T0227922

The
Global Forest
Sector

Changes, Practices, and Prospects

The
Global Forest Sector

Changes, Practices, and Prospects

Edited by
Eric Hansen • Rajat Panwar • Richard Vlosky

CRC Press
Taylor & Francis Group
Boca Raton London New York

CRC Press is an imprint of the
Taylor & Francis Group, an **informa** business

CRC Press
Taylor & Francis Group
6000 Broken Sound Parkway NW, Suite 300
Boca Raton, FL 33487-2742

© 2014 by Taylor & Francis Group, LLC
CRC Press is an imprint of Taylor & Francis Group, an Informa business

First issued in paperback 2017

No claim to original U.S. Government works
Version Date: 20131025

ISBN 13: 978-1-138-07581-8 (pbk)
ISBN 13: 978-1-4398-7927-6 (hbk)

Library of Congress Cataloging-in-Publication Data

The global forest sector : changes, practices, and prospects / edited by Eric Hansen, Rajat Panwar, and Richard Vlosky.
    pages cm
    "A CRC title, part of the Taylor & Francis imprint, a member of the Taylor & Francis Group, the academic division of T&F Informa plc."
    Includes bibliographical references and index.
    ISBN 978-1-4398-7927-6 (hardcover : acid-free paper)
    1. Forest products industry. 2. Globalization. 3. Forest products industry--Forecasting. I. Hansen, Eric, 1968- II. Panwar, Rajat. III. Vlosky, Richard P.

HD9750.5.G57 2013
338.1'7498--dc23   2013031804

Visit the Taylor & Francis Web site at
http://www.taylorandfrancis.com

and the CRC Press Web site at
http://www.crcpress.com

Contents

Section IV Capability Development and Strategic Imperatives for the Forest Sector

Section V Bringing It All Together

Preface

The impetus for this book came from conversations between Rajat Panwar and Eric Hansen while Rajat was pursuing his PhD under Eric's supervision at Oregon State University. It took several years before we began to refine our thoughts and convert them into a project plan. As we began to solidify the concept and talk with potential contributors, Richard Vlosky with the Louisiana State University Agricultural Center joined the editorial team.

Our overall goal in producing this book is twofold. First, we want to provide a means to support and advance teaching in forest sector business fields. Second, we want to consolidate current knowledge on various business management themes in the forest sector that would be useful to established and emerging scholars, business managers, and policy makers. Extant knowledge has largely remained dispersed across disparate academic sources albeit comprehensive books are available for specific functional areas such as forest products marketing and competitive strategy. Thus, we identified a gap in the literature that we believed was preventing students, scholars, policy makers, and others from developing a timely, structured, big-picture view of forest sector business. Furthermore, there was no single resource available for reviewing current thinking on a wide variety of business management issues in the forest sector. This gap looked even wider and deeper given the dramatic changes that have taken place in the global forest sector following the US housing crisis and global recession. How successful we have been in filling this gap is for the reader to judge, but we feel that this book is an important step toward the integration and further development of the forest sector business field.

The book is divided into five parts: Section I—Changing Context of the Global Forest Sector; Section II—Changes in Product Categories and Individual Markets; Section III—The Forest Sector within an Environmental Paradigm; Section IV—Capability Development and Strategic Imperatives for the Forest Sector; and Section V—Bringing It All Together. Each of these parts represents a separate theme within the book. Except for Section V, each part consists of several chapters. As a whole, the content provides a picture of the current and changing forest sector, including the state of forests, the nature of markets, the newly emerged patterns of stakeholder impact, and the evolution of key business practices.

This book could be used to support an entire academic course on the nature of forest sector business. For example, at Oregon State University it will be used for a three-credit, graduate-level course entitled The Context of the Forest Sector, a required course for all wood science graduate students, and at Louisiana State University for all graduate students in the forest products marketing program. Individual parts may be useful as supplemental readings for other courses at both graduate and undergraduate levels. For example, Section I could augment course content in global forestry and forest policy courses. Section II could fit in with forest products marketing courses. Section III could supplement a wide range of courses within forestry and support environmental studies fields. Section IV may be relevant to courses in forest-based recreation and nontimber forest products areas. Individual chapters may be of interest to advanced, focused scholars. Finally, Section V should be of interest to anyone contemplating the future of the sector.

We anticipate that practitioners will also find this volume valuable. Sections III and IV, in particular, might further stimulate managerial thinking on various topics of practical application. Also, because the book captures concepts from a broad swath of the forest industry, it can help business managers and policy makers to update their knowledge base, especially in areas in which they are less familiar or outside of their core responsibilities.

Happy reading.

Eric Hansen

Rajat Panwar

Richard Vlosky

Acknowledgment

Edited volumes such as this are successful to the extent they receive content from many scholars who are highly respected for their knowledge within their individual disciplines. We cannot thank enough each individual who contributed to the creation of this book. The support we received from colleagues, in fact, echoes the need for this book. In an ideal world, we would have liked to create a comprehensive book and title it "Forest products of the world, and the world of forest products," but the innumerous topics, product categories, and geographic areas that exist precluded their inclusion in this book. Whatever the reason for those omissions, we sincerely hope that colleagues and emerging scholars will find this book useful as a pedagogical tool—a basis to identify research areas and fill research gaps to make the discipline of forest products business management even more interesting, richer, and relevant to changing times.

Editors

Eric Hansen, PhD, currently serves as a professor of forest products marketing at the College of Forestry at Oregon State University. He earned his BS at the University of Idaho and his PhD at Virginia Tech. His research interests include organizational innovation, strategic marketing, and corporate social responsibility. At Oregon State University, he teaches courses in marketing; innovation, conservation, and design with natural resources; and forest sector business practices.

Dr. Hansen currently leads the Forest Products Marketing and Business Management Research Group within the International Union of Forestry Research Organizations.

Rajat Panwar, PhD, is currently an assistant professor of management and the Chapple Chair of Corporate Social Responsibility at Northland College in Ashland, Wisconsin. He earned his MBA at Lucknow University and his PhD at Oregon State University. His research interests include understanding the various interactions between commercial interests and sustainability issues, particularly from a strategic management perspective. At Northland College, he teaches courses on business and sustainability, entrepreneurship, and global business management.

Dr. Panwar is currently the editor of the *Journal of Forest Products Business Research* and also an affiliate faculty member at the College of Forestry, Oregon State University.

Richard Vlosky, PhD, is currently the Crosby Land and Resources Endowed Professor of Forest Products Marketing and Business Development at the Louisiana Forest Products Development Center at the Louisiana State University Agricultural Center. He earned his BS at Colorado State University, his MS at the University of Washington, and his PhD at Penn State University. His areas of research include biofuels, bioprocessing and bioenergy, domestic and international forest products marketing and business development, certification and green marketing, and e-business/e-commerce. At Louisiana State University, he teaches courses on forest products marketing.

Dr. Vlosky is a sector leader–wood products for the Louisiana Institute for Biofuels and Bioprocessing and a member of the board of directors for the Louisiana Forestry Association. Internationally, he is a US representative for the International Union of Forest Research Organizations Research Group on Forest Products Marketing and Business Development and an executive board member of the International Association for Economics and Management in Wood Processing and Furniture Manufacturing in Zagreb, Croatia. Dr. Vlosky is also an adjunct faculty member in the Faculty of Forestry at the University of Timisoara, Romania.

Contributors

Francisco X. Aguilar
Department of Forestry
School of Natural Resources
University of Missouri
Columbia, Missouri

Eduard L. Akim
Institute of Bio-Refinery of Wood and
    Nano-Technology in Forest Complex
and
Department of Technology of Pulp and
    Composite Material
Saint Petersburg State Technological
    University of Plant Polymer
Saint Petersburg, Russia

Leonid Akim
Wesco Distribution Canada, LP
Dorval, Quebec, Canada

Janaki Alavalapati
Department of Forest Resources and
    Environmental Conservation
Virginia Tech
Blacksburg, Virginia
and
Energy and Climate Partnership of the
    Americas
US Department of State
Washington, DC

Britta M. Anderson
University of Minnesota
and
E3 Environmental
Saint Paul, Minnesota

Jeffrey Biggs
Canadian Forest Service
Natural Resources Canada
Ottawa, Ontario, Canada

Michael A. Blazier
AgCenter Hill Farm Research Station
Louisiana State University
Baton Rouge, Louisiana

Mark Boyland
Canadian Forest Service
Natural Resources Canada
Ottawa, Ontario, Canada

Urs Buehlmann
Department of Sustainable Biomaterials
Virginia Tech
Blacksburg, Virginia

Lyndall Bull
Fenner School of Environment and Society
Australian National University
Canberra, Australian Capital Territory,
    Australia

Matt Bumgardner
US Forest Service
US Department of Agriculture
Madison, Wisconsin

Nikolay Burdin
Research and Design Institute on
    Economics and Information for Forest,
    Pulp and Paper and Woodworking
    Industries
Moscow, Russia

Zhiyong Cai
Forest Product Laboratory
US Forest Service
US Department of Agriculture
Madison, Wisconsin

Xiaozhi (Jeff) Cao
Forestry Asia Hub
Bureau Veritas
Shanghai, China

Benjamin Cashore
Yale School of Forestry & Environmental
    Studies
New Haven, Connecticut

David Cohen
Faculty of Forestry
Department of Wood Science
University of British Columbia
Vancouver, British Columbia, Canada

Ivan Eastin
Center for International Trade in Forest
    Products
School of Forest Resources
University of Washington
Seattle, Washington

Riitta Hänninen
Finnish Forest Research Institute
Vantaa, Finland

Eric Hansen
College of Forestry
Oregon State University
Corvallis, Oregon

Lauri Hetemäki
Foresight and Policy Support Programme
European Forest Institute
and
University of Eastern Finland
Joensuu, Finland

Steven Johnson
International Tropical Timber
    Organization
Yokohama, Japan

Chris Knowles
Department of Wood Science and
    Engineering
Oregon State University
Corvallis, Oregon

Robert A. Kozak
Faculty of Forestry
Department of Wood Science
University of British Columbia
Vancouver, British Columbia, Canada

Katja Lähtinen
Department of Forest Sciences
University of Helsinki
Helsinki, Finland

Pankaj Lal
Department of Earth and Environmental
    Studies
Montclair State University
Montclair, New Jersey

Scott Leavengood
Wood Innovation Center
Oregon State University
Corvallis, Oregon

William Luppold
US Forest Service
US Department of Agriculture
Madison, Wisconsin

Frances Maplesden
International Tropical Timber
    Organization
Auckland, New Zealand

Anne-Hélène Mathey
Canadian Forest Service
Natural Resources Canada
Ottawa, Ontario, Canada

Alexander Moiseyev
The Norwegian University of Life Sciences
Ås, Norway

Sergio A. Molina Murillo
School of Environmental Sciences
National University of Costa Rica
Heredia, Costa Rica

Rajat Panwar
Department of Management
Northland College
Ashland, Wisconsin

Ed Pepke
European Union Forest Law Enforcement,
    Governance and Trade Facility
European Forest Institute
Barcelona, Spain

Anatoly Petrov
All-Russian Institute of Continuous
    Education in Forestry
Federal Forest Agency
Pushkino, Russia

Erica Pohnan
Yale School of Forestry & Environmental
    Studies
New Haven, Connecticut

Sally A. Ralph
Forest Product Laboratory
US Forest Service
US Department of Agriculture
Madison, Wisconsin

Alan W. Rudie
Forest Product Laboratory
US Forest Service
US Department of Agriculture
Madison, Wisconsin

Ronald C. Sabo
Forest Product Laboratory
US Forest Service
US Department of Agriculture
Madison, Wisconsin

Jacki Schirmer
Fenner School of Environment and Society
Australian National University
and
Centre for Research and Action in Public
    Health
University of Canberra
Canberra, Australian Capital Territory,
    Australia

Al Schuler
US Forest Service
US Department of Agriculture
Princeton, West Virginia

Arijit Sinha
Department of Wood Science and
    Engineering
Oregon State University
Corvallis, Oregon

Timothy M. Smith
NorthStar Initiative for Sustainable
    Enterprise
and
Institute on the Environment
College of Food, Agricultural and Natural
    Resource Sciences
University of Minnesota
St. Paul, Minnesota

Taraneh Sowlati
Faculty of Forestry
Department of Wood Science
University of British Columbia
Vancouver, British Columbia, Canada

Nicole M. Stark
Forest Product Laboratory
US Forest Service
US Department of Agriculture
Madison, Wisconsin

Michael W. Stone
Yale School of Forestry & Environmental
    Studies
New Haven, Connecticut

Xiufang Sun
Forest Trends
Washington, DC

Anne Toppinen
University of Helsinki
Helsinki, Finland

Richard Vlosky
Louisiana State University
Baton Rouge, Louisiana

Ernesto Wagner
Tromen Enterprises
Santiago, Chile

Minli Wan
University of Helsinki
Helsinki, Finland

Section I

Changing Context of the Global Forest Sector

1

Understanding and Managing Change in the Global Forest Sector

Eric Hansen, Rajat Panwar, and Richard Vlosky

CONTENTS

1.1 Introduction

Globally, the forest sector employs over 18 million people in the formal sector. Inclusion of the informal sector estimates pushes the number closer to 50 million (Nair and Rutt 2009). Other experts suggest this number is much higher, only considering small- and medium-sized enterprises. Thus, it is a critical economic sector in many countries, especially for rural income and employment generation. This significant economic role has, understandably, attracted economists for many years making forest economics one of the well-established and mainstream fields within the forest sector academy.

In a sharp contrast, however, business management scholars largely did not address the forest sector until the 1950s. As a result, the academic contribution to business sophistication and entrepreneurial development within the forest industry has remained at a modest level. It is only during the past few decades that there has been a cadre of researchers around the world with specialized training in forest products marketing and related business development in the sector. For example, the Forest Products Marketing and Business Management group of the International Union of Forestry Research Organizations (IUFRO) and the team of specialists on forest products marketing within the auspices of the United Nations Economic Commission for Europe (UNECE) were formed in the early 1990s.

Because of its relatively recent history, the limited research-based knowledge of the business management aspects of the forest sector largely exists in journal articles, meeting proceedings, and other disparate sources. Our fundamental motivation to edit this book was to provide a consolidated, updated, and comprehensive resource encompassing the various facets of business management in the forest sector.

We trace the roots of forest industry management academics to the first coherent academic program focused on the managerial aspects of the forest sector at the University of Helsinki started during WWII. Most of the early work there was available only in the Finnish language and therefore not readily available to researchers in the rest of the world. The focus of most early work was to produce descriptive and highly applied studies, with little intention to develop fundamental understanding about the nature and underpinnings of business practices. The first area to see systematic study was business strategy with initial work at the University of Oregon (1980s), the University of Helsinki (1990s), and Virginia Tech (1990s). The emergence of forest certification and eco-labeling in the 1990s focused the academic attention of multiple researchers and thus promulgated systematic investigation and development of a strong knowledge base on one topic. Currently, organizational innovation, organizational capability development (specifically those related to information technology (IT)), and corporate social responsibility (CSR) (or sustainability) are examples of areas with multiple active researchers around the world.

A combined effect of accelerating globalization and the global recession of 2008 has produced dramatic changes in the forest sector. Consumer preferences have evolved and their purchasing power has increased. Accordingly, there have been changes in product offerings and prices, changes in technology, and changes in competition among industry players. More notably, however, the nature and extent of industry engagement in social and environmental issues have transformed fundamentally, especially with evolving environmental awareness. Emergence of low-cost producers from developing countries and resulting shifts in trade flows are other major factors driving change in the industry. Many argue that the forest industries now operate in a "green economy." This transition presents challenges and opportunities for forest sector companies as they navigate into the future.

When French poet Paul Valery said, "The trouble with our times is that the future is not what it used to be," he was not referring to the forest sector. But if he were, it would be apropos. Such is the extent of changes in the forest sector making its future both unknown and uncertain. The best we can do is to project into it by better comprehending the present. This is precisely what we have attempted through this book: to develop an understanding of the context that makes up the present operating environment for the global forest sector. As readers will note later, each chapter provides the state of the knowledge for the various topics covered. Before we describe the structure and contents of this book, it is, however, foremost to have an aerial view of the nature of changes in the global forest sector and develop a foundational understanding of the factors driving this change.

The balance of this chapter is organized as follows: we first provide a general characterization of the global forest sector to familiarize readers with its larger, primarily historical context. Section 1.3 portrays the changing picture of the forest sector wherein we identify and introduce a number of mega-trends in the external environment that, we believe, have produced widespread and penetrating changes in the forest sector. Where possible, we have tied these mega-trends with the chapters in this book (Figure 1.1). Section 1.4 outlines briefly the general structure and contents of the book. Section 1.5 concludes this chapter.

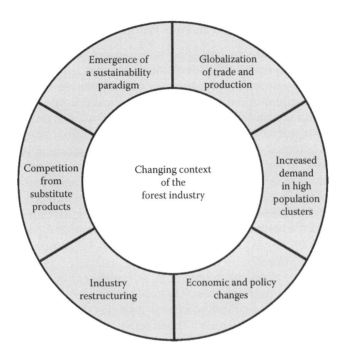

FIGURE 1.1
Major changes occurring in the operational environment of the forest sector.

1.2 A Look at the Historical Characteristics of the Industry

It is now almost cliché to say that forest products companies are undergoing a multidimensional change in the very fundamental nature of their business. In order to provide a context to where the industry is going, it is important to develop an understanding of the historical context of the industry. Later we identify two overarching features that have traditionally characterized the forest sector.

1.2.1 Mature Industry with Mature Products

Manufacturing of wood products started centuries ago and, in this sense, the forest sector represents a range of mature industries producing goods with extended life cycles. The industry has long been characterized as having a commodity and production orientation at the expense of a customer or market orientation (Hansen and Juslin 2011). Within this dominant business logic, firms within an industry tend to focus on reducing costs and gaining process efficiency and believe that low cost is a key to succeed in the marketplace.

Apart from a low-cost orientation, forest sector companies' focus on process efficiency is indeed attributable to other reasons as well that are industry specific. Within the sawmilling sector, for example, log costs can easily exceed 70% of total operating costs, making resource efficiency a logical imperative. Regardless of the underlying reason, a low-cost production orientation has arguably prevented industry from sufficiently investing in the development of new or value-added products. Following Prahalad (2004), a low-cost focus has created a blinder for forest sector companies.

1.2.2 Highly Fragmented Industry with Poor Profitability

Many sectors within the industry have had low barriers to entry, including relatively low-capital investment requirements. This characteristic has led to the presence of a large number of smaller operations with high levels of competitive intensity. The presence of a large number of small-size enterprises has historically limited the development of efficient, cohesive, and concentrated supply chains. Overall, the presence of small-scale actors in the forest sector and supporting industries has ultimately limited industry's capacity to reap economies of scale and efficiency-related gains (FAO 2011). Although significant consolidation has taken place during the last two decades, the nature of the forest sector continues to remain largely fragmented. This fragmentation is also responsible, in part, for relatively poor profitability levels within the industry. Poor profitability is illustrated in a PricewaterhouseCoopers report suggesting that the average return on capital employed (ROCE) of the 100 largest global pulp, paper, and packaging companies between 1999 and 2008 ranged between 2.3% and 6.5%, as compared to management goals of approximately 10%–12%.

While we label the previous characterization as traditional or historical, it is equally important to recognize that the line between history and the present is blurred and elusive. History is reflected in the present, more so in some domains than others. Today's forest sector presents a complex blend of historical characteristics and changed attributes. On one hand, the industry continues to produce mature products. Still, many new products have also been introduced to the market. Similarly, while the traditional, low-tech moniker of the forest sector continues to hold good especially relative to hi-tech or knowledge-based industries, the forest sector is also aggressively adopting technology to increase efficiencies throughout the full range of manufacturing and business processes. So, even though the final products may not often reveal the evolving technological sophistication within the industry, the tools and technologies employed to extract raw material and manufacture and deliver products are becoming increasingly hi-tech. Further, the traditional industry orientation within the sector is evolving from a production to a market orientation. You will read more about this transition in Chapter 17 where the authors argue that the global forest sector has made a shift from having a raw material-focused "upstream strategy" to adopting a market-focused "downstream strategy." Overall, while the vestiges of its historical characteristics are conspicuously visible, there are multidimensional changes occurring in the forest sector.

1.3 Changing Context of the Forest Sector

Changes in forest sector operations are much more pervasive and penetrating than the changes that happen within an organically evolving industry where the nature and pace of change is much less drastic. A large number of forest products companies face the need for actively exploring new markets in traditionally neglected regions and countries as well as integrating new business practices and technology in current markets. Accordingly, both large and small companies are investing in developing new processes and new products. Similarly, suppliers of raw material have found buyers in newly emerged segments, and they no longer rely exclusively on demand for raw material from forest products manufacturers. Additionally, companies are paying closer attention to social and environmental implications of not only their own actions but also those of their supply-chain partners. Reasons for these wide-ranging changes can often be directly traced to reactions to changes in the industry's macro-environment and the broad effects of the global economy. Figure 1.1 captures major changes in the operational environment of the forest sector, and each are discussed below.

1.3.1 Globalization of Trade and Production

Globalization has become the norm for all industrial sectors. Finished products, component parts, and intermediate manufacturing occur in widely dispersed regions, while final consumers or supply-chain partners are often geographically distant from their suppliers. The forest sector is no different and has become increasingly globalized over recent decades. Global trade of forest products has grown steadily in recent years. According to FAO data, the value of total world exports of forest products grew 1.7 times between 1997 and 2010. The growth in overall value of trade is associated with dramatic shifts in trade flows. Illustratively, China is currently the largest importer of hardwood lumber from the United States, and the European Union is the major importer of wood pellets for power generation. In addition to increased world trade, offshore manufacturing and emergence of a number of new economies have contributed to the globalization of the forest sector. For example, over the past decade, countries with lower costs of production and relatively minimal safety and environmental constraints have become major manufacturers of finished and semifinished wood products. Over time, low-cost producers lose their low-cost advantage due to increased labor costs and general economic development, making manufacturing units shift to places with newly gained lower-cost advantage. Shift in furniture production from China to Vietnam illustrates this phenomenon. Readers will note additional manifestations of the impact of globalization in the forest sector in many of the chapters, but especially in Chapters 3 through 5, 7, and 14.

1.3.2 Increased Demand in High-Population Clusters

The first decade of the new century can be termed an era of emerging economies. Fast economic growth within emerging economies phenomenally raised their purchasing power. Notably, some of these fastest growing economies, such as China, are also the largest population clusters of the world. A middle-class boom in these countries drove further the demand for various forest products. A similar phenomenon has taken place outside the typical "emerging economies" cluster. For example, the size of the middle class in the Asia Pacific region, which stood at 525 million people in 2009, is expected to climb to 1.7 billion in 2020 and 3.2 billion in 2030 (Kharas 2010). Globally, the middle-class population that consisted of 1.8 billion people in 2009 will grow to 4.9 billion people by 2030 (Kharas 2010). Readers will note the effect of the "emerging economies" phenomenon on global forest products trade diffused across Chapters 3 through 5, but Chapters 7 and 8 specifically focus on China and Russia for providing a magnified view of this phenomenon. Brazil, India, and several recently emerging economies also competed for inclusion in the book, but we were ultimately unsuccessful in including these countries.

1.3.3 Economic and Policy Changes

A number of trade, monetary, and public policy changes have happened in recent years and have had a profound impact on forest sector industries. In particular, energy policies in a number of countries have created new markets for forest products. Commitments by the various national governments to the Kyoto Protocol have also created major markets, for example, wood pellets in Europe. Resultantly, an idea, such as setting up a mill for manufacturing and exporting wood pellets, which might have sounded laughable a decade ago, now presents a business opportunity. Introduction of the Russian log export tax was a significant shock to global markets and resulted in market-level changes that extended beyond just the countries trading with Russia. Further, currency fluctuations have impacted global trade. The fall of the US dollar since the Great Recession has contributed to US success in exporting such as softwood logs and lumber shipped from the West Coast of the United States to China. European countries that have remained outside of the Euro Zone (e.g., Sweden) have maintained more currency flexibility that has contributed to their competitiveness in international trade. Readers will note that discussion and implications of the various policy changes primarily appear in Chapters 2 and 10 but are also diffused across Chapters 3 through 5.

1.3.4 Industry Restructuring

The historically fragmented nature of the industry has often been held responsible for overcapacity, lack of price discipline, and little ability to influence industry trends—all of which lead to poor profitability (PWC 2000). Therefore, calls for consolidation within and restructuring of the forest sector have received traction, and many mergers and acquisitions have taken place over the past couple of decades. Consolidation has spanned the entire value chain—from forestland to forest products retailers. On one hand, forestland ownership changes have profoundly impacted forest products availability. On the other hand, however, shortening of value chain between manufacturers and retailers, which was primarily driven by the emergence of mega-retailers and was touted as a cost-cutting mechanism, has led to a disintermediation effect in industry.

Chapter 2 provides a fuller picture of landownership changes, but the effects of disintermediation have not been well studied in the sector. We hold a view that disintermediation has certainly changed the "nature" of the buying experience, but, in the absence of any supporting research, we are not convinced that it has changed overall demand. Nonetheless, readers will be able to make a connection between disintermediation effect and developments outlined in Chapter 13.

1.3.5 Competition from Substitute Products

Technological advances in other industry sectors may create significant competition for an industry. We observe two features characterizing the nature of competition from substitute products. First, it seems to emerge in waves. In the 1970s, in the United States, for example, wall-to-wall carpeting became the popular floor covering, and wood flooring lost market share. Recent years have seen a resurgence of wood floor coverings. Second, the nature of competition from substitute products within some sectors is complex. For example, while the pulp and paper sector faces threats from digitalization of media, the same digitalization has opened new markets for packaging and various paper grades. Similarly, while the wood furniture sector has not faced any special threat from substitute products, the outdoor decking market has recently been invaded by plastics and wood–plastic composites, threatening the traditional wood-based outdoor decking segment of the industry. Overall, it is difficult to isolate the effect of the various substitutes on forest sector products and offer a comprehensive discussion, but readers will find this aspect sprinkled at various instances in Chapters 4 and 6.

1.3.6 Emergence of a Sustainability Paradigm

Unarguably, the single most prominent development influencing the global forest sector is what could broadly be labeled as "the emergence of a sustainability paradigm." Sustainability is a much more complex concept than can be summarized here. For the sake of brevity, therefore, we eschew the task of defining a sustainability paradigm but rather attempt to capture its manifestation in the forest sector in a tripartite fashion.

First of all, as economic development matured in the West, environmental awareness and considerations have been rising and have fostered the shift in societal views about natural resources, which our colleague Dr. Robert Kozak once aptly labeled as "a pure exploitation to a conservation-based paradigm" (see Chapter 18). While this shift spans a gamut of natural resources, it is especially pronounced in case of forest resources because linking environmental well-being with forest abundance is intuitively simpler. An increasingly larger proportion of population in the Western world now views forests as providing aesthetical value rather than just the raw material for manufacturing various goods. These changing values, through a variety of mechanisms, have impacted forest sector companies' operations in a number of ways. For example, harvesting and management activities have become more environmentally friendly and have shifted to lands that are viewed as being less sensitive. Intensive plantations now provide almost 2/3 of the global use of industrial wood (PWC 2011). Although not without their own socio-environmental controversies, harvesting from plantation "forests" appears to be more accepted by society than harvesting from "natural" forests. Readers will find Chapters 2, 10, and 15 helpful in further understanding those changes in forestry and forest products companies' actions that can be attributed to evolving societal considerations about the environment.

Companies' response to evolving environmental awareness, however, may not be as spontaneous as the foregoing text may suggest. A variety of "push and pull" factors contribute to a movement in companies' policies and practices. The role of environmental nongovernmental organizations (ENGOs) cannot be overestimated in creating a number of push and pull strategies. Chapter 9 offers a comprehensive analysis of the interaction of ENGOs and forest sector companies within an environmental paradigm. Notably, collaborative models between forest sector companies and ENGOs are emerging for joint problem-solving.

Emergence of environmental awareness has also created new market opportunities for companies so much so that a subfield within marketing called "environmental" or "green" marketing has emerged. Emergence of environmental marketing opens up unique opportunities for forest sector companies as these products have a potential to be viewed as more environmentally friendly than many competing products. Forest certification, triple-bottom line (or sustainability) reporting, and life-cycle analysis (LCA) are further helping companies to support their "green" claims. In particular, the recent emergence of two fields, namely, green building and bioenergy/biochemicals production, can directly be tied to greening markets. Detailed notes on these fields appear in Chapters 11 and 12, respectively.

1.4 This Book

We attempt in this book to capture how the changes, brought about by a host of factors mentioned earlier and many more, have manifested in the global forest sector. Some of these chapters focus on documenting the changes; others also encompass the various

TABLE 1.1

Organization of the Book

Section Number	Thematic Focus	Chapters within a Part
Section I	*Changing Context of the Global Forest Sector*	Chapter 1: Understanding and Managing Change in the Global Forest Industry Chapter 2: Impact of Globalization on Forest Users: Trends and Opportunities
Section II	*Changes in Product Categories and Individual Markets*	Chapter 3: Markets and Market Forces for Lumber Chapter 4: Markets and Market Forces for Secondary Wood Products Chapter 5: Markets and Market Forces for Pulp and Paper Products Chapter 6: New Products and Product Categories in the Forest Sector Chapter 7: Chinese Era Chapter 8: Russia in the Global Forest Sector
Section III	*Forest Sector within an Environmental Paradigm*	Chapter 9: Environmental Activism and the Global Forest Sector Chapter 10: Implementing Sustainability in the Forestry Sector: Toward the Convergence of Public and Private Forest Policy Chapter 11: Green Building and the Forest Sector Chapter 12: Assessment of Global Wood-Based Bioenergy
Section IV	*Capability Development and Strategic Imperatives for the Forest Sector*	Chapter 13: Current and Future Role of Information Technology in the Forest Sector Chapter 14: Cross-Cultural Sales, Marketing, and Management Issues in the Global Forest Sector Chapter 15: Corporate Social Responsibility in the Global Forest Sector Chapter 16: Innovation in the Global Forest Sector Chapter 17: Strategic Orientations in the Global Forest Sector
Section V	*Bringing It All Together*	Chapter 18: What Now, Mr. Jones? Some Thoughts about Today's Forest Sector and Tomorrow's Great Leap Forward

management techniques for leveraging these changes. The book, thus, has both a descriptive and a prescriptive tone (Table 1.1).

This book is organized into five sections. *Section I* (*Changing Context of the Global Forest Sector*), which the present chapter is also a part of, provides a macro-view of the forest sector. It would be tautological to introduce the present chapter here. Chapter 2 within Section I is titled "The Impact of Globalization on Forest Users: Trends and Opportunities." In this chapter, Ben Cashore, Erica Pohnan, and Michael Stone, all from the Yale School of Forestry and Environmental Studies, provide a broad outline of the state of world forests. They document and analyze the transition in forests in recent years and contemplate the future of global forests. The authors then suggest that, on a global scale, forest products markets are changing based on (1) the growth of global trade, (2) an increasing role of emerging economies in global trade, (3) the influence of domestic forest policies on policies in other countries, (4) globalization driven innovation, and (5) a host of actors being involved in policy making. Finally, the authors offer their vision for the future of global forests. In many ways, this chapter is a more formal, structured, and detailed analysis of the mega-changes surrounding the forest sector than the one we attempted earlier.

Section II (*Changes in Product Categories and Individual Markets*) contains six chapters (Chapters 3 through 8) all covering changes occurring either at the specific product level or at an individual market level.

Chapter 3, titled "Markets and Market Forces for Lumber," has three sections. In Section 3.1, Ed Pepke from the European Forest Institute focuses on softwood lumber that albeit being primarily used for building construction has a multitude of other uses. The chapter argues that new uses are being developed that extend its product life cycle beyond that of a simple commodity. In Section 3.2, William Luppold and Matt Bumgardner with the USDA Forest Service provide production, demand, and trade-related information of hardwood lumber. They also outline and discuss a number of trends affecting the global hardwood trade that have continued during the pre- and post-recession periods. In Section 3.3, Frances Maplesden and Steven Johnson of the International Tropical Timber Organization (ITTO) focus on tropical timber. The authors discuss how production from natural tropical forests is becoming increasingly supply constrained, affected by log export restrictions to encourage downstream processing, reductions in logging that achieve national sustainable forest management (SFM) targets, and crackdowns on illegal logging.

In Chapter 4, which is entitled "Markets and Market Forces for Secondary Wood Products," Urs Buehlmann of Virginia Tech and Al Schuler, recently retired from the USDA Forest Service, provide an overview of the global secondary wood products industry. They describe a sector that has recently experienced extensive globalization with major sectors such as furniture production moving from one low-cost region to another. They describe phenomena underlying global shifts in manufacturing locations as well as describe trends in raw materials, use patterns, and the long-term outlook for the sector.

In Chapter 5, titled "Markets and Market Forces for Pulp and Paper Products," Lauri Hetemäki, European Forest Institute and Professor at the University of Eastern Finland; Riitta Hänninen, Finnish Forest Research Institute; and Alexander Moiseyev, University of Ås, Norway, discuss the ongoing structural changes in the pulp and paper sector and project what the sector might look like in the next 10–20 years. They suggest that in order to grow in the long run, the sector needs to actively innovate and develop new value-added products. The authors believe that the focus will be on forest biorefineries producing bioenergy, biochemicals, and intelligent packaging products.

In Chapter 6, which is titled "New Products and Product Categories in the Forest Sector," authors Zhiyong Cai, Alan W. Rudie, Nicole M. Stark, Ronald C. Sabo, and Sally A. Ralph, all from the USDA Forest Service Forest Products Lab, discuss new products and applications, all of which go beyond traditional solid wood products. The first section of this chapter focuses on forest product-derived nanomaterials and provides a brief description of technologies and applications of nanotechnology for forest-based products. The second section explores research and development trends in wood–plastic composites. The authors argue that these products will increasingly penetrate the construction and value-added wood product markets in a variety of forms. In the third section, the authors cover chemicals derived from wood.

It is hard to overestimate the importance of China to the dynamics of the global forest sector moving into the heart of the twenty-first century. Xiaozhi (Jeff) Cao from Bureau Veritas, Xiufang Sun from Forest Trends, and Ivan Eastin from the University of Washington provide an overview of the forest sector in China in Chapter 7 titled "Chinese Era." With annual GDP in China typically reaching 8% or more, the window of growth is wide open and it is expected that the importance of China will only increase over time. China has risen to dominate a number of sectors in the global forest industry. The role of the country as a raw material supplier is overviewed followed by the role of China as a global supply-chain partner and as a consumer of forest products. Finally, the "green" drivers that are changing the industry in China and an overview of the significant FDI being made by China and Chinese companies in order to assure long-term supply for their manufacturing juggernaut are discussed.

In Chapter 8, which is titled "Russia in the Global Forest Sector," Eduard L. Akim of Saint Petersburg State Technological University of Plant Polymer; Nikolay Burdin of Research and Design Institute on Economics and Information for Forest, Pulp and Paper and Woodworking Industries; Anatloy Petrov of All-Russian Institute of Continuous Education of Forestry; and Leonid Akim of Wesco Distribution overview the country that today is the largest global exporter of roundwood. After explaining the current status of Russia's forest sector, the authors conclude that there are many challenges facing the sector based on the decline after the demise of the Soviet Union. However, structural transformation of the overall economy is largely complete, and the forest sector is primarily governed by market forces.

Section III (*Forest Sector within an Environmental Paradigm*) consists of four chapters that cover the various industry-level changes that can be attributed to changing environmental or sustainability concerns.

The first of these, Chapter 9 titled "Environmental Activism and the Global Forest Sector," is a comprehensive treaty on this topic. This chapter is contributed by Jackie Schirmer who holds a dual affiliation with the Australian National University and the University of Canberra. The chapter begins by documenting the history of and emerging trends in environmental activism with a particular focus on environmental campaigns in the forest sector. Replete with examples, this chapter also examines the factors that influence the success of ENGO campaigns. The author also contemplates the future of environmental activism surrounding the global forest sector.

In Chapter 10, entitled "Implementing Sustainability in the Forestry Sector: Toward the Convergence of Public and Private Forest Policy," Timothy M. Smith (University of Minnesota), Sergio A. Molina Murillo (National University of Costa Rica), and Britta M. Anderson (University of Minnesota) describe forestry and forest product-level policies, especially those aimed at implementing sustainability oriented principles. They first provide a typology of various SFM-focused mechanisms before discussing interaction among these various governance mechanisms. The authors further observe an evolving convergence and synergy between the policies within private and public realms.

Chapter 11, "Green Building and the Forest Sector," is authored by Chris Knowles and Arijit Sinha, both from Oregon State University. Writing in this chapter is a commentary of the evolution of the green building sector globally where the authors have displayed discipline in critically analyzing the role of wood in the green building sector. They explicitly recognize the need for more research before increased use of wood in green building can be emphasized. We hope this chapter would leave readers with questions and curiosities about the future interaction between this emerging field and the forest sector.

Chapter 12, "Assessment of Global Wood-Based Bioenergy," is authored by Francisco X. Aguilar (University of Missouri), Michael A. Blazier (Louisiana State University), Janaki Alavalapati (Virginia Tech), and Pankaj Lal (Montclair State University). The authors recognize the nascent stage of the sector but see promise triggered by recent developments in technology, improvements in energy conversion, and the potential net benefits of bioenergy use. The chapter discusses the role of government agencies in proactively promoting the use of woody biomass as feedstock for energy and biofuel production. The authors suggest that with improving technology and increasing economic viability of conversion processes, the evolution of wood-based bio-products development and production will likely continue on an upward trajectory.

Section IV (*Capability Development and Strategic Imperatives for the Forest Sector*) contains five chapters, each focusing on either a capability or a strategic imperative that is important for the forest sector in order to adapt to and succeed through the turbulence brought about by the various mega-changes.

Chapter 13, "Current and Future Role of Information Technology in the Forest Sector," is authored by Taraneh Sowlati from the University of British Columbia. IT has advanced rapidly during the past two decades and has impacted all aspects of our lives. New developments in IT, especially the advent of the Internet and cloud computing, have created significant opportunities for business competitiveness. IT has been recognized as a major enabler of business growth and innovation. In this chapter, she addresses IT applications in forestry, including forest management and forest sector and e-business adoption in the forest sector.

In what is written with a practitioner's eclectic style, Chapter 14 "Cross-Cultural Sales, Marketing, and Management Issues in the Global Forest Sector", Ernesto Wagner shares his observations from the field. This chapter blends theoretical input but remains heavy on the practical side. This chapter provides readers, particularly students, an opportunity to identify skill areas for development, were they interested to enter an ever-exciting field of international business. The author provides several examples to make this chapter informative and consumable.

"Corporate Social Responsibility in the Global Forest Sector" (Chapter 15) is authored by David Cohen (University of British Columbia), Anne-Hélène Mathey, Jeffrey Biggs, and Mark Boyland (all from the Canadian Forest Service). At the most basic level, the authors explicate the meaning of CSR, its application, and evolution in the forest sector. Not only does the chapter cover these foundational aspects but it also gets deeper into topics such as CSR strategies and approaches to measure CSR. The chapter generally vouches for corporate adoption of social responsibilities but does not fail to recognize the presence of a political skepticism that emanates from a larger debate concerning private provisions of public goods.

Chapter 16, titled "Innovation in the Global Forest Sector," is contributed by Scott Leavengood, Director of the Oregon Wood Innovation Center at Oregon State University, and Lyndall Bull from Australia National University. They provide an in-depth description of innovation within the forest sector. Innovation has been a hot topic for researchers and industry practitioners in the last decade, because it is generally viewed as one of the most promising propositions by industries in the developed world for maintaining competitiveness vis-a-vis low-cost manufacturers. The chapter provides a working definition of innovation and innovativeness and explains the context for innovation in the forest sector including an explanation of innovation systems. The authors show that these innovation systems encourage incremental innovations instead of big ticket, new ideas. This chapter also covers new product development in the sector, innovation management, etc.

In Chapter 17, titled "Strategic Orientations in the Global Forest Sector," Anne Toppinen, Katja Lähtinen and Minli Wan (all from the University of Helsinki) provide an in-depth exploration of the evolution of forest sector business strategies. Strategic management research has evolved significantly during the past several decades, as have the strategies pursued by forest sector companies. This chapter describes the changes taking place in the study of strategy, outlining the changes from the well-known Porter paradigm to the resource-based view of the firm. The authors further maintain that companies in the future may increasingly adopt a stakeholder orientation. The authors also articulate how forest sector companies may leverage their service capabilities and collaborate with customers for value cocreation.

Last but not the least, Robert A. Kozak from the University of British Columbia provides a grand canvas to this book in the solo chapter "What Now, Mr. Jones? Some Thoughts about Today's Forest Sector and Tomorrow's Great Leap Forward" within *Section V (Bringing It All Together)*. Indeed, this chapter wraps together the various pieces of this book, but more importantly marks the beginning of the future by raising many pertinent yet inconvenient questions. Thus, this chapter is both a conclusion to the present and an introduction to the future. Speculating industry's future is a complex task. As we stated before, the future is both uncertain and unknown. Being cognizant of this uncertainty makes Dr. Kozak cautious; being projective makes him audacious. Many readers will find this chapter a snapshot of what exists, others a glimpse of what may be. In either case, readers will find it insightful and thought provoking.

1.5 Conclusion

Overall, we present this book with the hope that it will help readers develop a bird's-eye view of the changes surrounding the forest sector as well as have a magnified view of numerous managerial issues associated with these changes. We did not intend to advance any scientific knowledge through this volume but rather aimed to create a consolidated resource of the state of knowledge in the global forest sector. Hopefully, this knowledge consolidation will help create new students and scholars in the field while helping existing scholars to identify and address knowledge gaps.

References

FAO. 2011. State of the world's forests, 2011. Food and Agriculture Organization of the United Nations, Rome, Italy. 164pp.

Hansen, E. and H. Juslin. 2011. *Strategic Marketing in the Global Forest Industries*. 2nd edn. Corvallis, OR. 327pp.

Kharas, H. 2010. The emerging middle class in developing countries. OECD Development Centre Working Paper No. 285. Paris, France. 61pp.

Nair, C.T.S. and R. Rutt. 2009. Creating forestry jobs to boost the economy and build a green future. *Unasylva*. 60(3): 3–10.

Prahalad, C.K. 2004. The blinders of dominant logic. *Long Range Planning*. 37(2): 171–179.

PWC. 2000. *Global Forest & Paper Industry Survey*. PricewaterhouseCoopers, Vancouver, British Columbia, Canada. 42pp.

PWC. 2011. *Growing the Future*. PricewaterhouseCoopers, Vancouver, British Columbia, Canada. 44pp.

2

Impact of Globalization on Forest Users: Trends and Opportunities

Benjamin Cashore, Erica Pohnan, and Michael W. Stone

CONTENTS

2.1 Introduction

For those forest owners, forest professionals, government agencies, and nongovernmental organizations seeking to promote sustainability and manage forest resources around the world, the last two decades have presented an exhilarating, sometimes frustrating, and complicated period. The emergence of China as an importer of raw materials, exporter of manufactured products, and consumer; the reawakening of Eastern Europe's forest sector; and the reorientation of African raw materials to Asian markets have presented obstacles and opportunities for a range of forest users.*

The purpose of this chapter is to shed light on such trends in the context of the economic global integration of the forest sector and, in so doing, better inform forest user strategies within a complex and changing world. We argue that while the economic globalization of world markets can facilitate forest loss and degradation, it also fosters technological innovation and a greater role for international efforts to promote environmental, social, and economic stewardship. As the chapter by Smith et al. details, there exist international influences, including economic incentives, such as eco-labeling "certification" and legality verification as well as norms regarding forest livelihoods such as "rights to resources," "subsidiarity," and "free and prior informed consent" (FPIC), which provide opportunities for forest users to engage with, and help shape, synergistic opportunities to address environmental and social concerns amidst increasing economic globalization. However, achieving such results is not preordained nor easy, rather it requires careful attention to four distinct pathways through which influence occurs.

We detail this argument in the following analytical steps. Following this introduction, Section II reviews the different configurations of "forest users" around the world through a review of country level approaches to ownership and tenure allocation. This section includes a review of the emergence of Timber Investment Management Organizations (TIMOs) in the United States that have reduced "vertical integration" within the industry, the role of the fall of communism in affecting Europe's forests, China's growth in providing new roles for community forestry, and the strong push for greater local resource access in Africa and Central and South America.

Section III reviews key domestic indicators that globalization of the forest sector, and forest users, attempt to shape and influence, including formal designations of "protected areas," extent and rates of deforestation/reforestation, and emergence of new environmental regulatory structures and governance of domestic forest practices. Section IV presents an analytical framework for understanding economic globalization and pathways of "internationalization" that forest users and stakeholders can travel to champion and influence strategic policy objectives.

2.2 Forest Users: Ownership and Tenure Patterns

There exists a complex and changing mosaic of forest ownership rights and responsibilities across the globe. For example, in much of Africa, national governments own most of the land, but in the last generation local communities have been granted increased

---

* We define forest users broadly to include any individual or organization that draws on, or seeks to influence, forest management and land use including forest-dependent communities, companies along the timber supply chain, and forest product consumers.

management responsibilities over public and private forests. Similarly, in Latin America, community forestry is a dominant mode of resource management. Postcommunist Eastern Europe has seen rapid growth of private forest ownership as land is repatriated to pre-communist owners. In the United States, there has been a decline in private land owner-ship by vertically integrated forest products companies (VIFPCs) to Timber Investment Management Organizations (TIMOs) and Real Estate Investment Trusts (REITs) that sell timber to other firms. In China, there have been substantial institutional reforms that involve the disassembling of community-owned forestlands into smaller plots managed by local individual owners, as well as logging bans that have converted state-owned tim-ber-producing companies into state-owned forest stewards.

2.2.1 Ownership

While the previous examples imply a global devolution of forest authority toward more local levels, it is important to note that, as detailed in Figure 2.1, national and state/ provincial governments still maintain ultimate ownership responsibilities governing 80% of the world's primary forests. Public ownership currently remains the most common form of ownership in many countries with a high percentage of forest cover, such as Brazil, Democratic Republic of the Congo, Indonesia, and the Russian Federation (FAO 2010b, 122). At the same time, the total area of forest under public ownership decreased by 141 million hectares (ha) from 1990 to 2005, while 113 million ha shifted into private ownership, most notably in China and Columbia (FAO 2010b, 124).

2.2.1.1 North America

Presently in the United States, 57% of forestland is privately owned (Butler 2008), but this figure masks an important change in the type of ownership over the last 20 years with

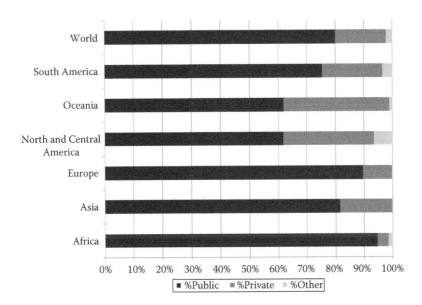

FIGURE 2.1
Forest ownership by region, 2005. (Based on data from Food and Agricultural Organization of the United Nations (FAO), *Global Forest Resources Assessment*, Food and Agricultural Organization of the United Nations, Rome, Italy, 2010b.)

the rise of investor-owned forestland through TIMOs and similar REITs. Investment in forestland by TIMOs and REITs has grown from less than $1 billion in 1985 to more than $25 billion in 27.3 million acres in 2005 constituting 5% of US timberland (Hickman 2007).

TIMOs are groups of private investors such as insurance companies, pension funds, endowments, and foundations that invest in forestland as a low-risk investment with returns from harvests and capital appreciation. The TIMO orchestrates sales and forest management, but the investors directly own the land. In contrast, REITs directly own the land and investors invest in the REIT. TIMOs generally purchase land from VIPCs such as Weyerhaeuser, whereas REITs often emerge following vertically integrated companies restructuring their assets. The result is that previous distinctions of "industrial" versus "private" ownership are less useful; private investors now own significant commercially harvested land (McDermott et al. 2010).

The change in ownership structure has prompted questions about the future management of forests, support for local communities, and funding for forest research (Hickman 2007). Whereas vertically integrated firms were known to make contributions to the local community and employ local professionals, there is the worry that private investors may be less attuned to, and integrated with, community concerns. Concerns have also been raised that these ownership patterns could lead to more mineral extraction and parcelization for land development (Bliss et al. 2009). Research findings to date indicate, however, that while ecological disturbance on TIMO and REIT managed forests to be higher than those on publicly owned land, their impacts are comparable to forestland managed by integrated forest management regimes (Noone et al. 2012).

The trend toward investor ownership in US forestland must be seen as the intersection of institutional changes beyond the US forest sector. The financial crisis in 2008 resulted in decreased housing starts, which caused financially troubled timber companies to liquidate some of their assets (Bliss et al. 2009). At the same time, US corporate accounting and tax policy favoring capital gains also provided incentives to investors in TIMOs and REITs (Clutter et al. 2005). Finally, international competition has driven US consolidation as expanded global markets meant that there was less need for timber companies to secure supply by owning forests (Clutter et al. 2005; Ince et al. 2007).

In contrast to the United States, the vast majority of forestland in Canada is publicly held by provincial and federal governments, although there is significant variation across provinces and territories. For example, 70% of Nova Scotia's forestland is privately held while the province of British Columbia owns 94% of land within its jurisdiction. The Canadian federal government also owns vast tracts of land in the sparsely populated northern territories. The most significant changes in land ownership have not been public to private but, reflecting similar trends elsewhere in the world, have been about granting, or ceding, greater authority to Aboriginal peoples. These efforts are most pronounced in the Canadian province of British Columbia; after 100 years of refusals, the government has embarked on a treaty negotiation process that will transfer vast tracks of land to First Nations (Cashore et al. 2001).

2.2.1.2 Latin America and Africa

Communities have been increasingly granted rights to own and/or manage both public and private forests in Africa and Latin America. In Latin America, there has been significant demand for recognition of indigenous land rights, support of which has been the main driver behind increasing community control over forests (Sunderlin et al. 2008). These norms were given further traction with the adoption of the United Nations Declaration on the Rights of Indigenous Peoples (UNDRIP) in 2007. In response to this demand, many countries have passed comprehensive legislation that award or expand community rights over forest resources. For example, Bolivia, which made specific reference to UNDRIP in its national enabling legislation, is in the process of clarifying and issuing land titles to indigenous communities (ibid.). Similarly, Brazil's 2007 Law on Public Forest Management granted communities the right to hold forest concessions and provided additional recognition of local community rights (ibid.).* Argentina's 2007 Forest Law mandated that forests used by local people and indigenous communities be protected, which can be seen as an effort to prioritizing the rights of many local communities and indigenous peoples over forest company interests (ibid.).†

In Africa, there has been a similar transition from state to community management of public countries in Cameroon, Sudan, and Tanzania (Sunderlin et al. 2008: 18). However, the vast majority of countries in Africa have made little progress in devolving forest management (ibid.). Still, it is noteworthy that of the 5% privately held forests in sub-Saharan Africa, 80% are managed by communities and/or indigenous peoples (FAO 2010b).

2.2.1.3 Europe

The decline of communism in the 1990s resulted in a major shift away from public ownership of forests in Eastern Europe. Communal forestlands were restituted to their pre-communist era owners, effectively returning them to private ownership, which coincided with similar efforts to privatize state forest management. Poland, Slovenia, and other Eastern European countries began privatizing and restructuring their forestry sectors as early as 1991, whereas other countries like Bulgaria did not begin until 1999 (Staddon 2001). In the Russian Federation, the private sector was granted management of 137 million ha of largely heavily deforested lands between 1990 and 2005, which have since been replanted (FAO 2010b: 125). In part owing to such incentives, forest cover in many former communist countries increased, including 850,000 ha in Belarus, 600,000 ha in Bulgaria, 456,000 ha in Poland, and 431,000 ha in Ukraine.‡

2.2.1.4 China

As Chapter 7 details, China began undertaking land tenure reforms in 2002, which was one of the most significant changes in direction in China's forest policy since collectivization of Chinese forests began in the 1950s. The reforms were primarily motivated by the combination of a severe drought in 1997 in the Yellow River Basin followed by a catastrophic flood in 1998 in the Yangtze River Basin and much of Northeastern China, which led to national attention to the problems of deforestation and erosion (Bennett et al. 2008). After 4 years of policy negotiations, the Chinese State Forest Administration initiated a five-pillar reform agenda aimed at transforming China's ecological and rural development through (1) forest-related activities (Zhou 2006) improving forestry financing, (2) reduction in taxes and fees, (3) encouraging better management practices, (4) liberalizing pricing while reducing power of timber-processing monopoly, and (5) contracts (Liu 2008).

---

* Government of Brazil. 2006. Lei No. 11.284. March 2, 2006. http://faolex.fao.org/docs/pdf/bra62562.pdf
† Government of Argentina. 2007. Ley de presupuestos mínimos de protección ambiental de los bosques nativas. December 19, 2007. http://faolex.fao.org/docs/texts/arg76156.doc
‡ Figures calculated from FAO Global Forest Assessment data. "Table 3: Trends in Extent of Forest, 1990–2010."

Taken as a whole, the tenure reforms represented a strong commitment to community/private forestry because they were designed to convince farmers on collective lands that it would be profitable to invest in forestry. Reducing fees and strengthening ownership not only provided such financial incentives, they also transformed the relationship between the state and forests in China. Prior to the reforms, collective forestry fees allotted substantial operational costs to forestry departments. In Jiangxi Province, for example, an annual operating budget of 300 million Chinese renminbi (USD 47.7 million) was largely financed by collective forestry fees (SFA 2007: 232). However, this funding relationship changed following a 2004 government decree stating that forestry departments would be funded by the central government and that local forest departments' mission would be focused on education (SFA 2007: 80, 124, 232).

While the fee reductions were an important incentive, the core of the institutional reforms was the expansion of ownership rights to farmers. These rights were embodied in the 2003 "Decision of the Chinese Communist Party Central Committee and the State Council on Accelerating the Development of Forestry" decree, which provided an overarching national rule: "the person who planted owns it, the investor is the beneficiary" (SFA 2007: 35,339). These reforms permitted trees to be inheritable and to be used as collateral for bank loans, which combined to bring much needed capital into the most rural areas of China (Kung and Liu 1997). Following these institutional reforms, decisions made within market transactions would have a much greater role in determining forest allocations.

Despite these changes, certain elements of forestry remain strictly in the hands of the government. For instance, local government forestry departments still retain the ability to decide when logging can occur and who will be able to do it. Forestry government officials hold lotteries for the distribution of a yearly quota of cutting, issue permits that prescribe a selective cut for farmers to undertake, and then audit for compliance.

The Chinese state has also heavily influenced forest management through its allocation of permits across China in national forests, which has served to widely restrict control over the amount of cutting (Hyde et al. 2008). In sum, the Chinese decentralization of rights is a mixed approach designed to bring access to capital to rural areas but imposes substantial limits on the decision-making power granted to forest users.

2.3 State of Global Forests

Efforts to address biodiversity conservation and forest degradation while enhancing livelihoods and economic opportunities have tended to focus policy makers' attention on three related approaches: promoting higher yields on existing forestlands and/or turning to higher production in degraded forests; designating high conservation value areas as "protected" from commercial extraction; and developing regulations to promote environmentally sensitive harvesting practices, such as delineation of buffer zones, selective harvesting, and eco-sensitive road building.

We review these trends later, which pave the way for assessing how economic globalization of the forest sector has both placed pressure on these efforts and fostered the development of global coalitions of domestic and international organizations attempting to promote economically viable, socially acceptable, and environmentally sensitive harvesting practices.

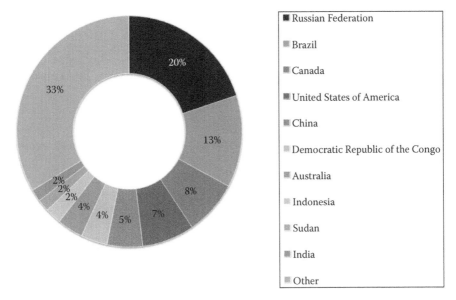

FIGURE 2.2
Ten countries with the largest forest area, 2010. (Based on data from Food and Agricultural Organization of the United Nations (FAO), *Global Forest Resources Assessment*, Food and Agricultural Organization of the United Nations, Rome, Italy, 2010b.)

2.3.1 Forest Area

Forest cover is distributed relatively evenly among the five main continents of Africa, Asia, Europe, and North and South America, with each hosting 15%–21% of the world's total. However, these general trends mask important subregional differences. In fact, just 15 countries contain 75% of the world's forest area (Figure 2.2). In Russia, 809 million ha of forests covers 49% of the country's land mass and constitutes 20% of total global forest area. Brazil hosts 510 million ha of forest, more than the combined forest cover of the United States (304 million ha) and China (207 million ha). Together, Russia, Brazil, United States, Canada, and China account for 53% of the world's forested area. Not including Russia, Europe has approximately 196 million ha, or 5% of the world's total, similar to China.

The proportion of forest area to total land area also varies (Table 2.1). Nearly 50% of South America is forested, while the Caribbean and Central America fall into the category of nations with high forest cover but accounts for a tiny fraction of the global total. Similarly, the world's most forested nations, as a percentage of land cover, are generally small countries, such as French Guiana and Gabon in Africa and Suriname in South America, and island archipelagos such as the Federated States of Micronesia, American Samoa, Palau, and the Solomon Islands.

2.3.2 Forest Losses and Gains

Although 2010 was the first instance in which the global rate of deforestation slowed, this can almost entirely be attributed to ambitious large-scale reforestation programs in China, India, and Vietnam. Deforestation still constitutes a major threat to forest areas throughout the world. This section overviews which regions are gaining and losing the most forest cover and the drivers behind these gains and losses.

TABLE 2.1

Distribution of Forests by Region and Subregion, 2010

Country/Region	Forest Area 2010 (1000 ha)	% of Total Land Area	% Global Forest Area
Africa	**674,419**	23	17
Eastern and Southern Africa	267,517	27	7
Northern Africa	78,814	8	2
Western and Central Africa	328,088	32	8
Asia	**592,512**	19	15
East Asia	254,626	22	6
South and Southeast Asia	294,373	35	7
Western and Central Asia	43,513	4	1
Europe	**1,005,001**	45	25
North and Central America	**705,393**	33	17
Caribbean	6,933	30	0
Central America	19,499	38	0
North America	678,961	33	17
Oceania	**191,384**	23	5
South America	**864,351**	49	21

Source: Food and Agricultural Organization of the United Nations (FAO), *Global Forest Resources Assessment*, Food and Agricultural Organization of the United Nations, Rome, Italy, 2010b.

2.3.2.1 Forest Losses

Unsustainable forest harvesting and land conversion to agriculture continue to be the main drivers behind global deforestation. From 2000 to 2010, 13 million ha of forest were lost every year (see Table 2.2), a decrease from the 16 million ha lost per year from 1990 to 2000 (FAO 2010b). Forests are also lost through natural causes, such as forest fires and disease. Forest fires are particularly a problem in Australia, Canada, Russia, and the United States. It is also believed that forest fires are a major problem in Africa, but quantitative data on the magnitude of area burned or number of incidents each year is lacking.

Countries experiencing the highest amount of deforestation over the last 20 years include Indonesia, which had the highest net forest loss in the 1990s but which slowed dramatically from 2000 to 2010. These changes can largely be attributed to the end of the Suharto regime and the Ministry of Forestry, which sanctioned large-scale deforestation. In the decade since Suharto fell from power, Indonesia's rate of deforestation decreased by over 70% and the magnitude of forest loss fell by nearly 75%.

Australia has recently emerged as high on the deforestation list, but this is owing to an increase in wildfires over the past decade, caused by extensive drought (FAO 2010b: 180). The devastating pandemic of bushfires Australia experienced in February 2009 certainly contributed to this as well with fires killing over 170 people and burning an area of 450,000 ha.

Between 2005 and 2010, Africa and South America experienced the highest magnitude of forest cover loss in the world (Table 2.3), together accounting for most of the world's forest loss (FAO 2010b: 19).

If we probe deeper, we see that rates and magnitude of deforestation have been decreasing in both regions. However, most of the improvement in Africa occurred in the northern region. Western and Central Africa and Eastern and Southern Africa continue to lose forests at roughly the same rate in 2010 as experienced in 1990. Sub-Saharan Africa is currently

TABLE 2.2

Ten Countries with Largest Annual Net Loss of Forest Area, 1990–2010

Country	Annual Change 1990–2000		Country	Annual Change 2000–2010	
	1000 ha/Year	%		1000 ha/Year	%
Brazil	−2890	−0.51	Brazil	−2642	−0.49
Indonesia	−1914	−1.75	Australia	−562	−0.37
Sudan	−589	−0.8	Indonesia	−498	−0.51
Myanmar	−435	−1.17	Nigeria	−410	−3.67
Nigeria	−410	−2.68	United Republic of Tanzania	−403	−1.13
United Republic of Tanzania	−403	−1.02	Zimbabwe	−327	−1.88
Mexico	−354	−0.52	Democratic Republic of Congo	−311	−0.2
Zimbabwe	−327	−1.58	Myanmar	−310	−0.93
Democratic Republic of Congo	−311	−0.2	Bolivia	−290	−0.49
Argentina	−293	−0.88	Venezuela	−288	−0.6

Source: Food and Agricultural Organization of the United Nations (FAO), *Global Forest Resources Assessment*, Food and Agricultural Organization of the United Nations, Rome, Italy, 2010b.

experiencing some of the highest rates of forest loss in the world. Of countries in this region, Nigeria and Zimbabwe are experiencing the greatest number of areas, and rate of loss. In South America, Brazil alone accounted for 61% of forest loss in 2010. Bolivia, Venezuela, and Argentina together accounted for an additional 23% of this total. However, as a region, Central America has the highest rate of forest loss of any region, at 1.2% per year (FAO 2010b: 18).

2.3.2.2 Forest Gains

Many countries have developed internal approaches to promoting gains in forest area. For example, China's concerns over desertification, Thailand's desire to avoid a recurrence of the severe 2011 floods in its central plains, and the Philippines' concerns to prevent land-slides have led each of these countries to initiate programs aimed at stabilizing soil and hilly slopes through forest regeneration.

Overall, the greatest gains in forest cover have been made through deliberate reforesta-tion (replacing forests that were logged or succumbed to fire) and afforestation efforts (planting in regions that previously had no trees). Afforestation plantings can generate natural forests or can be used to grow exotic and/or targeted species intensively, com-monly referred to as "plantation forestry." The Food and Agricultural Organization of the United Nations (FAO) estimated in 2005 (Figure 2.3) that timber production is the pri-mary function of around 76% of planted forests, but this share could be in decline with the advent of its large-scale tree planting campaigns, especially in China (FAO 2010b). Intensively managed forest plantations still constitute less than 3% of the world's forests.

The area of artificial regeneration forests has been increasing since 1990 in virtually all regions, with Asia leading the way (Figure 2.3). In 2005, East Asia accounted for 78% of the world's afforestation at 4.4 million ha. Thirty-three countries host artificial regeneration for-ests area above 1 million ha, accounting for 90% of the global artificial regeneration forests area. Five of these countries, China, the United States, the Russian Federation, Japan, and

TABLE 2.3

Annual Change in Forest Area by Region and Subregion, 1990–2010

Region	1990–2000		2000–2005		2005–2010	
	1000 ha/Year	% Change	1000 ha/Year	% Change	1000 ha/ Year	% Change
Africa	**−4067**	**−0.56**	**−3419**	**−0.49**	**−3410**	**−0.50**
Eastern and Southern Africa	−1841	−0.62	−1845	−0.65	−1832	−0.67
Northern Africa	−590	−0.72	−41	−0.05	−41	−0.05
Western and Central Africa	−1637	−0.46	−1533	−0.45	−1536	−0.46
Asia	**−595**	**−0.10**	**2777**	**0.48**	**1693**	**0.29**
East Asia	1762	0.81	3005	1.29	2557	1.04
South and Southeast Asia	−2428	−0.77	−363	−0.12	−991	−0.33
Western and Central Asia	72	0.17	135	0.32	127	0.29
Europe	**877**	**0.09**	**582**	**0.06**	**770**	**0.08**
North and Central America	**−289**	**−0.04**	**−40**	**−0.01**	**19**	**n.s.**
Caribbean	53	0.87	59	0.90	41	0.60
Central America	−374	−1.56	−247	−1.15	−249	−1.23
North America	32	−0.01	148	0.02	228	0.03
Oceania	**−36**	**−0.02**	**−327**	**−0.17**	**−1072**	**−0.55**
South America	**−4213**	**−0.45**	**−4413**	**−0.49**	**−3581**	**−0.41**

Source: Food and Agricultural Organization of the United Nations (FAO), *Global Forest Resources Assessment,* Food and Agricultural Organization of the United Nations, Rome, Italy, 2010b.

Note: The bold entries denote total for regions, e.g. the data for "Africa" refers to the totaled amount for Eastern and Southern Africa, West and Central Africa, and Northern Africa.

India, together account for more than half the world's planted forests (FAO 2010b: 91). The greatest proportion of planted forest is found in just a few countries within each region. For instance, in Northern Africa, 75% of the planted forest area is located in Sudan; in East Asia, 86% is found in China; and in South and Southeast Asia, 90% is in India, Indonesia, Malaysia, Thailand, and Vietnam. Afforestation is also increasing in arid zone countries in North Africa and the Middle East, such as Egypt, Kuwait, Libya, Oman, and the United Arab Emirates (FAO 2010b: 91). Table 2.4 lists countries with the largest net gain in forest area.

Overall, reforestation has increased, as Figure 2.4 shows, for over 20 years with China continuing to dominate the world's regreening efforts, planting at a rate that easily outpaced the rest of the world's combined reforestation effort between 2000 and 2010. India is also aggressively reforesting, with afforestation programs adding 304,000 ha per year of forests over the past decade.

India, and the South and Southeast Asian region, leads the world in reforestation. As of 2005, the region had rehabilitated 2.1 million ha of forest cover or 39% of the world's total.

2.3.3 Protected Areas

As of 2010, 13% of the world's forests (460 million ha) were located within legally protected areas such as national parks, game reserves, and designated wilderness areas (FAO 2010b). Figure 2.4 shows that the area under these forest classifications has increased by 94 million

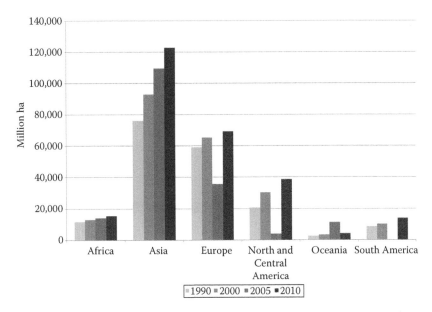

FIGURE 2.3
Planted forest area, 1990–2010. (Based on data from Food and Agricultural Organization of the United Nations (FAO), *Global Forest Resources Assessment*, Food and Agricultural Organization of the United Nations, Rome, Italy, 2010b.)

TABLE 2.4

Ten Countries with Largest Net Gain in Forest Area, 1990–2010

Country	Annual Change 1990–2000		Country	Annual Change 2000–2010	
	1000 ha/ Year	%		1000 ha/ Year	%
China	1986	1.2	China	2986	1.57
United States	386	0.13	United States	383	0.13
Spain	317	2.09	India	304	0.46
Vietnam	236	2.28	Vietnam	207	1.64
India	145	0.22	Turkey	119	1.11
France	82	0.55	Spain	119	0.68
Italy	78	0.98	Sweden	81	0.29
Chile	57	0.37	Italy	78	0.9
Finland	57	0.26	Norway	76	0.79
Philippines	55	0.8	France	60	0.38

Source: Food and Agricultural Organization of the United Nations (FAO), *Global Forest Resources Assessment*, Food and Agricultural Organization of the United Nations, Rome, Italy, 2010b.

ha since 1990 and continues to increase as more of the world's remaining primary forest is locked away for conservation of biodiversity, water, soil, cultural heritage, or other ecological services. Asia hosts the greatest extent of protected areas, at 125 million ha or 27% of the world's total. Central America has the highest extent of protected areas, covering nearly 55% of its forests, while Europe has the lowest proportion with less than 5%. However, there are significant challenges to enforcing protected area status "on the ground," given

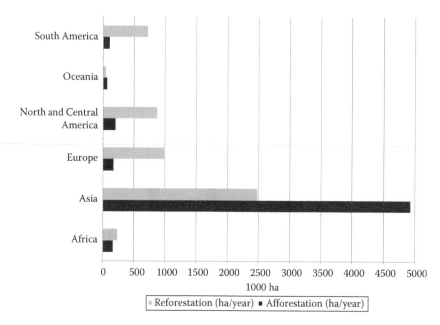

FIGURE 2.4
Reforestation and afforestation by region, 2010. (Based on data from Food and Agricultural Organization of the United Nations (FAO), *Global Forest Resources Assessment*, Food and Agricultural Organization of the United Nations, Rome, Italy, 2010b.)

the lack of financial and human resources currently earmarked for safeguarding these areas. Likewise even if fully enforced, research to date questions whether the current extent of protected areas is sufficient for conserving the world's biodiversity (Laurance et al. 2012).

2.3.3.1 Forest Regulations

Governments can foster the creation of an enabling environment by granting and enforcing legal access to forest resources, curbing illegal logging to reduce unfair competition, simplifying bureaucratic procedures for small to large forest enterprises (e.g., permitting processes for harvesting or transport), or providing financial incentives for start-ups (Donovan et al. 2007).

While it is beyond the scope of a single chapter to review the myriad of regulatory approaches to addressing and promoting sustainable and responsible forest management, a review of McDermott et al. (2010) comparative study of environmental practices and policies in 17 representative countries throughout the world provides a comprehensive snapshot of how different countries decide to develop policies and regulations governing environmental forest practices.

Challenging commonly held assumptions, McDermott et al. found that the lowest environmental prescriptions were found within developed countries in those jurisdictions regulating private land (most of these were found in the US southeast). While many developing countries had, on their books, relatively prescriptive regulations (see Table 2.5), they have difficulty being enforced owing to lack of resources, capacity, and training (Pacheco and Kaimowitz 1998).

There is also strong subnational variation. For example, in contrast to the low prescriptions in the US southeast, the *highest* levels of prescriptive regulations were also found in

TABLE 2.5

Average Prescriptiveness, Thresholds, and Enforcement: By "Development" and Tenure (Scale 0–10)

Category	Public	Private	All	Thresholds	Enforcement
Developing	6.7	6.0	**6.6**	High	Low
Developed	8.8	3.4	**6.1**	Mod (pvt) High (pub)	Mod (pvt) High (pub)
All	**7.9**	**3.5**	**5.6**		

Source: McDermott, C.L. et al., *Global Environmental Forest Policies: An International Comparison*, Earthscan, London, U.K., 2010.

Note: The bold entries denote the totals for all categories (e.g. total private for developed and developing countries, total public for developed and developing countries, etc.).

the United States, specifically governing public lands in the US Pacific Northwest region (McDermott et al. 2010).

These findings raise the need for nuanced deliberations about the ability of private forest management to promote environmental values that transcend timber sustainability and the appropriate mix of public and private management. At the very least, the assertion by post-Hardin economists that private land is better managed is simply too blunt to capture more complex incentives and disincentives facing forest managers, and policy makers (Cashore et al. 2010). This is not to say that higher regulations are always the most appropriate approach. Chavarria (2010) asserts that rules on the books often fail to be translated into on-the-ground practice, resulting in "illegal" activity. For example, obtaining legal harvesting permits in Honduras can take up to 8 months and require more than 40 steps, involving approximately 20 officials and foresters (Chavarria 2010). What we do know is that, depending on their design and approach, regulations can encourage forestland conversion and/or degradation, but they can also help foster sustainable forest management (SFM) and conservation, promote equitable access to forest users who depend upon forest resources for their livelihoods, and generate economically viable forestry sectors.

2.4 Opportunities for Forest User Engagement: Globalization and Four Pathways of Internationalization

To assess how forest users are shaped by, and also might be able to shape the increasing role of influences from beyond their borders, we draw on Bernstein and Cashore (2000, 2012a) to first distinguish economic globalization from internationalization, which reveals four different pathways through which strategists and users might travel. Bernstein and Cashore focus economic globalization on increasing levels of foreign trade, investment, and finance that influence and shape economic life (Bernstein and Cashore 2000). They use "internationalization," to capture "nondomestic" sources of influence on domestic policies and problems from environment, trade, development, and other organizations and agencies. These distinctions are important because some pathways may work to reverse asserted "downward" pressures of economic globalization on the ability of domestic governments to develop environmental and social stewardship, but only if strategic decisions follow the "logic" of the different pathways. To assess and elaborate on this framework, we first review economic globalization of the world's forest sector and then turn to review four pathways through which forest users might travel.

2.4.1 Economic Globalization and the Forest Sector

Economic globalization brought global market forces to bear on forestry management in all parts of the world, opening up new areas to market forces and bringing more actors into the policy-making process than ever before. This is driven by several key trends, which are outlined later.

2.4.1.1 Growth of Global Trade in Forest Products

Since 2000 there has been more than a 40% increase in the value of global trade in forest products (see Figure 2.5). In 2005, total international trade in wood products (including pulp, paper, solid wood, and secondary processed wood products) was valued at USD $257 billion and is expected to reach USD $450 billion by the year 2020 (ACPWP 2007).

Growth is expected to continue given fiber deficits in emerging markets such as China and India and because of improvements in transportation technology (Roberts 2008). These improvements are creating access to new markets and sources of raw material, enabling some countries' forest industries to enter the global market (Nair 2003). Some forest-related industries depend entirely upon global trade. In 2008, 80% or about 4.5 million cubic meters of the raw materials needed for the global furniture industry were imported from other countries around the world (Forest Trends 2009).

2.4.1.2 Emerging Economies' More Prominent Role in International Trade

Until recently, developed countries such as the United States, Canada, and the European Union (EU) have dominated international trade in forest products. However, wood production and manufacturing shifted from these developed countries to emerging economies

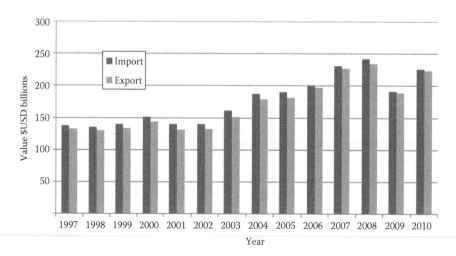

FIGURE 2.5
World import and export of forest products, 1997–2010. (From Food and Agriculture Organization of the United Nations (FAO) 2012. FAOSTAT. http://faostat.fao.org/site/626/DesktopDefault.aspx?PageID=626#ancor, accessed January 25, 2012.)

such as Brazil, Russia, and China, the latter of which primarily imports raw material from tropical Asia and Africa. In a decade, the percentage of US sales of wood household furniture from imports increased from 20% to more than 50% and continues to expand. This has significantly impacted sawmills and wrought drastic structural changes to US domestic industry (Ince et al. 2007). Many domestic forest sectors in the global North are now facing increasing consolidation, decreasing forest commodity prices, and increasing wood supply from competitors in the global south.

This brief review of economic globalization of the forest sector reveals the interdependence of global forest product markets, creating significant competition across countries and regions. These relationships mean that policies taken in one country will have significant, and perhaps countervailing, effects in other countries. For example, Thailand, Vietnam, India, and China—four countries with significant domestic timber-processing sectors—have partial or complete logging bans in place. But such policies end up creating pressures for extraction elsewhere. In fact, these countries rely on timber from neighboring countries, such as Cambodia and Laos, to meet their fiber needs (Luttrell and Brown 2006; EIA 2008a). One study of the international implications of Vietnam's forest policy calculated that approximately 39% of the volume of wood regrowth that took place in Vietnam's forests was achieved by the displacement of forest extraction to other countries (Meyfroidt and Lambin 2008).

As environmental governance arrangements have evolved as a function of multiple actors, institutions, and issues interacting with each other, so has the global forest products industry itself as it becomes increasingly integrated into a globalized world economy. For example, several countries in Latin America experienced resurgence in forest cover that was driven by the globalization of their agricultural sectors. Their domestic agricultural production was replaced by imports and by foreign subsidies, leading to an increase in available forestland. In Indonesia, the major cause of deforestation is expansion of an agricultural crop, namely, oil palm. It is, thus, becoming increasingly apparent that the major drivers of change in the forest sector come from external sources, such as economic, demographic, technological, and environmental changes (FAO 2010a).

What is certain is that the flow of capital investment to new regions and growing global competition in wood-based commodity product markets are forcing a reassessment of long-established manufacturing and marketing strategies in developed countries, particularly in North America (Bowyer 2004). This has resulted in a fundamentally altered structure of domestic forest sectors in developed countries, in which wood-based industries are forced to innovate and adapt for short-term survival and long-term growth (UNECE 2010). The result is that traditional, sector-oriented forecasting techniques have difficulty capturing the dramatic system-wide changes affecting forestry (Essmann et al. 2007). Hence, economic globalization has brought global market forces to bear on forest management in all parts of the world, opening up new areas to market forces and bringing more actors into the policy-making process than ever before. In order to understand these dynamics, as well as the potentially economic, social, and environmental stewardship synergies that scholars and practitioners might trigger, we turn to explore, and identify, four pathways that might be traveled.

2.5 Four Pathways of Internationalization

Bernstein and Cashore (2000, 2012a) argue that while much scholarly and practitioner attention has been placed on assessing and understanding the influences of economic globalization, much less attention has been placed on assessing the four pathways of internationalization: international rules, norms, markets, and "direct access," which may provide forest stakeholders in general and forest users in particular with new ways to engage in policy making and problem solving.

2.5.1 International Rules

For many years, multilateral environmental treaties were viewed as the leading way to address globally relevant environment and resource challenges. This "international rules" pathway focused strategist efforts, including governments and forest users, to create globally binding obligations on states through international law (Franck 1990).

Examples of international rules include single-issue treaties to address broad problems such as protection of biodiversity and endangered species through the Convention on Biological Diversity (CBD), Convention on International Trade in Endangered Species of Wild Fauna and Flora (CITES), protecting ecological integrity of landscapes through the Ramsar Convention, United Nations Convention to Combat Desertification (UNCCD), International Timber Trade Agreement (ITTA), regulation of chemicals or emissions through the Montreal Protocol, and the United Nations Framework Convention on Climate Change (UNFCCC). Most of these agreements were not initiated to address forests directly but have provisions relevant for forests. Such attention is even found in trade-based treaties at times, such as the United States–Peru Trade Promotion Agreement requires Peru to put several regulatory and control measures in place (Bernstein et al. 2010). This agreement, in turn, has driven change to Peruvian Forest Law and related endangered species legislation (Tomaselli and Hirakuri 2008).

Two possibilities exist for a forest user seeking to influence domestic policy through this pathway. They can work to shape and promote some type of international agreement, or they can pressure domestic governments who are not yet complying. Such efforts may be easier when the country in question is dependent on foreign markets, which leads to the importance in thinking about linking strategies across pathways as well.

2.5.2 International Norms and Discourse

Norms can also be an important source of influence. For example, there is now general agreement among many forestry users that three principles must be part of efforts to build "good forest governance": inclusiveness, transparency, and accountability (Cashore 2009a). These dominant norms, which have emerged through international multi-stakeholder deliberations and intergovernmental negotiations, are often championed by powerful international organizations such as the World Bank and large environmental NGOs.

A most prominent example of norm diffusion from the international to the domestic arena is the promotion of SFM (Singer 2008), which in practice attempts to capture environmental, social, and economic objectives into a single goal. Inspired by Brundtland's famous definition of sustainability as "meeting the needs of current generations without compromising the ability of future generations to meet their needs," SFM has become a powerful tool for framing debates about appropriate types of behavior in country-level contexts. The concept itself is strongly supported by a range of actors from the International Tropical Timber Organization (ITTO), which has played a major role in the development of criteria and indicator (C&I) for SFM processes. However, unlike hard law, C&I processes aim to define and assess rather than mandate SFM. The hope is that such processes will help states to develop their own internal approaches, such as the African Timber Organization, published a set of principles, criteria, and indicators that were inspired by the ITTO C&I in 2003 (Cashore et al. 2010).

There are a few factors that dictate the level of influence the norms pathway ultimately has on domestic forest policy. For one, the target state must be sensitive to challenges to its international reputation. This often arises following NGO campaigns that raise awareness and attempt to shame particular governments within the global community (Bernstein and Cashore 2012). Another route this can take is through international organizations, such as the World Bank, that often influences domestic policy change following engagement with governments seeking financial assistance. Bernstein and Cashore argue that users seeking to travel this path are often successful when they seek ways to make new norms resonate with domestic ideology, culture, and broader policy goals. For example, forest law reforms in Africa were aimed at promoting concepts deemed essential to SFM that resonated with local stakeholders, such as the need for greater participation of local people in forest resource management, subsidiarity, and benefit-sharing.

2.5.3 Markets

The market pathway encompasses processes or tactics that attempt to work with or leverage markets to create domestic policy change. Experience over the past 30 years demonstrates that market influence on domestic policy making in the forest sector is characterized by three key trends. First, there is rising awareness among consumers of how the actions of prominent companies along the timber supply chain negatively affect forests (e.g., environmental NGO boycott campaigns against retailers such as Home Depot, B&Q, as well as forest companies such as Domtar). Second, international organizations, such as the World Bank and the International Monetary fund, often require domestic policy reforms in exchange for financial assistance. Third, market-driven eco-labeling programs have emerged to "reward" firms that abide by preestablished standards.

The experience of global forest certification over the last 20 years provides an excellent example of how the market pathway can influence domestic policy making. Following the Rio Earth Summit in 1992, the World Wide Fund for Nature (WWF), and a coalition of environmental groups, retailers, and forest companies created the Forest Stewardship Council (FSC) in order to certify forestland owners and companies that practiced "sustainable forestry" in accordance with preestablished performance criteria. The idea of forest certification expanded upon the traditional "stick" approach of NGO-led boycott campaigns by offering "carrots" in the form of market access, firm recognition, and price premiums. The combination of the "carrot" and "stick" approach of economic incentives and disincentives thus helped expand the market's influence over domestic forestry practices (Auld and Cashore 2012).

These dynamics have been important for shaping market and supply chains, as support for forest certification took different shapes according to different national contexts across the world. However, the extent and impact of forest certification as an illustration of the potential of the market pathways provides some mixed results. The concept of forest certification has garnered widespread support in Europe and North America, although strong debates exist still among supporters of the environmental group-initiated FSC program, and the industry/landowner initiated Programme for the Endorsement of Forest Certification (PEFC) (Auld and Cashore 2012).

However, one of the key criticisms of forest certification is that it has failed to take hold in both tropical forests and developing countries. The progress of certification in these regions has been slow and uneven, reflecting challenges with poor infrastructure, corrupt institutions, a lack of resources, and environmentally insensitive domestic and foreign markets (Cashore et al. 2006a,b). Furthermore, rather limited economic incentives have hindered certification attempts to make inroads in the tropics, where these challenges are arguably more costly and difficult to address (Cashore et al. 2006a,b).

For these reasons, market incentives have brought through more modest efforts to promote "legal compliance," which can be seen as both illustrating the difficulty for users in traveling the market pathway in isolation and potential synergies by focusing on other pathways such as "direct access." In Africa, the dependence of the timber sector on European markets has been a catalyst for governments, including Ghana and the Republic of the Congo, to engage in "voluntary partnership agreement" (VPA) negotiations with the EU (Cashore et al. 2010).

2.5.4 Direct Access to Domestic Policy-Making Processes

As Bernstein and Cashore elaborate, "this pathway captures those processes in which non-domestic financial resources, technical knowledge, expertise, training, and learning can dramatically shape domestic politics. It works by mobilizing and funding societal interests, generating new coalitions, and providing resources for enduring impacts on domestic governance in ways that reinforce domestic sovereignty, rather than attempt to challenge from outside." This pathway can often result in large impact because it provides resources to civil society organizations, changing the relative influence of different actors and domestic policy networks (Bernstein and Cashore 2012a).

The key challenge for forest users is that traveling this pathway requires a high degree of sensitivity about concerns domestic players might have about being viewed as inappropriate sources of international interference. To address this, travelers must focus on helping improve capacity, knowledge, and training of domestic actors, in essence empowering them to be more productive members of domestic policy networks.

There are many examples of this type of influence within forestry. For instance, in Indonesia, international organizations, government and nongovernmental, have devoted considerable time, resources, and technology aimed at empowering community and marginalized groups following the fall of the Suharto regime. These include the World Resources Institute, The Nature Conservancy, and even Global Witness, to name but a few (Barr et al. 2006). In addition, donor agencies, including the UK Department for International Development (DFID) and Norway's Partnerships to GIZ, the German Aid Agency, have actively worked to build networks within specific countries. Similarly, informal policy networks such as the Association of Southeast Asian Nations' regional knowledge network region (Göhler et al. 2012) seem to be particularly effective in Asia because of "a cultural aversion to formal institutional arrangements and a reflection of an Asian style of governance and diplomacy" (Nesadurai and Stone 2000). Unquestionably, a range of international aid agencies, institutions, NGOs, and academic institutions have traveled this pathway in the last 20 years under the auspices of "capacity building," which often works to reinforce rather than to challenge domestic sovereign authority.

2.5.5 Case Study Example: Timber Legality Verification and the Problem of Illegal Logging

The previous discussion has revealed the importance of four pathways through which nondomestic influences are felt but also opportunities for domestic forest users to engage international opportunities and shape domestic policy responses. It is also the case that strategists can simultaneously travel more than one pathway. However, this requires understanding when traveling multiple pathways is synergistic and when to avoid contradictory or conflicting strategies. The case of global efforts to weed out illegal logging by reinforcing, rather than challenging, domestic rules illustrates these trends.

2.5.5.1 Case of Illegal Logging

Illegal logging is among the worst forest practices around the world (Kaimowitz 2005), especially in tropical developing countries where biodiversity loss poses a global challenge (Tacconi 2007). The prominence of illegal logging can be traced back, in large part, to widespread corruption, weak governance capacity, and enforcement gaps (Cashore 2009b). In Vietnam, forest protection officers uncovered some 300,000 violations of forest regulations between 2001 and 2006 (Sikor and Phuc 2011). The Government of Vietnam reported to the United Nations Forum on Forests in 2005 that illegal harvests accounted for more than half of national roundwood supply (ibid.). In Ghana, the annual production of timber by illegal chainsaw milling outside the forest reserves is reported to be as high as 2.5 million cubic meters, five times the total annual allowable cut (Marfo 2010). Illegal chainsaw milling is estimated to provide jobs for about 130,000 Ghanaians and livelihood support for about 650,000 people. In Indonesia, the rate of illegal logging was estimated to have halved from roughly 90% to 40% by 2009 (EIA 2008a). In 2011, the ITTO estimated that the volume of illegal logging was roughly half of the official harvest (Blaser et al. 2011).

In an in-depth, multi-country study of illegal logging, Seneca Creek Associates (2004) estimated that about USD $5 billion worth of illegal timber enters world markets each year, representing as much as 10% of the value of global trade of primary wood products. In developing countries, illegal logging in public lands alone causes estimated losses in assets and revenue in excess of USD $10 billion annually, more than eight times the total official development assistance dedicated to the sustainable management of forests in developing countries (World Bank 2006). Two studies of the global economic implications of eliminating illegal logging (Seneca Creek Associates 2004; Li et al. 2008) found that the losses due to reducing illegal logging are concentrated in developing countries or countries in transition to market economies. In almost all countries without illegal harvests, however, the elimination of illegal logging is predicted to lead to significant increases in the price and production of wood products. One economic simulation predicted that the prices for industrial roundwood would increase by 1.5%–3.5% and up to 2% for processed products (Li et al. 2008), while other simulations estimated this figure to be between 7% and 16% on average and 2% and 4% in the United States (Seneca Creek Associates 2004). This prediction suggests that an economic incentive exists for legitimate producers in all countries to support measures to reduce illegal logging.

Partly as a result of concerns regarding the effectiveness of global certification systems and good forest governance initiatives, "legality verification" is now emerging as a leading policy instrument with which to combat illegal logging. Like forest certification, legality verification relies on third-party verification. However, unlike certification, it does not rely on altruistic customers' support of eco-friendly practices. Several developed countries have enacted public procurement policies aimed at excluding illegal and unsustainable timber products. It is argued that these procurement policies are an effective tool because they can be developed and implemented more rapidly than most other policy options and the evidence suggests that they can have a much broader impact on consumer markets than simply through the direct effect of government purchases (Brack 2008).

Two primary policy instruments have emerged to promote legal compliance. In Europe, a Forest Law Enforcement and Governance and Trade (FLEGT) program was developed to facilitate VPA negotiations between the EU and producer country designed to promote legal compliance and "good forest governance." The second instrument consists of changes to domestic trade legislation forbidding the importation of illegally harvested wood products. This took the form in the United States of amendments to the Lacey Act to require that importers show "due care"* not to import illegal wood, while the European Parliament passed EU Timber Regulation (EUTR) requiring that importers exercise "due diligence" to avoid illegal wood imports. The Lacey Act and the EUTR are both designed to ask companies to record and be liable for the wood products they bring across national borders.† They risk not only fines of up to USD $200,000 but also having shipments seized that could run into millions of dollars of lost material (EIA 2008b).

This example of illegal logging illustrates how all four pathways might be traveled. The international rules pathway is traveled through VPA processes that lead to some type of intergovernmental agreement with agreed upon rules and approaches for compliance. Likewise the EUTR and Lacey Act draw on legislation that affects international trade. However, the market pathways are also traveled because instead of challenging sovereignty, these rules reward with market access countries who enforce their own domestic policies. Furthermore, governments benefit from the higher tax revenues associated with higher forms of legal compliance. Moreover, legally operating firms benefit economically since prices will rise as supply is restricted (Cashore and Stone 2012).

The norms pathway appears to be important as well. Here, coalitions of timber producers and environmentalist NGOs eventually coalesced around the norm/idea/problem definition of "illegal logging," which arguably paved the way for the championing of legislative amendments. Such a problem definition then helped entrench a coalition of domestic environmental group and industry interests‡ but who targeted forest practices located elsewhere.§ Finally, the "direct access" pathway is also being traveled with these efforts, as the legality verification process shows that major corporations, international environmental networks, and foreign governments all can show interest in timber-producing countries' political economy and work to influence it through infiltrating their policy-making processes. In the FLEGT process, for example, multi-stakeholder dialogues are required as part of the VPAs. Multi-stakeholder dialogue provides avenues for domestic interest groups and foreign allies to coordinate their activity to potentially shift power dynamics within the timber-producing countries such that environmental concerns can be moved to the foreground. Furthermore, because the EU acts as the judge of how representative these multi-stakeholder dialogues are, the VPA process becomes a way for EU political actors to directly participate in and influence the political discourse in timber-producing countries. Thus, in the case of legality verification, all of the pathways are intersecting at a powerful nexus of shared commercial and conservation incentives. The evolution of this policy change could not be said to originate in purely economic or ideational reasons, but rather the interplay between these motives leading to broad support, evolving norms, and strengthening legal mechanisms.

---

* The Ninth Circuit defined due care as "Due care means that degree of care which a reasonably prudent person would exercise under the same or similar circumstances (APHIS 2009)."
† The Lacey Act also applies to state borders within the United States as well as bringing wood through customs, though it is rarely enforced in that capacity.
‡ Interview with official at Environmental Investigation Agency. January 20, 2011.
§ Interview with official at American Forest and Paper Association. March 15, 2011.

2.6 Conclusion

The continued and accelerating economic integration of the global forest sector, and changing tenure arrangements, have profoundly affected the role and ability of forest users across the world. On the one hand, tenure reforms are giving increased access to local communities and forest livelihoods. On the other hand, increasing trade and investment means that many users are forced to adapt to increasing competition. As supply in Asian, Africa, Russia, and East Europe continue to grow, as does demand in Asia, many forest users have been forced to adapt to these competitive pressures. Likewise, those users promoting environmental and social stewardship worry that they are fighting an uphill battle as forests face increasing pressures from extraction and conversion to others uses, such as palm oil.

Despite these trends that appear to put so many users on the defensive and responding to overwhelming global forces, we have also reviewed how concerted efforts on the part of globally focused actors might be able to help shape social, environmental, and economic stewardship in the global era. Such an approach requires careful attention to the myriad of ways in which international influence might shape domestic policies and practices. We have drawn on Bernstein and Cashore's concept of four pathways with which to understand and identify how forest users proactively engage in shaping policy outcomes and forest management approaches.

To be sure, the ultimate choices made about forest management are still made by governments and actors within particular countries. However, the forces that influence these policy decisions—from economic globalization to consumption—extend from beyond their own borders. Recognition of this means that the world's forest users, and the governments that preside over them, must decide whether they are going to respond simply to pressures associated with economic integration and increased demand, or whether they are going to be part of proactive solutions that bind and shape their individual and collective identities.

Such possibilities include "Bootleggers and Baptists" coalitions of environmental, social, and business interests (Cashore and Stone 2012) that can, in turn, champion new requirements and norms of open, deliberative approaches to improve global forest policy making. This chapter has offered one way through which such efforts might be built that tend not only to the structural impacts of economic globalization but also towards enabling a range of forest users to develop and shape policy solutions.

References

ACPWP. 2007. Global wood and wood products flow: Trends and perspectives. Shanghai, People's Republic of China: Advisory Committee on Paper and Wood Products; Food and Agriculture Organization of the United Nations.

Auld, G., and B. Cashore. 2012. The forest stewardship council. In *Business Regulation and Non-State Actors: Whose Standards? Whose Development?* P. Utting, D. Reed, and A. M. Reed (eds.), London, U.K.: Routledge.

Barr, C., I. A. P. Resosudarmo, A. Dermawan, and B. Setiono. 2006. Decentralization's effects on forest concessions and timber production. In *Decentralization of Forest Administration in Indonesia: Implications for Forest Sustainability, Economic Development and Community Livelihoods.* C. Barr, I. A. P. Resosudarmo, A. Dermawan, and J. McCarthy (eds.), Bogor, Indonesia: CIFOR.

Bennett, M., A. Mehta, and J. Xu. 2008. Incomplete property rights, exposure to markets and the provision of ecosystem services in China. Working Paper: Environmental Economics Program in China.

Bernstein, S., and B. Cashore. 2000. Globalization, four paths of internationalization and domestic policy change: The case of eco-forestry in British Columbia, Canada. *Canadian Journal of Political Science* 33 (1):67–99.

Bernstein, S., and B. Cashore. 2012a. Complex global governance and domestic policies: Four pathways of influence. *International Affairs* 88 (3):585–604.

Bernstein, S., and B. Cashore. 2012b. Re-thinking environmental 'effectiveness': Complex global governance and influence on domestic policies. *International Affairs* (Under review).

Bernstein, S., B. Cashore, R. Eba'a Atyi, A. Maryudi, K. McGinley, T. Cadman, L. Gulbrandsen et al. 2010. Examination of the influences of global forest governance arrangements at the domestic level. In *Embracing Complexity: Meeting the Challenges of International Forest Governance*, J. Rayner, A. Buck, and P. Katila (eds.), Vienna, Austria: International Union of Forest Research Organizations (IUFRO).

Blaser, J., A. Sarre, D. Poore, and S. Johnson. 2011. Status of tropical forest management 2011. *ITTO Technical Series*, Yokohama, Japan: International Tropical Timber Organization.

Bliss, J. C., E. C. Kelly, J. Abrams, C. Bailey, and J. Dyer. 2009. Disintegration of the U.S. industrial forest estate: Dynamics, trajectories, and questions. *Small Scale Forestry* 9 (1):53–66.

Bowyer, J. L. 2004. Changing realities in forest sector markets. *FAO Unasylva* 55 (219): 59–64.

Brack, D. 2008. Controlling illegal logging: Using public procurement policy. In *Briefing Paper*, C. House (ed.), London, U.K.: C. House.

Butler, B. 2008. Private forest owners of the United States. Edited by G. T. R. NRS-27. Newtown Square, PA: Northern Research Station, USDA Forest Service.

Cashore, B. 2009a. Key components of good forest governance in ASEAN Part I: Overarching Principles and Criteria. Edited by Exlibris: ASEAN-German ReFOP project, the analysis and making of regional public policy, www.aseanforest-chm.org

Cashore, B. 2009b. Key Components of Good Forest Governance in ASEAN Part II: Institutional Fit, Policy Substance, Policy Instruments, and Evaluation. Edited by Exlibris: ASEAN-German ReFOP project, the analysis and making of regional public policy, www.aseanforest-chm.org

Cashore, B., F. Gale, E. Meidinger, and D. Newsom. 2006a. Forest certification in developing and transitioning countries: Part of a sustainable future? *Environment* 48 (9):6–25.

Cashore, B., F. Gale, E. Meidinger, and D. Newsom (eds.). 2006b. Confronting sustainability: Forest certification in developing and transition countries, Report Number 8. New Haven, CT: Yale F&ES Publication Series.

Cashore, B., G. Galloway, F. Cubbage, D. Humphreys, P. Katila, K. Levin, C. McDermott, A. Maryudi, and K. McGinley. 2010. The ability of institutions to address new challenges. In *World Forests, Society and Environment*, M. Gerardo (ed.), International Union of Forest Research Organizations, Vienna, Austria.

Cashore, B., G. Hoberg, M. Howlett, J. Rayner, and J. Wilson. 2001. *In Search of Sustainability: British Columbia Forest Policy in the 1990s*. Policy venues, policy spillovers, and policy change: The courts, aboriginal rights, and British Columbia Forest Policy. Vancouver, British Columbia, Canada: University of British Columbia Press.

Cashore, B., and M. W. Stone. 2012. Can legality verification rescue global forest governance? Analyzing the potential of public and private policy intersection to ameliorate forest challenges in Southeast Asia. *Forest Policy and Economics* 18:13–22.

Chavarria, O. A. 2010. *Incidencia de la legislacion forestal en el recurso maderable de fincas agroforestales con enfasis en sistemas silvopastorales de Copan, Honduras*. Turrialba, Costa Rica: CATIE.

Clutter, M., B. Mendell, D. Newman, D. Wear, and J. Greis. 2005. Strategic factors driving timberland ownership changes in the US South. S. R. Station, USDA Forest Service (ed.), Athens, GA: S. R. Station, USDA Forest Service.

Donovan, J., D. Stoian, S. Grouwels, D. Macqueen, A. Van Leeuwen, G. Boetekees, and K. Nicholson. 2007. Towards an enabling environment for small and medium forest enterprise development. In *Policy brief edited by F. CATIE, IIED, SNV, ICCO*.

Environmental Investigation Agency (EIA). 2008a. Borderlines: Vietnam's booming furniture industry and timber smuggling in the Mekong Region.

Environmental Investigation Agency (EIA). 2008b. The U.S. lacey act frequently asked questions. www.eia-global.org/lacey/P6.EIA.LaceyReport.pdf: EIA.

Essmann, H. F., G. Andrian, D. Pattenella, and P. Vantomme. 2007. Influence of globalization on forests and forestry. In *Allgemeine Forst und Jagdzeitung*. 178(4): 59–67. Frankfurt, Germany.

FAO. 2010a. Forest policies, legislation and institutions in Asia and the Pacific. Trends and emerging needs for 2010. In *Asia-Pacific Forestry Sector Outlook Study II. Working Paper Series*, FAO (ed.), Rome, Italy: Food and Agricultural Organization of the United Nations.

FAO. 2010b. *Global Forest Resources Assessment*. Rome, Italy: Food and Agricultural Organization of the United Nations.

Food and Agriculture Organization of the United Nations (FAO) 2012. FAOSTAT. http://faostat.fao.org/site/626/DesktopDefault.aspx?PageID=626#ancor, accessed January 25, 2012.)

Franck, T. 1990. *The Power of Legitimacy among Nations*. New York: Oxford University Press.

Göhler, D., B. Cashore, and B. Blom. 2012. Forest governance. In *Forests and Rural Development*, D. P. Darr (ed.), Berlin, Germany: Springer Verlag.

Hickman, C. 2007. TIMOs and REITs. Edited by R. a. Development: USDA Forest Service.

Hyde, W. F., J. Wei, and J. Xu. 2008. Economic growth and the natural environment: The example of China and its forests since 1978. Environment for Development.

Ince, P., A. Schuler, H. Spelter, and W. Luppold. 2007. Globalization and structural change in the US forest sector: An evolving context for sustainable forest management. In *A Technical Document Supporting the USDA Forest Service Interim Update of the 2000 RPA Assessment*, F. S. USDA (ed.), Fort Collins, CO: F. S. USDA.

Kaimowitz, D. 2005. Illegal logging: Causes and consequences. Paper read at the forests dialogue on illegal logging, Hong Kong, China.

Kung, J. K., and S. Liu. 1997. Farmers' preferences regarding ownership and land tenure in Post-Mao China: Unexpected evidence from eight counties. *The China Journal* 38:33–63.

Laurance, W. F., D. C. Useche, J. Rendeiro, M. Kalka, C. J. A. Bradshaw, S. P. Sloan, S. G. Laurance, M. Campbell, K. Abernethy, P. Alvarez, V. Arroyo-Rodriguez, P. Ashton, J. Benitez-Malvido, A. Blom, K. S. Bobo, C. H. Cannon, M. Cao, R. Carroll, C. Chapman, R. Coates, M. Cords, F. Danielsen, B. De Dijn, E. Dinerstein, and M. A. Donnelly. 2012. Averting biodiversity collapse in tropical forest protected areas. *Nature* 489 (September 13):290–294.

Li, R. H., J. Buongiorno, J. A. Turner, S. Zhu, and J. Prestemon. 2008. Long-term effects of eliminating illegal logging on the world forest industries, trade, and inventory. *Forest Policy and Economics* 10 (7–8):480–490.

Liu, C. 2008. *Collective Forestry Institutions and Forestry Development in China*. Beijing, China: Economic Science Press.

Luttrell, C., and D. Brown. 2006. VERIFOR country case study 4: The experience of independent forest monitoring in Cambodia. London, U.K.: VERIFOR-OD1.

Marfo, E. 2010. *Chainsaw Milling in Ghana: Context Drivers and Impacts*. Wageningen, the Netherlands: Tropenbos International.

McDermott, C. L., B. Cashore, and P. Kanowski. 2010. *Global Environmental Forest Policies: An International Comparison*. London, U.K.: Earthscan.

Meyfroidt, P., and E. F. Lambin. 2008. Forest transition in Vietnam and its environmental impacts. *Global Change Biology* 14:1–18.

Nair, C. T. S. 2003. Forests and forestry in the future: What can we expect in the next 50 years? In *XII World Forestry Congress 2003*, Montreal, Quebec, Canada.

Nesadurai, H., and D. Stone. 2000. Southeast Asian research institutes and regional cooperation. In *Banking on Knowledge: The Genesis of the Global Development Network*, D. Stone (ed.), New York: Routledge.

Noone, M. D., S. A. Sader, and K. R. Legaard. 2012. Are forest disturbance rates and composition influenced by changing ownerships, conservation easements, and land certification? *Forest Science* 58 (2):119–129.

Pacheco, P., and D. Kaimowitz. 1998. Municipios y gestacion forestal en al tropico boliviano. La Paz: Center for International Forestry Research, Centro de Estudios para el Desarrollo Laboral y Agrario, Fundacion TIERRA, Project for Sustainable Forest Management BOLFOR.

Roberts, D. G. 2008. Globalization and its implications for the Indian forestry sector. *International Forestry Review* 10 (2):401–413.

Seneca Creek Associates. 2004. *Illegal Logging and Global Wood Markets: The Competitive Impacts on the U.S. Wood Products Industry*. Poolesville, MD: Seneca Creek Associates, LLC.

Sikor, T., and X. T. Phuc. 2011. Illegal logging in Vietnam: Lam Toc (Forest Hijackers) in practice and talk. *Society and Natural Resources* 24 (7):688–701.

Singer, B. 2008. Putting the national back into forest-related policies: The international forests regime and national policies in Brazil and Indonesia. *International Forestry Review* 10 (3):523–537.

Staddon, C. 2001. Restitution of forest property in post-communist Bulgaria. *Natural Resources Forum* 24:237–246.

Sunderlin, W. D., S. Dewi, A. Puntodewo, D. Muller, A. Angelsen, and M. Epprecht. 2008. Why forests are important for global poverty alleviations: A spatial explanation. *Ecology and Society* 13 (2):24.

Tacconi, L. (ed.) 2007. *Illegal Logging: Law Enforcement, Livelihoods and the Timber Trade*. London, U.K.: Earthscan.

Tomaselli, I., and S. R. Hirakuri. 2008. Converting mahogany. *ITTO Tropical Forest Update* 18 (4):12–15.

UNECE. 2010. Forest products annual market review 2009–2010: Innovation for structural change recovery. Edited by G. a. S. United Nations Economic Commission for Europe.

World Bank. 2006. *Strengthening Forest Law Enforcement and Governance: Addressing a Systemic Constraint to Sustainable Development*. Washington, DC: The World Bank.

Zhou, S. 2006. *Forestry in China*. Beijing, China: China Forestry Press.

Section II

Changes in Product Categories and Individual Markets

3

Markets and Market Forces for Lumber*

Matt Bumgardner, Steven Johnson, William Luppold,
Frances Maplesden, and Ed Pepke

CONTENTS

---

* © European Forest Institute.

Lumber, also called sawnwood, is an important historical antecedent of the wood products industry. The Hierapolis sawmill, a Roman waterpowered stone sawmill at Hierapolis, Asia Minor (modern-day Turkey), dating to the second half of the third century AD, is the earliest known sawmill. It is also the earliest known machine to incorporate a crank and connecting rod mechanism. Waterpowered stone sawmills working with cranks and connecting rods, but without gear train, are archaeologically attested for the sixth century AD at the Eastern Roman cities Gerasa and Ephesus (Ritti et al. 2007).

The earliest literary reference to a working sawmill comes from a Roman poet, Ausonius, who wrote an epic poem about the river Moselle in Germany in the late fourth century AD. At one point in the poem, he describes the shrieking sound of a watermill cutting marble. Marble sawmills also seem to be indicated by the Christian saint Gregory of Nyssa from Anatolia around 370/390 AD, demonstrating a diversified use of waterpower in many parts of the Roman Empire (Wilson 2002). Sawmills became widespread in medieval Europe again, as one was sketched by Villard de Honnecourt in c. 1250 (Singer et al. 1956). They are claimed to have been introduced to Madeira following its discovery in c. 1420 and spread widely in Europe in the sixteenth century (Petersen 1973). By the eleventh century, hydro-powered sawmills were in widespread use in the medieval Islamic world, from Islamic Spain and North Africa in the west to Central Asia in the east (Lucas 2005).

The continued history of sawmilling includes dramatic new applications of technology to increase efficiencies and improve lumber product quality. These events accelerated with the advent of the industrial revolution that included the conversion from circular saws to band saws, the use of the steam engine to power sawmills, and the shift to gasoline and electricity. Today, technology and business practices and globalization have created a truly international marketplace for lumber.

Although lumber is ubiquitous on the international stage, not all lumber is similar or interchangeable in properties, applications, and in the production of potential downstream value-added or secondary wood products. As such, this chapter is divided into the three broadest generally recognized categories of lumber. Section 3.1 focuses on softwood lumber that is mainly used for building construction but also a multitude of other uses. Section 3.2 focuses on temperate hardwood lumber used in products ranging from industrial pallets to the finest custom furniture. Section 3.3 focuses on tropical hardwood lumber. Tropical lumber and other wood products have evolved on a different timeline than softwood or temperate hardwood products. Many tropical wood species are among the most valuable in the world while at the same time are often facing extinction with international efforts to protect them. The three parts of this chapter generally follow a uniform organization with an introduction, a discussion of global production and consumption, an analysis of international trade, and perspectives on influences on the sector and future markets.

3.1 Softwood Lumber Products and Markets

Ed Pepke

3.1.1 Introduction

Softwood (coniferous) lumber, also known as sawn softwood, has been an internationally traded commodity for centuries, with production and trade in softwood lumber intensifying in the twenty-first century. Softwood lumber has been used mainly for building

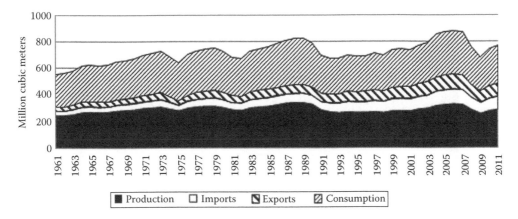

FIGURE 3.1
Softwood lumber consumption, production, and trade evolution.

construction, although it also has a multitude of other valuable uses. New uses continue to be developed that extend its product life cycle beyond that of a simple commodity. Softwood lumber production, consumption, and trade typically follow global economic cycles and, because softwood lumber is most used for home construction, are often strongly impacted by such cycles. In addition to general economic cycles, major global changes such as the fall of the Soviet Union in 1990 and the global economic recession in 2008–2009 have drastically affected the global production, consumption, and trade of softwood lumber (Figure 3.1).

Softwood lumber is the term used in North America, whereas in Europe it is referred to as sawnwood and sawn timber. Regardless of the terminology, lumber is sawn from logs or produced when cants (squared logs or parts of logs) are resawn into planks. Europeans distinguish sawnwood so that it includes cants, beams, and railroad ties (called sleepers in Europe). In this chapter, the term lumber is used to include all squared, sawn timber especially because there is no distinction in the international trade statistics between boards and thicker cants.

3.1.1.1 Where Are the Softwoods?

The geographical range of softwood forests is large and has been extended by the expansion of plantations of fast-growing species (Figure 3.2). The northern boreal forests have many species of softwoods, and often due to less favorable climatic conditions, their slow growth produces high-quality timber. This is due to narrow growth rings that produce dense, strong, and stable timber. This occurs, for example, in Russia, the country with the largest softwood resource. Softwoods extend southward to the tips of Africa and South America and are prevalent in Oceania. The Food and Agriculture Organization (FAO) of the United Nations (UN) shows softwood lumber production in over 100 countries, which is roughly half of the world's countries. Table 3.1 indicates the global softwood growing stock by region.

Softwood genetic research has taken place for centuries in search of creating fast-growing, straight trees to produce high-quality lumber. Plantations in Europe contain many softwood species from North America, for example, Douglas fir (*Pseudotsuga menziesii*), Sitka spruce (*Picea sitchensis*), loblolly pine (*Pinus taeda*), and lodgepole pine (*Pinus contorta*). Another American species, radiata pine (*Pinus radiata*) dominates the softwood plantations in Australia, Chile, New Zealand, and South Africa. Radiata pine has undergone considerable genetic improvements from its origins in California, and growing conditions in the southern

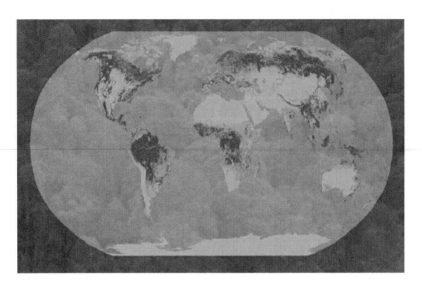

FIGURE 3.2
Global forest resources.

TABLE 3.1

Global Softwood Growing Stock by Region, 2010 (million m³)

CIS[a]	63,754
North America	59,618
Europe	13,672
Asia	11,143
Caribbean and Central America	1,224
South America	1,184
Oceania	968
Africa	217
World	**152,000**

Sources: Food and Agriculture Organization of the United Nations (FAO), Available at: www.fao.org/forestry/fra/fra2010; Food and Agriculture Organization of the United Nations (FAO), *Southeast Asian Forests and Forestry to 2020*, Subregional report of the second Asia-Pacific Forestry Sector Outlook Study, RAP Publication 2010/20, FAO, Bangkok, Thailand, 199pp, 2011.

Note: Growing stock is the volume of standing timber. In the CIS, the Russian Federation has 61,570 million m³ of softwoods. North America includes Canada and the United States.

[a] CIS includes Armenia, Azerbaijan, Belarus, Kazakhstan, Kyrgyzstan, Moldova, Russia, Tajikistan, Turkmenistan, Ukraine, and Uzbekistan.

hemisphere have been more suitable for the species than in the north. Scotch (or Scots) pine (*Pinus sylvestris*) is widely planted in Europe and provides the bulk of pine lumber.

Despite vast areas of softwood forests being converted to other uses, compensation owing to plantations has resulted in total softwood forest area expanding. Where softwood forests have flourished, a culture of using softwood lumber has developed. While softwoods were originally hewn into beams and sawn into lumber where they grew,

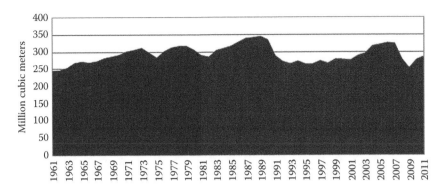

FIGURE 3.3
Global softwood lumber consumption, 1961–2011.

demand for softwoods has also developed in other areas of the world. Softwood timber, both round and sawn, was used originally to build housing and other structures in North America, the Nordic Countries, and Russia. Softwoods have always been used for a multitude of other uses, both interior and exterior. With proper design, protection, and treatments, softwood lumber has been developed to withstand potentially damaging effects of weather (precipitation and sun), insects, and disease (bacteria and fungi).

3.1.1.2 Softwood Lumber Consumption

Softwood lumber consumption is the direct indicator of demand for softwood lumber, and it can be derived from production and trade volumes and values (apparent consumption*). For the remainder of this section, consumption has been calculated as apparent consumption.

Globally, softwood lumber consumption has increased only slightly during the past 50 years, by 16.5% (Figure 3.3), although world population and gross domestic product (GDP) have increased at much greater rates. This disparity can be attributed to global consumption growth having suffered significant periodic setbacks during the oil crises of the mid-1970s, mid-1980s, the fall of the Soviet Union in the early 1990s, and more recently during the economic and financial crisis in the mid- to late 2000s. In 2012, global consumption was again increasing but, at 287 million m^3, is far from the historical high of 346 million m^3 in 1989.

Most consumption of softwood lumber occurs in regions with wood building culture, for example, in the United States (Figure 3.4 and Table 3.2). Second to the United States in consumption, China uses softwood lumber in construction but usually in conjunction with concrete forming. Substantial volumes in China are consumed for packaging, as China exports significant quantities of goods on pallets and in crates.

3.1.2 Softwood Lumber Uses

Softwoods have been utilized historically in a multitude of uses, including structural and nonstructural components of building, shipbuilding, furniture, energy, pulp, and particleboard manufacturing. This section describes lumber usage in two broad categories, structural and nonstructural.

---

* Apparent consumption (derived from production plus imports minus exports) excludes inventory as stocks either in the producing sawmill or with the customer and the volume of sawnwood in transportation.

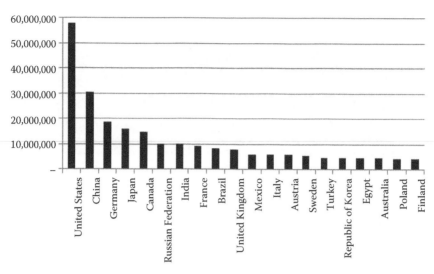

FIGURE 3.4
Top 20 softwood lumber–consuming countries (in m³), 2011.

TABLE 3.2

Top 20 Softwood Lumber–Consuming Countries (m³), 2011

United States	58,058,236
China	30,458,594
Germany	18,805,566
Japan	15,794,751
Canada	14,968,707
Russian Federation	10,223,750
India	10,019,231
France	9,175,142
Brazil	8,229,206
United Kingdom	7,609,345
Mexico	5,650,765
Italy	5,629,000
Austria	5,605,727
Sweden	5,380,589
Turkey	4,741,000
Republic of Korea	4,636,330
Egypt	4,484,553
Australia	4,467,000
Poland	4,358,402
Finland	4,058,384

Sources: FAO, 2013a; Food and Agriculture Organization of the United Nations (FAO), FAOSTAT, 2013a, http://faostat. fao.org/site/626/DesktopDefault. aspx?PageID=626#ancor, accessed March 11, 2013.

3.1.3 Structural Uses

The greatest demand for softwood-based housing is in North America. When the first settlers arrived, they found the continent covered with virgin forests, of which softwood species often predominated. The earliest buildings were constructed from softwood logs, and today there is a continued demand for log homes and buildings. Log construction, whether sawn square or left round, has a natural insulating value and also an appearance preferred by many homeowners. As sawing became mechanized with waterpowered mills, softwood lumber became the standard building material. Eventually, 2 × 4 (2 in. thick by 4 in. wide) construction became the norm, and it still is today, although the techniques have evolved considerably to make buildings more energy efficient and comfortable.

In Asia, for example, in Japan, a different construction method for softwood lumber evolved, that is, post and beam. Rather than the lighter framing of 2 × 4 techniques, the post and beam method uses heavier cross sections to construct the exterior and interior walls and roofs. In Japan, 56% of the homes built in 2011 were wood-based construction. The share of wooden homes has been growing during the past decade, in part due to the strong wood culture. North American sawmillers have successfully introduced the 2 × 4 construction technique to Japan, and it accounted for approximately one-fifth of all wooden houses in 2011. Wooden housing got a boost in 2011 in Japan during the reconstruction from the tsunami and earthquakes. Wooden structures withstood the earthquake and its aftershocks better than many concrete buildings. Japan's traditional post and beam construction has also witnessed innovations, with engineered wood products, such as glulam beams and laminated veneer lumber (LVL) substituting for traditional custom-made components.

Consumption of softwood lumber for building is receiving strong promotion with the implementation of green building policies. Governments, corporations, organizations, and trade associations have recognized the environmental, as well as economic, value in using wood in construction. As a renewable resource with good insulating values, wood is increasing its market share in construction of residential, commercial, and industrial buildings. European governments at national and local levels have established green building policies that in turn have boosted softwood lumber consumption. National and regional trade associations have ceased the opportunity to promote wooden buildings, and the share of wooden structures is increasing. The market share for wood is increasing in many European countries, but it lags relative to masonry and concrete home and building construction. Restrictive fire codes and expensive insurance are additional obstacles for developing the wooden housing industry in Europe. As these restrictions diminish, which has been taking place in recent years, markets for softwood lumber will expand. Further market development will be promoted by trade associations, designers, and architects.

Research and development (R&D) has produced significant improvements in buildings in terms of strength, cost, and energy efficiency. Softwood lumber in different dimensions from the standard 2 × 4 enabled walls and roofs to have greater insulation capacity, for example, by framing walls in 2 × 6s. Engineered wood products can improve strength while using less wood, for example, wooden I-beams that can use lumber or LVL for their flanges. Glulam beams are built from softwood lumber and have successfully competed against concrete and steel with regard to strength properties and from an environmental perspective on a life cycle basis.

Softwood lumber is a globally traded product. But a most interesting development in the past decade has been the increasing trade of prefabricated softwood-based buildings. Entire houses can be constructed in factories from lumber and softwood-based engineered wood products, such as glulam and LVL. The walls and other structures are sized to be shipped by container across land and sea and to be erected relatively quickly. The importance of this development is that affordable, energy-efficient buildings are being constructed in countries far from the softwood resources and in some countries where few buildings have been constructed from wood. Another advantage is that costly onsite construction waste is reduced considerably.

3.1.4 Nonstructural Uses

Softwood lumber is used extensively for packaging applications. In Europe, standard pallets are usually constructed from softwoods. In the United States, one-way (one-use) pallets and skids, as well as other crating, are often made from softwood. In contrast to Europe, most US multiple-use pallets are made from hardwoods.

Softwood lumber is also used for millwork, for example, window and door manufacturing. For windows, boards with straight grain and no knots are resawn into rails and sills, impregnated with antifungal preservatives, assembled into window frames, and often factory-painted. The technology for designing energy-efficient, wooden windows has improved dramatically, making wooden windows highly competitive with plastic and metal alternatives.

In another unique application of softwood lumber, the USDA Forest Service Forest Products Laboratory has developed wooden bridge technology and demonstrated its cost advantages, especially for shorter spans. One of the most popular uses of wooden bridges is in municipal settings for pedestrian crossings. While wooden road bridges may not appear as wood construction when an asphalt surface is applied, pedestrian bridges are often aesthetically pleasing structures.

Softwood lumber is also used for furniture manufacturing using both clear and knotty lumber. Tight knots, called red knots in the trade, provide attractive character marking. Softwood furniture can be finished with clear coats, or attractively stained or painted, with most softwoods having excellent staining and finishing characteristics. Most softwoods are easy to work, mill without grain problems, and produce products that given the proper care become heirlooms. In Nordic Countries where pine, spruce, and fir are plentiful, furniture is designed to exhibit the beauty of softwood. Some plantation-grown softwoods, which are becoming more prevalent, have stability, resin pockets, and other wood quality problems that are reducing recoveries of higher grades of sawnwood; these affect consumer perceptions of the quality of softwoods in some applications, particularly those that require a stable product.

3.1.5 Softwood Lumber Consumption

The development of demand for softwood lumber shows periods of growth, followed by sharp drops due to global economic downturns. Consumption was highest in 1989, just before a downward spiral due to the collapse of the Soviet Union. The decline is real, but the extent is not precise, due to a simultaneous collapse in statistical record keeping and reporting.*

---

* Following the fall of the USSR, the statistics on softwood lumber production by the Russian statistical agency, Rostat, were inaccurately reported. Rostat collected only production records from the largest sawmills, which were the ones exporting most of the lumber. This meant that considerable volumes produced from small- and medium-sized mills, which were the most numerous, were not collected and reported. This problem came to light in the rebound of demand in the CIS, and especially Russia, when rising exports did not keep pace with increases in production—the result was declining apparent consumption, something inconsistent with rising construction-related demand. Hence, prudence is required in looking at the figures for the USSR and the CIS.

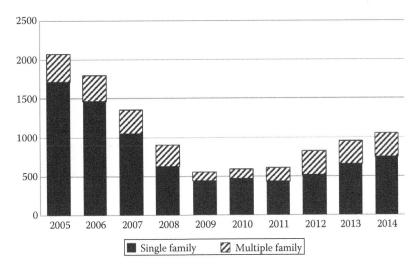

FIGURE 3.5
US housing starts (thousand units), 2005–2014.

Despite statistical concerns for data from the Commonwealth of Independent States (CIS), the global trend is cyclical, indicated by long periods of growth, followed by sharp declines. Consumption has only risen 16.5% in 50 years since 1961. Softwood lumber consumption in 2012 is rebounding from the global economic and financial crisis of 2008–2009. Before the crisis, consumption had risen to 327 million m^3 in 2006 but declined in response to the rapid decline in global housing demand, particularly in the United States (Figure 3.5). When the market collapsed, home values fell dramatically, foreclosures ensued, and the demand for housing, and lumber, collapsed. What had been a 16-year growth in softwood lumber consumption came to a halt.

Not only were Canadian and US lumber manufacturers tied to the US housing market but also offshore exporters. European, Asian, South American, and Oceanian suppliers were exporting significant quantities of lumber, and other wood products, to the United States. The downturn in demand not only resulted in sawmill closures and rationalization of capacity in the United States but also for its importers. Full recovery of global market demand for sawnwood is dependent on a recovery of the US housing market.

In contrast to the important rises in US housing starts, housing starts in Europe are not forecast to increase significantly in the coming years (Figure 3.6). The weak growth forecast led to sawmillers forecasting lower consumption levels, by 5.6% in 2012 and almost no growth in 2013 (International Softwood Conference 2012). This weak market situation has led sawmillers and traders to seek to strengthen positions in other markets, for example, Japan and the Middle East. Europeans are also regaining market share in the US market that was lost with the downturn in US housing and the weaker US dollar (meaning euro-based sawnwood was too expensive).

In 2011, the world consumed nearly 300 million m^3 of softwood lumber. This was the same volume as in 1971, 1983, and 2003 (with global statistics initiated in 1961). The peaks in softwood lumber were in 1973 (312 million m^3), 1978 (317 million m^3), 1989 (346 million m^3), and 2007 (325 million m^3). Globally, demand has not grown at the same rate as population and GDP, two factors with which lumber consumption is correlated. This can be

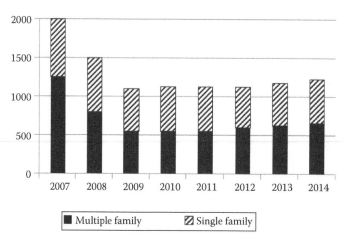

FIGURE 3.6
European housing starts (thousand units), 2007–2014.

attributed in part to substitution by other wood products such as composite panel products and engineered wood products and in part to substitution by non-wood products that compete for market share, such as plastic, steel, aluminum, and concrete.

Softwood lumber consumption is cyclical and despite regular downturns each decade has increased by 16.5% over the last 50 years. New market drivers for wood—such as green building, public procurement policies, and certification of sustainable forest management (SFM)—are expected to boost softwood lumber consumption. Other policies, for example, the European Union (EU) Timber Regulation, could create additional new demand. Chemical and heat treatments, which are expanding softwood end uses, as well as improved building systems, could boost consumption back to its historical high and drive growth in the future.

3.1.6 Softwood Lumber Production

As softwoods grow in temperate climates, in the past most softwood lumber was produced where it grew. Although it still is produced in mainly temperate regions, Asia has become the third largest softwood lumber–producing region, in part using imported logs from Russia, North America, and Oceania (Figure 3.7). China is the world's largest importer of softwood logs. The FAO of the UN indicates that softwood lumber is produced in over 100 countries. The 20 largest producers are shown in Figure 3.8 and Table 3.3.

Depending on conversion factors used, China may be the largest producer of softwood lumber (Wood Markets 2012). However, the phenomenal growth of China's consumption of all wood products has slowed in 2012 as housing started and related wood consumption slowed. Most Chinese housing is multifamily and concrete-based, but wood is used in concrete forming and in millwork and joinery. Chinese softwood lumber consumption nearly doubled from 2009 to 2011, increasing from 24.4 to 39.7 million m^3 (Wood Markets 2012).

It must be noted that the volumes produced in 2011 are in some cases far from the peak volumes produced. For example, the United States produced over 69 million m^3 in 2005, at the height of its housing boom. Russia, combined with the other countries of the former USSR, produced over 103 million m^3 in 1972. Long-term series are sometimes difficult to

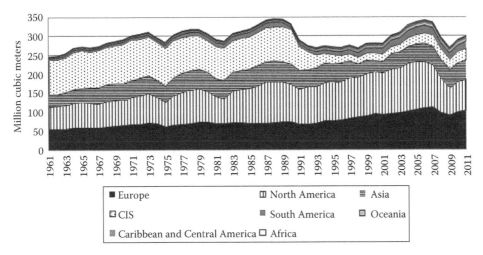

FIGURE 3.7
Softwood lumber production by region, 1961–2011.

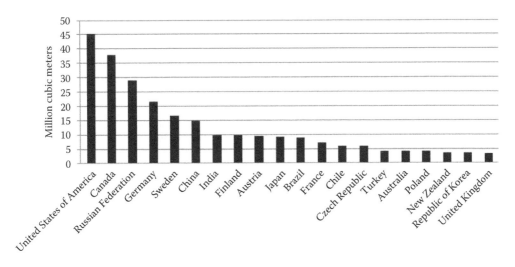

FIGURE 3.8
Top 20 softwood lumber producers (in m³), 2011.

accurately analyze due to changes of country groupings such as the formation of the CIS. Specifically, the current 11-country CIS (which replaced the 15-country USSR) does not contain the 3 Baltic countries, which are important softwood producers. Nevertheless, the global trends are evident in Figure 3.9.

The long-term production trend shows rising growth through the late 1980s, when the USSR collapsed. Although production had dropped in response to reduced demand from the oil crisis–induced economic shocks in the mid-1970s and mid-1980s, there was no precedence for the catastrophic fall in the early 1990s. Economic activity ground to a halt in Russia, the main lumber producer in the former USSR, and sawmills closed. When the economies of the new CIS started to improve in the mid-1990s, the systems for collecting and disseminating statistics were also reestablished. However, to date, Russian official

TABLE 3.3

Top 20 Softwood Lumber Producers, 2011

Country	Volume in m³
United States	45,410,400
Canada	37,991,503
Russian Federation	29,055,000
Germany	21,593,373
Sweden	16,700,000
China	14,920,000
India	9,900,000
Finland	9,700,000
Austria	9,485,000
Japan	9,277,000
Brazil	8,970,000
France	6,965,000
Chile	6,050,000
Czech Republic	5,783,665
Turkey	4,192,000
Australia	4,167,000
Poland	4,150,000
New Zealand	3,658,000
Republic of Korea	3,654,000
United Kingdom	3,227,334

Sources: FAO, 2013a; Food and Agriculture Organization of the United Nations (FAO), FAOSTAT, 2013a, http://faostat.fao.org/site/626/DesktopDefault.aspx?PageID=626#ancor, accessed March 11, 2013.

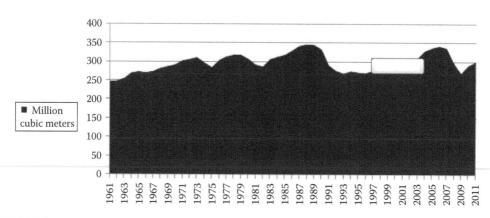

FIGURE 3.9
Long-term production volumes of softwood lumber, 1961–2011.

statistics underestimate lumber production because Rostat only records production from the large, export-oriented sawmills. Therefore, in the statistics in recent years, FAO has adjusted Russia's production upward to better account for the small- and medium-sized sawmills.

In 2012, global softwood lumber production has not regained the 1988 peak of 345 million m^3. In 2011, the most recent annual figure available, global production was 300 million m^3, 15% lower than the level of 1988. Since 1961, global production has grown 22%. However, the economic and financial crisis of the mid-2000s brought global production down to the low level of the early 1960s, which was also the production level in the mid-1990s.

Sawing logs into lumber has greatly improved in efficiency as sawmillers attempt to improve yields and production rates and hence profits. Circular saws have long been the favorite for log breakdown, and by reducing the kerf of saw teeth, more lumber is produced and less sawdust. Band saws typically have narrower kerfs, but the compromise is a slower throughput. In Europe, sash gang saws, which have kerfs equivalent to band saws, are common; however, production rates are less. In the past decades, chip-and-saw systems have improved production rates, especially where sawdust and chip demand enables sawmillers to achieve up to 40% of their revenue from residues. In a chip-and-saw system, logs are debarked, sorted by diameter, scanned for size and shape (and to avoid any metal), and then profiled into a rectangular cant that is sawn into boards. Chip-and-saw sawmills are sometimes owned by pulp producers, since the clean (debarked) chips are needed for pulp production and the sawdust can be used for energy production to heat the kilns and to generate electricity. After sawing, boards must be trimmed to standard lengths to remove unacceptable amounts of wane. Normally, edging is not necessary in softwood lumber since the widths produced at the headrig correspond to required dimensions. Visual and/or machine grading occurs after the next steps when lumber is further processed.

After sawing into lumber, mills can add value to the commodity product by drying in air or kilns, planing, and molding. Drying is critical for most applications as the lumber becomes stronger and more dimensionally stable. Planers give a smoother surface to the lumber and molding can ease the corners or give profiles for different applications. Planer and molding shavings are dry and valuable as a raw material for particleboard, fuel pellets, animal bedding, etc.

Many, but not all, softwoods are not naturally durable. However, species such as Cyprus cedar have good natural durability. For other species, treatment by chemical or heat can improve their durability and useful life in service. Chemical treatments have been changing to reduce their toxicity. Since the 1940s, copper–chrome–arsenate (CCA) was a favored pressure treating method which improved lumbers resistance to attack by insects and microbial agents (fungi and bacteria). The US Environmental Protection Agency banned CCA for most uses in residential settings, and further restrictions are to be imposed in 2013 (EPA 2012). Alternatives exist including alkaline copper quaternary (ACQ, a water-based preservative), borates, copper azole, cyproconazole, and propiconazole. Creosote is used for commercial, not residential, purposes such as railroad ties.

Pentachlorophenol (PCP or penta) was similarly banned by the EPA for residential uses due to its toxicity but is allowed for commercial uses such as walkways, docks, fences, and exterior glulam. Proper disposal of treated wood is critical as the preservatives can leach into the soil and water if buried or be volatized into the air if burned in an uncontrolled environment.

Heat treatments avoid the problem of chemical toxicity. New regulations for wooden packaging and pallets for international trade are the driver for most heat treating to reduce the possibility of transmission of insects and diseases. For packaging the heat treating temperatures are required to reach 56°C for a period of 30 min (IPPC 2012).

Higher temperature heat treatment, up to 230°C depending on the species, and for a 24–96 hour period, can render lumber more durable to attack by biological agents. It also changes the properties of wood by increasing its hardness. However, a drawback is that the modulus of rupture is diminished (Esteves and Pereira 2009). High-heat-treated lumber is resistant to water and can be used for outdoor residential requirements such as decking and siding. It also darkens the color of the wood, which enables it to have the appearance of some hardwoods, including some tropical species.

3.1.7 Softwood Lumber Trade

Trade of softwood lumber developed originally between regions where softwood resources were plentiful and those countries that had little or no softwood resources. However, over the past centuries, plantations of softwood species have enabled many countries to source their softwood needs domestically. Limited availability of resource has resulted in a major shift in the United States from being a net exporter of softwood lumber to becoming a net importer. Another remarkable development in the softwood trade is China's rise to become the world's largest softwood lumber importer, as well as producer, in a relatively short time period. While the previous sections showed slow developments in softwood lumber consumption and production, its trade has been dynamic, both positive and negative, in the past decade.

Global softwood lumber exports followed demand as shown in the consumption section mentioned earlier, with dips in the mid-1970s and mid-1980s due to economic weakness brought on by oil crises (Figure 3.10). Since the breakup of the USSR in the early 1990s, exports accelerated until the recent US housing collapse in the mid-2000s. Exports peaked in 2006 at 115.2 million m^3, in large part driven by US and Chinese imports for housing and nonresidential construction. The latest global export volume, 98.2 million m^3 in 2011, is tripling since 1961s 33.0 million m^3. Export growth has outstripped production and consumption growth at 22.0% and 16.5%, respectively.

Canada is by far the largest exporter of softwood lumber, mainly to the United States (Figure 3.11 and Table 3.4). During the housing slump in the United States, Canada also

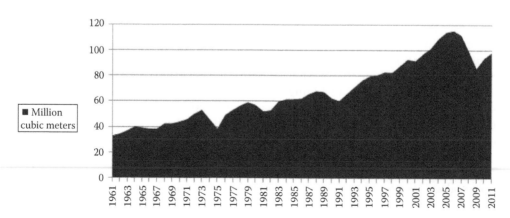

FIGURE 3.10
Global softwood lumber exports, 1961–2011.

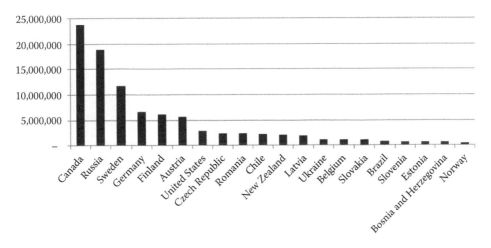

FIGURE 3.11
Top 20 softwood lumber exporters (in m³), 2011.

TABLE 3.4

Top 20 Softwood Lumber Exporters (m³), 2011

Canada	23,798,712
Russia	18,846,000
Sweden	11,656,400
Germany	6,712,198
Finland	6,102,293
Austria	5,591,703
United States	2,955,814
Czech Republic	2,401,809
Romania	2,324,297
Chile	2,289,000
New Zealand	2,023,217
Latvia	1,906,106
Ukraine	1,171,382
Belgium	1,097,474
Slovakia	1,012,797
Brazil	781,000
Slovenia	680,147
Estonia	631,230
Bosnia and Herzegovina	531,000
Norway	466,714

Sources: FAO, 2012; Food and Agriculture Organization of the United Nations (FAO), FAOSTAT, 2013a, http://faostat.fao.org/site/626/DesktopDefault.aspx?Page ID=626#ancor, accessed March 11, 2013.

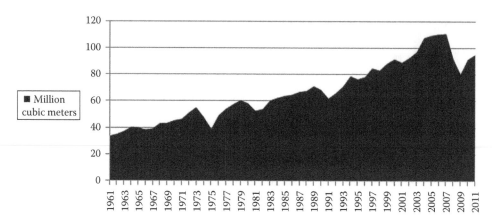

FIGURE 3.12
Global softwood lumber imports, 1961–2011.

developed markets in Asia. In the 1990s, an infestation of the pinewood nematode caused North American exports of softwood lumber to cease, unless it was kiln dried.

Globally, softwood lumber imports naturally follow exports and should be equal on a global level (Figure 3.12). However, there are always discrepancies in the statistics, which can be for valid reasons, or errors, or from other sources, for example, illegal trade. Discrepancies exist when trade data are collected at different periods, from different sources and by different methods. Some countries report lumber trade in volume, for example, in cubic meters or in board feet. Conversion between cubic meters and board feet rely on conversion factors, which are not standard. Conversion is also dependent upon whether the board feet were measured as actual or nominal. For example, a 2×4 cross section is a size in name only (nominal) because it is not 2 in. by 4 in. but rather, after drying and planing, 1.75 in. by 3.5 in. Discrepancies between importers' and exporters' records require validation, and it is not always clear which is correct. When countries report trade in units other than volume, for example, by value or weight, conversion to volume to enable comparison is challenging.

Sometimes, these discrepancies can indicate or suggest illegal trade. The elimination of illegal logging through prohibiting illegal imports is the aim of relatively new legislation in the United States, that is, the Lacey Act Amendment and the EU, via its EU Timber Regulation, as well as a growing number of other countries. Illegal logging deprives countries of revenues from their forest resources and is unsustainable. Illegal trade deprives the legal forest products industry of its rightful revenues as it competes unfairly with legal production and trade.

In the top 20 softwood lumber–producing countries, many countries have limited forest resources. Others, such as Japan, have rich forest resources, but the economics of harvesting the resource is prohibitive. Other countries are rich in forest resources, for example, the United States, Germany, France, and Austria, but their domestic production either does not meet their needs or, in the case of Germany, they import and export different species and qualities for particular end uses.

The two major softwood lumber importers, China and the United States, have had the greatest impact on global markets (Figure 3.13 and Table 3.5). Until 2010, the United States was the global leader in lumber imports. But with the downturn in demand, US imports fell and in 2011, while Chinese imports continued to rise they surpassed the United States for the first time. Most of the US imports come from Canada, while China sources its lumber needs from many countries. The maximum production for the United States,

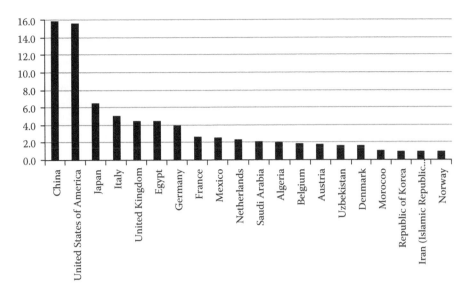

FIGURE 3.13
Top 20 sawn softwood importers (in million m³), 2011.

TABLE 3.5

Top 20 Sawn Softwood Importers (million m³), 2011

China	15.8
United States	15.6
Japan	6.6
Italy	5.0
United Kingdom	4.5
Egypt	4.5
Germany	3.9
France	2.7
Mexico	2.6
Netherlands	2.4
Saudi Arabia	2.1
Algeria	1.9
Belgium	1.8
Austria	1.7
Uzbekistan	1.7
Denmark	1.7
Morocco	1.0
Republic of Korea	1.0
Iran (Islamic Republic of)	0.9
Norway	0.9

Sources: FAO, 2012; Food and Agriculture Organization of the United Nations (FAO), FAOSTAT, 2013a, http://faostat.fao.org/site/626/DesktopDefault. aspx?PageID=626#ancor, accessed March 11, 2013.

69.2 million m^3 could have made the United States self-sufficient for earlier years, but when timber availability decreased substantially from federal lands in the 1980s, the United States became dependent on imports.

Countries that are members of the World Trade Organization (WTO) are committed to lowering and eliminating tariffs and resolving trade disputes. Canada and the United States are members of the WTO but have a long running battle over tariffs and taxes for softwood lumber. US lumber manufacturers and their associations claim that their Canadian counterparts are unfairly subsidized by the provincial and federal timberland owners. In Canada, stumpage prices are set by the government and not by market forces as in the United States. However, the Canadian governments, lumber manufacturers, and their trade associations dispute the assertion of subsidies. Since 1982, there have been a number of iterations of this trade dispute that resulted in 1996 in the Softwood Lumber Agreement (SLA). The Agreement has been renewed several times, despite being contested. It has measures to impose higher tariffs by the United States when lumber prices are low; conversely, when lumber prices are high, there is a lower tariff or quota, and when prices reach $355 per 1000 board feet (MBF), there is no quota or tariff (US Lumber Coalition 2012). The current SLA, established in 2006 for a period of 7–9 years, has resulted in stability and an end of trade litigation.

The drivers for lumber imports are those mentioned earlier, and trends are correlated with gross national product (GNP). Lumber imports are climbing back from the global economic and financial crisis in the mid-2000s and are expected to continue to climb when housing demand improves.

3.1.8 Future Softwood Lumber Market Trends

Softwood lumber continues to renew itself with improvements in its uses, preventing it from peaking in the product life cycle. Processing efficiency increases have enabled structural lumber to be produced from increasingly smaller diameter logs, which is especially important with plantation-grown timber being grown on shorter rotations.

Softwood structural lumber competes with alternative building materials, for example, concrete and steel. For windows and doors, softwoods compete with aluminum and plastic. On a life cycle basis, wood is a superior material, which is why some green building systems promote the use of lumber and associated wood-based engineered wood products. In countries such as Sweden, multistory wooden buildings are well established (Figure 3.14).

Production of softwood lumber is forecast to continue to grow slowly. The primary drivers of construction, residential and nonresidential, are trending upward in North America in 2012 and forecasts are positive for the coming years. Conversely, in Europe and Asia, housing construction has slowed in 2012 and is forecast to remain relatively flat in the coming years. When new housing demand slows, production will be oriented toward the repair and renovation markets that are massive due to the huge existing and aging housing stock. New products and treatments will also enable softwood lumber production to continue to increase.

Softwood lumber is currently used for a multitude of purposes, and demand will continue to grow as new uses are developed. But in order to compete with, both wood and non-wood competitors, softwood lumber will need to be produced in a cost-effective manner and to be able to be shipped to distant markets economically. Chemical (impregnation and coating) and heat treatments can improve lumber's properties for different uses. R&D will always be necessary to improve lumber production and processing efficiency and to improve wood longevity in use.

FIGURE 3.14
Eight-story wooden apartment house in Växjö, Sweden, 2012.

3.2 Temperate Hardwood Lumber Products and Markets

William Luppold and Matt Bumgardner

3.2.1 Introduction

Hardwood (non-coniferous) tree species are found on all continents other than Antarctica. The uses for the fibrous material derived from this resource range from wood pulp to architectural plywood, but hardwood lumber is the most common solid wood product. Hardwood lumber is used in appearance, industrial, and building framing applications. The major appearance applications include furniture, cabinets, flooring, and millwork. The most important industrial applications are pallets, cross ties or sleepers, scaffolding, and dunnage. While softwood (coniferous) tree species are preferred for building framing in developed countries, hardwood lumber can be used for this application in countries that do not have a large inventory of softwood timber.

This section focuses on hardwood lumber production, exports, imports, and apparent consumption for major producing and consuming regions and countries in temperate zones. The temperate zones are defined as north of the Tropic of Cancer and south of the Tropic of Capricorn.

3.2.2 Temperate Hardwood Lumber Production

Over 50% of worldwide hardwood lumber production is produced in the temperate regions. Temperate hardwood lumber production trended downward between 1990 and 2003, increased between 2003 and 2007, decreased in 2008 and 2009, and rebounded in 2010 and 2011 (Figure 3.15).

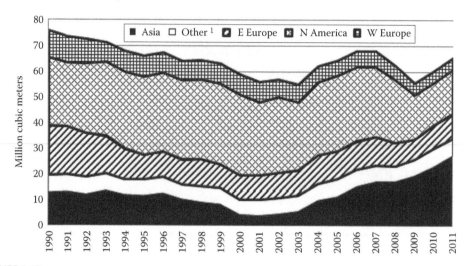

FIGURE 3.15

Hardwood lumber production by temperate region, 1990–2011. *Note*: [1]Other is primarily India, Turkey, and Australia. (From Food and Agriculture Organization of the United Nations (FAO), FAOSTAT, 2013a, http://faostat.fao.org/site/626/DesktopDefault.aspx?PageID=626#ancor, accessed March 11, 2013.)

TABLE 3.6

Top Five Hardwood Lumber–Producing Countries in the Temperate Region in 1990, 2000, 2005, and 2011

	1990		2000		2005		2011
Country	Market Share (%)	Country	Market Share (%)	Country	Market Share (%)	Country	Market Share (%)
United States	33.2	United States	50.8	United States	43.0	China	41.5
USSR	17.1	China	5.4	China	16.7	United States	24.9
China	11.1	France	5.0	Turkey	4.1	Russia	4.0
France	5.1	Russia	4.3	Russia	4.1	Turkey	3.5
Japan	4.4	Turkey	4.1	France	3.0	Romania	2.4
Top five[a]	71.1	Top five	69.6	Top five	63.4	Top five	76.3

Source: Food and Agriculture Organization of the United Nations (FAO), FAOSTAT, 2013a, http://faostat.fao.org/site/626/DesktopDefault.aspx?PageID=626#ancor, accessed March 11, 2013.

[a] May not add up due to rounding error.

In 1990, the United States, the former USSR, and China were the largest producers of temperate hardwood lumber accounting for 33.2%, 17.1%, and 11.1% of worldwide production, respectively (Table 3.6). Regionally, North America* was the largest producer of temperate hardwood accounting for over 25% of worldwide production (Figure 3.15). Temperate hardwood lumber production declined by over 50% in East Europe[†] between

---

* North American temperate lumber producers are located in the United States and Canada.

† East Europe includes Albania, Armenia, Belarus, Bosnia and Herzegovina, Bulgaria, Croatia, Czech Republic, Czechoslovakia, Estonia, Hungary, Latvia, Lithuania, Montenegro, Poland, Republic of Moldova, Romania, Russian Federation, Serbia, Serbia and Montenegro, Slovakia, Slovenia, the former Yugoslav Republic of Macedonia, Ukraine, USSR, and Yugoslav SFR.

1990 and 2000, with the greatest declines occurring in countries that were formerly the USSR* (FAO 2013a). In 1990, the USSR was reported to produce 13 million m^3 of hardwood lumber. In 2000, the combined production of the 15 countries that made up the former USSR was 5.1 million m^3, or a 61% decrease. Most of this decline occurred between 1992 and 1995. West Europe[†] had a 37% decline in hardwood lumber production between 1990 and 2003 (Figure 3.15). The decline in West Europe hardwood lumber production was relatively constant over time, with the greatest declines occurring in Germany and France.

Between 2003 and 2007, temperate hardwood lumber production increased by 24% but there were major shifts in the amount of lumber produced in specific countries, regions, and zones (Figure 3.15). Production in North America remained nearly constant during this period. West European production continued to decline between 2003 and 2007 with the greatest declines occurring in France. By contrast, production in East Europe increased by 14% and production in East Asia[‡] increased by 209%. China's production accounted for most of the increase in East Asia, soaring by 240%, coinciding with a 100% increase in imports of temperate hardwood roundwood on a volume basis (UN FAO 2013a). The major source of temperate roundwood imports was Russia (FAO 2013b).

Between 2007 and 2009, temperate hardwood lumber production declined by 18% (Figure 3.15) as a result of the global recession. The most affected regions were North America, East Europe, and West Europe where hardwood lumber production declined by 36%, 29%, and 26%, respectively. The countries with the greatest absolute declines were the United States, Russia, and France. By contrast, hardwood lumber production in China increased by 14% and by 2009, China had become the largest hardwood lumber producer in the world (UN FAO 2013a).

3.2.3 Temperate Hardwood Lumber Trade

3.2.3.1 Exports

In 1990, the United States accounted for 34% of international shipments of temperate hardwood lumber (Table 3.7). Other important exporters of temperate hardwood lumber in 1990 included France, the former Yugoslavia, Canada, and Germany. The United States remained the largest exporter of temperate hardwood lumber in 2000. China, Canada, France, and Romania were also major exporters of temperate hardwood.

The United States continued to be the largest exporter of temperate hardwood lumber in 2005. Thirty-three percent of US exports in 2005 went to Canada while another 17% went to China (USDA FAS 2012)[§]. Other major temperate hardwood lumber–exporting counties

---

[*] Countries that were formerly the USSR include Russia, Ukraine, Belarus, Uzbekistan, Kazakhstan, Georgia, Azerbaijan, Lithuania, Moldova, Latvia, Kyrgyzstan, Tajikistan, Armenia, Turkmenistan, and Estonia.

[†] West Europe includes Austria, Belgium, Belgium–Luxembourg, Denmark, Finland, France, Germany, Greece, Iceland, Ireland, Italy, Liechtenstein, Luxembourg, the Netherlands, Norway, Portugal, Spain, Sweden, Switzerland, and the United Kingdom.

[‡] Temperate East Asia includes China, Democratic People's Republic of Korea, Japan, and Republic of Korea.

[§] US sources (USDA FAS) consider China, Hong Kong SAR, and Taiwan POC as separate export destinations, whereas UN sources (FAO) only provide trade data for China. When US exports to Taiwan, Hong Kong, and China (USDA FAS 2012) are combined, they equal UN FAO estimates of Chinese imports from the United States in 1997 and 1998. These two series were poorly correlated between 1999 and 2004 but were in a similar range between 2005 and 2009. In 1997 and 1998, combined exports to Hong Kong and Taiwan from the United States exceeded exports to China by over 85%. Direct exports to China started to increase in 1999 and exceeded combined exports to Taiwan and Hong Kong in 2003. It is interesting to note that FAO (2013b) indicated that imports to China from the United States declined sharply in 2003, while USDA FAS (2012) indicated a steady increase.

were Canada, China, Germany, and Romania. Exports of temperate hardwood lumber declined by 34% between 2007 and 2009 (FAO 2013a) as a result of the global recession. Hardwood lumber exports from the United States declined by 46% mainly as a result of a 56% decline in shipments to Canada (USDA FAS 2012).

The United States remained the largest exporter of temperate hardwood lumber in 2011 with a 23% market share (Table 3.7). Belgium was a major exporter of hardwood lumber in 2011, but the lumber shipped from Belgium could have been produced in other countries since it imported nearly as much lumber as it exported that year. Russia became one of the five top exporters in 2011, displacing China.

The relative size of hardwood lumber exports as a percentage of regional production is shown in Table 3.8. East Europe exported only modest amounts of hardwood lumber in 1990, but by 2000, over 41% of the lumber produced in this region was exported. This increase was the result of a 252% increase in lumber exports and a 50% decrease in lumber production. In 1990, West Europe exported 17% of the lumber it produced,

TABLE 3.7

Top Five Hardwood Lumber–Exporting Countries in the Temperate Regions in 1990, 2000, 2005, and 2011

1990		2000		2005		2011	
Country	Market Share (%)	Country	Market Share (%)	Country	Market Share (%)	Country	Market Share (%)
United States	34.3	United States	23.6	United States	25.6	United States	22.8
France	12.1	China	13.0	Canada	11.5	Belgium	8.9
Yugoslavia	11.8	Canada	11.5	China	6.8	Russia	7.2
Canada	7.8	France	5.3	Germany	6.5	Romania	7.1
Germany	5.8	Romania	5.2	Romania	5.9	Germany	5.9
Top five[a]	71.8	Top five	58.5	Top five	56.3	Top five	51.9

Source: Food and Agriculture Organization of the United Nations (FAO), FAOSTAT, 2013a, http://faostat.fao.org/site/626/DesktopDefault.aspx?PageID=626#ancor, accessed March 11, 2013.
[a] May not add up due to rounding error.

TABLE 3.8

Exports as a Percentage of Hardwood Lumber Production by Temperate Region in 1990, 2000, 2005, and 2009

Region	1990 (%)	2000 (%)	2005 (%)	2011 (%)
East Asia	2.4	5.8	7.2	2.0
East Europe	5.8	41.4	37.6	43.9
West Europe	17.4	29.6	37.4	59.4
North America	9.6	14.2	14.7	15.9
Other	3.4	3.4	4.6	2.5
All temperate	7.9	21.2	18.2	15.8

Source: Food and Agriculture Organization of the United Nations (FAO), FAOSTAT, 2013a, http://faostat.fao.org/site/626/DesktopDefault.aspx?PageID=626#ancor, accessed March 11, 2013.
Note: Other is primarily India, Turkey, and Australia.

but by 2011 this proportion had increased to 59% as hardwood lumber production decreased and exports increased.

3.2.3.2 Imports

In 1990, Japan, Italy, and China accounted for nearly 34% of total hardwood lumber imports by countries in the temperate zone (Table 3.9). Between 1990 and 2000, worldwide imports of hardwood lumber (temperate and tropical) increased and a high proportion of this increase was the result of a 376% increase in imports by China (FAO 2013a). By contrast, imports by Japan declined by 34% during this period. In 2000, the major sources of hardwood lumber imported by China were Indonesia and Malaysia (tropical hardwoods), and the United States (temperate hardwoods) (FAO 2013b).

Imports of hardwood lumber by countries in the temperate region remained constant between 2000 and 2007, but imports into the United States, Italy, and Japan declined sharply (FAO 2013a). US hardwood lumber imports declined by 27% between 2000 and 2007, and nearly all of this decrease resulted from reduced imports from Canada (USDA FAS 2012). This decline occurred at the same time hardwood lumber consumption by the furniture industry in the United States declined by 60% (HMR 2009). Italian imports declined by 23% during this period with the greatest decline being shipments from East Europe and West Africa (FAO 2013a, b).

The recession of 2008 and 2009 caused worldwide imports of hardwood lumber to decline with the largest absolute declines in China, Italy, the United States, and Spain (FAO 2013a). China reduced its hardwood lumber imports from the United States, Thailand, and Brazil while increasing its imports from Russia (FAO 2013b). The overall reduction in Chinese imports was the apparent result of increased hardwood lumber production in China and the shift in furniture manufacturing from China to Vietnam (Luppold and Bumgardner 2011). Imports of hardwood lumber by temperate regions rebounded to near 2007 levels by 2011, with the greatest increases occurring in the United States and China. In 2011, 43% of the hardwood lumber imported by temperate region countries went to China.

TABLE 3.9

Top Five Hardwood Lumber–Importing Countries in the Temperate Regions in 1990, 2000, 2005, and 2011

1990		2000		2005		2011	
Country	Market Share (%)	Country	Market Share (%)	Country	Market Share (%)	Country	Market Share (%)
Japan	13.2	China	24.2	China	24.9	China	43.1
Italy	12.5	Italy	10.3	United States	9.3	Italy	5.8
China	8.1	United States	8.3	Canada	7.8	Belgium	4.7
Netherlands	7.2	Canada	5.9	Italy	7.6	United States	4.1
Spain	6.7	Japan	5.7	Spain	4.9	Egypt	3.6
Top five[a]	47.6	Top five	54.4	Top five	54.5	Top five	61.4

Source: Food and Agriculture Organization of the United Nations (FAO), FAOSTAT, 2013a, http://faostat.fao.org/site/626/DesktopDefault.aspx?PageID=626#ancor, accessed March 11, 2013.

[a] May not add up due to rounding error.

3.2.4 Temperate Hardwood Lumber Consumption*

Hardwood lumber consumption is affected by numerous factors including lumber availability and price, population size and age, wealth, and the existence of industries that use
hardwood lumber as a major input. Population size is important because the greater the
number of people the greater the demand for furniture, cabinets, flooring, millwork, and
industrial products. The age of the population is also important because people in their 60s
and older consume fewer durable products than younger people. Normally, countries with
median population in the mid-40s have a considerably greater proportion of older people
than countries with median population in the mid-20s. In addition, the greater levels of
wealth, as defined by GDP, the more goods consumers will purchase. Countries that have
established wood-consuming industries as a part of their overall economic development
plan or have historically had a secondary hardwood processing industry will consume
more lumber than countries that have few such manufacturers. Similarly, countries that
have lost their secondary manufacturing industries due to international competition will
reduce hardwood lumber consumption.

In 1990, 62% of the hardwood lumber consumed in the temperate region was by four
countries, the United States, the former USSR, China, and Japan (Table 3.10). These four
countries were also highly populated, collectively accounting for 34% of world population
in 1990 (Table 3.11). The United States was by far the largest worldwide consumer of hardwood lumber in 1990 even though it was the fourth most populated country in the world.[†]
The number one ranking of the United States in 1990 was primarily the result of the size of
its economy, which accounted for 25% of the world GDP in that year (Table 3.12), as well as
the presence of numerous secondary processing industries. The United States was also the
largest producer of hardwood lumber (FAO 2013a). The large consumption by the former

TABLE 3.10

Top Five Hardwood Lumber–Consuming Countries in the Temperate Regions in 1990, 2000,
2005, and 2011

1990		2000		2005		2011	
Country	Market Share (%)	Country	Market Share (%)	Country	Market Share (%)	Country	Market Share (%)
United States	28.8	United States	43.0	United States	36.5	China	47.1
USSR	15.8	China	9.7	China	20.6	United States	20.4
China	11.2	France	4.5	Turkey	3.7	Turkey	3.2
Japan	6.1	Italy	4.2	Italy	3.1	Russia	2.7
France	4.5	Turkey	3.7	Russia	3.0	France	2.0
Top five[a]	66.4	Top five	65.1	Top five	66.8	Top five	75.4

Source: Food and Agriculture Organization of the United Nations (FAO), FAOSTAT, 2013a, http://faostat.
fao.org/site/626/DesktopDefault.aspx?PageID=626#ancor, accessed March 11, 2013.

[a] May not add up due to rounding error.

---

* Consumption figures used in this section are approximate estimates of consumption (apparent consumption)
and have been derived from production plus imports minus exports. Hardwood lumber inventories have not
been considered.

[†] While the USSR state that is now the nation of Russia was the sixth most populous country in 2003, the
combined population of the 15 countries in the former USSR was the third largest country in 1990.

TABLE 3.11

Population of Countries That Were Top Five Consumers in Temperate Regions One or More Times during 1990, 2000, and 2011 and the Median Age of Population in 2011

Country	Population 1990 (Million)	Population 2000 (Million)	Population 2011 (Million)	Median Age 2011 (Years)
United Sates	249.6	282.2	311.7	36.9
USSR	289.1	NA	NA	NA
China	1135.2	1262.6	1336.7	35.5
Japan	123.5	126.9	126.5	44.8
France	56.7	58.9	65.1	39.9
Turkey	56.2	67.4	78.8	28.5
Italy	56.7	56.9	61.0	43.5
Russia	NA	146.3	138.7	38.7
World	5222.8	5991.3	6882.9	28.4

Sources: US Central Intelligence Agency (US CIA), The World Factbook, December 18, 2003–March 28, 2011, NationMaster.com 2013Population (1990) by country, 2013, http://www.nationmaster.com/graph/peo_pop-people-population&date=1990, accessed on March 11, 2013; U.S. Central Intelligence Agency (US CIA), The World Factbook, Median age, 2012, https://www.cia.gov/library/publications/the-world-factbook/fields/2177.html# 133, accessed March 11, 2013.

TABLE 3.12

GDP in Terms of Purchasing Power Parity[a] of Countries in Temperate Regions That Were Top 10 Consumers One or More Times during 1990, 2000, 2005, and 2011

Country	1990 Billions of US Dollars	2000 Billions of US Dollars	2005 Billions of US Dollars	2011 Billions of US Dollars
United Sates	5,800	9,951	12,623	15,076
USSR	NA	NA	NA	NA
China	910	3,015	5,364	11,300
Japan	2,370	3,256	3,890	4,444
France	1,027	1,532	1,862	2,214
Turkey	292	513	747	1,076
Italy	976	1,404	1,642	1,847
Russia	NA	1,205	1,894	2,512
World	23,490	42,310	61,705	78,970

Source: Knoema, GDP statistics by country, 2013, http://knoema.com/gdp-by-country?gclid=CMOu4ImPha8CFcMbQgod62dn4w#United%20States, accessed March 13, 2013.

[a] Purchasing power parity is the rate at which the currency of one country would have to be converted into that of another country to buy the same amount of goods and services in each country (Callen 2012).

USSR in 1990 appears to be the result of high populations and high volumes of hardwood lumber production.

The combined populations of Japan, Germany, and France were similar to that of the United States in 1990 (Table 3.11), and the combined GDP was 83% of the United States (Table 3.12), yet lumber consumption was half that of the United States (Table 3.10).

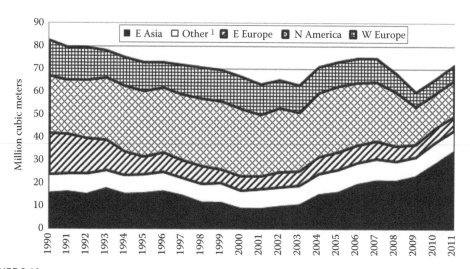

FIGURE 3.16

Apparent hardwood lumber consumption by temperate region, 1990–2011. *Note*: [1]Other is primarily India, Turkey, and Australia. (From Food and Agriculture Organization of the United Nations (FAO), FAOSTAT, 2013a http://faostat.fao.org/site/626/DesktopDefault.aspx?PageID=626#ancor, accessed March 11, 2013.)

Two potential reasons for combined lower lumber consumption in these countries are lower levels of hardwood lumber production (Table 3.6) and older populations (Table 3.11).

Hardwood lumber consumption in the temperate regions declined between 1990 and 2000 (Figure 3.16). The greatest declines occurred in East Europe and East Asia. The decline in East Europe has been affected by a slowly growing and aging population (US CIA 2013). The decline in East Asia was primarily caused by reduced consumption by Japan. Japan had slow GDP growth between 1990 and 2000 (Table 3.12) and an aging population (Table 3.11). The decline (though relatively small) in West Europe consumption is attributed to aging populations and relatively low GDP growth in larger countries in this region.

Chinese consumption of hardwood lumber declined 30% between 1990 and 2000. As noted earlier, hardwood lumber production in this country decreased, as a result of redirecting logs to plywood production. North American consumption increased as a result of increased demand for hardwood lumber in the United States (Luppold and Bumgardner 2008). The secondary hardwood processing industries that increased their consumption in the United States during this period were kitchen cabinet, millwork, and flooring, as greater volumes of hardwood products were used in home construction.

For the 2000–2005 period, hardwood lumber consumption increased in the temperate zone countries primarily as a result of increased demand by China. The worldwide recession caused a reduction in hardwood lumber consumption between 2007 and 2009. Consumption in North America fell by 53% as the housing market and overall economic activity declined. East European consumption also declined 25%, while West European consumption declined by 40%. By contrast, consumption by China increased (FAO 2013a).

3.2.5 Future Temperate Hardwood Market Trends

The worldwide economic downturn that began in 2008 has caused continued uncertainty, making it difficult to project the future of international hardwood lumber markets. However, there are a number of trends that have continued during the pre- and post recession periods. These include increased production and consumption of hardwood lumber in China, the continued decline in West European and Japanese production and consumption, and the stagnation of the US market for higher grade lumber.

China became the largest market for hardwood lumber in 2009, while US production and consumption declined by over 36% between 2007 and 2009. China sources lumber imports from nearly every part of the world and was the top export market in 2009 for hardwood lumber from the United States, Indonesia, Malaysia, Cameroon, and Germany. While a portion of the hardwood lumber consumed in China is remanufactured for export products such as furniture, the majority of lumber consumed in China is for domestic markets. Given China's rapidly growing GDP, large population, and rising incomes, this country will probably remain a major consumer if not the largest consumer of hardwood lumber for years to come. However, there are some potential issues that could curtail future increases in hardwood lumber production and consumption in China. The expansion of Chinese hardwood lumber consumption since 2003 has been predicated on imports of logs primarily from the Pacific Rim nations and Russia and lumber from nearly every corner of the world.

Hardwood lumber consumption in West Europe has been declining since 1990. West European hardwood lumber production has also been declining since 1990 but at a lower rate than consumption. The apparent causes of the declines in hardwood lumber consumption are economic and demographic. The major West European hardwood lumber-consuming countries listed in Table 3.12 had lower GDP growth than any country other than Japan. The population growth of these countries between 1990 and 2011 was below or near 10% and the median ages were among the highest in the world (US CIA 2013 and Table 3.11).

In 1990, Japan was the world's largest importer of hardwood lumber and the seventh largest consumer. It also was the seventh most populous country and the second largest economy in the world (US CIA 2013 and Table 3.12). Between 1990 and 2000, the Japanese economy and population rate grew at extremely slow rates relative to other industrial countries as it went through a series of economic structural shocks. The median age of the population reached 44 years in 2011, comparable to Germany but with a lower immigration rate of guest workers. This series of events led to a continued reduction in hardwood lumber production, consumption, and imports of 93%, 88%, and 76%, respectively, between 1990 and 2007. Japan also increased its imports of secondary processed hardwood products such as furniture during this period. The recession years of 2008 and 2009 brought even greater declines in hardwood lumber production, consumption, and imports of 29%, 40%, and 46%, respectively. Japanese consumption probably will increase in the near future because of the size of their economy and the rebuilding efforts after the 2011 earthquake and tsunami, but long-term increases appear to be improbable.

For most of the time period examined in this chapter, the United States was the dominant producer and consumer of temperate hardwood lumber and continued to be the most important exporter. Since 2000, US hardwood lumber production and consumption have been trending downward, before declining by over 35% between 2007 and 2009. The decline in the US hardwood lumber market began as domestic furniture manufacturers could no longer compete with East Asian imports (Luppold and Bumgardner 2011). Initially, the decline in consumption by the furniture industry was offset by increased consumption by industries associated with home construction. The crash in the US housing market in 2009 caused hardwood lumber consumption and production to decline to its lowest levels since the early 1960s. The probability that US hardwood lumber consumption will reach 2000 levels in the near future is low. In contrast, the outlook for hardwood lumber production is better because the United States has a large and sustainable hardwood resource base and the infrastructure to access this base allowing it to remain the world's largest exporter of hardwood lumber.

Since 2000, East European hardwood lumber production and consumption has been relatively stable. While some countries in East Europe have experienced slow population growth, Russia (the largest) has experienced negative population growth. Similar to West Europe, the median age of the East Europe population is relatively high but with a lower rate of immigration. Similar to North America, there is little change expected in production and consumption in East Europe in the short term.

3.3 Tropical Lumber Products and Markets

Frances Maplesden and Steven Johnson

3.3.1 Introduction

Tropical timber products, including lumber, are derived from timber that is grown or produced in the countries situated between the Tropic of Cancer and the Tropic of Capricorn. Tropical industrial roundwood production, used in the production of sawnwood and veneer for plywood, has been more or less stable in the 16 years from 1995 to 2010 in each of the three tropical regions (Figure 3.17).* However, production in natural forests has become increasingly supply constrained, affected by log export restrictions to encourage downstream processing, reductions in logging quotas to achieve national SFM targets, and crackdowns on illegal logging in the major supplying countries. Illegal logging and trade in illegally sourced products is widely perceived to be more a more serious issue in tropical supplying countries, where forest governance is often poor. Hence, data on forest

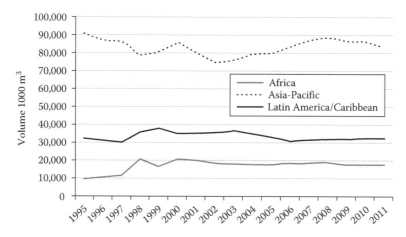

FIGURE 3.17
Tropical log production by region, 1995–2011.

---

* Unless otherwise stated, the data on tropical production, consumption, and trade presented in this section are for ITTO member countries only. ITTO's 60 member countries together constitute 95% of all tropical timber trade and over 80% of tropical forest area.

resources and roundwood production in the tropical zone are often uncertain with the problem exacerbated by poor monitoring and reporting capacities in many tropical countries. Although declines in production in natural forests in some countries have been offset by increases in production from planted forests, plantation development in the tropics has not kept pace with development in the temperate zone, with only 15.6 million hectares (14%) of the world's plantations in countries located in the tropical zone and concentrated in relatively few countries (ITTO 2011).

3.3.2 Production of Tropical Sawnwood

Production of tropical sawnwood in International Tropical Timber Organization (ITTO) producing countries totaled about 41.0 million m^3 in 2010, with the Asia-Pacific and Latin America/Caribbean regions each accounting for approximately 44% of production while Africa accounted for the remainder (ITTO 2012). There have been longstanding regional disparities in the proportion of log production utilized domestically, with Latin America converting almost all of its domestically produced logs to at least primary products over the last 5 years, while Asia-Pacific producers domestically consumed about 95% of their log production. Both regions have had rising domestic demand for wood-based products resulting from population and economic growth, as well as emphasis on producing and exporting value-added products.

In the African region, the proportion of all logs produced that were converted domestically to further processed products is relatively small compared with the other regions, although it has increased to an estimated 84% in 2011, reflecting increasing government restrictions on log exports in many ITTO African member countries (ITTO 2012). Africa's tropical log production is more dependent on exports, and EU markets, than the other regions, with over 16% of log production exported as logs in 2010. Compared to Asia and Latin America, the African region was more sensitive to the depressed wood products demand in traditional markets caused by the global economic downturn in 2008 and 2009. Many of the major producing countries relaxed log export restrictions during the economic crisis, to assist their forestry sectors to improve profitability (particularly Gabon, Cameroon, and the Republic of Congo), but in 2010, many countries reimposed these restrictions to assist the recovery of their sawmilling and other wood processing industries (ITTO 2012, ITTO MIS).

Brazil's tropical sawnwood production, totaling 15.5 million m^3 in 2010, constitutes over 85% of production in the Latin American region and it remains the largest global producer and consumer of tropical sawnwood, with high economic growth and an increase in construction activity fuelling an increase in domestic sawnwood demand over the last few years (Figure 3.18). India (4.9 million m^3), Malaysia (4.3 million m^3), Indonesia (4.2 million m^3), and Thailand (2.9 million m^3) were other major producers of tropical sawnwood in 2010, although the accuracy of aggregate data for the Asian region may be impaired by the lack of official data on sawnwood production available for three of the major producing countries, India, Indonesia, and Thailand. In 2010, sawnwood production in the Asia-Pacific region recovered from a low in 2009, increasing 3% to reach 18.2 million m^3.

The top five tropical sawnwood producing countries produced over 77% of tropical sawnwood in 2010, although there were eight other countries (Nigeria, China, Lao PDR, Myanmar, Cameroon, Peru, Côte d'Ivoire, and Ghana) that produced over 500,000 m^3 of tropical sawnwood in 2010. China imports more tropical sawnwood than it produces from its considerable tropical log imports. China's domestic sawmills supplied only 33% of

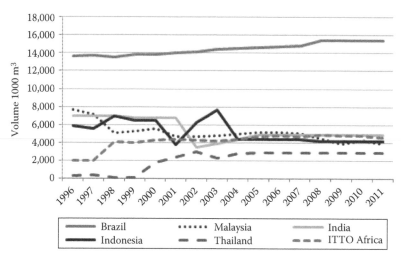

FIGURE 3.18
Major tropical sawnwood producers, 1996–2011.

tropical sawnwood demand in 2011, compared with 67% of coniferous sawnwood demand. This reflects the introduction of log export restrictions in tropical supplying countries (e.g., Gabon) and increases in China's labor costs, reducing competitive advantage in primary processing compared with moving up the value chain. China's sawmilling industry is dominated by small- and medium-sized enterprises and production figures from such numerous, small-scale operations is likely to be underestimated (ITTO 2012).

3.3.3 Consumption of Tropical Sawnwood

Brazil remains the largest ITTO tropical sawnwood consumer at over 14.8 million m³ in 2010. Domestic consumption has been relatively stable since 2009, supported by strong sawn timber demand in the growing construction sector. India, China, and Indonesia were the next most important consumers in 2010, with tropical sawnwood consumption of 4.9, 4.9, and 3.6 million m³, respectively (ITTO 2012). There has been a developing trend toward domestic consumption within tropical producer countries. This has been driven by relatively high population growth and rising GDP/capita in tropical producer (developing) countries, with economic growth generally being higher in developing economies than in developed economies.

China's consumption increased 33% in 2010 as domestic demand grew strongly and as China's wooden furniture and flooring exports recovered from the effects of the global economic downturn on its major markets—the United States and EU countries. EU consumption of tropical sawnwood picked up in 2010 to reach 1.5 million m³, although this was significantly less than precrisis levels. Consumption dropped again in 2011 to 1.4 million m³ with the outlook for a further decline in consumption in 2012. There have been a number of other factors contributing to declining EU consumption, including a loss of secondary processed manufacturing capacity as a result of strong competition from Asian manufacturers (particularly China and Vietnam), substitution by nontropical sawnwood in furniture and joinery manufacture, and more recently a lack of availability of certified tropical sawnwood that will become more critical when the new EU Timber Regulation is enforced in March 2013.

3.3.4 Tropical Sawnwood Trade

Although the share of tropical wood products in the global wood products trade has been declining, tropical sawnwood has maintained its share of global sawnwood imports at 10% between 1994 and 2010. By contrast, tropical logs as a share of global log imports declined from 30% to 13% over the same period. The downturn in the tropical plywood trade has been more pronounced, from 68% share of global plywood imports in 1994 to 37% in 2010. The tropical sawnwood trade is dominated by trade within the Asia-Pacific region, with over 75% of global imports and 65% of global exports of tropical sawnwood being between countries in the region (ITTO 2012). Intra-regional trade within the African region has also developed in response to growing demand within the region (ITTO 2010a).

3.3.4.1 Imports

Total imports of tropical sawnwood rebounded from a low in 2009 to reach 8.1 million m^3 in 2010, a year-on-year increase of 23%, as construction demand and consumer spending began to pick up in consumer countries. In 2011, imports moved downward again to 7.2 million m^3 as the economic situation in the euro zone deteriorated and the US economic outlook remained uncertain. The largest country importers are in Asia (China and Thailand) with other Asian (Malaysia, Taiwan Province of China (POC), and Japan) and EU importers (the Netherlands, France, Italy, and Belgium) also being important to the trade (Figure 3.19).

China's imports soared in 2010, reaching 3.3 million m^3, 50% more than the previous year. The reasons for this growth include rising demand for sawnwood in China's furniture and flooring industries; increases in log export restrictions from supplying countries (Gabon and Russia) creating a substantial log supply gap; and increasing labor costs, rising domestic sawnwood prices, and appreciation of the Chinese currency that has eroded the competitiveness of tropical sawnwood manufactured in China. During the period of the global financial and economic crisis (2008–2009), China's economy was assisted by aggressive fiscal stimulus packages and the subsequent growth in domestic consumption, including demand for tropical sawnwood, more than compensated for depressed demand

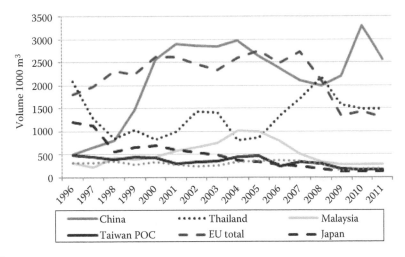

FIGURE 3.19
Major tropical sawnwood importers, 1996–2011.

in China's export-oriented wood remanufacturing industries. China's tropical sawnwood imports are mainly used in furniture, interior decoration, and home improvement and are more sensitive to export market conditions than is the case for softwoods, which are used predominantly in domestic construction. In contrast to Thailand (the second largest importer and also a major exporter), China has a larger range of tropical sawnwood suppliers, the main suppliers in 2010 being Thailand (43%), the Philippines (16%), Indonesia (12%), Malaysia (7%), Myanmar (3%), and Brazil (3%). China's imports from Thailand are predominantly of lower priced rubberwood that has become popular in the production of low-cost furniture products. Imports from African countries (Gabon, Cameroon, the Republic of Congo, Côte d'Ivoire, and Ghana) were less than 3% of China's tropical sawnwood imports in 2010 but have the potential to rise given China's investment in many infrastructure projects such as roads and ports to facilitate transport and trade of basic commodities such as logs and sawnwood in the African region. In 2011, China's tropical sawnwood imports declined but remained substantial at 2.6 million m^3.

Thailand's tropical sawnwood imports are mainly structural grade material, with 92% of imports from Malaysia and Lao People's Democratic Republic (PDR). However, Thailand's reported imports from Lao PDR of over 672,000 m^3 in 2010 were not corroborated by Lao PDR export statistics. Although the Government of Lao PDR has committed to SFM, high demand levels from neighboring countries such as Thailand and Vietnam and a suspected high incidence of illegal logging and poor governance mean that its export figures may be underestimated (Forest Trends 2011).

Taiwan POC's imports rebounded in 2010, increasing nearly 90% to 333,000 m^3, with most of the supply (nearly 80%) from Malaysia. Malaysia's imports recovered slightly to 282,000 m^3 in 2010 but were still nearly half the 2007 level. Malaysia's suppliers were mostly from the Asian region, with 37% of imports in 2010 from Thailand and most of the remainder from Indonesia, the Philippines, and Myanmar. In Japan, imports and consumption of tropical sawnwood have declined in recent years because the use of solid wood for shop renovations and housing renovations is declining, and the use of substitute products, such as MDF with printed wood grain patterns, has grown because of their low prices and ease of installation/workability (ITTO 2010b).

Total tropical sawnwood imports by EU countries remained at a very low level in 2010, increasing slightly to 1.4 million m^3, nearly half the peak level of 2007. In 2011, as economic uncertainty mounted, imports dropped to 1.3 million m^3, the lowest level in ITTO's statistical records. Many EU member countries are experiencing government austerity measures, sluggish construction activity, and a continuing tendency for importers to maintain low stocks. There have also been some major structural changes in the sawnwood market, with the EU temperate hardwood sawnwood industry weathering the demand crisis better than most external suppliers. Its share of the EU hardwood sawn timber market increased from 66% to 74% during the period 2006–2010 while tropical hardwood's market share declined from 18% to 12%.

The Netherlands is the largest EU importer although a significant proportion is reexported to other EU destinations. It is mainly supplied by Cameroon, Brazil, and Malaysia. France is also a major importer, supplied mostly from Cameroon, Côte d'Ivoire, and Ghana, as well as significant imports from Brazil. A decline in furniture manufacturing in France and Belgium in recent years implies that there is a limited prospect for tropical sawnwood consumption and imports returning to the high volume import levels before the economic crisis.

Italy's tropical sawnwood imports, which have been significant in past years, have been declining continuously since 2007. The Italian hardwood furniture sector has undergone significant structural change, with demand for tropical sawnwood declining due to the shift by larger manufacturers to lower-cost locations including in the tropics. Italy's imports were mainly from countries within Africa—Cameroon, Côte d'Ivoire, and Gabon. Spain's imports, which have been most affected by significant setbacks in the construction sector from 2008, remained depressed in 2010 and 2011 as the construction sector continued to decline and the important door manufacturing sector remaining depressed. Although economic conditions remained relatively positive in Germany in 2011, with new residential construction and renovations remaining strong, tropical sawnwood imports declined. This partly reflects changing fashion trends in Germany favoring character and grain, demand for which has been increasingly met by the application of stain or heat treatment to oak and ash (ITTO MIS).

3.3.4.2 Exports

Figure 3.20 shows the major tropical sawnwood exporters over the last 15 years. Thailand's exports of tropical sawnwood jumped to 2.8 million m³ in 2010, well in excess of pre crisis levels. The growth is attributed to China's surge in demand for low-cost raw materials (particularly rubberwood) for its export furniture and flooring industries with consumers in end-use markets demanding lower-priced furniture and flooring products during the economic crisis. Thai exports were predominantly to China (74%) and Malaysia (23%). Malaysia is also a significant exporter, recording exports of 2.6 million m³ in 2010, up to 32% on the previous year. In contrast to Thailand, Malaysia has more diverse market destinations, with Thailand importing the largest share (27%), but there were also a significant number of other important destinations, including Taiwan POC, China, Singapore, the Philippines, the United Arab Emirates, Maldives, Yemen, and Sri Lanka.

With some stability returning to global markets in 2010, Cameroon's tropical sawnwood exports totaled 738,000 m³, the same level as 2009. Exports were mainly to European destinations—Italy, the Netherlands, Belgium, and France—with Cameroon as the largest

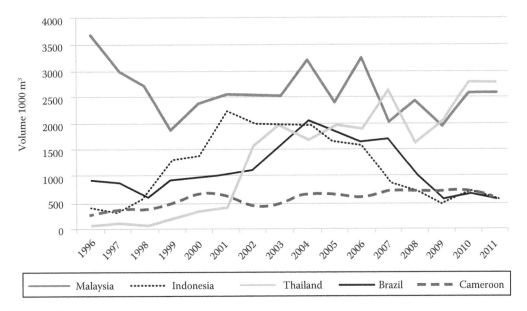

FIGURE 3.20
Major tropical sawnwood exporters, 1996–2011.

single supplier of tropical sawnwood to Europe. In 2010, Cameroon's sawnwood exports were assisted by Gabon's log export ban that reduced tropical log supplies and strengthened demand for tropical sawnwood imports, particularly okoumé, from other African sources. In 2010, the Cameroon government reimposed log export quotas on several species, and demand for iroko sawnwood (one of the major species exported) had strengthened. However, in 2011 and early 2012, with an uncertain economic outlook for the euro zone countries, exports were declining. West and Central African exporters have the advantage of shorter lead times and lower freight costs to Europe compared to competitors from Southeast Asia.

Brazil's exports have plunged in recent years. In 2007 Brazil exported 1.7 million m³, declining steadily to 571,000 m³ in 2009 and increasing slightly to 668,000 m³ in 2010. Over this period, Brazil's export competitiveness declined as the currency continued to appreciate relative to the US dollar, domestic demand grew, and sawnwood demand in Brazil's major sawnwood export markets, particularly the United States, declined. Brazil's major markets in 2010 were China (which has grown in importance in recent years), the Netherlands, France, and the United States. Indonesia's exports of tropical sawnwood increased in 2010 to 635,000 m³, although this was still less than before the global economic crisis. Estimates for Indonesia's exports of tropical sawnwood have underestimated total trade in previous years, particularly with China.

3.3.5 Future Tropical Sawnwood Market Trends

International tropical timber markets are continually undergoing dynamic structural changes that threaten the competitive position of tropical timber products compared with other wood and non-wood products. Changes in factors such as price, product availability, technology, shipping and freight costs, manufacturing costs, environmental concern, and consumer taste have been impacting on the relative competitiveness of tropical wood products. Substitution of tropical sawnwood in outdoor decking applications, for example, has occurred from both new wood-based products and plastics. Tropical hardwoods such as meranti and sapele, which have been highly favored for window manufacture for their technical and aesthetic attributes at the high end of European window markets, are losing market share to engineered wood products. Innovations in product development and processing have generally benefited softwoods and other materials rather than tropical hardwoods, with global research focusing on improving the ability of softwoods to match the technical performance of tropical hardwoods rather than the development of new and innovative products and applications to extend the market share of tropical hardwoods. In EU markets, European hardwood products with targeted performance attributes are being developed and marketed as alternatives to tropical hardwoods in the external joinery and furniture sectors.

Market perceptions of the environmental credentials of tropical wood products and emerging public and private timber procurement policies also represent a major challenge for tropical timber exporters, impacting their competitiveness in some markets. In EU markets, environmental concerns have benefited FSC and PEFC certified hardwoods, the majority of which are sourced from Europe rather than tropical supplying regions. The lack of availability of certified tropical sawnwood remains a concern given the numerous trade policy measures now being implemented with the aim of improving forest law enforcement in tropical supplying countries and countering the trade in illegal harvested timber. There is widespread expectation in the EU that demand for certified tropical wood products will pick up strongly as the EU moves toward full implementation of the EU Timber Regulation in 2013. Tropical producer countries, particularly in the African region, are underrepresented in the global supply of environmentally certified wood products that are derived overwhelmingly from forests in Europe and North America. However, these policy measures will increasingly need to accommodate legality verification procedures in more complex value-added wood product supply chains, given the shift in tropical sawnwood trade toward Asian countries, particularly China, which further process and reexport products to EU and North American markets.

References

Callen, T. 2012. Purchasing power parity: Weights matter. International Monetary Fund. http://www.imf.org/external/pubs/ft/fandd/basics/ppp.html (Accessed March 28, 2012).

Environmental Protection Agency (EPA). 2012. Chromated copper arsenate (CCA). Available at: www.epa.gov/oppad001/reregistration/cca/ (Accessed March 19, 2013).

Esteves, B.M. and H.M. Pereira. 2009. Wood modification by heat treatment: A review. *Bio Resources.* 4(1): 370–404.

Food and Agriculture Organization of the United Nations (FAO). 2011a. Forest resources assessment 2010. Available at: www.fao.org/forestry/fra/fra2010 (Accessed March 19, 2013).

Food and Agriculture Organization of the United Nations (FAO). 2011b. *Southeast Asian Forests and Forestry to 2020. Subregional Report of the Second Asia-Pacific Forestry Sector Outlook Study.* RAP Publication 2010/20. 199pp. FAO, Bangkok, Thailand.

Food and Agriculture Organization of the United Nations (FAO). 2013a. FAOSTAT. http://faostat.fao.org/site/626/DesktopDefault.aspx?PageID=626#ancor (Accessed March 11, 2013.)

Food and Agriculture Organization of the United Nations (FAO). 2013b. FAOSTAT. http://faostat.fao.org/site/628/DesktopDefault.aspx?PageID=628 (Accessed March 11, 2013.)

Forest Trends. 2011. Baseline study 2, Lao PDR: Overview of forest governance, markets and trade. July 2011. Available at: http://www.forest-trends.org/documents/files/doc_2920.pdf. (Accessed 1 June 2012.)

Hardwood Market Report (HMR). 2009. 2008: The year at a glance, 12th annual statistical analysis of the North American hardwood marketplace. Hardwood Market Report, Memphis, TN.

International Plant Protection Convention (IPPC) 2012. International standards for phytosanitary measures. Available at: www.ippc.int (Accessed March 19, 2013).

International Softwood Conference. 2012. Available at: www.isc2012.se (Accessed March 19, 2013).

International Tropical Timber Organization (ITTO). 2010a. Good neighbours. Promoting intra-African markets for timber and timber products. *ITTO Technical Series #35.* June 2010. 112pp. ITTO, Yokohama, Japan.

International Tropical Timber Organization (ITTO). 2010b. Leveling the playing field. Options for boosting the competitiveness of tropical hardwoods against substitute products. *ITTO Technical Series #36.* November 2010. 164pp. ITTO, Yokohama, Japan.

International Tropical Timber Organization (ITTO). 2011. Status of tropical forest management 2011. *ITTO Technical Series #38.* July 2011. 418pp. ITTO, Yokohama, Japan.

International Tropical Timber Organization (ITTO). 2012. *Annual Review and Assessment of the World Timber Situation 2011.* ITTO, Yokohama, Japan.

International Tropical Timber Organization Market Information Service (ITTO MIS). Biweekly Publication of ITTO. Available at: http://www.itto.int/market_information_service/ (Accessed March 1, 2013).

Japan Wood-Products Information and Research Center (JAWIC). 2012. *Japan Wood Market Statistics.* January 2012. 26pp. Tokyo, Japan.

Knoema. 2013. GDP statistics by country. http://knoema.com/gdp-by-country?gclid=CMOu4ImPh a8CFcMbQgod62dn4w#United%20States (Accessed March 13, 2013.)

Lucas, A.R. 2005. Industrial milling in the ancient and medieval worlds: A survey of the evidence for an industrial revolution in medieval Europe. *Technology and Culture.* 46(1): 1–30. [10–11].

Luppold, W. and M. Bumgardner. 2008. Forty years of hardwood lumber consumption: 1963 to 2002. *Forest Products Journal.* 58(5): 7–12.

Luppold, W.G. and M.S. Bumgardner. 2011. Thirty-nine years of US wood furniture importing: Sources and products. *Bioresources.* 6(4): 4895–4908.

Peterson, C.E. 1973. Sawdust trail: Annals of sawmilling and the lumber trade. *Bulletin of the Association for Preservation Technology.* 5(2): 84–85.

Ritti, T. K. Grewe, and P. Kessener. 2007, A relief of a water-powered stone saw mill on a sarcophagus at Hierapolis and its implications. *Journal of Roman Archaeology.* 20: 138–163.

Singer, C., E.J. Holmyard, A.R. Hall, and T.I. Williams (eds.). 1956. *History of Technology* (Volume II of IV). Oxford, U.K., pp. 643–644.

US Central Intelligence Agency (US CIA). 2013a. The World Factbook, December 18, 2003–March 28, 2011, NationMaster.com 2013Population (1990) by country. http://www.nationmaster.com/graph/peo_pop-people-population&date=1990 (Accessed March 11, 2013).

US Central Intelligence Agency (US CIA). 2013b. The World Factbook, Median age. https://www.cia.gov/library/publications/the-world-factbook/fields/2177.html#133 (Accessed March 11, 2013).

US Department of Agriculture Foreign Agricultural Service (USDA FAS). 2012. Global agricultural trade system. http://www.fas.usda.gov/gats/default.aspx (Accessed March 7, 2012).

US Department of Commerce, Census Bureau 2013. New Residential Construction. www.census.gov/construction/nrc/ (Accessed March 19, 2013).

US Lumber Coalition. 2012. About the Softwood Lumber Agreement. Available at: http://www.uslumbercoalition.org/general.cfm?page=4 (Accessed March 19, 2013).

Virginia Tech. 2013. Virginia Tech Housing Reports. http://woodproducts.sbio.vt.edu/housing-report/ (Accessed March 19, 2013).

Wikipedia. 2012. Wood preservation. Available at: http://en.wikipedia.org/wiki/Wood_preservation#Heat_treatments

Wilson, A. 2002. Machines, power and the ancient economy, *The Journal of Roman Studies.* 92: 1–32.

Wood Markets. 2012. China at a crossroads: *Outlook. Monthly International Report.* Volume 17, Number 9, November 2012.

4

Markets and Market Forces for Secondary Wood Products

Urs Buehlmann and Al Schuler

CONTENTS

4.1 Introduction

Global demand for durable products, that is, goods that yield utility over time such as furniture, floors, refrigerators, cars, or mobile phones, depends on political and socioeconomic factors, such as trade patterns, exchange rates, resource availability, employment, and affluence (UN 2005; FAO 2009). At present, sizeable markets exist globally for these products (e.g., European Union [EU], China, India and Brazil) with the United States being the largest, single uniform market that includes some of the most affluent consumers. Despite recent rhetoric of domestic protectionism in the United States (Herbst 2009), the availability of such a unique and dynamic consumer-oriented market combined with liberal trade policies (WTO 2008) has made the United States an attractive outlet for goods and services created in all corners of the world. Thanks primarily to comparative cost advantages that some mostly developing countries have over the United States and historically favorable exchange rates, the United States has been the leading net importer of goods and services for the last 40 years with the nation's trade deficit in 2011 settling at $560 billion (down from a record $753 billion in 2006; U.S. Census Bureau 2012). However, in line with the shifting economic balance of the world's regions, where developing markets like China,

India, and Brazil are growing much faster than the developed economies in the West, the US currency has dropped to historical lows relative to other major currencies (€, CN¥, £, J¥, CHF, CA$, among others; Economist 2008, 2009; NZZ 2008; Elwell 2012). Since a powerful US recovery that would strengthen the value of the dollar is not in sight, adjustments in global trade patterns are occurring and are likely to continue (Economist 2008; IMF 2011; Elwell 2012).

The forest products industry has classically been divided into the primary and secondary forest (wood) products sectors (Smith et al. 2009). Secondary wood products, a loosely defined term, typically refer to products made from roundwood that has already undergone a primary conversion process to lumber or other intermediate products such as engineered wood composites. Examples of secondary wood products include, but are not limited to, furniture, windows, doors, fixtures, floors, cabinets, wooden pallets and wooden containers.

In a globalized economy, secondary wood products, like other commodities, are driven by global supply and demand. Centers of production of secondary wood products continue to shift with the changing patterns of the global economy in line with shifting comparative advantages. Such patterns can be observed with the US furniture industry, which had its manufacturing center originally in the Great Lakes region, then moved to the South of the United States, where cheaper labor and cheaper lumber were available after WW II, just to move on to Taiwan and then to China in the last decades of the twentieth century (Schuler and Buehlmann 2003; Buehlmann et al. 2004; Buehlmann and Schuler 2009). Taiwan, in turn, accounted for approximately 30% of all imported furniture to the United States in 1990, just to have its share shrink to less than 4% today (ITA 2012). Reasons for this decline of Taiwan's success can be found in "...*Labor shortages, ballooning wages, and soaring industrial land costs...* (Li 2011, 1)," e.g., declining comparative advantages in the business of manufacturing furniture. However, while Taiwan's exports of furniture declined dramatically after 1990, export of Taiwan's furniture marketing and manufacturing know-how increased, with the Taiwanese Furniture Manufacturers' Association (TFMA) estimating that more than 80% of the island's furniture makers moved to China and other destinations in Southeast Asia (Li 2011). The steep rise of furniture production in China thus is largely ascribed to Taiwanese entrepreneurs shifting their money, knowledge, and skills from a country with fast-rising production costs (e.g., declining comparative advantages) to a country offering lower production costs, a seemingly endless pool of cheap labor, and a government focused on economic growth. Thus, huge furniture manufacturing complexes of Kasen, Makor, or Lacquercraft in China were born where thousands of workers labor to fill the pipeline of global furniture demand (Champine and Krishnan 2004). However, as production costs in China begin to rise (Areddy 2008; Chowdhury 2011; Economist 2012a,b), the global furniture industry is beginning to move to lower cost destinations that offer the right mix of production costs, investment climate, and supplies (e.g., destinations that offer comparative advantages in the manufacture of furniture, Stickley 2005; Economist 2011; Li 2011; Bland 2012).

A large part of secondary wood products sold are commodities that are conceived, designed, engineered, manufactured, and sold globally; however, there also exists a sizeable market for secondary wood products that is geographically restricted. In other words, regional or local markets for secondary wood products exist. For example, the market for architectural millwork that includes receptions, bars, retail space, and private, commercial, and public interiors remains firmly in the hands of local, regional, or, to an extent, national businesses. Additionally, increased attention has been paid to mass customization in recent years by local, regional, and national manufacturers (Toffler 1970; Davis 1987;

Pine 1993), where customers expect products and services to be customized according to their needs and expectations (Lihra et al. 2008, 2011). Mass customization necessitates interactions between customers and producers and also calls for short lead times—the acceptable time between when an order is placed and when it is delivered. As short lead times are more easily feasible for local or regional producers, mass customization helps local, regional, or national businesses to retain manufacturing that otherwise could go global (Huyett and Viguerie 2005; Grant Thornton LLP 2006; Bernhardt et al. 2007; Wan and Bullard 2008).

Overall, the secondary wood products sector is undergoing numerous changes that are driven by economic, political, and social factors. In the following pages, we highlight trends in raw materials, present facts about the global and regional production and consumption, elaborate on data pertaining to trade flows, and discuss market characteristics and product use patterns and trends. We conclude by presenting a long-term outlook for the secondary wood products sector.

4.2 Trends in Raw Materials

The secondary wood products sector is arguably one of the largest consumers of primary forest products, especially hardwoods. While softwoods are of importance to the secondary wood products sector, construction markets consume more softwood globally, particularly on a volume basis. Historically, the secondary wood products sector has been using lumber as a raw material. However, due to technical (size limitations, shrinkage and swelling, warp), economic (due to decreasing supplies of saw logs, prices are increasing), and social reasons (design trends that do not ask for solid wood or demand certified lumber, Espinoza et al. 2012), lumber usage is shrinking and numerous wood fiber–based derivative materials such as wood composites, wood fiber–plastic, and other lightweight materials are becoming increasingly popular. Figure 4.1 shows the product life cycle for wood-based products and estimates selected wood products' position in the product life cycle.

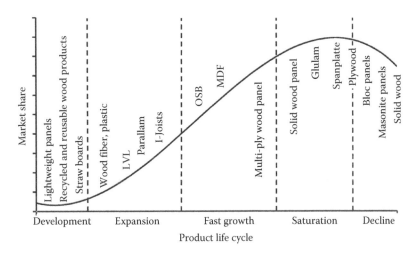

FIGURE 4.1
Selected wood products' position on product life cycle.

Lumber was the primary raw material for secondary wood products manufacturers throughout much of known history, but solid wood is losing its historic dominance for most sectors of the industry and has been replaced by engineered wood composites, veneers, or other engineered materials based on wood fibers. Also, the dependence of the secondary wood sector on solid wood varies greatly. While the solid hardwood flooring sector relies almost entirely on hardwood lumber for its production, the laminated flooring sector consumes no lumber whatsoever but uses high-density fiberboard and laminated materials. Furthermore, depending on the industry sector, auxiliary materials, such as metals, glass, plastics, and fabrics, are also widely used by the secondary wood products industry. For example, a manufacturer of architectural millwork may, in fact, rely as much on some of these auxiliary materials as on any type of wood product to manufacture its products.

While this characterization of forest raw material usage for secondary wood products may vary across countries, the trend away from the use of solid wood (lumber) is universal. Mostly, this trend is because solid wood is no longer available in sufficient quantities to meet global demand. Moreover, substitute products offer better quality at a competitive price compared to solid wood. Given the ever increasing choice of available materials, the secondary wood products sector continues to adopt new, wood-based, engineered materials. Furthermore, increasing raw material scarcity and increasing transportation costs continue to encourage the use of novel, lightweight materials (Buehlmann et al. 2008, 2012).

4.3 Global Production and Consumption

Wood has been used as material for energy production, construction, and many other practical purposes since early mankind (Buehlmann 2001). Sutton (1999) estimated the annual global per capita wood consumption (fuel and industrial wood) at 0.6 m³, with a split of 55%–45% for fuel wood and industrial wood.* However, annual per capita wood consumption varies widely by nation or region. For example, the average Chinese person consumes an estimated 0.3 m³ of wood annually, while the average US person consumes six times that much (1.8 m³, FAO 1999).

With the emergence of the global economy, the source point of the secondary wood products industry's raw material, the manufacturing location, and the point of product use have shifted and have become geographically dispersed. For example, a set of dining room furniture manufactured in China and sold in Europe may use hardwoods from the United States (Appalachia), wood composite panels made in China with logs imported from Russia (Siberia), and wooden drawer boxes made from plywood manufactured in Indonesia. The globalization of the wood products value chain, such as raw material procurement, product manufacturing, and use of the final good, has impacted entire regions' economic fortunes and has lead to sharply growing global trade in forest products. While world exports of primary forest products (defined as round wood, fuel wood, sawn wood, wood-based panels, and pulp and paperboard but not including secondary wood products; FAO 1982) was $5 billion in 1961, it increased to almost $250 billion in 2011 (Figure 4.2,

---

* FAO (2011) defines "industrial wood in the rough" as (1) saw and veneer logs, (2) pulpwood (round and split), and (3) other industrial roundwood. Fuel wood is not included in industrial roundwood.

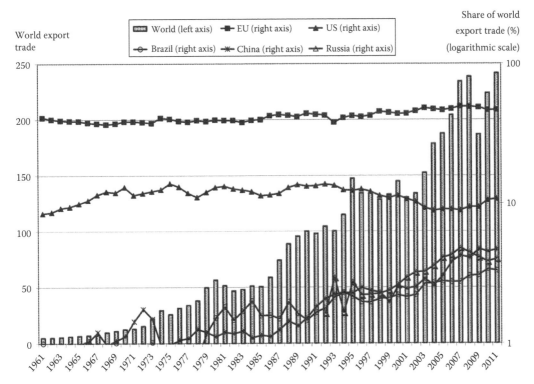

FIGURE 4.2
Primary forest products trade 1961–2011, overall and for selected countries. (From Food and Agricultural Organization of the United Nations (FAO), FAOSTAT, http://faostat.fao.org/DesktopDefault.aspx?PageID=626 &lang=en#ancor, accessed January 18, 2012, 2012.)

FAO 2012). During the recent global recession, world trade of primary forest products shrank to $187 billion but has since recovered and exceeded prerecession levels (FAO 2012). Historically, the EU, with its strong internal trade among member countries, accounts for almost half of global trade in forest products (47% in 2011, FAO 2012) while the United States is responsible for another 10% of world trade (Figure 4.2).

The emergence of three of the four BRIC countries (Brazil, Russia, India, and China) as notable exporters of primary forest products over the last few decades is also noteworthy (Figure 4.2). China, with 5% of world exports, is closely followed by Russia (4%) and Brazil (3%, FAO 2012). These three countries' share of global primary forest products exports is bigger than the exports of more widely recognized exporters of primary forest products like Chile (2%) or New Zealand (1%). Figure 4.3, using geographic data adjusted by primary forest products trade volume, shows the enormous primary forest products trade imbalances that exist globally (Pepke 2011). Pepke's (2011) world atlas, when adjusted for primary forest products trade activity, centers on Europe, which makes up nearly half of all trade, followed by Canada and the United States. China, Russia, and some Southeast Asian nations appear prominent on the right end of the map, while South America and Africa are shrunk to minimal proportions.

The evolution of global trade has impacted industries around the globe. For example, while the US share of global primary forest products trade as a percentage of total trade has declined only slightly over the past decades, exports have, for some segments of the US industry, grown considerably (Buehlmann et al. 2007, 2010a; Espinoza et al. 2011).

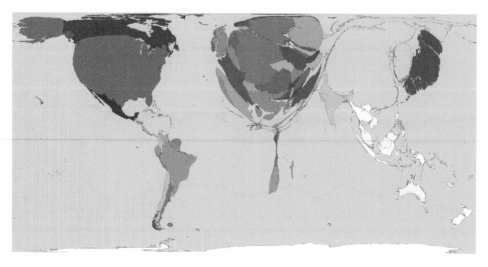

FIGURE 4.3
World atlas with country sizes adjusted for contribution to forest products trade. (From Pepke, E., European markets and certification, *International Scientific Conference on Hardwood Processing (ISCHP 2011)*, Blacksburg, VA, October 17, 2011; Worldmapper and FAOStat, 2009.)

For example, the US hardwood lumber industry today exports 17% of its total annual production, up from less than 2% in 1963 (Barford 2012; Luppold 2012).

4.4 Globalization and Secondary Wood Products Industry

Until the emergence of the global economy in the last quarter of the twentieth century, the secondary wood sector was national in nature. However, the relaxation of trade barriers (WTO 2008) combined with the existence of comparative advantages between countries and regions and the containerization of transportation made non-customized secondary wood products a globally sourced commodity. Comparative advantages of a country over another refer to a country's ability to produce a good or a service at lower marginal and opportunity costs. For example, China, over the past two decades, has benefited from comparative advantages in manufacturing goods and has thus become a major producer of manufactured goods globally. Specifically, the rise of China as a manufacturing hub has had a remarkable impact on the secondary wood products sector globally. Similarly, low-cost manufacturing locations in Eastern Europe, the former Soviet Bloc states in particular, have surfaced on the global secondary sector manufacturers' map. Notably, Germany's furniture industry has lost 70,000 positions from 1997 to 2007 (a 40% drop) mostly to Eastern European countries such as Poland, Czech Republic, Slovenia, and Hungary. However, over the past years, China has successfully taken market share from these Eastern European manufacturers and continues to consolidate its dominance as a secondary sector manufacturer (Klaas 2007). In addition to cost-based comparative advantages, other political decisions also help shape the overall manufacturing competitiveness of a region that may include aspects such as the ease of trade through customs and tariffs.

Globalization of the secondary wood products sector has impacted the industry in more than one way. Firstly, globalization has allowed investors to reap the benefits of comparative advantages by moving production to more advantageous locations. In management literature, this continuous search for and shift to cheaper manufacturing locations is often referred to as "industrial flight." In the secondary products sector, this phenomenon can be seen through the Taiwan experience over the decades. In 1987, Taiwan's furniture manufacturing was at its peak and exports rose to $2.4 billion. However, this manufacturing boom led to labor shortages, ballooning wages, and soaring industrial land costs—all impairing the Taiwanese cost advantage. In response, the Taiwanese furniture industry started to move offshore, taking advantage of the cultural and linguistic similarities with China. Since then, the TFMA estimates that more than 80% of Taiwan's furniture manufacturers have relocated their production to the mainland (Li 2011). At the same time, from 1998 to 2008, China's production of furniture increased almost sixfold (from $10.5 billion to $56 billion) while exports octupled (from $2.1 billion to $17 billion, China National Furniture Association 2010).

However, in spite of the secondary wood products manufacturing sector having become a global business, at least for commodity-type secondary wood products, the rise of the mass-customized economy counterbalances the global nature of secondary wood products manufacturing and creates opportunities for local/regional businesses. The simultaneous presence of globalization and localization within the same sector is an interesting characteristic evident in the furniture sector. While this trend is also observable elsewhere, in the following, we focus on the US furniture sector to illustrate this phenomenon.

4.4.1 How Does Globalization/Localization Play Out in the Secondary Sector?

Two subsegments of the furniture sector, namely, the non-upholstered wood household (hereafter referred to as wood household) furniture segment (NAICS 337122, U.S. Census Bureau 2010) and the wood kitchen cabinet and countertop segment (NAICS 337110, U.S. Census Bureau 2010), vividly exemplify a coexistence of globalization and localization. The US wood household furniture segment and the US wood kitchen cabinet and countertop segment differ in size. In 2005, wood household furniture consisted of 1,442 establishments, employed 84,172 people, and produced slightly more than $10 billion worth of products. In the same year, the kitchen cabinet and countertop manufacturing segment consisted of 9,373 establishments, employed 139,258 individuals, and produced more than $18 billion worth of products. The wood household furniture segment is more traditional and predominantly produces commodity-type furniture that offers little opportunity for consumers to customize the product to their tastes. Conversely, the kitchen cabinet sector offers customers extensive opportunities to customize their product with reasonable lead times and at a reasonable price. Consequently, while wood household furniture is produced mostly in low production cost countries like China or Vietnam, 95% of the kitchen cabinet and countertops sold in the United States are manufactured domestically. Figure 4.4 graphically illustrates the difference between these two segments, indicating that globalization and/or localization forces may play out differently even within the same industry.

As shown in Figure 4.4, in 2009, 69% of all non-upholstered wood household furniture sold in the United States was imported, while only 4% of all wood kitchen cabinets and countertops had been manufactured abroad. Thus, while the wood kitchen cabinet and countertop sector evolved into offering highly personalized cabinets made to order (MTO), the non-upholstered wood household furniture sector stuck with the traditional model

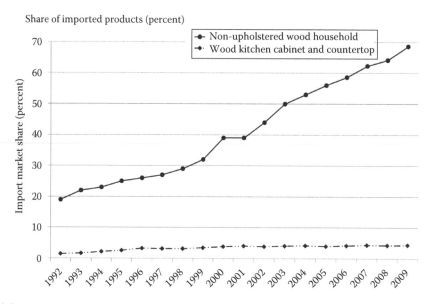

Share of imported products (percent)

FIGURE 4.4

US market share of imports 1992–2009 in percent. *Note*: These market share computations are conservative since some imported components and original equipment manufacturers (OEM) furniture is included in domestic shipments. This observation applies to all numbers regarding the value of domestic shipments throughout this manuscript. [consumption=shipments+imports−exports; import share=imports/consumption]. (From U.S. Census Bureau NAICS 337: Furniture and related product manufacturing, http://www.census.gov/epcd/ec97/def/337.HTM, accessed November 12, 2009, 2009; International Trade Administration (ITA), Consumer goods industries, International Trade Administration (ITA), U.S. Department of Commerce, http://www.ita.doc.gov/td/ocg/index.htm, accessed November 15, 2011, 2011.)

of standardized, made to stock (MTS) furniture. MTS, standardized furniture enabled global manufacturers to supply retailers with inexpensive, quality products, thus competing one on one with domestically based manufacturers. Thus, in the non-upholstered wood household furniture sector, comparative advantages worked in favor of global manufacturers, displacing most of the traditional, domestic manufacturers. Although differences exist, the forces of localization and globalization are essentially applicable to all segments of the secondary wood products sector and, in the absence of trade barriers, apply globally. For example, manufacturers in Germany, where average hourly compensation costs in the manufacturing sector in 2011 were $43.76, which placed Germany as the sixth most costly country for manufacturing globally (BLS 2012), face fierce global competition for commodity-type products. As a result, German furniture imports from China grew by 19.1% between 2005 and 2006 (Klaas 2007). However, due to its geographic proximity with other low-cost manufacturing locations (e.g., Poland that has a $ 8.01 average hourly compensation costs in manufacturing sector, BLS 2012), even the customized, MTO products can be manufactured at low-cost locations since transportation costs, lead times, and even communication with customers is much less a challenge. Thus, customization as a strategy to combat globalization is not as realistic an option for countries that are in geographic proximity to low-cost producers. However, using advanced design capabilities and automation can provide a buffer against the comparative advantage of low-cost producers. The German kitchen cabinet sector has succeeded by using this strategy, which has kept the imports of kitchens from low production cost countries at relatively low levels (Dierig 2012).

The law of comparative advantages, absent any trade barriers (such as duties, tariffs, or quotas), works in favor of the most efficient and effective global market participants. Thus, countries with favorable production economics, such as China and Vietnam, have benefited considerably from globalization over the past decades. Local producers of secondary wood products in locations with less favorable production economics, such as producers in the United States or the EU, have to rely on competitive factors other than price (such as customization, automation, and/or lead time) to remain viable.

As a rising tide lifts all boats, globalization has created sizeable new markets for secondary wood products. For example, according to the China National Furniture Association (2010), China consumes 70% of its furniture production domestically. Other countries, like India, Brazil, and Russia, have become important consumer markets as well. These fast-growing economies possess huge numbers of potential customers who are becoming more affluent. However, their annual per capita spending on furniture (and other secondary wood products) is still extremely low compared to affluent countries like the United States, Italy, or Germany ($2.67, $6.68, or $26.74 for India, China, or Russia, respectively, versus $280.75, $270.05, or $494.65 for the United States, Italy, and Germany, respectively, Klaas 2007).

4.5 Regional Production and Consumption

Since the definition of secondary wood products is vague and differs from country to country, the use of residential housing activity may serve as a proxy for understanding secondary sector developments. While around 70% of structural lumber and panel product sales in North America is attributable to the housing sector (Schuler and Adair 2003), the dependence between housing activity and secondary wood products sales such as furniture is not well understood. However, there is a general agreement that household furniture sales are derived primarily from new housing and remodeling activity (Majumdar 2004; AP 2007), and therefore the selection of housing as a proxy to secondary sector development is reasonable. In the following, we discuss these developments in different regions of the world.

4.5.1 Americas

The Americas encompass a wide variety of regions and countries, ranging from locations with annual per capita gross domestic product (GDP) below $5,000 per year, for example, Bolivia, Honduras, Guatemala, Nicaragua, and Haiti, (CIA 2012), to the United States and Canada with annual per capita GDPs approaching $50,000 (CIA 2012). Secondary wood products consumption not only depends on the affluence of the population but also traditions and availability of materials. Thus, in some regions, residential buildings, especially family homes, are made with concrete, stone, or steel, while in other places wood is the material of choice.

Latin America, encompassing South and Central America, is economically dominated by Brazil and Mexico, which together make up approximately 70% of the region's GDP (Garcia 2011). Brazil, the 10th largest economy globally, possesses large reservoirs of wood resources and has a fast-growing population (estimated at +1.1%/year) and a fast growing economy (forecasted +5%/year). Thus, the secondary wood products sector appears to have upside potential in the country. Also, Brazil is host to the 2014 World Cup and the

2016 Olympics, promising to further strengthen the demand for secondary wood products. Growth is also predicted in Mexico, the 13th largest economy and the 11th most populous country globally, but possibly at a somewhat slower pace than Brazil. Other large economies in Latin America, such as Colombia, Chile, or Peru, are also expected to do well and offer growth opportunities. Argentina, however, may remain subdued and see lackluster growth into the future as a consequence of its past problems and current government policies. Overall, the secondary wood products sector in Latin America can look to the future with confidence, as growing economies and populations will drive demand for products.

North America, consisting of Canada and the United States, both with mature and well-off economies, has been affected by the recent recession differently. Canada weathered the recent economic struggles, mainly thanks to having a more conservatively structured financial system that did not leverage home equity to the extent seen in the United States. Canada's housing market, with roughly 200,000 new housing starts (CMHC 2012) and 450,000 existing homes sold in 2012 (Bloomberg 2012), in fact has shown signs of overheating over the past years. Only lately, the market has cooled off somewhat, mainly because housing prices, especially in urban areas like Toronto or Vancouver, have reached unsustainable levels. However, despite the slowdown in housing, the Canadian secondary wood products industry should remain busy at a decent level by serving their home market and taking advantage of its closeness to the vast US market.

The United States, the world's largest economy and second largest construction market (Garcia 2011), is currently recovering from the worst recession since the great depression (Lazear 2012). This latest recession that started in 2008, caused by unabashed financial and housing speculation, has impacted the US secondary hardwood sector and its international suppliers profoundly. New single-family home sales in 2010 fell to 323,000 from their peak in 2005 (1,283,000, −75%) and existing home sales fell from 6,180,000 in 2005 to 4,309,000 in 2008 (−30%, NAHB 2012). Figure 4.5 displays the number of US single-family home sales from 2000 to 2012.

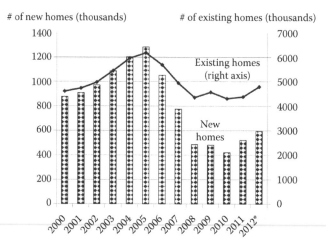

FIGURE 4.5
New and existing single-family home sales 2000–2012. *Note*: 2012 annualized through October 2012. (From National Association of Homebuilders (NAHB), Housing Economics, Washington, DC, http://www.nahb.org, accessed December 1, 2012, 2012.)

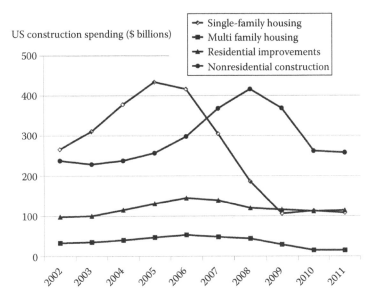

FIGURE 4.6
Value of private US construction. (U.S. Census Bureau, n.d. Annual value of private construction put in place 2002–2010, http://www.census.gov/const/C30/private.pdf, accessed April 22, 2011.)

The current challenges in the US residential housing market have had a detrimental impact on the country's secondary wood products sector. Between 2008 and 2009, 81% of the secondary wood products manufacturers participating in the annual housing markets and wood products industry survey (Buehlmann et al. 2010b) indicated that they lost sales, with 60% reporting sales volume declines of 20% or more. In 2010, only half of the respondents reported declines in sales, showing the stabilization, albeit on a low level, of the markets for secondary wood products in the United States (Buehlmann et al. 2011). Also, results from these studies indicate that over the past 3 years, the US secondary wood products sector has shifted business away from the single-family housing market. Figure 4.6 shows the value of US private construction from 2002 to 2011.

4.5.2 Europe

Europe, at present, displays a mixed basket of fortunes for secondary wood products manufacturers. Preferences for building materials vary within Europe by region. While the south relies mostly on stone buildings, the north is home to some of the most historic and modern wooden buildings. In fact, northern Europe is home of the most wood-centric culture globally and is on the forefront of tackling climate change through the use of renewable, carbon-neutral, or carbon-absorbing materials, creating considerable opportunities for wood products manufacturers. The combination of traditional craft skills and modern technologies has created some of the most stunning secondary wood products. Figure 4.7 shows a modern pedestrian bridge in Switzerland.

Nonetheless, overall, the European home construction market is sluggish (Ball 2011). European residential construction contracted more than 20% from 2008 to 2011 to approximately $796 billion/year (Alderman 2011), with Spain (−21.5%) and Ireland (−32.2%) suffering the most severe decreases (Euroconstruct 2010). A business located in the south of the EU serving local markets will find business difficult. The same business located in

FIGURE 4.7
Modern wooden bridge in Wimmis, Switzerland.

the north of the EU may experience a good business climate with opportunities. In particular, Scandinavia, the Benelux countries (Belgium, the Netherlands, and Luxembourg), and Germany are in the midst of a decent economic recovery from the global financial crisis, while Greece, Portugal, Spain, Ireland, and Italy are struggling to overcome their debt crisis making their economies sluggish or declining. While not part of the Euro zone and thus having more economic liberties, Great Britain and Hungary are dealing with economic problems of their own.

Whereas secondary wood products nowadays can be sold freely throughout most of the EU and associated nations, product and building codes still apply nationally. Even though the harmonization of these codes is ongoing, individual nations still offer shelter to some segments of their industries through administrative and legalistic procedures. Also, the definition of "imported product" has an ambiguous meaning in Europe as it is not always clear if imported refers to any product from outside a given nation state or outside the EU.

Production of secondary wood products moved, with the opening of the Eastern Bloc countries in the 1990s, from Western European countries east, a trend that continues and expands even more eastward, into former Soviet Republic countries. At first (starting in the early 1990s), this eastward shift included Poland, Czech Republic, Slovakia, and Hungary but has moved on to countries like Romania, Moldova, Ukraine, and Russia. At the same time, with the improved competitiveness of Asian manufacturers, mostly from China but also from countries like Vietnam, imports of secondary wood products from outside Europe are also increasing. However, some Western European countries with advanced secondary wood products manufacturing capabilities have successfully prevented parts of their industry from moving eastward. The German kitchen cabinet sector and the Italian furniture sector are two prime examples. Other countries, such as Portugal, are striving to regain international competitiveness as a consequence of their recent economic austerity that lowers wages and production costs (Sayare 2012).

Europe, especially Germany and Italy, is also home to many of the leading secondary wood products equipment manufacturers worldwide. Biesse, Homag, IMA, and SCM, to name just a few, are all European suppliers of highly advanced secondary wood products manufacturing equipment throughout the globe. While these manufacturers have started outsourcing some of their lower end equipment manufacturing to places like China, their most advanced equipment is still manufactured at their European home bases.

4.5.3 Asia

Asia hosts 60% of the global population but covers only 30% of the total global land area (UN 2012a). The second and third largest economies of the globe are in Asia (China and Japan). While China, the most populous country on Earth, experienced dramatic growth over the past decades, Japan, one of the most developed economies in Asia, has been mired in a decade-old deflationary period and is struggling to regain its footing (Nishiyama and Sekiguchi 2012). India, the world's second most populous country, while still being on a low level of economic development, is also growing rapidly. Other emerging economies such as the Philippines, Indonesia, Vietnam, and Thailand are growing quickly, adding large numbers of potential customers for the secondary wood products sector every year.

China, the world's largest importer and third largest consumer of logs (Wood Resources International 2012), is predicted to have its roundwood deficit grow to 150 million m^3 by 2015 (e.g., a deficit bigger than the entire Canadian timber harvest in 2009). China's wood supply problem partially exists due to the success of the country's secondary wood sector. The country has, after Nixon's visit in the 1970s and Deng Xiaoping's interweaving of capitalism and communism in the 1980s, become the world's leading exporter of manufactured goods, serving every conceivable market around the globe with inexpensive products. One of the most illustrative examples of China's prowess in manufacturing is its overtaking of the US non-upholstered wood household furniture market. Between 1992 and 2006, China increased its exports of non-upholstered wood household furniture to the United States by 7437% (from $69 million in 1992 to $5.2 billion in 2006, ITA 2011), thereby eliminating a substantial part of the existing US manufacturing capacity.

China is also home to a gigantic domestic market. With more Chinese people earning middle-class wages every year and a social yearning to display wealth with status symbol products including kitchens and furniture, China is a huge, growing market opportunity for businesses around the globe. China's government is committed to build 35 million housing units over the next 5 years (Stephens 2011). Although traditional Chinese buildings are not typically constructed with wood (Alderman 2011), the wood industry, mainly from the United States and Canada, is continuing to work on making the Chinese building market more amenable to wood as a construction material (IWMG 2011).

Japan presents a different context for secondary wood products manufacturers than does China. While China has been experiencing annual growth of 10% or more for the past decades, Japan is stuck in a period of stagnation and deflation (Nishiyama and Sekiguchi 2012). Wood is the traditional building material for houses in Japan (post and beam construction), but demand for housing in Japan is slated to be relatively modest. Eastin (2008) estimated that less than 200,000 housing units are being built per year and Cohen et al. (2001, 1) described several trends in wood products use in Japan, among them "... The growing importance of high-performance, aesthetically pleasing wood products, the adoption of platform-frame technologies and engineered wood products, and the emergence of a pre-cut component manufacturing sector." These authors also allude to the changing demographics, changes in the regulatory environment, and protectionism as being important for the future of Japanese wood products use. It appears that Japan, with an aging and declining population and little exports of secondary wood products, will offer a challenging environment for secondary wood products businesses into the foreseeable future.

India, the world's second most populous country, is slated to grow its GDP by 7%–8% annually for the coming decade (Garcia 2011) and is expected to invest considerable sums into infrastructure. However, while the residential construction sector in India is small, it should, with a shortage of 31.1 million housing units, increase its size considerably over the next years (Mallela 2009). Scarcity of land, difficulties obtaining land titles, and problems in sourcing of materials make housing construction in India a challenge and prevent demand to drive the industry. Yet, given the large need and the increasing affluence of the average Indian, markets for secondary wood products are expected to grow considerably.

The Middle East, especially countries like Qatar, Kuwait, Oman, the United Arab Emirates, and Saudi Arabia, are experiencing considerable economic growth as a consequence of their enormous oil wealth and their strive to diversify their economies. Much of the construction materials and secondary wood products needed to sustain this construction boom is sourced from other countries. The increasing demand has wood products exporters from around the world considering these Middle Eastern countries as an important export market for the future, and they thus have started attending regional trade shows and are actively marketing their products.

4.5.4 Oceania

Oceania consists of over 20 independent states in the southern Pacific with considerable differences in size, wealth, and growth. Australia, where over 60% of Oceania's population lives, together with New Zealand, with roughly 20% of the population (UN 2012b), is also the economic center of the region. New Zealand is well known for a thriving forest products industry with forest products being the second most important export product of the country. However, while New Zealand excels in supplying world markets with forest products, it struggles to create a sizeable secondary wood products sector that would add additional value to their domestic raw material (Spetic 2009). Australia, the largest economy of Oceania, suffers from the global economic slowdown, which translates into lower demand for raw materials, which are Australia's biggest export. Thus, while the economy might be slowing, reports indicate that housing will improve over the next few years (Anonymous 2012). Thus, markets for secondary wood products in Australia should be positive for both, domestic producers and exporters to Australia.

4.5.5 Africa

Africa, the world's second largest continent with the second largest population (slightly more than one billion people) is, on average, the poorest continent on Earth. Average per capita GDP was only $1603 in 2010, about half of the next poorest (Asia, per capita GDP $2902, CIA 2012). However, while at low levels, per capita GDP varies widely, ranging from $12,248 in Libya to $185 in the Democratic Republic of Congo (CIA 2012). Thus, the markets for secondary wood products in Africa are unevenly distributed throughout the continent, and industrial-scale secondary wood products manufacturers may be rare in some regions of the continent.

South Africa, Africa's largest economy (World Bank 2012), has a sizeable wood furniture sector consisting of 830 enterprises with 39,000 employees (Pogue et al. 2008), most of them small- and medium-sized enterprises. While the South African government has encouraged the development of the secondary wood products industry due to its potential to employ individuals with a variety of skills and education, the industry's limited attention to research and development (R&D) limits the success of the industry, especially internationally (Pogue et al. 2008).

Ndiapi et al. (2011) analyzed the status of the secondary wood processing sector in Cameroon. They point out that the focus of the Cameroonian government has been the development of the primary wood sector to the disadvantage of the secondary sector. These authors, however, identified a comprehensive list of opportunities for the future of the secondary wood sector in Cameroon, the presence of potential markets (local, regional, and international), the availability of raw materials, and a strong desire of the government and NGOs to promote the secondary wood sector to create jobs and income. However, Ndiapi et al. (2011) also identified challenges facing the Cameroonian secondary wood sector, among them the difficulty to obtain quality raw materials, skilled labor, financing, and the strong competition from imported products (mainly from Asia). Ba (2009) described similar conditions for producers of non-timber forest products in Senegal, showing the widespread challenges facing industrial endeavors in large parts of Africa.

4.5.6 Global Threats to Regional Markets

Regional markets can, broadly speaking, face threats from two sources. First, protectionism, which refers to obstacles that prevent the free flow of goods and services, may affect regional markets. Second, is the ability of a global supplier to build a competitive business model that allows it to serve local or regional customers through a global supply chain. Both of these are at play in global markets and the outcome is not yet clear.

Protectionism, "… The policy a country has of helping its own industries by putting a tax on imported goods (Collins ELT 1987, 1155)," now is, to a large extent, overseen by the World Trade Organization (WTO 2008) to ensure that a country is following mutually agreed rules (for the 157 member nations of the WTO (2012a)). In the case of unfair trade practices, such as dumping (e.g., selling a product in a foreign market for less than in the domestic market or for less than fair value, WTO 2012b) of goods, however, tariffs are still applicable and can protect a national industry. A widely discussed example of this practice is the US non-upholstered wood bedroom furniture sector, which has been protected by tariffs since 2005 (Harris 2009). The WTO, in cases of dumping, not only allows the nation harmed by the dumping to levy a tariff on the goods imported in question but also allows compensatory payments to the industry hurt by the dumping (e.g., the duties collected by US customs is paid to US firms hurt by the dumping). However, politicians, when deciding on protectionist measures, also need to take into account that the emergence of global competitors implies lower prices and increased product choices for customers.

Global competition, because of comparative advantages, is supposed to raise standards of living in all places. Thus, for example, while globalization in the sector led to the demise of the US non-upholstered wood household furniture sector that ultimately brought economic plight to numerous furniture workers in the United States, the very globalization brought US consumers cheaper and, possibly, even better furniture with more product choices. Globalization thus is a mixed blessing. While proponents of globalization would argue that workers, who lost their jobs due to globalization, should have moved to higher value-added, more advanced, better paid employment (Anonymous 2004), many found themselves unemployed or working in minimum-wage jobs. Globalization has proven efficient in "constructive destruction," but in the case of the US secondary wood sector, no constructive replacement of the lost opportunities has been found at regional levels where the jobs were lost. Such unfortunate outcomes of trade liberalization bear the potential to revive domestic protectionism (Herbst 2009) and threaten the free trade of goods and services, creating a problem for global suppliers while offering an opportunity to local or regional industries.

4.6 Regional Market Characteristics, Use Patterns, and Trends

Both markets and end-use patterns differ for secondary wood products throughout the world. However, given the continuing globalization of the sector, product specifications, market characteristics, and end-use patterns are slowly converging. Also, because most secondary wood products like furniture, kitchen cabinets, or flooring are fashion products, they are highly dependent on larger trends created by actors influencing fashion symbols globally.

Alongside the strong wave of globalization, some regional preferences continue to have their stronghold. For example, while the kitchen cabinet sector has eliminated cabinetry with face frames for many decades now in Europe, face-framed cabinetry still makes up the majority of all kitchen cabinets in the United States and Canada. This preference is mostly based on customer preferences, as kitchen cabinetry without face frames is easier to produce and thus would be easy for North American manufacturers to switch to, if customer demand asked for it. Thus, manufacturers who want to compete in regional markets must be sensitive to the practices and preferences that prevail at local levels.

The increasing consumer awareness, mostly in developed economies such as the United States, Europe, or Japan, about the environmental impact of their purchasing decisions, has implications for secondary wood products manufacturers, too. Given increasing customer interest in gauging the environmental merit of the products they buy, numerous systems have evolved to gauge product "greenness" and may cover a wide range of products—from furniture to an entire building (Espinoza et al. 2012). Wood, a carbon-absorbing, natural material, has been suggested as having a more benign environmental impact than most other building materials (CEI-bois 2006), and as discussed in Chapter 11 of this book, developments in the green building policies and practices may have significant impacts on the secondary wood products sector.

Perhaps most importantly, demographic shifts around the globe will create changes in markets for secondary wood products. Peter Drucker, a renowned management sage, is quoted as saying that "Demography is the future that already happened (Siegel 2006)." Global demographic trends show a two-part world, with regions where the population is aging fast and regions where the population is growing fast. Europe, Japan, and, to a lesser degree, the United States, among others, are faced with an increasing share of older citizens, putting strain on these countries' social safety net and increased burdens on their younger citizens. Conversely, in countries like India, much of Southeast Asia, South America, or Africa, a fast-growing population creates more future customers. Either way, an aging population or a fast-growing one creates opportunities for secondary wood products producers. For example, in the United States, one can observe several notable secondary wood products manufacturers shifting their product lines away from traditional mass markets to more specialized, higher-end products serving the aging population.

4.7 Long-Term Outlook

Demand for secondary wood products is, to a great extent, tied to housing and remodeling activity in a country or region. Thus, the future demand for secondary wood products should be strongest in countries with high growth, such as the BRIC countries or other fast-growing nations. Some of these emerging countries (Brazil, Russia) have good quality domestic wood supplies, while others (India, China) must rely on imported fiber to cover their needs.

The United States, Europe, Japan, and other developed countries will continue to remain important markets for secondary wood products despite their slow economic growth. Time will tell if these countries' secondary wood industries can benefit from the fast-growing economies in developing countries through export of secondary wood products or if trade flows will generally go in the opposite direction. Much will depend on customer preferences, international trade conditions, raw material availability, and production economics in individual countries and regions.

In any case, over the past two decades, the secondary wood products sector has evolved into a global industry consisting of two types of market participants. On the one hand are companies that are pursuing market opportunities in every corner of the world where their products can be sold and that continue to move their manufacturing facilities to places where new comparative advantages emerge. On the other hand are companies that operate locally/regionally and live off products that are MTO and are highly customized and specialized to fit the needs of particular customers. As discussed, manufacturing of such products cannot be easily moved to low-cost locations. The global companies will compete in markets selling standardized products where demand forecasts allow for long lead time and mass production, while the customizers will serve markets where customers expect or need customized solutions using mass customization techniques.

Companies competing globally will supply commodity products or components for standard furniture, flooring, or kitchen cabinets that require no customization and allow for extended lead times. These companies will also compete heavily for standardized components used in the local/regional business. For example, while the US upholstered household furniture is mostly selling standardized products, customers can still select the fabric. Thus, some suppliers produce the standard frame and the upholstery without fabric in a low production cost location, ship the semifinished product to the United States, and install the fabric demanded by the customer in a small assembly operation somewhere locally/regionally. We anticipate more such collaborations will emerge in the future.

Also, while customization currently limits globalization of some secondary wood products, future technologies will again shift these limits. Currently, communicating individual customer preferences and speedy delivery of the product is challenging. However, with the ongoing revolution in communication and future improvements in transportation and technology, new business models might evolve and make possible what we today think as far-fetched. Human progress has always happened around new ideas, technologies, and concepts. The future of secondary wood products manufacturing will evolve with these new opportunities.

References

Alderman, D. 2011. International housing construction developments—Implications for hardwood utilization. *Proceedings of the International Scientific Conference on Hardwood Processing 2011 (ISCHP 2011)*. Blacksburg, VA, pp. 223–234.

Anonymous. 2004. Shaking up trade theory. *Bloomberg Businessweek Magazine*. December 5, 2004. http://www.businessweek.com/stories/2004-12-05/shaking-up-trade-theory. Accessed October 2, 2012.

Anonymous. 2012. Housing market forecast to improve. *Herald Sun*, Melbourne, Australia. http://www.heraldsun.com.au/money/property/housing-market-forecast-to-improve/story-fnb-pe2tv-1226407174692. Accessed October 1, 2012.

AP. 2007. Housing market's next victim? Furniture sales. Consumer worried about selling, mortgages are deferring purchases. *MSNBC Business/U.S. Business*. October 7, 2007. http://www.msnbc.msn.com/id/21182357/. Accessed April 6, 2008.

Areddy, J. T. 2008. China's export machine threatened by rising costs. *Wall Street Journal*. June 30, 2008. A1 and A9.

Ba, B. 2009. Poverty alleviation and non-timber forest products in Senegal: The role of small-scale enterprises. Research paper submitted to the Faculty of Virginia Tech., Blacksburg, VA, April 28, 2009. 78pp.

Ball, M. 2011. RICS European housing review 2011. Royal Institution of Chartered Surveyors (RICS). January 2011. Brussels, Belgium, 90pp.

Barford, M. 2012. *The Future of North American Hardwood Lumber*. National Hardwood Lumber Association (NHLA), Memphis, TN, 18pp.

Bernhardt, D., Q. Liu, and K. Serfes. 2007. Product customization. *European Economic Review*. 51(2007):1396–1422.

Bland, B. 2012. Vietnam offers companies China alternative. *Financial Times*. http://www.ft.com/intl/cms/s/0/46d052b8-6446-11e1-b30e-00144feabdc0.html#axzz25VnstQx3. Accessed September 4, 2012.

Bloomberg. 2012. Canada November existing home sales fall 1.7% on month. Bloomberg.com. http://www.bloomberg.com/news/2012-12-17/canada-november-existing-home-sales-fall-1-7-on-month.html. Accessed December 21, 2012.

BLS. 2012. Charting international labor comparisons (2012 edition)—hourly compensation costs in manufacturing, selected countries, in U.S. dollars, 2010. Bureau of Labor Statistics, U.S. Department of Labor, Washington, DC. http://www.bls.gov/fls/chartbook/2012/section3a.htm#chart3.1. Accessed September 26, 2012.

Buehlmann, U. 2001. Entwicklung von Holzeinschlag und Holzverbrauch [Future development of wood supply and demand]. *Holz-Zentralblatt*. September 14, 2001, pp. 1373–1374.

Buehlmann, U., M. Bumgardner, and K. D. Forth. 2012. Lightweight panel usage, acceptance grow: Survey. CabinetMaker+FDM. August 2012. pp. 36–41.

Buehlmann, U., M. Bumgardner, A. Schuler, and M. Barford. 2007. Assessing the impacts of global competition on the Appalachian hardwood industry. *Forest Products Journal*. 57(3):89–93.

Buehlmann, U., M. Bumgardner, A. Schuler, and K. Koenig. 2010b. Housing market's impact on the secondary wood industry. *Wood & Wood Products*. July 2010. 21–29.

Buehlmann, U., M. Bumgardner, A. Schuler, and K. Koenig. 2011. Housing and the wood industry: trends and market conditions. *Wood & Wood Products*. July 2011. 24–29.

Buehlmann, U., O. Espinoza, M. Bumgardner, and R. Smith. 2010a. Trends in the U.S. hardwood lumber distribution industry: Changing products, customers, and services. *Forest Products Journal*. 60(6):547–553.

Buehlmann, U., K. D. Forth, and M. Bumgardner. 2008. Office furniture, fixtures promising for lightweight panels. FDMOnline. http://www.fdmonline.com/ViewArticle.aspx?id=30586. Accessed November 16, 2008.

Buehlmann, U. and A. Schuler. 2009. The U.S. household furniture industry: Status and opportunities. *Forest Products Journal*. 59(9):20–29.

Buehlmann, U., A. Schuler, and D. Merz. 2004. Reinventing the U.S. furniture industry—facts and ideas. *Proceedings of the Annual Meeting of the Forest Products Society–Industry Day. Forest Products Society*. Madison, WI.

CEI-bois. 2006. *Tackle Climate Change: Use Wood*. European Confederation of Woodworking Industries, Brussels, Belgium, 86pp.

Champine, L. A. and A. Krishnan. 2004. *Asia's Impact on the Residential Furniture Industry*. Morgan Keegan & Company, Inc., New York, 34pp.

China National Furniture Association as cited in: Snow, M. 2010. *Penetrating International Markets*. American Hardwood Council, Reston, VA, May 2010.

Chowdhury, N. 2011. The China effect. *Time Magazine Business*. http://www.time.com/time/magazine/article/0,9171,2065153,00.html. Accessed September 4, 2012.

CIA. 2012. The world factbook—Country comparison: GDP—per capita (PPP). Central Intelligence Agency (CIA). https://www.cia.gov/library/publications/the-world-factbook/rankorder/2004rank.html. Accessed September 28, 2012.

CMHC. 2012. Preliminary housing start data released December 2012. Canada Mortgage and Housing Corporation. 22pp. http://www.cmhc-schl.gc.ca/odpub/esub/64695/64695_2012_M12.pdf?fr=1356093865312. Accessed December 21, 2012.

Cohen, D., C. Gaston, and R. Kozak. 2001. Influences on Japanese demand for wood products. ECE/FAO Forest Products Annual Market Review, 2000–2001. http://www.unece.org/forests/docs/rev-01/rev01.html. Accessed October 2, 2012.

Collins, ELT. 1987. *English Language Dictionary*. Collins ELT, London, U.K., 1703pp.

Davis, S. 1987. *Future Perfect*. Addison-Wesley Publishing Company, Reading, MA, 243 pp.

Dierig, C. 2012. Deutsche räkeln sich gern auf chinesischen Sofas. *Die Welt*. http://www.welt.de/wirtschaft/article13823039/Deutsche-raekeln-sich-gern-auf-chinesischen-Sofas.html. Accessed September 25, 2012.

Eastin, I. 2008. *Japanese Green Building Program and the Domestic Woodworking Program. CINTRAFOR*, University of Washington, Seattle, WA.

Economist. 2008. The dollar—The fear of falling. *The Economist*, June 21, 2008. Online: http://www.economist.com/finance/displaystory.cfm?story_id=11599161. Accessed June 21, 2008.

Economist. 2009. Denial or acceptance—Dollar appreciation. *The Economist*, October 22, 2009. Online: http://www.economist.com/node/14700644. Accessed September 3, 2012.

Economist. 2011. Moving back to America—The dwindling allure of building factories offshore. *The Economist*, May 12, 2011. Online: http://www.economist.com/node/18682182. Accessed September 4, 2012.

Economist. 2012a. The end of cheap China. *The Economist*, March 10, 2012. Online: http://www.economist.com/node/21549956. Accessed September 4, 2012.

Economist. 2012b. The boomerang effect—As wages rise, some production is moving back to the rich world. *The Economist*, April 21, 2012. Online: http://www.economist.com/node/21552898. Accessed September 4, 2012.

Elwell, C. K. 2012. The depreciating dollar: Economic effects and policy response. Congressional Research Service. RL34582. February 23, 2012. 24 pp.

Espinoza, O., U. Buehlmann, M. Bumgardner, and B. Smith. 2011. Assessing changes in the U.S. hardwood sawmill industry with a focus on markets and distribution. *BioResources*. 6(3):2676–2689.

Espinoza, O., U. Buehlmann, and B. Smith. 2012. Forest certification and green building standards: Knowledge, use and perceptions in the US hardwood industry. *Journal of Cleaner Production*. 33(2012):30–41.

Euroconstruct. 2010. Summary report. 70th euroconstruct summary report. *70th Euroconstruct Conference*. Budapest, Hungary, December 2–3, 2010, 170pp.

FAO. 1982. Classification and definitions of forest products. Food and Agricultural Organization of the United Nations, Rome, Italy. http://www.fao.org/docrep/015/an647e/an647e00.pdf. Accessed December 21, 2012.

FAO. 1999. State of the World's forests 1999. Food and Agricultural Organization of the United Nations, Rome, Italy. http://www.fao.org/docrep/w9950e/w9950e00.htm. Accessed November 15, 2012.

FAO. 2009. State of the World's Forests 2009. Food and Agricultural Organization of the United Nations, Rome, Italy. http://www.fao.org/docrep/011/i0350e/i0350e00.htm. Accessed June 17, 2010. 4pp.

FAO. 2012. FAOSTAT. Food and Agricultural Organization of the United Nations. http://faostat.fao.org/DesktopDefault.aspx?PageID=626&lang=en#ancor. Accessed January 18, 2012.

Garcia, T. 2011. The global construction industry—What can engineers expect in the coming years? *Plumbing Systems & Designs*. December 2011. pp. 22–25. http://www.aspe.org/sites/default/files/webfm/ArchivedIssues/2011/201112/TheGlobalConstructionIndustry.pdf. Accessed December 31, 2012.

Grant Thornton, LLP. 2006. *Survey of U.S. Business Leaders*, 12th edn. Grant Thornton International, Chicago, IL, 16pp.

Harris, J. 2009. U.S. Department of commerce industry report: Furniture and related products NAICS code 337. Office of Health and Consumer Goods, US Department of Commerce, Washington, DC. http://ita.doc.gov/td/ocg/outlook09_furniture.pdf. Accessed October 2, 2012.

Herbst, M. 2009. Jobs and protectionism in the stimulus package. *Business Week*. February 16, 2009. http://www.businessweek.com¬/bwdaily/¬dnflash/¬content/¬feb2009/¬db20090216_¬920561.htm?chan=top+news_top+news+index+-+temp_top+story. Accessed April 14, 2009.

Huyett, W. I. and S. P. Viguerie. 2005. Extreme competition. *The McKinsey Quarterly Number*. 1: 46–57.

IMF. 2011. Changing patterns of global trade. International Monetary Fund. http://www.imf.org/external/np/pp/eng/2011/061511.pdf. http://www.imf.org/external/np/pp/eng/2011/061511.pdf. Accessed September 3, 2012. 69pp.

ITA. 2011. Consumer goods industries. International Trade Administration (ITA), U.S. Department of Commerce Washington, DC. http://www.ita.doc.gov/td/ocg/index.htm. Accessed November 15, 2011.

ITA. 2012. Global patterns of U.S. merchandise trade. International Trade Administration (ITA), U.S. Department of Commerce Washington, DC. http://tse.export.gov/TSE/MapDisplay.aspx. Accessed September 4, 2012.

IWMG. 2011. China's imports soar in 2010–led by logs (+22%) and lumber (+49%). International Wood Markets Group, Inc., Vancouver, British Columbia, Canada, February 8, 2pp.

Klaas, D.-U. 2007. Künftige Entwicklung der Möbelindustrie—ein Ausblick auf die nächsten 3–5 Jahre. Hauptverband der deutschen Holzindustrie, e.V. und Verband der deutschen Möbelindustrie, e.V. Tappi-Symposium zur interzum 2007.

Lazear, E. P. 2012. The worst economic recovery in history. *Wall Street Journal*. April 2, 2012. http://online.wsj.com/article/SB10001424052702303816504577311470997904292.html. Accessed September 26, 2012.

Li, J. 2011. Taiwans's furniture industry: Yesterday, today, and tomorrow. *China Economic News Service*. http://cens.com/cens/html/en/news/news_inner_37386.html. Accessed September 4, 2012.

Lihra, T., U. Buehlmann, and R. Beauregard. 2008. Mass customization of wood furniture as a competitive strategy. *International Journal of Mass Customization*. 2(3/4):200–215.

Lihra, T., U. Buehlmann, and R. Graf. 2011. Customer preferences for customized household furniture. *Journal of Forest Economics*. 18(2012):94–112.

Luppold, W. G. in: Buehlmann, U., A. Schuler, and D. Alderman. 2012. Forest products markets in 2012: Background and opportunities. WVFA presentation. Kanaan Valley, WV, July 2012.

Majumdar, R. 2004. The effect of changes in housing wealth on retail sales. *Wharton Research Scholars Journal*. University of Pennsylvania, Philadelphia, PA, 27 pp.

Mallela, K. 2009. The housing construction sector in India—growth, development and outlook. YahooVoices. http://voices.yahoo.com/the-housing-construction-sector-india-growth-development-2591878.html. Accessed October 1, 2012.

NAHB. 2012. *National Association of Homebuilders*. Housing Economics, Washington, DC. http://www.nahb.org. Accessed December 1, 2012.

Ndiapi, O., J. M. Njankouo, and L. M. Aynia Ohandja. 2011. Technical diagnosis of secondary wood processing in Cameroon. *Proceedings of the International Scientific Conference on Hardwood Processing (ISCHP2011)* Virginia Tech., Blacksburg, VA, II: Publications and abstracts. 167 pp.

Nishiyama, G., and T. Sekiguchi. 2012. Japan's Noda vows quick end to deflation. *Wall Street Journal*. http://online.wsj.com/article/SB10000872396390444358804578015651997380798.html. Accessed September 26, 2012.

NZZ. 2008. Der Dollar erreicht Parität zum Franken–Sorge um die amerikanische Wirtschaft lastet auf der Währung. Neue Zürcher Zeitung online. March 14, 2008. http://www.nzz.ch/nachrichten/wirtschaft/boersen_und_maerkte/der_dollar_erreicht_paritaet_zum_franken_1.688872.html. Accessed March 24, 2008.

Pepke, E. 2011. European markets and certification. *International Scientific Conference on Hardwood Processing (ISCHP 2011)*. Blacksburg, VA, October 17, 2011.

Pine, J. B. 1993. *Mass Customization: The New Frontier in Business Competition*. Harvard Business School Press, Boston, MA, 333pp.

Pogue, T. E., L. Jordan, S. Mabitsela, K. Morobane, M. Sumbula, and L. Wang. 2008. A sectoral analysis of wood, paper and pulp industries in South Africa. Institute for Economic Research on Innovation (IERI), Pretoria, South Africa 128pp.

Sayare, S. 2012. Portuguese just shrug and go on in the face of cuts and job losses. *New York Times*. http://www.nytimes.com/2012/06/08/world/europe/portugal-shrugs-at-austerity.html?pagewanted=all&_r=0. Accessed October 1, 2012.

Schuler, A. and C. Adair. 2003. Demographics, the housing market and demand for building materials. *Forest Product Journal*. 53(5):8–17.

Schuler, A. and U. Buehlmann. 2003. Benchmarking the wood household furniture industry: A basis for identifying competitive business strategies for today's global economy. USDA Forest Service General Technical Report. GTR-NE 304. Princeton, West Virginia. 18pp.

Siegel, J. J. 2006. Gray world. *Wall Street Journal*. http://online.wsj.com/article/SB115871347054768375.html. Accessed October 7, 2012.

Smith, M., M. J. Fannin, and R. P. Vlosky. 2009. Forest industry supply chain mapping: An application in Louisiana. *Forest Products Journal*. 59(6):7–16.

Spetic, W. C. 2009. Competitiveness and sustainability: Perspective from the secondary wood industry of British Columbia, the forest industries of New Zealand, Chile, and Brazil, and the sugarcane-based ethanol industry of Brazil. Doctoral Dissertation. University of British Columbia, Vancouver, British Columbia, Canada, 154pp.

Stephens, D. 2011. The west looks to the east: Exports to China play key role in western forest products industry. Pallet Enterprise, Ashland, VA. www.palletenterprise.com. Accessed April 3, 2011.

Stickley. 2005. Vietnam Expansion. L. & J.G. *Stickley News*. May 18, 2005. http://www.stickley.com/NewsAndEvents.cfm?SubPgName=StickleyNews. Accessed April 28, 2008.

Sutton, R. J. 1999. Does the world need plantations? *International Forestry Symposium*. Santiago, Chile. April 1999.

Toffler, A. 1970. *Future Shock*. Bantam Books, New York, 545pp.

UN. 2005. European forest sector outlook study main report. United Nations. http://www.unece.org/timber/docs/sp/sp-20.pdf. Accessed June 25, 2010. 265pp.

UN. 2012a. Country/region profiles. United Nations, Department of Economic and Social Affairs. http://esa.un.org/unpd/wpp/unpp/Panel_profiles.htm. Accessed December 21, 2012.

UN. 2012b. Composition of macro geographical (continental) regions, geographical sub-regions, and selected economic and other groupings. United Nations, Department of Economic and Social Affairs. http://millenniumindicators.un.org/unsd/methods/m49/m49regin.htm#oceania. Accessed December 21, 2012.

U.S. Census Bureau. 2009. NAICS 337: Furniture and related product manufacturing. http://www.census.gov/epcd/ec97/def/337.HTM. Accessed November 12, 2009.

U.S. Census Bureau. 2010. NAICS – North American Industry Classification System–Definition. http://www.census.gov/eos/www/naics/index.html. Accessed June 22, 2010.

U.S. Census Bureau. 2012. Foreign Trade. U.S. Department of Commerce, U.S. Census Bureau, Washington, DC. http://www.census.gov/foreign-trade/statistics/historical/. Accessed September 3, 2012.

U.S. Census Bureau. n.d. Annual value of private construction put in place 2002–2010. http://www.census.gov/const/C30/private.pdf. Accessed April 22, 2011.

Wan, Z. and S. H. Bullard. 2008. Firm size and competitive advantage in the U.S. upholstered, wood household furniture industry. *Forest Products Journal.* 58(1/2):91–96.

Wood Resources International. 2012. China is now the world's largest importer of softwood lumber and logs despite a slowdown in imports during the 4Q/11. Wood Resources International LLC Company. http://www.cisionwire.com/wood-resources-international-llc-company/r/china-is-now-the-world-s-largest-importer-of-softwood-lumber-and-logs-despite-a-slowdown-in-imports-,c9213679. Accessed September 26, 2012.

World Bank. 2012. World Development Indicators (WDI). World Data Bank. http://databank.worldbank.org/ddp/home.do?Step=12&id=4&CNO=2. Accessed December 21, 2012.

WTO. 2008. What is the WTO? World Trade Organization Geneva, Switzerland. http://www.wto.org/. Accessed April 5, 2008.

WTO. 2012a. Members and observers. World Trade Organization Geneva, Switzerland. http://www.wto.org/english/thewto_e/whatis_e/tif_e/org6_e.htm. Accessed October 2, 2012.

WTO. 2012b. Glossary. World Trade Organization. http://www.wto.org/english/thewto_e/glossary_e/dumping_e.htm. Accessed December 21, 2012.

5

Markets and Market Forces for Pulp and Paper Products

Lauri Hetemäki, Riitta Hänninen, and Alexander Moiseyev*

CONTENTS

5.1 Background and the State of the Sector

5.1.1 Background

The pulp and paper industry is highly diversified in terms of products, raw materials, product qualities, distribution channels, and end uses. For instance, tissue, carton board,

---

* Lauri Hetemäki is head of the Foresight and Policy Support Programme at the European Forest Institute and professor at the University of Eastern Finland. Riitta Hänninen is a senior researcher at the Finnish Forest Research Institute. Alexander Moiseyev is a senior researcher at the Norwegian University of Life Sciences Ås, Norway. Hetemäki is responsible for Sections 5.1 through 5.2 and 5.5 through 5.7, Hänninen for Section 5.3, and Moiseyev for Section 5.4.

and newsprint have very little in common, apart from their basic production processes and being capital intensive to manufacture. Pulp, paper, and packaging boards are typically intermediate products, used as inputs in the production of downstream value-added products, while some products, such as tissue and office papers, are generally distributed to consumers without further conversion.

Global pulp and paper markets are experiencing large-scale structural changes due to several factors. First, after having increased for over a century, at the turn of the twenty-first century, for the first time in history, communication paper consumption (printing and writing paper + newsprint) started to decline in many industrialized Organization for Economic Co-operation and Development (OECD)* countries. There are many reasons for this decline in consumption; however, the single most important reason is the substitution impact from digital media. Second, there is a clear dichotomy in global markets: Paper and paperboard consumption and production are increasing significantly in non-OECD countries, whereas they are stagnating or declining in many OECD countries. Third, real prices of paper and paperboard products have been declining in the past decade. Fourth, the share of hardwood pulp in total pulp production based on fast-growing plantation forests in South America, Asia, and Oceania is increasing rapidly. Fifth, international trade in pulp and paper products is increasing, both absolutely and relatively (share of exports in total production). Sixth, pulp and paper industry companies in OECD countries that are suffering stagnating or declining markets are innovating their business models by diversifying into new products, particularly in the areas of bioenergy and biorefining. Finally, the pulp and paper sector has become more global, with increased multinational companies and industry consolidation.

The main purpose of this chapter is to discuss the ongoing structural changes in the pulp and paper sector and to project what the sector might look like in the next 10–20 years. We start by first presenting recent trends and the current state of the global paper and paperboard sector. Next, the communication (or graphics) paper sector is discussed, after which the packaging and carton board products sector is analyzed.

Changes in paper and paperboard production and consumption have significant implications for wood raw material demand; demand for pulp and industrial wood is derived from the demand for forest products, (e.g., Klemperer 1996; Uutela 1987). Therefore, we also analyze the implications on raw material demand, that is, for pulp, recycled paper, pulpwood, and chips. The chapter concludes with a broad discussion and conclusions regarding major changes in the global pulp and paper industry structure, implications for companies in the sector, and an outlook for new product development.

5.1.2 Significance, Structure, and Outlook of the Global Paper and Paperboard Sector

In 2010, the world total quantity of paper and paperboard products produced was 400 million metric tons, with an estimated value of US $ 360 billion (using the 2010 average export unit value as a basis of valuation).[†] For comparison, the aggregate sales of Apple Inc.,

---

* The mission of the OECD is to "promote policies that will improve the economic and social well-being of people around the world" (www.oecd.org). It is comprised of 34 member countries spanning the globe from North and South America to Europe, and the Asia-Pacific region. The organization includes many of the world's high-income countries, but also some emerging countries like Mexico, Chile, and Turkey.

† According to FAO, the average world export value of paper and paperboard in 2010 was US$ 902 per ton (FAOSTAT). The world production of paper and paperboard in 2010 was 400 million tons. If we value this production by the export value, the total world paper and paperboard value was 902 × 400 = 360, 800 million US dollars or 360.8 billion. All the data related to forest industry given in this chapter are either from FAOSTAT or from RISI.

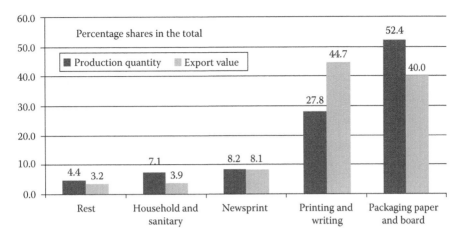

FIGURE 5.1
The production quantity and export value shares (%) of paper product groups in 2010. (Computed as percentage of the total world paper and paperboard production and export value in 2010; http://faostat.org.)

Microsoft Inc., and Nokia Inc. in 2011 was about US $ 228 billion, or 63% of the production value of the entire global paper and paperboard industry.*

In terms of the quantity produced or consumed, packaging and board products are the largest paper product sector. They accounted for 52% of the total paper and paperboard production in 2010, whereas the share of printing and writing paper and newsprint together was 36% (Figure 5.1). However, the pattern of export value of these products is reversed; communication papers accounted for 53%, while packaging and board just under 40% in 2010.

At the global level, aggregate paper and paperboard consumption has continued to grow, on average, approximately 2% per annum during the past decade. If this trend continues in the current decade, global consumption will increase by 83 million tons from 2010 to 2020, that is, about the same amount as North American consumption was in 2010 (81.5 million tons). However, this growth pace is unlikely to continue in the future. As indicated by the 5-year moving average (MA) of the paper consumption growth rate, the trend has been declining since the end of the 1980s (Figure 5.2). In the 1980s, the average growth rate of world paper and paperboard consumption was 3.75%, in the 1990s it was 3.11%, and in the 2000s it has slowed to 2.12%, on average.

The major reason behind declining overall global growth appears to be driven by saturated or declining consumption of some major paper products in high-income OECD countries. If global paper and paperboard markets were divided into two regions, *industrialized high-income regions/countries* and *primarily low-income countries*, the differences in consumption patterns become more striking. We define *high-income regions/countries* to be North America, Western Europe,[†] Japan, Australia, and New Zealand, and the *low-income countries* are all others.

---

* *Sources*: Annual reports for 2011 in company websites: Nokia = 50.1, Apple = 108, Microsoft = 69.9.
† *Western Europe* is here defined to consist of Austria, Belgium, Denmark, Finland, France, Germany, Greece, Iceland, Ireland, Italy, the Netherlands, Norway, Portugal, Spain, Sweden, Switzerland, and the United Kingdom.

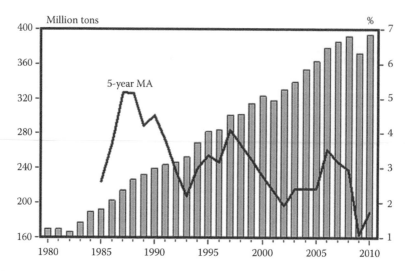

FIGURE 5.2
World paper and paperboard consumption and 5-year MA change in consumption growth rate, 1980–2010. (Data from FAOSTAT, http://faostatifao.org/.)

An interesting observation is that in 2000 paper and paperboard consumption in the high-income region/countries was twice that of low-income countries; but already in 2010, aggregate consumption was higher in the low-income countries. Thus, within a decade, there has been a striking change in these markets. A second observation is that, at the turn of this century, consumption has become saturated and even declined in high-income countries. If consumption trends of the last decade were to continue in the coming decade, consumption in the high-income regions/countries would decline by 2020 to the level it was in the late 1980s, while in the low-income countries, consumption would be over 50% higher than it was in 2010.

One of the major factors influencing growth in low-income countries has been the extraordinary development in China. Between 2000 and 2010, Chinese paper and paperboard consumption grew by 143%. However, this was outstripped by a 182% increase in Chinese production. This dramatic influence has global implications for the paper and wood fiber markets.

This global paper consumption perspective does not reveal large differences between major regions and various paper grades. What are these differences and the factors behind the different patterns? Moreover, what is the outlook for the next 10–20 years? These are the questions we next analyze in more detail. Before doing this, a short description of the approach used for the analysis is helpful. The market outlook prospects are described by the use of simple trend models that can be viewed as "base scenarios" where markets and market structures are assumed to follow the same patterns for the next 20 years as they have over the 2000–2010 period. The year 2000 has been chosen as the base year for the trend analyses, because for many paper grades and regions, data from the twentieth century reflect a different structure than what has been experienced in the past decade in the global paper markets.

However, the longer to the future we project, the more likely it is that even the most recent trend patterns will not continue as such. Inevitably, there will be new structural changes emerging that will create dislocations for the trend of the past decade. Still, the trend projection is a helpful baseline against which we can reflect and speculate possible structural changes and how they would impact the projections.

5.1.3 Regional Pulp and Paper Markets and Trade

There are large regional differences in pulp and paper consumption and production patterns, as seen in Table 5.1. Asia is clearly the largest region in terms of paper consumption and production, about twice as big as the next region, that is, North America. Perhaps the most striking fact is that Africa's consumption and production are so extremely low compared to other regions. Africa's population (one billion) is roughly equal to that of the total of North America and Europe, but its consumption of paper and paperboard is only about 4% of consumption in these continents.

In examining global development of paper markets between 2000 and 2010, the highest consumption growth has been in Asia, both in absolute volumes and in terms of the rate of growth. Latin America and Eastern Europe also show high growth rates, but in absolute volume they are below Western Europe and North America consumption.

Table 5.1 also provides trend projections to 2020 and 2030. The projections show that clearly the most significant paper consumption and production growth would take place in Asia, doubling by 2030 from 2010 levels. Although this is clearly a possibility, as mentioned earlier, this projection has a high level of uncertainty. Paper consumption and production is quite likely to increase in the emerging economies of Asia (but stagnate or decline in

TABLE 5.1

Paper and Paperboard Consumption, Production, and Net Imports in 2000 and 2010 and Projections to 2020 and 2030 (in Million Metric Tons)

		2000	2010	2020	2030
Asia	Consumption	109.4	178.1	250.7	323.7
	Production	96.0	164.4	238.8	313.3
	Net imports	13.4	13.7	11.9	10.4
North America	Consumption	100.7	81.5	67.0	49.4
	Production	106.8	88.6	75.2	59.4
	Net imports	−6.1	−7.1	−8.2	−10.0
Western Europe	Consumption	81.4	75.9	75.3	72.2
	Production	88.6	90.5	96.0	99.0
	Net imports	−7.2	−14.6	−20.7	−26.8
Eastern Europe	Consumption	13.1	24.0	37.0	48.9
	Production	12.8	19.0	25.9	32.1
	Net imports	0.3	5.0	11.1	16.8
Latin America	Consumption	19.4	27.1	35.3	43.7
	Production	14.9	20.3	26.3	32.0
	Net imports	4.5	6.8	9.0	11.7
Africa	Consumption	4.8	8.0	11.1	14.2
	Production	3.3	4.3	5.6	6.7
	Net imports	1.6	3.6	5.5	7.5

Source: Data from RISI, 15-Year North American graphic paper forecast, June 2012, www.risi.com, 2012.

Japan) up to around 2015–2020. However, the further the time horizon, the more likely it is that consumption growth will face market saturation and consumption may decline (see the discussion in Section 5.2).

The other important message of Table 5.1 is that North American paper consumption and production have declined significantly during the past decade, and logically, the trend projection forecasts this pattern to continue to 2030 barring a shock to the supply/ demand system. Similarly, Western European consumption has started to decline during the past decade, but at a slower rate than North America. Although Western European production has increased from 2000 to 2010, one should be cautious to expect this trend to continue. This is because there have been significant fluctuations (ups and downs) in the production during this period, and also recent data (since 2007) indicate that production has been declining. Many Western European companies have been reducing capacity, which appears to be continuing in the near term.

The share of international trade in paper and paperboard markets relative to global production has increased slightly in the last decade. Globally, the share of exports to production was on average 27% in the 1990s and increased to 30% on average in the 2000s. Regionally, there are significant differences in trade, as seen by the regional net import figures in Table 5.1. North America and Western Europe are the only regions that have been, and are projected to be, net exporters of paper products.

Historically, Asia has been the biggest global importer of paper, and this trend is likely to continue in the coming decade. However, in China, paper production has expanded in recent years more rapidly than consumption. As a result, Chinese exports of paper and paperboard almost tripled from 2005 to 2010 as exports increased by nearly 2 million tons. According to China's 12th Five-Year Plan for the pulp and paper industry, the country targets total paper and board consumption and production to grow at an annual rate of 4.6% to 2015 (Yao 2012). As there are already significant ongoing projects or investment plans for paper and paperboard capacity increases in China, the balanced growth of production and consumption may require closure of outdated production facilities (Ou 2011; Yao 2012).

Assuming that the trend of the recent decade continues, Africa, Eastern Europe, and South America will continue to increase their imports of paper and paperboard. The most significant importer in 2030 is projected to be Eastern Europe. North America and Western Europe will most likely remain the main global exporters of paper and paperboard. However, these trends may be altered by a continuing global economic slowdown, which has significantly reduced paper consumption and production both in North America and in Western Europe. In addition, large global paper companies are redirecting investment to emerging Asian and South American regions.

5.2 Impact of Digital Media in the Graphics Paper Sector

5.2.1 Background

Paper grades used for communication purposes are called *communication papers* or *graphics papers*. They consist of two main paper grade types: *printing and writing papers* and *newsprint*. Printing and writing papers are often disaggregated into four major grades: coated woodfree (freesheet), uncoated woodfree (freesheet), coated mechanical, and

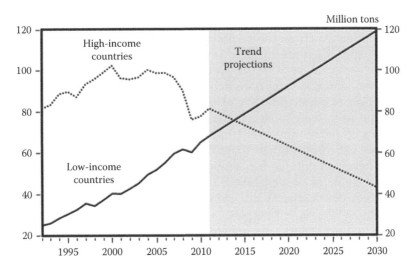

FIGURE 5.3
World graphics paper consumption in high-income and low-income countries (1992–2010) and trend (2000–2010) projections to 2030. (Data from RISI, 2000–2010. www.risi.com.)

uncoated mechanical papers.[*] In terms of world consumption of graphics papers, the most significant grade is the uncoated woodfree, accounting for over 37% of the total graphics paper consumption, followed by newsprint (23%).[†]

One of the most significant features of global graphics paper markets during the past decade has been the significant rise of low-income emerging economies as consumers and producers of paper products and the simultaneous decline of many of the high-income OECD countries. As a result, a shift from West to the East has taken place in the global forest products markets. For example, in 2000, the consumption of graphics papers was 2.5 times as high in high-income countries compared to the low-income countries, while levels are almost equal now, and in 2014 the latter countries are projected to have a higher consumption level (Figure 5.3).

Although Figure 5.3 provides projections to 2030, it is unlikely that the trends would continue without any changes for the next 20 years or so. For example, it could very well be possible that, say after 10 years, emerging paper-consuming economies, such as India, China, Brazil, and Russia, could hit the saturation point in communication paper consumption (e.g., due to consumers increasingly adopting digital media), and consumption would start to stagnate, and even decline. Thus, the trend projection has a high degree of uncertainty, but is still useful as a baseline.

5.2.2 Drivers of Market Changes

The basic structure of the models used to project forest products demand has not changed significantly over time (see, e.g., Buongiorno et al. 2003; FAO 1999; Hetemäki

---

[*] Uncoated freesheet papers are used, for example, for office and business printing (copiers, computer printers, facsimiles), business forms and envelopes, and commercial printing and writing (stationery). Coated woodfree and coated mechanical papers are used, for example, for magazines and catalog. The uncoated mechanical papers are used, for example, for inserts, flyers, directories, and books.

[†] Coated woodfree papers accounted for 19.1, coated mechanical for 11.5, and uncoated mechanical for 9.2 percent of the total world graphics paper consumption in 2010.

2005; HermeKoski and Hetemäki 2013; Zhang and Buongiorno 1997). Typically, these are empirical models, such as the Global Forest Products Model (Buongiorno et al. 2003), in which paper consumption is a function of economic activity (usually GDP or GDP per capita), paper demand in the previous year, and the price of the paper commodity. One of the central assumptions behind these models and projections is that per capita consumption of paper products is directly and positively related to per capita income (GDP) and negatively related to the price of the paper product. This is assumed to be valid across countries and over time. Researchers, industry firms, analysts, government agencies, etc., typically use these drivers when considering the long-term outlook for market development (see, e.g., Buongiorno et al. 2003; FAO 1999; Hetemäki 2005; HermeKoski and Hetemäki 2013; UNECE 2011; Zhang and Buongiorno 1997).

Looking at recent developments in *low-income countries*, these assumptions appear to be valid. In many "emerging" economies, rapid economic growth, along with increasing urbanization and educational levels, is generating increasing demand also for communication papers. For example, in China, India, Indonesia, Poland, Russia, Turkey, and all populous countries, communication paper consumption grew from 60% to 100% between 2000 and 2010, depending on the country (FAOSTAT). On the basis of long-term economic growth projections (e.g., Consensus Economics 2012) and the population projections by the United Nations (*World Population Prospects: The 2010 Revision*), we would expect this trend to continue, at least in the coming decade (Figure 5.3).

However, in case of *high-income countries*, it is much more difficult to use economic growth as a primary driver for communication paper consumption (UNECE 2011). For example, for North America and Western Europe, it would be problematic to project communication paper consumption to grow as GDP grows, except during the short-run business cycles (Gordon et al. 2007; Hetemäki 2005; Hetemäki and Obersteiner 2001; Hujala 2012; HermeKoski and Hetemäki 2013; Soirinsuo 2010).

Figure 5.4 is illustrative of this situation. It shows US newsprint consumption, real GDP, and population data from 1939 to 2010. All these variables were increasing until 1987,

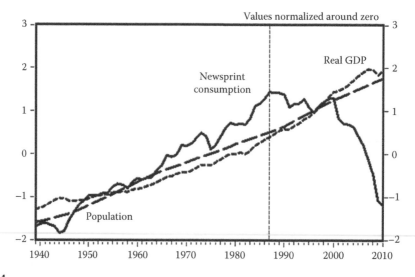

FIGURE 5.4
US newsprint consumption, population, and real GDP in 1939–2010 (values are scaled by normalizing the data series around zero). (Data from RISI; US Census Bureau; US Bureau of Economic Analysis.)

after which newsprint consumption started first to stagnate and later to decline rapidly. The market behaved before 1987 very much in a way that forest economists and industry analysts expected (Buongiorno et al. 2003; FAO 1999; HermeKoski and Hetemäki 2013; Zhang and Buongiorno 1997). Using this type of "classical" model in 1999, the FAO (1999) projected that the US newsprint consumption would continuously increase up to 2010, when it was projected to be 16.4 million tons. However, as Figure 5.4 shows, there has been a drastic drop in consumption, and according to FAO statistics, it had dropped to 4.6 million tons in 2010, that is, almost 12 million tons less than projected by FAO (1999). Viewing this "projection error" from the industry perspective, it equals the annual production of approximately 79 newsprint machines in North America (average size of a newsprint machine being 150,000 tons). Also, the "official" US RPA projection in 2002 forecasted that the consumption in 2010 would be over 11 million tons, about 2.5 times higher than the actual figure (see Haynes 2002).

The drastic structural change is also reflected in the correlation coefficient between newsprint consumption and GDP: The correlation for period 1939–1987 was +0.97, but for period 1988–2010 it was −0.73. The positive relationship between economic growth and newsprint consumption has ended and has now turned to a negative relationship.

Clearly, there are many reasons behind the structural change that resulted in widely different projections relative to actual consumption figures. For example, many commercial printers have switched from newsprint to other paper grades (SC paper); the weight of newsprint has declined from 60 to 48 mg; and there have been changes from broadsheet to tabloid newspaper formats (Hetemäki 2005). Yet the most important factor has been the fact that fewer people are reading newspapers and newspaper circulation has thus started to decline markedly, as the Newspaper Association of America (NAA) statistics have shown. But, in addition, due to the declining circulation, both business and classified advertisements began to abandon newspapers for digital venues. Thus, fewer pages were needed and therefore also less newsprint.

The major reasons for declining circulation and newspaper readership appear to be twofold: firstly, people increasingly are reading the news on the Internet, and secondly, an increasing number of people do not read newspapers at all. The latter may be a result of many things, but one important factor, as household media surveys point out, is that they spend more time on electronic media (the Internet, TV, video games, mobile phones, tablets, etc.) and have less time and interest for reading newspapers.

It appears that the US newsprint market development is an anticipatory example of what is expected to happen to other communication paper grades and in other regions. For example, according to RISI (2012) projections for US paper consumption up to 2027, the consumption of magazine paper is projected to decline on average 4.6% per year. This is combined with a 2% per year loss in magazine circulation and a 3% per year loss in ad pages. At this rate, magazine paper usage will be cut in half from 3.0 million tons in 2011 to 1.4 million tons by 2027, a net loss of 1.6 million tons of demand.

Moreover, the impacts of digital media on print media and the paper industry are universal. Electronic communication supersedes print media in New York, Moscow, Peking, or Nairobi in exactly the same way (Hetemäki 2010; Hujala 2012; PricewaterCoopers 2011). However, in emerging economies, due to rapid economic growth and urbanization, there is still a clear *net increase* in communication paper consumption. In addition, digital media impacts all printed communication forms, such as magazines and company annual reports (mainly coated mechanical and woodfree papers), business and office forms (uncoated woodfree paper), and home-delivered advertisements (mainly uncoated mechanical paper).

However, there are large differences in the timing and magnitude of the impacts between countries and paper grades, as shown in the differences in the two biggest Western Europe communication paper markets, Germany and the United Kingdom. According to RISI data, newsprint consumption in Germany has declined only slightly from 2000, and printing and writing paper consumption is practically equal (although clearly lower than at the height of 2007). In contrast, in the United Kingdom, newsprint consumption and printing and writing paper consumption have declined 24% and 22% from 2000 to 2010, respectively.

If consumption in high-income countries follows the trend of 2000–2010 into the next decade, consumption would *decline* by 38% by 2020 from its maximum level in 2000. In contrast, in the same period, in low-income countries, it would *increase* by 128%. The *world net increase* would therefore be 9% from 2000.

One significant unknown is when and to what extent electronic media will start to replace print media in the low-income region. In 2010, the low-income region population-weighted average Internet penetration rate was still only 17%, which is what it was in the United States in 1997. But in some major low-income countries, change is taking place rapidly, as in China. According to *Internet World Stats,* in China there were 538 million Internet users in June 2012, which is the largest number for any country. However, the Internet penetration rate is still only about 40%, whereas in the United States it is estimated to be 78%. But the Internet penetration in China grows very rapidly—if the trend of the last 5 years continues, China's penetration rate will reach in 2018 the same level the United States has currently.

In summary, given the rapid spread of the Internet and electronic media also in the low-income countries, it may be that the current rapid consumption growth may weaken already in the coming decade. Indeed, there are already indications of this happening. For example, the Chinese newsprint consumption growth rate has already started to decline: In 1995–2004, the consumption grew on average by 15.9% per annum, whereas in 2005–2011 this figure was only 3.6% (RISI data).

5.2.3 Declining Price Trend

The discussion of the impacts of digital media on the graphics paper sector is very much focused on what happens to paper consumption. However, from the perspective of paper company profitability, it is important to draw attention to the potential *price impacts* that arise from changes in the marketplace. The increasing competition between print and digital media has led to a reduction in pricing power for the paper sector. Companies in the paper industry are not simply competing against other paper companies, but increasingly they are also competing against digital media companies, who provide alternative platforms for information dissemination and publishing (Green 2012). In the face of this increasing competition between print and electronic media, publishers of print products seek to cut operating, materials, and other costs, which is intended to lead to lower, more competitive prices of their paper products. In short, *communication paper prices* are also increasingly determined by digital media development. Indeed, the real price of communication papers has been declining in the past decade. In recent years, the average real price has been around 30% lower than in the beginning of the century.*

---

* Based on FAOSTAT data and computing the world price as an average of the export and import unit prices and deflating it with world commodity input price index from IMF.

As a result of the competition from digital media, the pulp and paper industry most implement strategies to adjust to structural changes in communication paper markets. First, the industry can continue cutting production costs and increase productivity through investment in modernization. Another pathway to competitiveness is the application of information technology for intracompany business processes and intercompany connectivity with exchange partners. Third, companies can reduce capacity (close mills and paper machines) in order to maintain the supply/demand balance and maintain/gain pricing power within the markets they serve. This is what companies have been doing in recent years in North America and Western Europe and has been an essential tactic to keep their current businesses profitable. Of course, for some companies the capacity closures may not be a result of well-planned strategic decision making, but simply a *force majeure*, in that they have no other possibility. Also, companies merge to increase market share and gain market and pricing power. Finally, the paper industry can innovate new products for which there will also be growing markets in the high-income industrialized countries (see Section 5.5).

5.3 Packaging Sector Increases Paperboard Consumption

5.3.1 Overview of Packaging and Paperboard Products

In the previous section it was shown how the development of information technology and digital media is resulting in substitution impacts and declining consumption and price trends for graphics paper products. In contrast, information technology is not expected to have such a negative effect on paperboard consumption. For example, a rapidly growing consumer Internet trade increases the need for packaging, which translates to growth in paper products used for packaging. In this section, we focus on the current state and outlook of the global paperboard and packaging markets.

One of the most important driving forces in determining the success of paperboard is how well it can compete against other packaging materials. Currently, packaging paper and paperboard are the most important packaging materials in terms of market share. Their value share of the total global packaging products is 38%, while the second largest is plastics with a 34% share (WPO 2008). As these two product groups dominate global packaging, their relative competitiveness determines how the paperboard sector will develop in the future.

Global consumption growth of packaging paper and paperboard has been stable during the last two decades with an average annual growth rate of 3.4% (based on the RISI data). Although packaging markets are affected by global economic changes, economic recessions have generally had smaller negative impacts on paperboard markets than on graphics papers markets (Finnish Forest Sector Economic Outlook 2011). Global growth of paperboard production is mainly correlated to the rapid consumption growth of containerboard that is used for bulk packaging of industrial commodities. According to the RISI data, carton board consumption has been growing slower, but the growth rate has increased during the end of the 2000s. The carton board product group includes folding boxboard, liquid packaging board, solid bleached sulfate board, and white lined chipboard. These grades are used for many kinds of consumer packaging such as food, liquor, light industrial products, medicine, health-care products, cosmetics, and electronics. The World

Packaging Organization (WPO 2008) expects that growth opportunities exist for packaging in such areas as fresh food and ready-to-eat meals especially in emerging markets in developing countries. Additional opportunities exist for suppliers in beer and mineral water consumption especially in Eastern Europe, the Middle East, and Asia. Health care and cosmetics are also fast-growing end-use areas for carton board. Despite the significant growth possibilities for carton board, plastic packaging is a challenging competitor. For example, according to WPO (2008), rigid plastics have been, and will be in the future, the fastest growing packaging material.

5.3.2 Regional Developments in Paperboard Markets

The largest consumer and producer countries of paperboard are the United States, China, and Japan. With declining production, Japan has become more dependent on paperboard imports during the last decade. An important structural change that has affected the global paperboard markets is the remarkable consumption and production growth in Asia resulting from the rapid economic development of China (Tables 5.2 and 5.3). In contrast, consumption has declined in the United States and in Western Europe. An important reason for this was that production of consumer and industrial goods has increasingly been transferred from OECD countries to emerging economies, such as China. As a result, packaging also has migrated to these regions. However, a significant portion of packaging board produced and "consumed" in China actually ends up in the United States and Western Europe through Chinese exporters (Hetemäki and Hänninen 2009). North America and Western Europe clearly produce more paperboard than they consume and have been important exporters (Table 5.3). The production growth in China and overall decline in demand during the recession of 2008–2010 have led to capacity cuts in the paperboard industry in North America and Western Europe. Africa, Eastern Europe, and Latin America are net importers of paperboard. In particular, Africa's consumption and production volumes are very low compared to the other regions.

International trade of paperboard in terms of export and import volumes doubled between 1992 and 2010. However, the volume of trade relative to production has been rather stable: the share of exports to production has been around 20% on average during this period. A number of important changes have occurred in the regional net trade (exports–imports), of which the most important is the volume growth of Western

TABLE 5.2

Paperboard Consumption and Production by Regions
(in Million Metric Tons)

Region	Consumption			Production		
	1992	2000	2010	1992	2000	2010
Africa	2	2	4	1	2	2
Asia	26	41	84	24	39	82
North America	37	44	40	41	48	46
Latin America	5	9	14	5	7	10
Eastern Europe	4	6	12	5	6	10
Western Europe	23	30	30	24	31	34

Source: Data from RISI, www.risi.com, includes containerboard and carton board.

TABLE 5.3

Net Trade Volumes of Paperboard
by Regions (in Million Metric Tons)

Region	Net Trade Volumes		
	1992	2000	2010
Africa	−0.3	−0.3	−1.3
Asia	−2.0	−2.3	−2.0
North America	4.4	4.6	5.8
Latin America	−0.6	−2.0	−3.1
Eastern Europe	0.3	0.2	−2.0
Western Europe	0.3	1.5	4.6

Source: Data from RISI, www.risi.com,
includes containerboard and
carton board.

European net exports between 1992 and 2010. In 2010, Western European exports were 4.6 million tons larger than imports. North America is the other region where exports have increased in relation to imports. Western Europe and North America will continue to be important exporters of paperboard in the future, due to the stagnating consumption in these regions.

Eastern Europe, Latin America, and Africa have become more dependent on imports. In Eastern Europe, the development of the Russian market is important, as it covers about one-third of Eastern European production and consumption. Russia's paperboard imports have risen quickly, boosted by domestic consumption. Import growth will continue in Eastern Europe (particularly Russia), Latin America, and Africa. Asia will continue investing in new capacity in order to meet rapidly growing demand in the region.

5.3.3 Changes in Paperboard Prices

The rapid increase in the production of paperboard in the low-cost emerging countries (e.g., China) in the past decade seems to have changed the paperboard world price pattern (Figure 5.5).* In the 1990s, there was significant cyclical variation in paperboard prices, but no clear declining trend. However, during the last decade, when rapid production enlargements started in new Asian low-cost countries, a clear declining price trend is evident. This changing world price pattern has been a particular challenge for the profitability of the North American and Western European producers.

5.3.4 Drivers of Paperboard and Packaging Markets

What are the main drivers that help to explain past developments and anticipate the future of the paperboard and packaging sector? These are questions that have not been

---

* The price development of paperboard is described by the average of world import and export unit values in US dollars (FAOSTAT). Prices were transformed to real prices by deflating nominal prices by world commodity industrial inputs price index (IMF). Prices are for wrapping and packaging paper and paperboard (FAOSTAT code 1681).

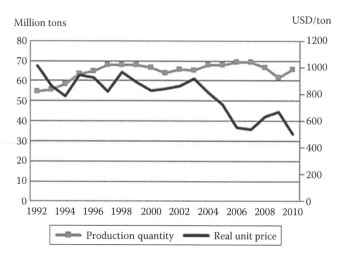

FIGURE 5.5
Production of wrapping and packaging paper and board in high-income countries and world real price, 1992–2010. (Data from FAOSTAT, http://faostat.fao.org/; IMF, http://www.imf.org/.)

thoroughly addressed by academic researchers. Previous studies on the paper industry have focused mainly on graphics papers (e.g., Bolkesjö et al. 2003; Hänninen and Toppinen 1999; Hetemäki 1999, 2008; Hetemäki and Obersteiner 2001; Laaksonen-Craig 1998; Zhang and Buongiorno 1997). On the other hand, academic research on paperboard markets has been relatively scarce.

The studies that do exist have focused on different aspects of the paperboard sector. For example, Li and Luo (2008) examined consolidation of the paperboard industry. Their results for the US linerboard industry suggested that consolidation has not necessarily resulted in higher market prices. One reason for this was suggested to be a low concentration ratio. In another study, Löfgren and Witell (2005) suggest that quality attributes of packaging, such as recyclability, influence the decisions to buy and use packaging products. Also, the findings of Rokka and Uusitalo (2008) emphasize the increasing importance of environmental dimensions of packaging in product choices. In contrast to the scarce academic research on paperboard sector, there are many empirical surveys and reports made in industry organizations and consulting companies working in the packaging sector. According to a WPO (2008) study, the most important demand drivers of packaging and packaging materials are economic development, population growth, consumption habits, Internet trade, and technological development.

In the following section, we present two alternative trend projections for paperboard consumption in five regions. Consumption is analyzed based on the figures for total consumption volumes as well as for volumes per capita (consumption/inhabitant). According to the United Nations estimates (UN 2011), population growth has been shrinking gradually in all five regions during the 1990s and 2000s and is anticipated to continue to 2030 (Table 5.4). For Eastern Europe, the UN (2011) estimates indicate negative population growth also in the future.

In the first projection, we keep the consumption/per capita at 2010 levels and assume that only population growth (UN 2011) will determine future total paperboard consumption. Figure 5.6 indicates growth rate changes for paperboard consumption in different regions based on this scenario.

TABLE 5.4

Consumption of Paperboard Per Capita and Projections for Population Growth by Regions

Apparent Consumption of Paperboard and Population	Asia	Western Europe	Eastern Europe	Latin America	North America
2010: consumption kg/capita	2.2	7.3	3.5	2.0	11.6
2010: population, millions	4164.3	189.1	294.8	590.1	344.5
Average annual growth rate of population by 5-year periods					
2010–2015, %	0.99	0.15	−0.17	1.07	0.86
2015–2020, %	0.85	0.16	−0.21	0.93	0.80
2020–2025, %	0.71	0.12	−0.29	0.80	0.74
2025–2030, %	0.57	0.08	−0.38	0.66	0.67

Sources: Data from FAOSTAT, http://faostat.fao.org; the United Nations http://www.un.en/development/desa/population/.
Note: Population growth, medium-fertility variant 2010–2100.

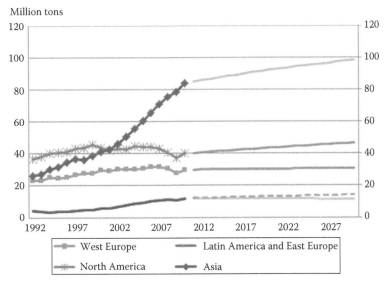

FIGURE 5.6
Paperboard consumption by regions in 1992–2010 and projections based on population growth for 2011–2030 (dotted trend line is for Latin America).

For emerging markets, Asia, Eastern Europe, and Latin America, the projections probably underestimate future development. In these regions, current consumption of paperboard per capita is clearly lower than in North America or Western Europe (Table 5.4). The increasing trend in industrial investments and production of industrial commodities in emerging countries will probably increase their packaging demand and consumption of containerboard and carton board per capita. WPO (2008) estimates that the emerging markets are also areas where especially food and fresh products packaging is growing, which are possible new geographical markets for the carton board industry. For North America, the projection shows continued growth in consumption in the future but is premised on sluggish growth observed in the 2000s.

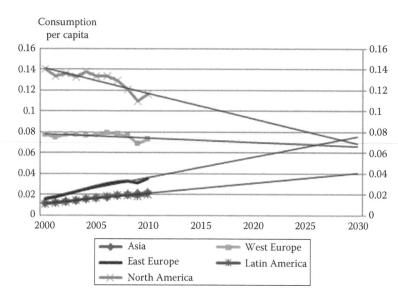

FIGURE 5.7
Regional paperboard consumption (per capita in 2000–2010 and trend projections to 2030).

The second projections are based on linear trends calculated for consumption per capita for the period 2000–2010. As mentioned earlier, globalization has rapidly changed the structure of the global forest industry and product markets during the 2000s. Paperboard consumption per capita started to decrease in North America and to stagnate in Western Europe. In the other areas, where consumption per capita is clearly lower than in the last mentioned regions, the figures show steady growth: the fastest growing area is Eastern Europe. Even small changes in consumption per capita will have a very large effect on the total absolute volume of consumption. The second projection (Figure 5.7) indicates 77% lower paperboard consumption for North America than the first projection (Figure 5.6) in 2030. For Eastern Europe, the second projection indicates consumption levels that are two times larger, and for Asia the forecast is three times larger than the first projection.

In summary, the two projections differ considerably, but they may help in assessing possible future developments in the paperboard sector. Industrial production and export packaging will continue to grow in Asia with a concurrent relatively lower consumption of paperboard for packaging in traditionally large producer regions in North America and Western Europe in the future.

An important source of uncertainty in global paperboard markets is China's rapid economic growth rate and concurrent packaging consumption. This is an issue that is very difficult to project. Another important source of uncertainty and potential opportunity relates to the development of new packaging materials and the ability to innovate new packaging products in reaction to changing needs and habits of consumers. Substitution from alternative materials, particularly plastic, will influence the development of new wood-based packaging materials. An example of an emerging product/market is intelligent packaging that combines wood fibers with modern digital information technology, such as interactive pharmaceutical packages that remind people to take their pills with a programmed frequency. Another example is incorporating new technology in packaging, providing information about food spoilage, which could prevent huge volumes of food waste in the chain of food retailers, wholesalers, or consumers.

Although current paperboard product consumption is decreasing in Western Europe and North America, global industrial packaging is concentrating in Asia. The growth in packaging related to everyday life, such as foods, may offset this decline. Larger production volumes will be needed also to satisfy the growing consumption of tissue paper.

Packaging materials and tissue paper are typically the most profitable to produce near their end use and final customers, because of high unit transportation costs. For example, carton board production for food packaging is typically based on coniferous virgin fiber that is available in the traditional producer countries of North America and Western Europe. On the other hand, tissue products are relatively expensive to transport, and therefore, they tend to be produced near the consuming markets.

Finally, environmental concerns are likely to be an important determinant of packaging sector development in the future. This is positive for the paperboard sector that uses renewable raw materials relative to the main competing fossil-based plastic products.

5.4 Implications of Paper Markets on Wood Fiber Demand

What would be the implications of trends and outlooks for paper and paperboard consumption and production on the markets and trade flows for wood fiber raw material (wood pulp, recycled paper, and pulpwood)? In order to shed light on this question, we first start by analyzing the recovered paper and wood pulp inputs at the global level. Then, we analyze the recycled paper recovery and utilization trends at the regional level. Finally, we discuss the outlook for pulpwood demand.

Recovered paper (Figure 5.8) has become by far the largest fiber type used in papermaking.* Its input share in papermaking has grown from below 40% in the beginning of 1990 to 57% in 2010, while wood pulp share has declined from 60% to 43%. Among the different wood pulp grades, bleached hardwood kraft pulp (BHKP) has increased substantially in absolute volume and has also gained input share marginally from 14% in 1992 to 15% in 2010. All other grades of wood pulp show decline in absolute volumes except unbleached kraft pulp, which has been relatively stable during the last 20 years.

Western European *paper recovery rate* has recently reached 75% of the paper consumed, which is assumed to be close to the practical limit for the region.† The recovery rate in Germany was 80% in 2010; however, a more conservative assumption was applied for 2010–2030 period for the region as a whole. The North American paper recovery rate is second highest at around 63% in 2010. It is projected that North American paper recovery rate will increase following the trend from the past decade until reaching 70% soon after 2015 and then remain stable. In Asia, the paper recovery rate has been steadily increasing and was 53% in 2010, the third highest rate after Western Europe and North America. It's assumed that the Asian recovery rate will increase until it reaches 65% near 2025.

It is also expected that some regional differences in paper recovery rate will remain in the medium term. Eastern European and Latin American recovery rates are assumed to be increasing more gradually, but in line with the last decade's linear trend. The African paper recovery rate remains stagnant at around 30%.

---

* The statistics used in this section is based on RISI data.
† *Paper recovery rate* is the same as paper recycling rate, and it can be defined as the total amount of paper and paperboard recovered (collected) as a percentage of the total amount of paper and paperboard consumed.

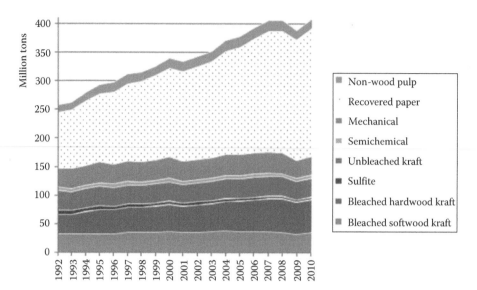

FIGURE 5.8
Global paper and paperboard fiber input, 1992–2010. (Data from RISI, www.risi.com.)

Recovered paper *utilization rates* are projected in a similar way—based on the last decade's regional trends.* Asia has reached the highest recovered paper utilization rate of around 69% in 2010. That is, out of the total fiber used for papermaking, 69% was based on recovered paper and 31% on pulp. It is expected that this utilization rate is close to its upper limit, and it will grow only slightly from this level in the future.

Based on the projected regional recovered paper utilization rates and the projected regional paper production volumes (shown in Table 5.1), we projected the use of recovered paper in papermaking up to 2030 (results not shown here). According to the projections, Asia's use of recycled paper would double from 2010 to 2030. In Western Europe, Eastern Europe, Latin America, and Africa, recycled paper usage is projected to grow moderately. Only in North America would the recovered paper usage decline in the coming decades. Given these trend projections, in 2030 Asia's share of the global recovered paper use would reach nearly two-thirds.

On the basis of the projected recovered paper utilization rates, regional paper production volumes, and the projected paper consumption (Table 5.1), regional supplies of recovered paper were estimated. The balance of recovered paper used for paper production and recovered paper supply results in the regional *net trade* of recovered paper (Figure 5.9). Therefore, in contrast to previous projections, the recovered paper net trade projection is not purely a trend projection. Asia has been the largest and the fastest growing importer of recovered paper, but this trend cannot continue much longer, given that Western European and North American exports of recovered paper are not likely to increase in the long run. The reasons for this are the declining Western European and North American paper consumption and paper recovery rates (they can increase only by a relatively small margin without a resulting very high price level).

---

* Recovered paper *utilization rate* can be defined as the share of recovered paper in the total fiber input in paper production.

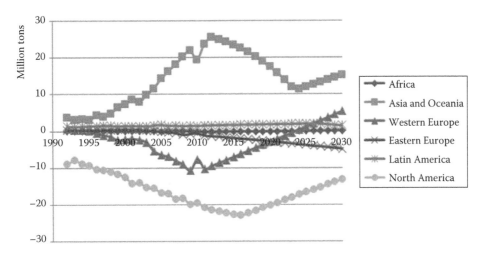

FIGURE 5.9
Recovered paper net trade in 1992–2010 and projections to 2030. (Data from RISI, www.risi.com.)

On the other hand, growing paper consumption and paper recovery rates in Asia will increase the overall supply of Asian recovered paper.

The wood pulp production trends for the next two decades are linearly projected from the past decade (Table 5.5). Given these projections, Asian and Latin American wood pulp production will exceed North American production, which will decline substantially from its current level. In order to supply pulpwood for increasing pulp production in Asia and Latin America, new plantations would need to be established at the same rate as has been observed over the past two decades (FAO 2006). In addition to the need to maintain the rate of new plantations established, plantation productivity would have to improve in Asia.

Table 5.5 shows what the projected changes could imply for regional wood pulp production and pulpwood consumption during 2010–2030.* The most important projected changes are significant increases in pulp production in Asia and Latin America and the decline of pulpwood production in North America by 38% from 2010 to 2030. In Western Europe, the projection anticipates a slight decline in pulp production in the coming decades. If the projected wood pulp production trends were to be realized, Asia would become the largest and fastest growing net importer of wood pulp and Latin America the largest and fastest growing net exporter of pulp.

These changes in pulpwood production would naturally impact regional pulpwood consumption. Table 5.5 shows that pulpwood consumption would increase from 2010 to 2030 in Latin America and Asia roughly 131 million m³ accompanied by a decrease in North America and Western Europe of 121 million m³. However, demand increases in Latin America and Asia would be mostly in hardwood pulpwood, whereas the decline in North America and Western Europe would be mainly in softwood pulpwood.

---

* It was assumed that to produce one metric ton of wood pulp in Asia, Africa, and Latin America, three cubic meters of hardwood pulpwood is required. In North America and Western and Eastern Europe, where pulpwood is largely softwood coniferous requiring higher input, this figure was assumed to be 4.5 m³.

TABLE 5.5

Change in Wood Pulp Production and Pulpwood
Consumption

	Wood Pulp Production (Million Tons)			Pulpwood Consumption (Million m³)
	2010	2030	Change 2010–2030	Change 2010–2030
Africa	1.8	1.7	−0.1	0
Asia and Oceania	32.1	51.0	19.0	57
Latin America	20.6	45.1	24.6	74
North America	67.8	42.2	−25.5	−115
Western Europe	36.0	34.6	−1.3	−6
Eastern Europe	10.4	12.2	1.8	8
World	168.6	186.9	18.3	17

The projections would most likely imply that Asia and Latin America would need to increase their wood pulp production significantly from present levels. This could be achieved by increasing the volume of fast-growing plantations, as well as increasing productivity. According to Carle and Holmgren (2008), in their "higher productivity scenario" (assuming 2% annual productivity increase for planted forests), Asia will be able to increase pulpwood supply by 63 million m³ over the 2005–2030 period. This would be sufficient to sustain Asian wood pulp production growth. However, questions remain as to whether newly established plantations and associated productivity increases will be realized to the extent required for the projected levels of growth to take place. On the other hand, if Asia can achieve higher than assumed paper recovery rates, the region can potentially sustain the same level of paper production with lower use of wood pulp. Clearly, if plantation forest productivity growth is not achieved and the paper recovery rates do not increase, Asia would need to import more pulpwood or wood pulp from other regions.

One interesting implication from Table 5.5 is that the global demand for pulpwood will grow only marginally due to higher efficiency of using tropical hardwood for pulping compared to using softwood fiber from the Northern Hemisphere. To a large extent, an increase of wood fiber use in Asia and Latin America is offset by a significant reduction in North America due to a decline of wood pulp production. Wood fiber deficits could exist regionally. Should Asia or Latin America fail with plans for expected growth of wood pulp production and increasing wood fiber supply, North America may try to reverse the declining trend for wood pulping.

Another possible option is to attract more investments to the Russian Siberian forest sector. In terms of potential large raw material supply to the global pulp and paper industry, there is particular uncertainty with regard to such development in Russia. Russia has the largest coniferous forest resource in the world, but currently the utilization rate of this resource is low. For example, the *annual allowable cut utilization rate* has typically been under 25% (UNECE 2003). There are many obstacles related to the utilization of this resource, such as poor infrastructure (logistics and forest roads), ambiguous forest ownership legislation, and security of wood supplies. However, if the global demand for pulpwood resources increases markedly, the potential pulpwood supply that exists in Russia could start to materialize to a much greater extent than today.

5.5 Pulp and Paper Industry Is Still Consolidating and Globalizing

The pulp and paper industry has been steadily consolidating as a result of a number of waves of mergers and acquisitions during the 1980s, the 1990s, and in the past decade. However, the industry remains fragmented compared to many other industries (Kroeger et al. 2008; Uronen 2010). In 2010, the 10 biggest companies produced 21% of the world's paper and paperboard output (FAOSTAT 2012; Rushton et al. 2011). However, there are significant differences in the market shares of the top few producers across regions and across paper and board grades. For example, in North America and Western Europe, the top five companies typically have a 60%–85% share of a particular paper grade (see Uronen 2010, 79). Specifically, the markets are significantly consolidated for coated woodfree paper in North America and mechanical papers in Europe.

Theoretically, higher consolidation should lead to better capacity management, control of supply, and improved pricing power. According to Uronen (2010), "despite the high degree of consolidation in some product groups within the main production regions, there is inconclusive evidence on its impact on improved pricing." Also, this has not seemed to have led to better corporate profitability. Uronen (2010) argues that the situation may be the result of intense rivalry between the top five producers, strong bargaining power of the biggest customers, limits set by competition authorities, the role of imports, and the remaining smaller producers taking advantage of the actions of market leaders. In the communication paper sector, competition between the digital and print media is also exacerbating the situation.

With regard to industry consolidation in the European Union and North America, it is unclear to what extent antitrust legislation will allow it to develop much further. However, the sector may still experience bigger companies acquiring their smaller competitors.

The paper industry has been called a *home market industry*, indicating that the products have been traditionally produced and consumed in the same region (Lindqvist 2009). Thus, regional strategies have been more common. When studying the paper companies' regionalization and globalization strategies in the 1990s, Siitonen (2003) identified only 4 truly global and 10 globalizing producers among the top 100 pulp and paper industry companies. Also, in the recent decade, some paper companies have transitioned their strategy from globalization to inter- and intraregional concentrations after experiencing disappointing globalization results (Lindqvist 2009). For example, many Nordic companies have disinvested their North American operations to a significant degree.

Table 5.6 and Figure 5.10 provide some basic statistics on the magnitude of operations and profitability of the pulp and paper industry and how they differ across the major regions. When interpreting the results, it should be kept in mind that many of the companies are already global and may have operations in several continents. Table 5.6 shows that three continents dominated the pulp and paper industry in terms of sales and production volumes in 2010: Europe, North America, and Asia. Europe and North America are almost of equal size according to these indicators and Asia somewhat smaller. The statistics for 1 year does not, however, show the dynamics of the industry's development. For example, Chinese, Brazilian, and Chilean companies have grown rapidly and have increased in importance over the past decade.

In Figure 5.10, the earnings before interest, taxes, depreciation, and amortization (EBITDA) as percentage of company sales are shown for year 2010 for different regions (PricewaterhouseCoopers 2011). The companies included are those that are in the top

TABLE 5.6

Comparison of Geographic Distribution of Top 100 Pulp and Paper Companies in the World in 2011

Region	No. of Companies	2010 Sales ($US Billion)	% of Total 2010	P&B (Million Tons)	% of Total 2010
Europe	34	95.4	31.4	72.6	36.4
North America	32	115.2	37.9	59.2	30.0
Asia	24	65.3	21.5	53.8	27.0
Latin America	7	14.9	4.9	6.8	3.4
Oceania	1	5.2	1.7	0	0.0
Africa	2	7.8	2.6	6.9	3.5
	100	304.0	100	199.3	100

Source: Data from Pulp and Paper International Magazine, August 30, 2011; Rushton, M. et al., 2011.

Note: Not all of the North American, Asian, and South American companies listed in the PPI 100 report their financial results.

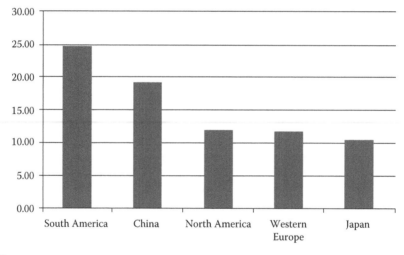

FIGURE 5.10
Top 100 global forest, paper, and packaging industry companies' EBITDA as a percentage of sales by regions in 2010. (Data from PricewaterhouseCoopers, *Global Forest, Paper and Packaging Industry Survey*, 2011 ed., 2011.)

100 list of forest, paper, and packaging industry companies in terms of value of sales. The EBITDA is an indicator of the company profitability. The regional figures are averages of the companies' EBITDA in a specific region. The statistics clearly indicate that in 2010 companies with most of their operations in emerging regions have been more profitable than the companies based in OECD countries (or having most of their operations in this region). In South American companies, EBITDA is over twice as high as that of companies in North America, Western Europe, or Japan. Although the data are for 1 year, they reflect the general situation in recent years. It is also an indication of the change that is taking place in the global forest and pulp and paper industry—the companies based in emerging regions are becoming more significant players and are likely to increase their role in the top 100 list in the coming years.

EBITDA figures can be very different from the return on capital employed (ROCE) statistics. For example, according to Roberts (2012), the average ROCE in between 1999 and 2009 was under 9% in all of the regions and under 5% in North America, Europe, and Japan. In capital markets, these are not considered to be robust figures and are particularly alarming in North America, Europe, and Japan, where the ROCE has been lower than the cost of capital. This is one indication of the current poor situation of the industry in OECD countries and the great challenges that lie ahead. In order to solve these challenges, it appears evident that there will be further restructuring and mergers among the pulp and paper companies in North America and Western Europe, through which they will seek to increase productivity and gain pricing power. On the other hand, many companies will simultaneously grow their pulp and paper operations in the growing markets, particularly in Asia, Latin America, and Eastern Europe. Moreover, North American and Nordic country companies in particular are engaged in product development efforts to produce new wood fiber–based products in regional and international markets. These new products are discussed in the next section.

5.6 New Products

5.6.1 Background: Saturation of Old Products, Birth of New Products

Pulp and paper products can be considered to be standardized and at the maturity stage of the product life cycle. They are commoditized, which is reflected in that "market growth in North America and Europe is below GDP growth, price differentials between suppliers are marginal at best, switching costs are low and negotiating power lies with the customers. Radical changes are needed, if a supplier wants to break away from this situation" (Uronen 2010, 25–26).

However, this does not mean that technological and product development in pulp and paper products has stopped. Continued product refinement and advances in production technology are being conducted in the sector. As times change, innovation in business processes and products is taking place.

In 1854, *The Times of London* offered to pay £1000 to anyone who could develop a method of using rags as the raw material in papermaking. While the announcement did not produce the desired result, it clearly demonstrated the need to resolve the shortage of raw materials that was causing problems for papermakers and their customers. Today, 150 years later, the big challenge in the paper industry is not so much the raw materials, but rather that besides paper, there is a need to develop new products. Indeed, the pulp and paper industry is investing increasingly on R&D to develop new products, such as second-generation biofuels, smart packaging, biocomposites, biochemicals, and products derived from nanofibers. In many of the major and traditional pulp and paper-producing countries, such as Canada, the United States, Finland, and Sweden, there are *internal* and *external* factors in the forest sector that drive this development of new forest products.

The internal factors relate mainly to the structural difficulties of the forest industry in the traditional big forest sector countries, such as Canada, Finland, Sweden, and the United States. As discussed earlier, the industry has been suffering from continuing profitability problems, and new investments are going mainly to Asia and South America. In this context, the industry is rethinking its strategies, and new products are seen as one important development that could enhance the profitability and viability of operations in big forest sector countries.

The external factors are likely to turn out to be also important ones. Not least because the external factors can attract to the forest sector new industries and operators and diversify the forest sector. These external factors are particularly related to increasing energy consumption, greenhouse gas emissions, and concerns over energy import dependence. These are, in turn, prompting changes in the sources from which energy is expected to be derived in the coming decades. In this context, forests and biorefineries are seen as potentially important sources and producers of carbon-neutral energy. The primary benefits generated by the biorefinery development to the forest industry, and society in general, are often seen to be the following (Hetemäki 2010):

1. Through new technology and synergies with current operations, increased production efficiency and profitability of current forest products
2. Production of new products that are increasingly needed by society (e.g., bioenergy, bio-based polymers, and chemicals)
3. Help to meet regional policy targets (preserves and creates jobs in rural forest-based communities)
4. Help to meet climate change policy targets (replaces fossil fuel–based energy)
5. Help to meet energy security policy targets (replaces imported energy)

Reflecting these objectives, there have been large-scale R&D efforts in developing technologies and pilot projects that promise to open up new and more efficient ways to utilize forests and wood fibers in energy production. In these efforts, forest biorefineries play a central role.

5.6.2 Toward a Forest-Based Biorefinery

A *forest-based biorefinery* (FBB) is a facility that integrates biomass conversion processes and equipment to produce fuels (e.g., ethanol and biodiesel), electricity, power, and chemicals (e.g., polymers and acids), along with the conventional forest products (pulp, paper, sawnwood, etc.) (Hetemäki 2010; Ragauskas et al. 2006; Söderholm and Lundmark 2009). An FBB can use multiple feedstocks, including harvesting residues, extracts from effluents, fractions of pulping liquors, as well as agri-biomass, recycled paper, and municipal and industrial wastes. It can be a large-scale industrial facility, integrated into a pulp and paper mill, or a medium- or small-scale facility integrated into a sawmill or plywood mill, or a stand-alone bioenergy or chemical plant. Most of the discussions have so far focused on the pulp and paper–integrated biorefineries.

Within the FBB platform, there are a number of different output mix and technology possibilities (Hetemäki 2010; Ragauskas et al. 2006; Söderholm and Lundmark 2009). Therefore, the number of investment opportunities and risk factors related to forest biorefineries are many. The viability of each specific forest biorefinery product–technology mix depends on end markets (demand, supply, and prices), substitute markets (e.g., oil and plastics), biomass markets, and global, national, and regional policies. These may vary between countries, and even within countries. Also, the policies to support biorefinery development depend on the goals that biorefineries hope to target. For example, depending on the degree that the policy goal emphasizes, climate change mitigation, domestic energy production, rural employment, energy efficiency, or some combination of these, the optimal biorefinery concept may differ (Kangas et al. 2011). In short, there is no single best uniform solution for FBB, rather a large number of different concepts, raw material options, production processes, and output mixes, each tailored to be optimal for the local conditions and objectives.

An essential part of FBB is the objective to more efficiently utilize the various fractions of woody biomass. This biomass is lignocelluloses material, which is made up of three primary chemical fractions: hemicelluloses, cellulose, and lignin. All of these can be converted to carbon-neutral renewable energy or chemicals. Some of the conversion technologies are already mature and commercial; others require development to move to commercial applications.

The interest in forest biorefineries is still a rather recent phenomenon, which means that societies and the forest sector have not yet had much time to reflect on the issues associated with the concept. Thus, although the biorefinery landscape is promising, it is also broad, complex, and even confusing. More research is needed to understand the implications of FBBs to society and to the forest sector.

So far, the research on biorefineries has been very much technology driven and specialized. This is natural, since advances in the technology have been recent, and the possibilities for moving this technology to practice are only just opening up. However, now that the technology is close to the stage where it can be moved to commercial application, there is a need for synthesis on current knowledge and analytical assessment of future environmental, economic, and policy prospects. The work of Kangas et al. (2011) is one of the few economic analyses of the profitability of the forest biorefinery investments. There is clearly a need for further studies in this field. Also, research is lacking on the socioeconomic implications of FBB to the forest sector, as well as the environmental impacts.

Pulp and paper industry associations and industry R&D organizations in Western Europe and North America have in recent years published ambitious strategies and visions that set targets to launch new "bioeconomy" products that are set to significantly increase the value added of the industry and at the same time help to cut CO_2 emissions. For example, the Confederation of European Paper Industries (CEPI 2011) roadmap report lays out the future of the forest fiber industry in Europe up to 2050. The main targets are to meet future consumer demands by inventing new products that are produced sustainably and which help to reduce the use of nonrenewable, fossil-based raw materials. Indeed, the industry targets to reduce its CO_2 emissions by 80% over the next 40 years. One important means by which this vision is to be achieved is the FBB concept.

However, as promising and necessary as these new products are, developing them into commercially viable production processes will inevitably take many years. Until then, the pulp and paper industry will have to operate in the very challenging mature or declining segment of the product life cycle. Also, although the activities in developing these new products have so far very much centered in Western Europe and North America, it is to be expected that the pulp and paper industry (and energy and chemical industries) in emerging economies will also be increasingly interested in these new business opportunities.

5.7 Conclusions

Some of the most important characteristics and global trends of pulp and paper markets and industry have been discussed in this chapter. We have focused on the "Markets and Market Forces for Pulp and Paper Products." Many of the issues we have touched are very wide topics, and thus, the focus has been on reviewing some of the major recent trends, rather than providing an exhaustive account of the issues.

One of the main aims of this chapter has been to analyze the structural changes that are taking place in global pulp and paper markets and the industry. In short, the landscape of pulp and paper markets and industry is shifting from North to South (mostly emerging economies). The Asian and South American emerging economies are becoming the biggest paper and board consumers and producers, while the traditional big consumer producer regions in North America and Western Europe are stagnating or declining in the traditional paper products and at the same time trying to reinnovate their businesses with new products. In Asia, South America, Russia, and Central Eastern Europe, consumption and production of current paper products is still growing, while in North America and Western Europe, they will to a large extent remain close to the current level or decline, depending on paper grades.

In analyzing the future outlook of paper markets, we based the projections mainly on trends in the past decade (2000–2010). That is, we looked to the future of paper and paperboard markets, as they would develop in way they have done recently. This approach is both simplistic and transparent at the same time. Clearly, the longer in the future we make our projections, the more likely it is that the trend projections fail to capture the development. On the other hand, due to its simplicity and transparency, it is easy to use these projections as baseline "scenarios," against which one can make alternative assumptions and scenarios of possible developments. Indeed, we supplemented the trend projections with commenting and assumptions that helped to point out alternative possibilities.

According to the projections, the ongoing global structural change of growing "Southern" and declining "Northern" paper markets appears to continue in the coming decades. If world paper and paperboard production would continue as it has in the past decade, world production would grow from about 394 million tons in 2010 to about 553 million tons in 2030, that is, by about 40%. Almost all of this growth (159 million tons) would come from Asia (149 million tons). At the same time, North American paper and paperboard production would decline by 29 million tons, or by 33%. Although, these figures have very high uncertainty in them, and most probably they will not be realized as such, they give indications of the possible relative trends and magnitudes.

Due to the facts that paper and paperboard products are at a maturity stage of their product life cycle, consumption is stagnating or declining in many OECD countries, digital media is increasingly competing with print publishing, and new emerging low-cost regions are increasing their market share of paper, the world prices of paper products have been declining already for a decade or so. This trend may continue in the long term, or, at least, it is unrealistic to expect it to reverse and start to increase. The higher cost region in the "North" is particularly suffering from this trend, and intense restructuring and cost efficiency programs have been underway in North America and Western Europe.

Prospects for paperboard are somewhat different than for printing and writing paper, because the need for packaging is increasing globally. Important drivers are rapidly growing consumer Internet trade and the future need to improve food packaging. One of the uncertainties affecting wood-based packaging materials is substitution from alternative materials, particularly plastics. In the future, increasing industrial production and international trade supports the growth of Asian paperboard consumption and production. In contrast, consumption will decrease or stagnate in North America and Western Europe.

In order to grow in the long run, the industries in North America and Western Europe have also now started to actively reinnovate their businesses, by developing new and more value-added products. The focus is on forest biorefineries producing, for example, bioenergy, biochemicals, and intelligent packaging products. These biorefineries are based on new and more efficient technologies, which also can utilize a wider raw material basis than the current pulp and paper plants. However, despite some new promising products and technologies such as smart packaging and biofuels, the industry will have to improve and maintain the competitiveness of its existing product portfolio for many years to come.

References

Bolkesjø, T.F., Obersteiner, M., and Solberg, B. 2003. Information technology and the newsprint demand in Western Europe: A Bayesian approach. *Canadian Journal of Forest Research*, 33(9):1644–1652.

Buongiorno, J., Zhu, S., Zhang, D., Turner, J., and Tomberlin, D. 2003. *The Global Forest Products Model (GFPM): Structure, Estimation, and Applications*. Academic Press, San Diego, CA. 301 pp.

Carle, J. and Holmgren, P. 2008. Wood from planted forests. A global outlook 2005–2030. *Forest Products Journal*, 58(12):6–18.

Confederation of European Paper Industries (CEPI). 2011. The Forest Fibre Industry 2050: Roadmap to a low-carbon bio-economy. Burssels, Belgium. 46 pp.

Consensus Economics. 2012. Consensus forecasts global outlook 2011–2021. http://www.consensuseconomics.com/global_economic_outlook.htm (Accessed June 15, 2012).

FAO. 1999. Global forest products consumption, production, trade and prices: Global forest products model projections to 2010. Working Paper GFPOS/WP/01. Rome, Italy.

FAO. 2006. Global planted forests thematic study: Results and analysis. By A. Del Lungo, J. Ball and J. Carle. Planted Forests and trees working paper 38. Rome, Italy.

FAOSTAT. 2012. Databank for foresty statistics. http://faostat.fao.org/ (Accessed August 11, 2013).

Gordon, P., Gjerstad, K., Clark, A., Lange, K., Andersson, H., and Eliertsen, P. 2007. The prospects for graphic paper: The impact of substitution, the outlook for demand. Report. Boston Consulting Group Inc., Boston, MA 32 pp.

Green, F. 2012. Do print this e-mail. *Pulp and Paper International (PPI)*. April 2012, pp. 20–22.

Haynes, R.W. 2002. *An Analysis of the Timber Situation in the United States: 1952 to 2050*. Technical document supporting the 2000. USDA, Forest Service RPA Assessment, Pacific Northwest Research Station. Portland, OR.

Hänninen, R. and Toppinen, A. 1999. Long-run price effects of exchange rate changes in finnish pulp and paper exports. *Applied Economics*, 31:947–956.

Hetemäki, L. 1999. Information technology and paper demand scenarios. In *World Forests, Society and Environment. World Forests*, (eds.) Palo, M. and Uusivuori, J. Vol. 1. Kluwer Academic Publishers, Dordrecht, the Netherlands. pp. 31–40.

Hetemäki, L. 2005. ICT and communication paper markets. In *Information Technology and the Forest Sector*, (eds.) Hetemäki, L. and Nilsson, S. Vol. 18. IUFRO World Series, Vienna, Austria. pp. 76–104.

Hetemäki, L. and Hänninen, R. 2009. Arvio Suomen puunjalostuksen tuotannosta ja puunkäytöstä vuosina 2015 ja 2020. English summary and conclusions: Outlook for Finland's Forest Industry and Wood Consumption for 2015 and 2020. Working Papers of the Finnish Forest Research Institute 122. 63 p.

Hetemäki, L. 2010. Forest biorefinery: An example of policy driven technology. In *Forests and Society—Responding to Global Drivers of Change*, (eds.) Mery, G., Katila, P., Galloway, G., Alfaro, R.I., Kanninen, M., Lobovikov, M., and Varjo, J. Vol. 25. IUFRO World Series, Vienna, Austria. pp. 160–161.

Hetemäki, L. and Nilsson, S. (eds.) 2005. *Information Technology and the Forest Sector*. Vol. 18. IUFRO World Series, Vienna, Austria.

Hetemäki, L. and Obersteiner, M. 2001. U.S. newsprint demand forecasts to 2020. Interim Report IR-01-070. International Institute for Applied Systems Analysis.

Hujala, M. 2012. Structural dynamics in the global pulp and paper industry. Ph.D. thesis, Lappeenranta University of Technology, School of Business, Business Economics and Law, Finance, Lappeenranta, Finland.

Hurmekoski, E. and Hetemäki, L. 2013. Studying the future of the forest sector: Review and implications for long-term outlook studies. *Forest Policy and Economics*, 34 (2013):17–29.

Kallio, M., Dykstra, D., and Binkley, C. (eds.) 1987. *The Global Forest Sector. An Analytical Perspective*. John Wiley & Sons, New York.

Kangas, H.-L., Lintunen, J., Pohjola, J., Hetemäki, L., and Uusivuori, J. 2011. Investments into forest biorefineries under different price and policy structures. *Energy Economics*, 33(6):1165–1176.

Klemperer, D.W. 1996. *Forest Resource Economics and Finance*. McGraw-Hill, New York.

Kroeger, F., Vizjak, A., and Moriarty, M. 2008. *Beating the Global Consolidation Endgame*. McGraw-Hill, New York.

Laaksonen-Craig, S. 1998. Price adjustment for forest products under fixed and floating exchange rate regimes. Ph.D. dissertation, College of Natural Resources, University of California, Berkeley, CA, 128 pp.

Li, H. and Luo, J. 2008. Industry consolidation and price in U.S. linerboard industry. *Journal of Forest Economics*, 14:53–115.

Lindqvist, A. 2009. Engendering group support based foresight for capital intensive manufacturing industries–case paper and steel industry scenarios by 2018. Ph.D. thesis, Lappeenranta University of Technology. Lappeenranta, Finland, May 2009.

Löfgren, M. and Witell, L. 2005. Kano's theory of attractive quality and packaging. *Quality Management Journal*, 12:7–20.

Ou, V. 2011. China—Shaping the flow of paper and board around the world. *PaperAge*, November/December 2011, 127(6):22–24.

PricewaterhouseCoopers. 2011. *Global Forest, Paper and Packaging Industry Survey*, 2011 ed. http://www.pwc.com/ox/on/forest–paper–packaging/assets/global–forest–survey–2011.pdf

Ragauskas, A.J., Nagy, M., Kim, D.H., Eckert, C.A., Hallett, J.P., and Liotta, C.L. 2006. From wood to fuels: Integrating biofuels and pulp production. *Industrial Biotechnology*, 2(1):55–65.

RISI. 2012. 15-Year North American graphic paper forecast. June 2012, www.risi.com

Roberts, D. 2012. The forest sector's status and opportunities from the perspective of an investor. Presentation at *The Transformation of the Canadian Forest Sector and Swedish Experiences–Seminar*, Stockholm, May 28. http://www.ksla.se/wp-content/uploads/2012/04/Don-Roberts.pdf (Accessed June 15, 2012).

Rokka, J. and Uusitalo, L. 2008. Preference for green packaging in consumer product choices—Do consumers care? *International Journal of Consumer Studies*, 32:516–525.

Rushton, M., Rodden, G., James-Van Beuningen, R., and Lees, R. 2011. The PPI top 100–most companies in the black. RISI, August 30, 2011. http://www.risiinfo.com/techchannels/papermaking/The-PPI-Top-100-2010-stayed-the-course.html (Accessed June 15, 2012).

Siitonen, S. 2003. Impact of globalisation and regionalisation strategies on the performance of the world's pulp and paper companies. Ph.D. thesis, Acta Universitatis Oeconomicae Helsingiensis A-225, Helsinki School of Economics, Helsinki, Finland.

Söderholm, P. and Lundmark, R. 2009. Forest-based biorefineries: Implications for market behavior and policy. *Forest Products Journal*, January/February, 59:6–16.

Soirinsuo, J. 2010. *The Long-Term Consumption of Magazine Paper in the United States*. LAP Lambert Academic Publishing, U.K.

United Nations Economic Commission for Europe. 2003. *Russian Federation Forest Sector Outlook Study*. Geneva Timber and Forest Discussion Paper 27. United Nations, Geneva, Switzerland.

United Nations Economic Commission for Europe–FAO. 2011. *The European Forest Sector Outlook Study II: 2010–2030*. United Nations, Geneva, Switzerland.

United Nations UN (2011). Population statistics database. http://unstats.un.org/unsd/demographic/products/vitstats/. August 11, 2013.

Uronen, T. 2010. On the transformation processes of the global pulp and paper industry and their implications for corporate strategies. Doctoral dissertation, TKK Reports in Forest Products Technology, Series A14. Espuo, Finland.

Uutela, E. 1987. Demand for paper and board: Estimation of parameters for global models. In *The Global Forest Sector: An Analytical Perspective*, (eds.) Kallio, M., Dykstra, D., and Binkley, C.S. John Wiley, New York. pp. 328–354.

WPO. 2008. Market statistics and future trends in global packaging. World Packaging (WPO) Organization/Pira International Ltda. 44 pp. http://www.worldpackaging.org/publications/documents/market-statistics.pdf (Accessed June 15, 2012).

Yao, H. 2012. An analysis of china's new five-year plan for the pulp and paper industry and its impact on the paperboard sector. RISI Viewpoint, April 12, 2012, RISI.

Zhang, Y. and Buongiorno, J. 1997. Communication media and demand for printing and publishing papers in the United States. *Forest Science*, 43(3):362–377.

6

New Products and Product Categories
in the Global Forest Sector

Zhiyong Cai, Alan W. Rudie, Nicole M. Stark, Ronald C. Sabo, and Sally A. Ralph

CONTENTS

Forests, covering about 30% of the earth's land area, are a major component in the global ecosystem, influencing the carbon cycle, climate change, habitat protection, clean water supplies, and sustainable economies (FAO 2011). Globally, the vast cellulosic resource found in forests provides about half of all major industrial raw materials for renewable energy, chemical feedstock, and biocomposites (Winandy et al. 2008). Although forests and forest products have been used since the dawn of time, exciting new opportunities are emerging for sustainably meeting global energy needs and simultaneously creating new high-value biobased products from forests.

This chapter is organized into three sections, discussing new products and applications, all of which go beyond traditional solid wood products. The first section of this chapter focuses on forest productderived nanomaterials and provides a brief description of technologies and applications of nanotechnology for forest-based products. The second section explores research and development trends in wood–plastic composites (WPCs). These products will increasingly penetrate the construction and value-added wood product markets in the form of siding, fencing, bridge decking, foundation isolation elements, laminate flooring, residential furniture, utility poles, railroad ties, and exterior and interior molding and millwork. The third section covers chemicals derived from wood. It summarizes methods that use distinctive extraction processes that are attractive for new and higher value products and processes such as solvent extraction used to obtain phytosterols and hot water extraction used to obtain arabinogalactans. Books have been written about each of these categories; so in this chapter, we provide a general foundation of knowledge for the reader.

6.1 Nanoproducts

Less than two decades ago, the first reports of cellulose nanocomposites came out of France (Favier et al. 1995a,b), so the field is fairly new and there is still much to be learned, and great potential for the development of new materials with unique properties imparted with the aid of nanotechnology. The use of nanocellulose has been targeted for a number of unique effects, including reinforcement, barrier properties, and stimulus sensitivity.

Nanoparticles, with at least one dimension in the range of 1–100 nm, are materials that have been found to exhibit interesting physical and mechanical properties compared to coarse particles of the same composition (Wegner and Jones 2005). In the last few decades, many industrial and technological sectors have begun researching or adopting nanotechnology. In the forest products industry, nanotechnology is manifested in one of the two forms: nanomaterials derived from forest products or nanotechnology incorporated into traditional forest-based products. The focus of this section will be primarily on forest product—derived nanomaterials, but first a brief description of technologies and applications of nanotechnology into forest-based products is provided.

6.1.1 Incorporation of Nanotechnology into Forest Products

Nanoparticles are currently being added to wood-based products, primarily to enhance durability. Products are currently available on the market that employ nanoparticles to enhance ultraviolet (UV) resistance, scratch resistance, water repellency, fire retardancy, and microbial decay resistance. These nanoparticles are typically composed of inorganic materials, such as metals or clay. Nanoclays can add fire retardancy and barrier properties to materials, while metal nanoparticles are known best as antimicrobial agents. While the methods of incorporating these nanomaterials into products are varied and typically proprietary, they are often applied as thin exterior coatings.

6.1.2 Nanocellulose Introduction

As trees and plants are broken down into nanometer-scale fibers and particles (nanocellulose), the resulting materials begin to exhibit unique and interesting properties, which are resulting in new applications and industries in the twenty-first century. Some of the more unique properties of various nanocelluloses include remarkable strength, liquid crystal behavior in solution, transparency when cast as a film, low thermal expansion, capacity to absorb water, and piezoelectric and electroactive behavior.

Although commercial application of nanocellulose is currently in its infancy, applications are varied, covering many industries. Some applications include food additives (Turbak et al. 1982, 1983; Innami and Fukui 1987), medical and pharmaceutical applications (Innami and Fukui 1987; Pääkkö et al. 2008; Mathew and Oksman 2011), paper applications (Taipale et al. 2010; Klemm et al. 2011), automobile parts (Oksman et al. 2006), substrates for flexible displays (Nakagaito et al. 2010), electronic actuators (Olsson et al. 2010), battery membranes (Nyström et al. 2009), separation membranes (Pääkkö et al. 2008), barrier membranes (Paralikar et al. 2008), paints and coatings (Hoeger et al. 2011; Syverud 2011), and many more. Klemm et al. (2011) provide a good narrative of potential applications and list numerous nonpatents that have been filed.

In recent years, nanoscale cellulosic materials have received a great deal of attention because of their physical and mechanical properties and because of the abundance and renewability of sources of cellulose in forests and agricultural crops. Although numerous thorough reviews of cellulose nanomaterials have been recently compiled (Lima and Borsali 2004; Azizi Samir et al. 2005; Berglund 2005; Kamel 2007; Hubbe et al. 2008; Ioelovich 2008; Eichhorn et al. 2010; Habibi et al. 2010; Siró and Plackett 2010; Klemm et al. 2011; Moon et al. 2011; Peng et al. 2011), this section provides an introduction to the properties and potential applications of nanocellulose.

6.1.3 Nanocellulose Types, Sources, and Preparation Methods

Generally, nanocellulose is considered to be of three different types: cellulose nanocrystals (CNCs) (Figure 6.1), nanofibrillated cellulose (Figure 6.2), and microbial cellulose. CNCs consist of discrete, highly crystalline, rodlike cellulose particles, whereas nanofibrillated cellulose consists of more amorphous, high-aspect-ratio fiber networks. Microbial cellulose is excreted from bacteria (Berglund 2005) and has many similar properties to nanofibrillated cellulose. The source and production method both influence the size range and properties of the nanocellulose.

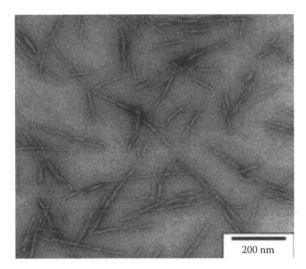

200 nm

FIGURE 6.1
Transmission electron microscopy image of CNCs from wood. (From Moon, R.J. et al., *Chem. Soc. Rev.*, 40, 3941, 2011.)

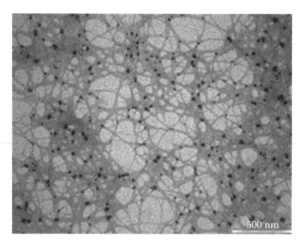

FIGURE 6.2
Cellulose nanofibers prepared by TEMPO-mediated oxidation of eucalyptus pulp.

Sources for nanocellulose include forests, plants, animals, and microbes, all of which yield somewhat different sizes and types of nanocellulose (Beck-Candanedo et al. 2005). The plant source material for producing nanocellulose is quite varied and includes hardwood (Fukuzumi et al. 2009; Saito et al. 2009; Stelte and Sanadi 2009; Zhu et al. 2011) and softwood (Nakagaito and Yano 2004; Zimmerman et al. 2004; Henriksson et al. 2007; Pääkkö et al. 2007; Stelte and Sanadi 2009; Syverud and Stenius 2009) pulps, potatoes (Dufresne et al. 2000), sugar beet (Azizi Samir et al. 2004), pea hulls (Chen et al. 2009), sisal (Moran et al. 2008), wheat straw (Alemdar and Sain 2008a,b), and banana crop residues (Zuluaga et al. 2007). Most reported studies on nanocellulose production used material processed as nearly pure cellulose (Andresen et al. 2006; Bondeson et al. 2006; Henriksson et al. 2007), such as bleached wood fibers. However, some work has been done on the preparation of nanocellulose from pulps containing significant amounts of lignin and hemicelluloses (Okita et al. 2009).

CNCs are typically produced by the acid hydrolysis of native crystalline cellulose using hydrochloric, sulfuric, or phosphoric acid. Reports of cellulose crystals from acid hydrolysis date back to work done in the middle of the twentieth century (Battista 1950; Ranby 1951). Processes for producing stable colloidal solutions of crystalline cellulose by hydrolysis with sulfuric acid were later patented (Battista and Smith 1962; Battista 1975). The yield of nanocrystals from natural plant fibers is found to be about 30% (Bondeson et al. 2006).

Nanofibrillated cellulose production generally requires some method of mechanical refining of coarse cellulose fibers, and various chemical and enzymatic pretreatments have been explored to aid in the refinement of cellulose fibers. Mechanical actions through shearing or grinding have been used to produce nanocellulose (Nakagaito and Yano 2004; Andresen et al. 2006; Iwamoto et al. 2007; Alemdar and Sain 2008a).

The method was initially demonstrated through homogenization (Herrick et al. 1983; Turbak et al. 1983). Although cellulose nanofibers (CNFs) were successfully produced using a homogenizer, other mechanical methods, such as stone grinding and disk refining, have also been found to be useful for producing nanofibrillated cellulose (Iwamoto et al. 2007; Okahisa et al. 2009). The purely mechanical method is extremely energy intensive, with the energy consumption as high as 20,000–30,000 kWh/ton (Siró and Plackett 2010). Because the high energy consumption of early techniques was cost prohibitive, research and development of nanofibrillated cellulose was slow to develop (Siró and Plackett 2010; Klemm et al. 2011). Recent advances in pretreatments and chemical methods for producing CNFs have reduced the energy requirements by more than an order of magnitude with energy consumption as low as 500 KWh/ton (Ankerfors and Lindström 2007, 2009).

6.1.4 Properties and Potential Applications of Nanocellulose and Cellulose Nanocomposites

The crystalline portion of cellulose has been characterized as having excellent mechanical properties. Researchers have both calculated and measured the strength and stiffness of cellulose building blocks to be quite high, making them potentially suitable for various applications, including composite reinforcement (Sakurada et al. 1962; Eichhorn et al. 2001; Azizi Samir et al. 2005; Lahiji et al. 2010; Moon et al. 2011; Postek et al. 2011). Moon et al. (2011) provide an excellent summary of the calculated and experimentally measured moduli of cellulose. A variety of other interesting properties of nanocellulose have been observed. For example, the crystalline portion of cellulose is reported to be piezoelectric and/or electroactive, making nanocellulose potentially suitable for actuators or smart devices (Fukada 2000; Kim and Yun 2006; Yun et al. 2008; Wegner and Jones 2009).

The suspensions of CNCs exhibit birefringence from the anisotropy of the crystals in solution either from their self-ordering at higher concentrations or from induced ordering (Revol et al. 1992; Revol and Marchessault 1993; Dong et al. 1996; Lima et al. 2003; Azizi Samir et al. 2005; Kimura et al. 2005; Beck-Candanedo et al. 2006; Teters et al. 2007). CNCs also align in the presence of an electric field (Lima et al. 2003; Teters et al. 2007). Such ordering properties of CNCs have potential applications in electronics and display devices.

CNF suspensions exhibit gel-like behavior at even low concentrations (Turbak et al. 1983; Pääkkö et al. 2007), and they are known to have shear thinning properties (Klemm et al. 2011). Such rheological properties make CNFs potentially suitable for thickening of foods and as additives in other industrial processes.

Natural fibers have been used extensively as reinforcement for composite materials, and because of their remarkable mechanical properties, nanocellulosic materials are expected to serve as excellent reinforcements. Nanofibrillated cellulose films tend to be transparent or at least translucent, have good mechanical properties, and are often used a starting point for making composites. Dewatering CNF solutions presents a practical challenge, but CNFs have been commonly cast or filtered into films and evaluated for their properties. These films have been coined "nanopaper" and have been touted as the "strongest cellulose-based material made by man" (Ankerfors and Lindström 2007). A survey of the literature reveals reported tensile strengths of nanofibrillated cellulose films ranging from about 70 MPa to over 300 MPa and tensile moduli ranging from about 3 GPa to nearly 20 GPa. The transparency and flexibility, along with low thermal expansion, of CNF films make them suitable for flexible displays and electronics.

6.1.5 Outlook

Different types of nanocellulose have distinctive properties, so the applications for the various forms are likely to be different. CNCs certainly have some interesting properties and potential for applications such as security printing, scratch-resistant coatings, or liquid crystals. While CNCs from the sea animal have provided good reinforcement for polymer composites, the results have not been as widely successful with CNCs from plants, especially wood. However, some promising success has been achieved using CNCs from plants such as ramie. Even so, most of the success of reinforcing polymers with CNCs uses solvent evaporation techniques that are not commercially viable. Given that tunicate (and even ramie) are not extremely abundant, it seems the widespread commercial adoption of CNCs as a polymer reinforcement is not likely. Perhaps some breakthrough research or technology will emerge that will change this perspective, but it seems that nanofibrillated cellulose is more likely than CNCs to act as a polymer reinforcement.

CNFs are somewhat analogous to pulp fibers used for producing paper products in that they exhibit entangled networks and can be formed into sheets using similar processes. Agglomeration of nanofibers does not seem to be as worrisome as for CNCs, and the sheets made from CNFs are much less brittle than CNCs and handle similar to paper made from coarse pulp fibers. Strength and other properties of CNF suspensions, sheets, and composites have led to adoption of CNFs in a variety of applications including rheology modification, paint emulsification, polymer reinforcement, coatings, and packaging. Therefore, CNFs are expected by many to prove as a useful material in a broad range of applications and markets in the future.

In the long term, nanotechnology is expected to add significant improvements in the performance and functionality of materials, including forest-based products. Nanomaterials are expected to dramatically enhance both mechanical and barrier properties of papers, packaging composites, and numerous other wood-derived materials. Furthermore, functionalized and tailored nanomaterials are expected to result in stimuli-sensitive products that will offer specific sensing capabilities, including the ability to detect contaminants and food spoilage. The unique properties of wood and other plants refined to the nanometer scale are expected to result in a new generation of renewable, forest-based products, including electronics and composites.

6.2 Trends in Wood–Plastic Composites

The term of WPCs is broad and encompasses the incorporation of wood and/or other lignocellulosic materials into thermoplastics. In the past decade, WPCs have become a widely recognized commercial product in construction, automotive, furniture, and other consumer applications. Commercialization of WPCs in North America has been primarily due to penetration into the construction industry. Current WPC applications include decking, railing, window and door lineals, roofing, picnic tables and benches, fencing, landscape timbers, patios, gazebos, pergolas, auto parts, and playground equipment (Smith and Wolcott 2006). The automotive industry in Europe has been a leader in using WPCs for interior panel parts and is leading the way in developing furniture applications. Manufacturers in Asia are targeting the furniture industry, in addition to interior construction and decorative applications.

Inorganic materials (e.g., glass, clays, and minerals) are used as reinforcements or fillers in the majority of reinforced or filled thermoplastics. Wood and other lignocellulosic materials offer some advantages over inorganic materials: they are lighter, much less abrasive, and renewable. Wood also may reinforce the thermoplastic by stiffening and strengthening and can improve thermal stability of the product compared with that of unfilled material. In typical WPCs, wood content is less than 60% by weight.

The manufacture of WPCs is usually a two-step process. The raw materials are first mixed together, and the composite blend is then formed into a product. The most common types of product-forming methods for WPCs involve forcing molten material through a die (sheet or profile extrusion) or into a cold mold (injection molding), and pressing in calenders (calendering) or between mold halves (thermoforming and compression molding).

Most WPCs in North America are formed using profile extrusion. Products such as decking, railings, and window profiles readily lend themselves to extrusion through a 2D die. Injection-molded applications such as consumer household goods and furniture parts are gaining importance. Thermoforming or compression molding is the forming method of choice for the automotive industry.

6.2.1 New Materials

Thermoplastics selected for WPCs traditionally have melt temperatures below 200°C (392°F) to maintain processing temperatures below the degradation point of the wood component. Higher processing temperatures can result in the release of volatiles, discoloration, odor, and embrittlement of the wood component. Thermoplastics commonly used in WPCs include polyethylene (PE), polypropylene (PP), and polyvinyl chloride (PVC). However, new manufacturing strategies allow for the use of engineering thermoplastics with melting temperatures higher than 230°C (446°F) such as polyethylene terephthalate (PET), polyamide (PA, nylon), and acrylonitrile butadiene styrene (ABS) (Gardner et al. 2008). The use of engineering thermoplastics for WPCs will grow as new applications requiring superior mechanical and thermal properties are introduced.

A driving force in the development of many new materials is to decrease our use of petroleum. Because the most common plastics used in WPCs, PE, and PP are typically derived from petroleum, there is growing interest in replacing the common WPC thermoplastics with bioplastics. Bioplastics, that is, plastics derived from renewable biomass, may also be biodegradable. Biodegradable bioplastics include polylactic acid (PLA) and starch acetate; nonbiodegradable bioplastics include cellulose acetate and bioderived PE and polyamide (Lampinen 2009). PLA has received the most attention from researchers as it is widely available commercially. Bioplastics will be particularly important in automotive and packaging applications.

The wood and/or lignocellulosics material used in WPCs can be derived from a variety of sources. Geographical location often dictates the material choice. In North America, wood is the most common raw material; in Europe, natural fibers such as jute, hemp, and kenaf are preferred, while rice hull flour and bamboo fiber are typical in Asia. The wood is incorporated as either fiber bundles with low aspect ratio (wood flour) or as single fibers with higher aspect ratio (wood fiber). Wood flour is processed commercially, often from postindustrial materials such as planer shavings, chips, and sawdust. Wood and lignocellulosic fibers are available from virgin and recycled sources. New lignocellulosic sources include paper mill sludge and biorefinery residues. Because wood and lignocellulosic fibers can lead to superior WPC properties by acting more as a reinforcement than a filler, in applications requiring additional strength, a trend is to move toward the use of wood and other lignocellulosic fibers.

The adaptation of nanotechnology in WPCs includes the use of nanofibers derived from wood and other lignocellulosics as a reinforcement for plastics. These new composites are termed cellulose nanocomposites. Cellulose nanocomposites are rapidly expected to open new markets in medicine, packaging, electronics, automotive, construction, and other sectors (Oksman et al. 2009). The emergence of nontraditional forest products markets is particularly exciting.

Other materials can be added to affect processing and product performance of WPCs. These additives can improve bonding between the thermoplastic and wood components (e.g., coupling agents), product performance (talc, impact modifiers, UV light stabilizers, and flame retardants), and ability to be processed (lubricants). Additives for WPCs are continually evolving, but nanotechnology will drive changes in additive technology as the use of nanomaterials in small amounts in WPCs improves performance. The incorporation of nanomaterials into WPCs is still in its infancy and in the research stage but includes the use of carbon nanotubes (Faruk and Matuana 2008a; Jin and Matuana 2010), nanoclays (Faruk and Matuana 2008b; Hetzer and DeKee 2008), and nanosized titanium dioxide (Stark and Matuana 2009). It is expected that as nano-WPC technology becomes better understood, the use of these and other nanomaterials in WPCs will experience tremendous growth.

6.2.2 New WPC Manufacturing Techniques

Improvements in and changes to manufacturing methods will help pave the way for the next generation of WPCs. New processes under development include coextrusion, foaming during extrusion, and in-line coating technologies. Coextrusion consists of the extrusion of multiple materials through a single die simultaneously. The most common application in WPCs is coextrusion of an opaque, unfilled plastic cap layer over a WPC core. This process became popular for fencing and is quickly becoming the manufacturing method of choice for decking. The unfilled cap layer enhances durability by improving moisture resistance. In addition, coextrusion allows manufacturers to concentrate expensive additives such as biocides, fungicides, and photostabilizers in the cap layer. Coextrusion is also being evaluated as a method to extrude a clear plastic cap layer over a WPC core (Stark and Matuana 2009; Matuana et al. 2011). This method allows for a more natural appearance while still providing enhanced moisture resistance and concentrated stabilization. Others are evaluating coextrusion as a method to extrude an all WPC profile with different additives in different layers as needed (Yao and Wu 2010).

Growth is also expected in the production of foamed WPCs. Creating a microcellular foamed structure in WPCs not only results in weight reduction but also improves impact resistance and allows for better surface definition and sharper contours (Faruk et al. 2007). Chemical and physical foaming agents are typically used to foam WPCs. Chemical foaming agents decompose at processing temperatures into gases, while physical foaming agents liberate gases as a result of evaporation or desorption at elevated temperatures. A new trend in foaming is to meter and dissolve inert gases in the polymer melt during processing. This is termed extrusion foaming when done during extrusion (Diaz and Matuana 2009) or gas-assist injection molding if done during injection molding. Another option is to use moisture in the wood to foam WPCs (Gardner et al. 2008).

Applying a coating to WPCs post processing but before they are made available to the consumer is a growing trend to provide increased durability and enhanced aesthetics. Coatings can include latex paints, polyurethanes, or acrylics. However, surface treatments are required to attain adequate adhesion between the WPC and the coating. Treatments that have been found effective include oxygen plasma, flame, chromic acid, and benzophenone/UV irradiation (Gupta and Laborie 2008). Another treatment being commercialized uses fluorooxidation to modify the surface of WPCs for improved coating adhesion. The advantage of this process is that the rapid chemistry allows for in-line processing. Fluorooxidation also modifies only the outer few molecular layers so embossed patterns are unaffected (Williams and Bauman 2007). New coatings being evaluated include UV-curable coatings and powder coating. Advantages of a UV-curable coating include virtually unlimited color choice and gloss level; long-term resistance to fading; increased scratch, stain, and mar resistance; and prevention of mold and mildew growth (Burton 2008).

6.2.3 Emerging Applications

Continued research and development will allow each new application to penetrate new global markets. For example, in North America, traditional wood applications that will see increased pressure from emerging technologies include siding, fencing, bridge decking, foundation isolation elements, marine structure (chocks, wales, and pier decking), laminate flooring, residential furniture (bathroom/kitchen cabinets and patio furniture), utility poles, railroad ties, and exterior and interior molding and millwork (Crespell and Vidal 2008). Emerging markets are expected as injection molding becomes more common for WPCs, including applications such as cosmetics packaging and toys. Markets in injection-molded automotive application are poised to grow in Europe and include parts such as glove boxes, fixing hooks, sound systems, and fan boxes (Carus et al. 2008).

Growth is also expected as improvements to WPCs allow them to be used as structural members. There are various methods being investigated to improve the strength, stiffness, and creep performance of WPCs. Cross-linking the PE polymer matrix in WPCs using silanes in a reactive extrusion process can improve toughness, reduce creep, and improve durability (Bengtsson et al. 2006, 2007). Combining WPCs with other non-WPC materials is also a trend that will allow for more structural composites. Currently, some WPC manufacturers in China are extruding WPCs over solid metal or solid wood. This allows for structural members in applications such as pergolas and gazebos to have WPC surfaces that match the nonstructural members.

6.3 Wood Chemicals

Although there are hundreds of species of trees, the chemical makeup is quite consistent and falls into two broad groups. Softwood trees, also called evergreens or technically conifers, contain 25%–35% lignin, 40%–50% cellulose, and 10%–20% hemicellulose (Figure 6.3).

Hardwoods or deciduous trees contain 15%–25% lignin and 15%–30% hemicellulose. The cellulose content of hardwoods is about 45%–50%. Woods contain a group of small molecules often referred to as extractives. For softwoods, these include volatile chemicals classed as turpentine, fats and fatty acids, resin acids, steroids, and several groups of phenolic compounds. Softwood extractives range from as little as 2% up to as much as 10% of dry wood mass. Hardwood extractives are generally limited to fats and phenols with no or very little turpentine and resin acids. There are exceptions with some tropical species containing as much or more resin than softwoods, but this is not common.

Cellulose is a linear polymer of glucose sugars. The degree of polymerization is about 100,000 monomers and the cellulose regions are of sufficient uniformity and purity that they form highly ordered (crystalline) regions that are very resistant to chemicals and biological degradation. Hemicellulose consists of three polymeric groups of mixed sugars: xylan in which the dominant sugar is xylose, mannans with the dominant sugar being

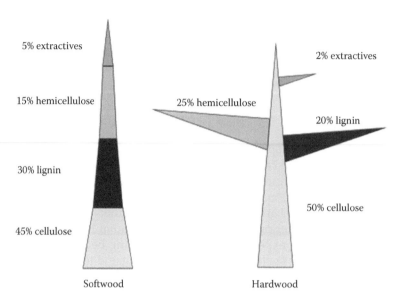

FIGURE 6.3
Typical composition of softwood and hardwood tree species.

mannose, and arabinogalactans where the dominant sugar is galactose (Timell 1957, 1967; McGinnis and Shafizadeh 1980; Pettersen 1984). Xylans contain glucuronic acid as a secondary sugar acid in hardwoods and also contain some arabinose in softwoods. Mannan contains glucose as a secondary sugar in hardwoods and also contains some galactose in softwoods. Arabinogalactan is generally found at low levels in softwoods—about 2% with the exception being larch where it can be 10%–20% of wood mass. Lignin is a cross-linked phenolic polymer that helps to bind the carbohydrate polymers in the tree into a rigid composite structure. Although composed of just one (softwoods) or two (hardwoods) monomers, the bonding pattern is varied resulting in a mixture of reactive and chemically resistant bonds that produces a wide mixture of oligomers and very few pure and useful chemicals when lignin is chemically depolymerized.

Just as wood was the primary construction material and the primary fuel for preindustrial societies, it was also a primary source of organic chemicals. Many early paints and wood finishes were wood derived using a specific group of wood resins that did not remain sticky and did not crystallize. Initially, amber, the fossilized resin of a number of extinct softwoods, was used as the carrier or binder in paints. Later, other wood extrudents including damar (*Dipterocarpaceae* spp.), copals (*Hymenaea* spp.), and kauri (*Agathis australis*) were also used for this purpose. Shellac and tung oil are tree-derived wood finishes still commonly used today. Many fragrances including frankincense, myrrh, sandalwood, and vanillin (Hocking 1997) and some adhesives including Canada balsam and latex are obtained from trees. Some medications were also originally found in trees including quinine (*Cinchona* spp.), salicylic acid (aspirin) from *Betula* and *Salix* spp., and *Taxol* (Pacific yew) as examples. Tannins from oaks and other wood species were critical for tanning leather, a process used for over 5000 years.

A huge historical use of wood chemicals was and still is Naval stores, so named because pine pitch (pine tar) was critical for waterproofing the hulls and rigging of wooden sailing ships (Zinkel and Russell 1989). Pitch is specifically mentioned in Genesis when describing the construction of Noah's ark and was an item of international commerce during the

FIGURE 6.4
Gum harvesting for naval stores.

Greek and Roman eras. There are a number of methods for collecting and processing pine tar, but large industries (for the times) developed around bleeding resin from pine trees by cutting off strips of bark (Figure 6.4) and recovering resin from pine trees by heating the wood and stumps. The Scandinavian countries became major suppliers for much in Europe, and the British colonies in the Carolinas and Georgia and Florida became the major suppliers for England before the Revolutionary War. Slash pine was the preferred species for pine pitch and is named after the process of cutting the trunk to collect pitch. Wood also provided two of the earliest postindustrial textile fabrics (rayon and acetate), early barrier and clear films (cellophane and cellulose nitrate), and smokeless powder (cellulose nitrate).

Obviously, developments in petroleum, chemical industries, and new synthetic polymers have replaced many of these historic uses for wood chemicals, but the naval stores business continues using turpentine, resin acids, and fatty acids recovered from pulping processes, as well as remnants of the traditional recovery processes using steaming and destructive distillation of pine stumps. Fatty acids are still used to make soap, detergents, and other surfactants. Resin acids are also used for soaps and are modified to improve the properties of paints and varnishes.

Although many of the traditional uses of wood-based chemicals are declining, markets, pharmaceutical, nutraceutical, and specialty chemical product applications are expanding. The litany of known chemicals available in wood extractives and in waste liquor from wood pulping is well beyond the scope of this chapter, with books devoted to the topic (Rowe 1989). There are a number of high-value target compounds that cannot be readily prepared from starting materials found in petroleum. The problem for the forest products and chemical industries has always been the large numbers of similar compounds found in the wood and the high cost of separation processes needed to isolate the more valuable of these compounds. It is said in the various commercial sectors of the industry that "we can make anything out of lignin or black liquor except money." As industry moves into other chemical treatment methods, in particular, the biorefinery for fuels and commodity chemicals, the desire to capitalize on other wood chemicals remains high. New markets for wood chemicals are focusing on methods to collect target chemicals without producing a myriad of other compounds that lower yields and increase isolation costs.

The remainder of this section discusses the production of higher-value chemicals in a biorefinery producing wood sugars and current processes and research to obtain higher-value chemicals from the unused parts of the tree. The discussion focuses on phytostanols for nutraceutical use, xylitol as a low-calorie sugar substitute, and hemicelluloses including current use of arabinogalactans and research on new uses for xylans. There are also intense efforts to derive higher value from lignin (Zakzeski et al. 2010), bark (Pietarinen et al. 2006; Liimatainen et al. 2012), and knots (Pietarinen et al. 2006).

6.3.1 Biorefinery

Biorefinery as we understand it today is geared toward replacing petroleum-based fuels or commodity chemicals. There are two broad methods for converting wood in most modern concepts of biorefinery: hydrolysis to sugars (Mosier et al. 2005; Wyman et al. 2005) and partial combustion to produce pyrolysis oil or producer gas. Pyrolysis oil is a very complex mixture, and research today is directed at improving its characteristics to serve directly as a fuel oil or transportation fuel. Research on producer gas (synthesis gas) is focused on Fischer–Tropsch conversion to fuel oil (Anderson et al. 1952), synthetic natural gas (methane), or other stable, high-energy density liquid fuels (Tijmensen 2000). These processes are not well suited to provide high-value chemicals. On the other hand, fermentation methods are capable of producing more complex types of chemicals with higher value. Potential products include alcohols, such as ethanol, propanol, or butanol; diols, such as ethylene, propylene, or butylene diol; polyhydroxyalkanoates; and lactic acid (Hajny 1981; Werpy and Petersen 2004). During World War II, the US Government constructed a dilute acid wood hydrolysis plant with the intention of fermenting the sugars to ethanol for conversion into butadiene for synthetic rubber (Katzen and Schell 2006). Currently, there are industrial-scale processes producing lactic acid (Sreekumar and Thomas 2008), 1,3-propanediol (Kurian 2005), and polyhydroxyalkanoates (Poirier et al. 1995; Sreekumar and Thomas 2008). In all of these cases, the source of sugar is a sugar or starch crop but these are product options for biorefineries using wood as a raw material as well.

6.3.2 Phytosterols

The phytosterols are a group of over a dozen compounds of which sitosterol and campesterol are the most prevalent (Conner 1989). They can be a significant component of crude tall oil, the mixture of fatty acids, resin acids, and sterols that survive the pulping process and can be recovered from the soluble waste products. The sterols present a problem in that they are an impurity that reduces both the yield and value of the resin and fatty acids. Traditionally, crude tall oil is processed by distillation to collect a relatively pure fatty acid fraction, a relatively pure resin acid fraction, and a mixture in the still bottoms referred to as tall oil pitch (Conner 1989). Sterols can end up as contaminants in both the lower and higher boiling fractions (Holmbom and Era 1978; Traitler and Kratzl 1980; Smith 1989). The tall oil pitch fraction is typically 15%–25% of the crude tall oil and the phytosterols can be as much as 25% of the tall oil pitch (Steiner and Fritz 1959; Holmbom and Era 1978). Again, they dilute the value of this fraction and discolor the product.

Several companies developed processes to remove the sterols before distillation, generally involving either solvent extraction or chemical derivitization followed by extraction (Conner 1989; Rouskova et al. 2011). A second method developed later is able to distill the steroids from the tall oil mixture by adding sodium hydroxide to raise the pH and convert the fatty and resin acids to less volatile salts. By 1980, Oy Kaukas AB in Finland was marketing the tall oil steroids as emulsifiers and viscosity modifiers for use in cosmetics. The ability of phytosterols to inhibit production of serum cholesterol was first reported by Pollak (1953). Sitosterol itself or the esters are used as a food additive to help people lower their cholesterol (Lichtenstein and Deckelbaum 2001; Maki et al. 2001). In addition, lower amounts of other steroids—sitostanol (also called stigmastanol) and campestanol—are available from crude tall oil (Murzin et al. 2007), and these are also used as food additives to inhibit the formation of cholesterol in humans.

These products have considerable market presence today as food-based sterols that help lower serum cholesterol. The Benecol® line of products originating from the Raisio Group in Finland includes margarine, yogurts, cream cheese, and bread, and the Becel brand of margarine (Promise—Smart Balance® Heart Right™) is produced by Unilever. As wood-based chemicals, they represent a new by-product with high value. The phytosterols also have use as starting chemicals for producing other steroid-based drugs including Rogaine®.

6.3.3 Xylitol

Xylitol is a low-calorie sweetener marketed by Danisko and now several other companies. Xylose, one of the five common sugars in wood (glucose, mannose, galactose, xylose, and arabinose), differs from glucose primarily in that it has just five carbons in the sugar molecule where glucose contains six carbons. Common table sugar produced from sugar cane is usually sucrose, a dimer containing one molecule of glucose and one molecule of a similar six-carbon sugar, fructose. Xylose is about twice the sweetness of sucrose (Robyt 1998) and is poorly adsorbed and metabolized by humans. It is not as chemically stable as glucose and sucrose and was not commonly marketed as a sugar substitute. A reduced form called xylitol is regarded as equally sweet as sucrose, stable, and has been used as a sugar substitute in Europe for many years. Xylitol penetration in the US market has been slower, possibly because it is toxic to dogs. Xylan is quite prominent in hardwoods, typically 20% or more of the dry wood mass.

Sugar is a relatively low-price commodity food product and this sets a price standard for low-calorie substitutes. Direct wood extraction of xylose has not been a suitable process primarily because of yield and cost. However, this option may work well in new wood-processing concepts, as the xylose is readily removed in the pretreatment processes needed to prepare the cellulose for enzyme hydrolysis. Xylan degrades in the kraft pulping process and is not available from this source. A less common pulping process uses acid to neutral pulping conditions with sulfurous acid or sodium sulfite as the active pulping chemical. Xylose is more stable under these conditions (Bryce 1980) and sulfite waste product or pulping liquor is a common source of xylose for xylitol (Heikkilä et al. 2005). The waste pulping liquor from hardwoods is obviously better than the liquor from softwoods. A single sulfite mill can supply 100 T of xylose per day as a valued product. As this market grows, Danisko and competitors may need to look for other sources. Biorefineries are clearly good candidates.

Xylitol represents a modest value use of a wood chemical. The higher concentration in sulfite waste liquors is fortuitous in that this process does not have a good chemical recovery process making the waste product less valuable as process fuel. There are also high concentrations of xylan in corn stover, and this option is equally valid for biorefinery concepts based on agricultural residuals, as it is for wood.

6.3.4 Hemicellulose

Xylitol is a sugar monomer produced from a hemicellulose, but two of the hemicellulose polymers have interesting product applications as polymers and oligomers. Arabinogalactan is a branched chain polymer of galactose monomers with short branches of one or two monomers of galactose or arabinose (Timell 1967). Galactose is a six-carbon sugar-like glucose, but the configuration of atoms around carbon 1 and carbon 4 are different. These seemingly minor changes make the sugar difficult for humans to digest and for microorganisms to ferment. It is not prevalent in most tree species, with Pettersen reporting 0.5%–2.7% arabinose and 1.0%–4.7% galactose in 19 North American softwood species. Because both arabinose and galactose are incorporated into other hemicellulose polymers in softwoods, the sugar analysis is not definitive of the polymer but does cap the maximum levels. Clearly, it cannot exceed about 7%, which Pettersen (1984) reported for Douglas fir. Missing from Pettersen's review is western larch (*Larix occidentalis*), which has an unusually high level of arabinogalactans, generally reported as between 8% and 17%. This is the commercial source of arabinogalactan (Schorger and Smith 1916; Wise and Peterson 1930; White 1941). Unique to arabinogalactan and somewhat to larch, the hemicelluloses can be extracted with water providing a relatively pure polymeric product (Schorger and Smith 1916; White 1941). Arabinogalactan extracted from larch has been an item of commerce for decades, formerly sold under the trade name Stractan by the St. Regis Paper Company and now sold by Lonza under various application specific trade names: ResistAid™ as an antioxidant food additive, LaraFeed™ as an animal nutrition supplement, and LaraCare™ as a cosmetic ingredient.

As stated in the section on xylitol, glucuronoxylan is the most abundant hemicellulose polymer in hardwoods. The polymer and monomer are of interest as by-products of biorefinery, dissolving pulp, and cellulose nanomaterial production processes. The xylan polymers are not really a product of commerce yet, but there is considerable interest in developing uses for this carbohydrate. Xylose is a less valuable sugar in sugar-based biorefinery applications where it is typically removed in the pretreatment and is hard to ferment using traditional organisms (Lee and McCaskey 1983). It is an impurity that must be removed for production of chemical dissolving pulp (Richter 1955; Simmonds et al. 1955; Bernardin 1958) and also is an impurity in the concentrated acid hydrolysis process used to isolate CNCs (Ranby 1952; Bondeson et al. 2006). It is expected that the polymer xylan and monomer sugar xylose are about to become either a valuable by-product or a costly waste product as cellulose-based biorefinery becomes a commercial reality. The current research interest is to find valuable uses.

Polymeric xylan is readily extracted from hardwoods, nearly intact, using 10% sodium hydroxide (Booker and Schuerch 1958; Sixta et al. 2011). This is an expensive process and will require a high-value use, but the residual wood can be used for dissolving pulp or other wood pulp applications (Lyytikäinen et al. 2011). The Finnish-sponsored BioRefine program is evaluating options to integrate this as a pretreatment into kraft pulping where the alkaline extraction liquor can be recovered after separating out the xylan and then used in the pulping process (Sixta et al. 2011).

In addition to xylitol discussed earlier, potential markets for xylan polymers include viscosity modifiers, cosmetics, and dietary fiber, similar to the arabinogalactans. Research groups are also working on modifying the polymer to produce clear barrier films (Gröndahl et al. 2004; Sixta et al. 2011; Escalante et al. 2012). As a packaging film, xylans are renewable and compostable, making them a green alternative for existing petroleum-based polymer films.

6.4 Conclusion

Renewable, recyclable, and compostable products sourced from trees have great potential as low ecological impact products and part of a new sustainable way of life. Wood is a green renewable material that has been accepted mostly as the structural members of choice for residential constructions. However, the potential for an array of wood-based building materials for application in sustainable buildings, both residential and commercial, has not been fully realized. With the incorporation of nanoscale science and engineering into wood-based building materials, new generations of multifunctional, high-performance, ultralow maintenance, and durable building materials and components can be achieved. Globally, the vast lignocellulosic and other wood chemical resources provide about half of all major industrial raw materials for renewable energy, chemical feedstock, and biocomposites. Conversion of woody biomass to biofuels is technically feasible, but this conversion process is marginally economical with the current technology and price of crude petroleum. An integrated utilization of biomass is needed to overcome economic shortcomings by optimizing biomass use and value for a wider array of products.

References

Alemdar, A. and M. Sain. 2008a. Isolation and characterization of nanofibers from agricultural residues–Wheat straw and soy hulls. *Bioresource Technology*, 99(6):1664–1671, April 2008.

Alemdar, A. and M. Sain. 2008b. Biocomposites from wheat straw nanofibers: Morphology, thermal and mechanical properties. *Composites Science and Technology*, 68(2):557–565, February 2008.

Anderson, R.B., B. Seligman, J.F. Shultz, and M.A. Elliott. 1952. Fischer-Tropsch synthesis: Some important variables of the synthesis on iron catalysts. *Industrial and Engineering Chemistry*, 44(2):391–397.

Andresen, M., L.S. Johansson, B.S. Tanem, and P. Stenius. 2006. Properties and characterization of hydrophobized microfibrillated cellulose. *Cellulose*, 13:665–677.

Ankerfors, M. and T. Lindström. 2007. On the manufacture and use of nanocellulose. Presented at the *Ninth International Conference on Wood and Biofiber Plastic Composites*, Madison, WI.

Ankerfors, M. and T. Lindström. 2009. *NanoCellulose Developments in Scandinavia*. A contribution to the *Paper and Coating Chemistry Symposium (PCCS)*, Hamilton, Ontario, Canada, June 2009.

Azizi Samir, M.A.S., F. Alloin, and A. Dufresne. 2005. Review of recent research into cellulosic whiskers, their properties and their application in nanocomposite field. *Biomacromolecules*, 6:612–626.

Azizi Samir, M.A.S., F. Alloin, M. Paillet, and A. Dufresne. 2004. Tangling effect in fibrillated cellulose reinforced nanocomposites. *Macromolecules*, 37:4313–4316.

Battista, O.A. 1950. Hydrolysis and crystallization of cellulose. *Industrial and Engineering Chemistry*, 42:502–507.

Battista, O.A. 1975. *Microcrystal Polymer Science.* McGraw-Hill Inc., New York.

Battista, O.A. and P.A. Smith. 1962. Microcrystalline cellulose. *Journal of Industrial and Enginerring Chemistry,* 54(9):20–29.

Beck-Candanedo, S., M. Roman, and D.G. Gray. 2005. Effect of reaction conditions on the properties and behavior of wood cellulose nanocrystal suspensions. *Biomacromolecules,* 6(2):1048–1054.

Beck-Candanedo, S., D. Viet, and D.G. Gray. 2006. Induced phase separation in cellulose nanocrystal suspensions containing ionic dye species. *Cellulose,* 13:629–635.

Bengtsson, M., K. Oksman, and N.M. Stark. 2006. Profile extrusion and mechanical properties of crosslinked wood-thermoplastic composites. *Polymer Composites,* 27(2):184–194.

Bengtsson, M., N.M. Stark, and K. Oksman. 2007. Durability and mechanical properties of silane cross-linked wood thermoplastic composites. *Composites Science and Technology,* 67:2728–2738.

Berglund, L. 2005. Cellulose-based nanocomposites. In A. K. Mohanty, M. Misra, and L. Drzal (eds.), *Natural Fibers, Biopolymers and Biocomposites.* CRC Press, Boca Raton, FL.

Bernardin, L.J. 1958. The nature of the polysaccharide hydrolysis in black gumwood treated with water at 160°C. *TAPPI,* 41(9):491–499.

Bondeson, D., A. Mathew, and K. Oksman. 2006. Optimization of the isolation of nanocrystals from microcrystalline cellulose by acid hydrolysis. *Cellulose,* 13:171–180.

Booker, E. and C. Schuerch. 1958. Extraction of pentosans from woody tissues, III. *TAPPI,* 41(11):650–654.

Bryce, J.R.G. 1980. Sulfite pulping. *In* J.P. Casey (ed.), *Pulp and Paper, Chemistry and Chemical Technology.* Wiley & Sons, New York, pp. 291–376.

Burton, K. 2008. A cure for wood alternatives: New UV coating technologies can enhance wood plastic composites, fiber cement siding and vinyl substrates used in the construction industry. *Finishing Today,* April 2008:32–35.

Carus, M., C. Gahle, and H. Korte. 2008. Market and future trends for wood-polymer composites in Europe: The example of Germany. In *Wood-Polymer Composites.* Woodhead Publishing Ltd., Cambridge, U.K. pp. 300–330.

Chen, Y., C. Liu, P.R. Chang, X. Cao, and D.P. Anderson. 2009. Bionanocomposites based on pea starch and cellulose nanowhiskers hydrolyzed from pea hull fibre: Effect of hydrolysis time. *Carbohydrate Polymers,* 76(4):607–615.

Conner, A.H. 1989. Chemistry of other components in naval stores. In D.F. Zinkel and J. Russell (eds.), *Naval Stores, Production, Chemistry, Utilization.* Pulp Chemicals Association, New York, pp. 440–475.

Crespell, P. and M. Vidal. 2008. Market and technology trends and challenges for wood plastic composites in North America. *In Proceedings of the 51st International Convention of Society of Wood Science and Technology,* November 10–12, 2008, Concepción, Chile.

Diaz, C.A. and L.M. Matuana. 2009. Continuous extrusion production of microcellular rigid PVC. *Journal of Vinyl and Additive Technology,* 15(4):211–208.

Dong, X.M., T. Kimura, J.F. Revol, and D.G. Gray. 1996. Effects of ionic strength on the isotropic-chiral nematic phase transition of suspensions of cellulose crystallites. *Langmuir,* 12:2076–2082.

Dufresne, A., D. Dupeyre, and M.P. Vignon. 2000. Cellulose microfibrils from potato tuber cells: Processing and characterization of starch-cellulose microfibril composites. *Journal of Applied Polymer Science,* 76(14):2080–2092.

Eichhorn, S.J., C.A. Baillie, N. Zafeiropoulos, L.Y. Mwaikambo, M.P. Ansell, A. Dufresne, K.M. Entwistle et al. 2001. Current international research into cellulosic fibres and composites. *Journal of Materials Science,* 36(2001):2107–2131.

Eichhorn, S.J., A. Dufresne, M. Aranguren, N.E. Marchovich, J.R. Capadona, S.J. Rowan, C. Weder et al. 2010. Review: Current international research into cellulose nanofibres and nano-composites. *Journal of Materials Science,* 45:1–33.

Escalante, A., A. Goncalves, A. Bodin, A. Stepan, C. Sandström, G. Toriz, and P. Gatenholm. 2012. Flexible oxygen barrier films from spruce xylan. *Carbohydrate Polymers,* 87:2381–2387.

Faruk, O., A.K. Bledzki, and L.M. Matuana. 2007. Microcellular foamed wood-plastic composites by different processes: A review. *Macromolecular Materials and Engineering,* 292:113–127.

Faruk, O. and L.M. Matuana. 2008a. Reinforcement of rigid PVC/wood-flour composites with multi-walled carbon nanotubes. *Journal of Vinyl and Additive Technology*, 14(2):60–64.

Faruk, O. and L.M. Matuana. 2008b. Nanoclay reinforced HDPE as a matrix for wood-plastic composites. *Composites Science and Technology*, 68(9):2073–2077.

Favier, V., G.R. Canova, J.Y. Cavaillé, H. Chanzy, A. Dufresne, and C. Gauthier. 1995a. Nanocomposite materials from latex and cellulose whiskers. *Polymer for Advanced Technologies*, 6:351–355.

Favier, V., H. Chanzy, and J.Y. Cavaillé. 1995b. Polymer nano-composite reinforced by cellulose whiskers. *Macromolecules*, 28:6365–6367.

Food and Agriculture Organization of the United Nations (FAO). 2011. *State of the World's Forests*. FAO, Rome, Italy.

Fukada, E. 2000. History and recent progress in piezoelectric polymers. *IEEE Transactions on Ultrasonics, Ferroelectrics, and Frequency Control*, 47(6):1277–1290.

Fukuzumi, H., T. Saito, T. Wata, Y. Kumamoto, and A. Isogai. 2009. Transparent and high gas barrier films of cellulose nanofibers prepared by TEMPO-mediated oxidation. *Biomacromolecules*, 10:162–165.

Gardner, D.J., Y. Han, and W. Song. 2008. Wood plastic composites technology trends. *Proceedings of the 51st International Convention of Society of Wood Science and Technology 2008*. November 10–12, Concepción, Chile.

Gröndahl, M., L. Erikson, and P. Gatenholm. 2004. Material properties of plasticized hardwood xylans for potential application as oxygen barrier films. *Biomacromolecules*, 5:1528–1535.

Gupta, B.S. and M.P. Laborie. 2008. Surface activation and adhesion properties of wood-fiber reinforced thermoplastic composites. *The Journal of Adhesion*, 83(11):939–955.

Habibi, Y., L.A. Lucia, and O.J. Rojas. 2010. Cellulose nanocrystals: Chemistry, self-assembly, and applications. *Chemical Reviews*, 110:3479–3500.

Hajny, G.J. 1981. *Biological Utilization of Wood for Production of Chemicals and Foodstuffs*. USDA Forest Service, Research Paper: FPL 385. Available on-line: Forest Products Laboratory, Madison, WI.

Heikkilä, H., M. Mänttäri, M. Lindroos, and M. Nyström. 2005. Recovery of Xylose. U.S. Patent 6872316B2.

Henriksson, M., G. Henriksson, L.A. Berglund, and T. Lindström. 2007. An environmentally friendly method for enzyme-assisted preparation of microfibrillated cellulose (MFC) nanofibers. *European Polymer Journal*, 43:3434–3441.

Herrick, F.W., R.L. Casebier, J.K. Hamilton, and K.R. Sandberg. 1983. Microfibrillated cellulose: Morphology and accessibility. *Journal of Applied Polymer Science*, 37:797–813.

Hetzer, M. and D. DeKee. 2008. Wood/polymer/nanoclay composites, environmentally friendly sustainable technology: A review. *Chemical Engineering Research and Design*, 86(10):1083–1093.

Hocking, M.B. 1997. Vanillin: Synthetic flavoring from spent sulfite liquor. *Journal of Chemical Education*, 74(9):1055–1059.

Hoeger, I., O.J. Rojas, K. Efimenko, O.D. Velev, and S.S. Kelley. 2011. Ultrathin film coatings of aligned cellulose nanocrystals from a convective-shear assembly system and their surface mechanical properties. *Soft Matter*, 7:1957–1967.

Holmbom, B. and E. Era. 1978. Composition of tall oil pitch. *Journal of the American Oil Chemists' Society*, 55(3):342–344.

Hubbe, M.A., O.J. Rojas, L.A. Lucia, and M. Sain. 2008. Cellulose nanocomposites: A review. *Bioresources*, 3(3):929–980.

Innami, S. and Y. Fukui. 1987. Additive composition for food or drugs. U.S. Patent 4,659,388.

Ioelovich, M. 2008. Cellulose as a nanostructured polymer: A short review. *Bioresources*, 3(4):1403–1418.

Iwamoto, S., A.N. Nakagaito, and H. Yano. 2007. Nano-fibrillation of pulp fibers for the processing of transparent nanocomposites. *Applied Physics A: Materials Science and Processing*, 89:461–466.

Jin, S. and L.M. Matuana. 2010. Wood/plastic composites co-extruded with multi-walled carbon nanotube-filled rigid poly(vinyl chloride) cap layer. *Polymer International*, 59(5):648–657.

Kamel, S. 2007. Nanotechnology and its applications in lignocellulosic composites, a mini review. *Express Polymer Letters*, 1(9):546–575. doi: 10.3144/expresspolymlett.2007.78.

Katzen, R. and D.J. Schell. 2006. *Lignocellulosic Feedstock Biorefinery: History and Plant Development for Biomass Hydrolysis in Biorefineries—Industrial Processes and Products.* Vol. 1, Kamm, B., Gruber, P.R. and Kamm, M. (eds.). Wiley-Vch, Weinheim, Germany, pp. 129–138.

Kim, J. and S. Yun. 2006. Discovery of cellulose as a smart material. *Macromolecules*, 39:4202–4206.

Kimura, F., T. Kimura, M. Tamura, A. Hirai, M. Ikuno, and F. Horii. 2005. Magnetic alignment of the chiral nematic phase of a cellulose microfibril suspension. *Langmuir*, 21:2034–2037.

Klemm, D., F. Kramer, S. Moritz, T. Lindstrom, M. Ankerfors, D. Gray, and A. Dorris. 2011. Nanocelluloses: A new family of nature-based materials. *Angewandte Chemie International Edition*, 50:5438–5466.

Kurian, J. 2005. A new polymer platform for the future—Sonora from corn derived 1,3-propanediol. *Journal of Polymers and the Environment*, 13(2):159–167.

Lahiji, R.R., X. Xu, R. Reifenberger, A. Raman, A. Rudie, and R.J. Moon. 2010. Atomic force microscopy characterization of cellulose nanocrystals. *Langmuir*, 26:4480–4488.

Lampinen, J. 2009. Biocomposite research as a way to add value and sustainability to composites. *Proceedings 4th Wood Fibre Polymer Composites International Symposium.* March 30–31, 2009, Bordeaux, France.

Lee, Y.Y. and T.A. McCaskey. 1983. Hemicellulose hydrolysis and fermentation of resulting pentoses to ethanol. *Tappi Journal*, 66(5):102–107.

Lichtenstein, A.H. and R.J. Deckelbaum. 2001. Stanol/Sterol ester—Containing foods and blood cholesterol levels. *Circulation*, 103:1177–1179.

Liimatainen, J., M. Karonen, J. Sinkkonen, M. Helander, and J.P. Salminen. 2012. Characterization of phenolic compounds from inner bark of Betula pendula. *Holzforschung*, 66:171–181.

Lima, M.M.D.S. and R. Borsali. 2004. Rodlike cellulose microcrystals: Structure, properties, and applications. *Macromolecular Rapid Communications*, 25:771–787.

Lima, M.M.D.S., J.T. Wong, M. Paillet, R. Borsali, and R. Pecora. 2003. Translational and rotational dynamics of rodlike cellulose whiskers. *Langmuir*, 19:24–29.

Lyytikäinen, K., E. Saukkonen, I. Kajanto, and J. Käyhkö. 2011. The effect of hemicelluloses extraction on fiber charge properties and retention behavior of kraft pulp fibers. *Bioresources*, 6(1):219–231.

Maki, K.C., M.H. Davidson, D.M. Umporowicz, E.J. Schaefer, M.R. Dicklin, K.A. Ingram, S. Chen et al. 2001. Lipid response to plant-sterol-enriched reduced fat spreads incorporated into a national cholesterol education program step 1 diet. *The American Journal of Clinical Nutrition*, 74:33–43.

Mathew, A.P. and K. Oksman Niska. 2011. Artificial ligaments/tendons based on cellulose nanofibers. *11th International Conference on Wood & Biofiber Plastic Composites and Nanotechnology in Wood Composites Symposium*, May 18, Madison, WI.

Matuana, L.M., S. Jin, and N.M. Stark. 2011. Ultraviolet weathering of HDPE/wood-flour composites coextruded with a clear HDE cap layer. *Polymer Degradation and Stability*, 96(1):97–106.

McGinnis, G.D. and F. Shafizadeh. 1980. *Cellulose and Hemicellulose in Pulp and Paper, Chemistry and Chemical Technology*, Casey, J.P. (ed.). Wiley & Sons, New York, Chapter 1: pp. 1–38.

Moon, R.J., A. Martini, J. Nairn, J. Simonsen, and J. Youngblood. 2011. Cellulose nanomaterials review: Structure, properties and nanocomposites. *Chemical Society Reviews*, 40:3941–3994.

Moran, J.L., V.A. Alvarez, V.P. Cyras, and A. Vazquez. 2008. Extraction of cellulose and preparation of nanocellulose from sisal fibers. *Cellulose*, 15:149–159.

Mosier, N., C. Wyman, B. Dale, R. Elander, Y.Y. Lee, M. Holtzapple, and M. Ladisch. 2005. Features of promising technologies for pretreatment of lignocellulosic biomass. *Bioresource Technology*, 96:673–686.

Murzin, D.Y., P. Mäki-Arvela, T. Salmi, and B. Holmbom. 2007. Catalytic transformation for production of fine chemicals and pharmaceuticals from wood-derived raw materials. *Chemical Engineering and Technology*, 30(5):569–576.

Nakagaito, A.N., M. Nogi, and H. Yano. 2010. Displays from transparent films of natural nanofibers. *MRS Bulletin*, 35:214–218.

Nakagaito, A.N. and H. Yano. 2004. The effect of morphological changes from pulp fiber towards nano-scale fibrillated cellulose on the mechanical properties of high-strength plant fiber based composites. *Applied Physics A: Materials Science and Processing*, 78:547–552.

Nyström, G., A. Razaq, M. Strømme, L. Nyholm, and A. Mihranyan. 2009. Ultrafast all-polymer paper-based batteries. *Nano Letters*, 9(10):3635–3639.

Okahisa, Y., A. Yoshida, S. Miyaguchi, and H. Yano. 2009. Optically transparent wood–cellulose nanocomposite as a base substrate for flexible organic light-emitting diode displays. *Composites Science and Technology*, 69:1958–1961.

Okita, Y., T. Saito, and A. Isogai. 2009. TEMPO-mediated oxidation of softwood thermomechanical pulp. *Holzforschung*, 63:529–535.

Oksman, K., A.P. Mathew, D. Bondeson, and I. Kvien. 2006. Manufacturing process of cellulose whiskers/polylactic acid nanocomposites. *Composites Science and Technology*, 66:2776–2784.

Oksman, K., A.P. Mathew, and M. Sain. 2009. Novel bioanocomposites: Processing, properties and potential applications. *Plastics, Rubber and Composites*, 38(9–10):396–405.

Olsson, R.T., M.A. Azizi Samir, G. Salazar-Alvarez, L. Belova, V. Ström, L.A. Berglund, O. Ikkala, J. Nogués, and U.W. Gedde. 2010. Making flexible magnetic aerogels and stiff magnetic nano-paper using cellulose nanofibrils as templates. *Nature Nanotechnology*, 5:584–588.

Pääkkö, M., M. Ankerfors, H. Kosonen, A. Nykänen, S. Ahola, M. Österberg, J. Ruokolainen et al. 2007. Enzymatic hydrolysis combined with mechanical shearing and high-pressure homogenization for nanoscale cellulose fibrils and strong gels. *Biomacromolecules*, 8(6):1934–1941.

Pääkkö, M., J. Vapaavuori, R. Silvennoinen, H. Kosonen, M. Ankerfors, T. Lindström, L.A. Berglund, and O. Ikkala. 2008. Long and entangled native cellulose I nanofibers allow flexible aerogels and hierarchically porous templates for functionalities. *Soft Matter*, 4:2492–2499.

Paralikar, S.A., J. Simonsen, and J. Lombardi. 2008. Poly(vinyl alcohol)/cellulose nanocrystal barrier membranes. *Journal of Membrane Science*, 320:248–258.

Peng, B.L., N. Dhar, H.L. Liu, and K.C. Tam. 2011. Chemistry and applications of nanocrystalline cellulose and its derivatives: A nanotechnology perspective. *The Canadian Journal of Chemical Engineering*, 89:1191–1206.

Pettersen, R.C. 1984. The chemical composition of wood. In R.M. Rowell (ed.), *The Chemistry of Solid Wood*. American Chemical Society, Washington, DC, Chapter 2: pp. 57–126.

Pietarinen, S.P., S.M. Willför, M.O. Ahotupa, J.E. Hemming, and B.R. Holmbom. 2006. Knotwood and bark extracts: Strong antioxidants from waste materials. *Journal of Wood Science*, 52:436–444.

Poirier, Y., C. Nawrath, and C. Somerville. 1995. Production of polyhydroxyalkanoates, a family of biodegradable plastics and elastomers, in bacteria and plants. *Biotechnology*, 13:142–150.

Pollak, O.J. 1953. Successful prevention of experimental hypercholesterolemia and cholesterol atherosclerosis in rabbit. *Circulation*, 7:696–701; Reduction of blood cholesterol in man. *Circulation*, 7:702–706.

Postek, M., A. András Vladár, J. Dagata, N. Farkas, B. Ming, R. Wagner, A. Raman et al. 2011. Development of the metrology and imaging of cellulose nanocrystals. *Measurement Science and Technology*, 22:024005.

Ranby, B.G. 1951. The colloidal properties of cellulose micelles. *Discussions of the Faraday Society*, 11:158–164.

Ranby, B.G. 1952. The cellulose micelles. *Tappi*, 35(2):53–58.

Revol, J.F., H. Bradford, J. Giasson, R.H. Marchessault, and D.G. Gray. 1992. Helicoidal self-ordering of cellulose microfibrils in aqueous suspension. *International Journal of Biological Macromolecules*, 14:170–172.

Revol, J.F. and R.H. Marchessault. 1993. In vitro chiral nematic ordering of chitin crystallites. *International Journal of Biological Macromolecules*, 15:329–335.

Richter, G.A. 1955. Production of high alpha-cellulose wood pulps and their properties. *Tappi*, 38(3):129–150.

Robyt, J.F. 1998. *Essentials of Carbohydrate Chemistry*. Springer-Verlag, New York, p. 143.

Rouskova, M., A. Heyberger, J. Triska, and M. Krticka. 2011. Extraction of phytosterols from tall oil soap using selected organic solvents. *Chemical Papers*, 65(6):805–812.

Rowe, J.W. 1989. *Natural Products of Woody Plants*, 2 vols. Springer-Verlag, Berlin, Germany.

Saito, T., M. Hirota, N. Tamura, S. Kimura, H. Fukuzumi, L. Heux, and A. Isogai A. 2009. Individualization of nano-sized plant cellulose fibrils by direct surface carboxylation using tempo catalyst under neutral conditions. *Biomacromolecules*, 10:1992–1996.

Sakurada, I., Y. Nukushina, and T. Ito. 1962. Experimental determination of the elastic modulus of crystalline regions in oriented polymers. *Journal of Polymer Science*, 57:651–659.

Schorger, A.W. and D.F. Smith. 1916. The galactan of *Larix occidentalis*. *Industrial and Engineering Chemistry*, 8(6):494–499.

Simmonds, F.A., R.M. Kingsbury, and J.S. Martin. 1955. Purified hardwood pulps for chemical conversion. *Tappi*, 38(3):178–186.

Siró, I. and D. Plackett. 2010. Microfibrillated cellulose and new nanocomposite materials: A review. *Cellulose*, 17:459–494.

Sixta, H., L. Testova, M. Alekhina, A. Roselli, O. Ershova, M. Tenkanen, H. Komu et al. 2011. Raw material pretreatment, fractionation and extraction. In *Bio Refine Yearbook* 2011, Tekes Review. Helsinki, Finland. pp. 17–29.

Smith, K.T. 1989. Vinsol resin and tall oil pitch, in naval stores. In D.F. Zinkel and J. Russell (eds.), *Naval Stores, Production, Chemistry, Utilization*. Pulp Chemicals Association, New York, pp. 715–737.

Smith, P.M. and M. Wolcott. 2006. Opportunities for wood/natural fiber-plastic composites in residential and industrial applications. *Forest Products Journal*, 56(3):4–11.

Sreekumar, P.A. and S. Thomas. 2008. Matrices for natural-fiber reinforced composites. In K.M. Pickering (ed.), *Properties and Performance of Natural-Fiber Composites*. Woodbridge Publishing Ltd., Cambridge, U.K. Chapter 2: pp. 67–126.

Stark, N.M. and L.M. Matuana. 2009. Co-extrusion of WPCs with a clear cap layer to improve color stability. *Proceedings of 4th Wood Fiber Polymer Composites International Symposium*, March 30–31, 2009, Bordeaux, France.

Steiner, C.S. and E. Fritz. 1959. Pharmaceutical-grade sterols from tall oil. *Journal of the American Oil Chemists Society*, 36(8):354–357.

Stelte, W. and A.R. Sanadi. 2009. Preparation and characterization of cellulose nanofibers from two commercial hardwood and softwood pulps. *Industrial and Engineering Chemistry Research*, 48:11211–11219.

Syverud, K. 2011. A novel application of cellulose nanofibrils for improving the water resistance of commercial paints. Presented at the *11th International Conference on Wood & Biofiber Plastic Composites and Nanotechnology in Wood Composites Symposium*, May 18, 2011, Madison, WI.

Syverud, K. and P. Stenius. 2009. Strength and barrier properties of MFC films. *Cellulose*, 16:75–85.

Taipale, T., M. Österberg, A. Nykänen, J. Ruokolainen, and L. Janne. 2010. Effect of microfibrillated cellulose and fines on the drainage of kraft pulp suspension and paper strength. *Cellulose*, 17:1005–1020.

Teters, C., W. Kong, M. Taylor, J. Simonsen, M. Lerner, T. Plant, and G. Evans. 2007. The cellulose nanocrystal electro-optic effect. Presented at the *International Conference on Nanotechnology for the Forest Products Industry*, June 13–15, Knoxville, TN.

Tijmensen, M.J.A. 2000. *The Production of Fischer-Tropsch Liquids and Power Through Biomass Gasification*. University of Utrecht, Utrecht, The Netherlands. NWS-E-2000-29, ISBN 90-73958-62-8. Doctoral thesis.

Timell, T.E. 1957. Carbohydrate composition of ten North American species of wood. *Tappi*, 40(7):568–572.

Timell, T.E. 1967. Recent progress in the chemistry of wood hemicelluloses. *Wood Science and Technology*, 1:45–70.

Traitler, H. and K. Kratzl. 1980. Investigations on tall oil of southern pine wood. Part 2: Tall oil pitch. *Wood Science and Technology*, 14:101–106.

Turbak, A.F., F.W. Snyder, and K.R. Sandberg. 1982. Food products containing microfibrillated cellulose. U.S. Patent 4,341,807.

Turbak, A.F., F.W. Snyder, and K.R. Sandberg. 1983. Microfibrillated cellulose, a new cellulose product: Properties, uses, and commercial potential. *Journal of Applied Polymer Science, Applied Polymer Symposium*, 37:815–827.

Wegner, T. and P. Jones. 2005. Nanotechnology for the forest products industry. *Wood and Fiber Science*, 37(4):549–551.

Wegner, T. and P. Jones. 2009. A fundamental review of the relationships between nanotechnology and lignocellulosic biomass. In Lucian, A.L. and Orlando, J.R. (eds.), *The Nanoscience and Technology of Renewable Biomaterials*. John Wiley & Sons Ltd, West Essex, U.K.

Werpy, T. and G. Petersen. 2004. *Top Value Added Chemicals from Biomass*. Vol. 1. US Department of Energy, National Renewable Energy Laboratory, Golden, CO, DOE/GO-102044-1992.

White, E.V. 1941. The constitution of arabo-galactan. I. The components and position linkage. *Journal of the American Chemical Society*, 63:2871–2875.

Williams, K. and B. Bauman. 2007. New technology for enhancing wood-plastic composites. *JCT CoatingsTech*, 4(8):52–57.

Winandy, J.E., A.W. Rudie, R.S. Williams, and T.H. Wegner. 2008. Integrated biomass technology: A future vision for optimally using. *Forest Products Journal*, 58(6):6–16.

Wise, L.E. and F.C. Peterson. 1930. The chemistry of wood, II—Water-soluble polysaccharide of western larch wood. *Industrial and Engineering Chemistry*, 22(4):362–365.

Wyman C.E., B.E. Dale, R.T. Elander, M. Holtzapple, M.P. Ladisch, and Y.Y. Lee. 2005. Coordinated development of leading biomass pretreatment technologies. *Bioresource Technology*, 96:1959–1966.

Yao, F. and Q. Wu. 2010. Coextruded polyethylene and wood-flour composite: Effect of shell thickness, wood loading, and core quality. *Journal of Applied Polymer Science*, 118:3594–3601.

Yun, G.Y., H.S. Kim, J. Kim, K. Kim, and C. Yang. 2008. Effect of aligned cellulose film to the performance of electro-active paper actuator. *Sensors and Actuators A*, 141(2008):530–535.

Zakzeski, J., P.C.A. Bruijnincx, A.L. Jongerius, and B.M. Weckhuysen. 2010. The catalytic valorization of lignin for the production of renewable chemicals. *Chemical Reviews*, 110:3552–3599.

Zhu, J.Y., R. Sabo, and X.L. Luo. 2011. Integrated production of nano-fibrillated cellulose and biofuel (ethanol) by enzymatic fractionation of wood fibers. *Green Chemistry*, 13(5):1339–1344.

Zimmermann, T., E. Pohler and T. Geiger. 2004. Cellulose fibrils for polymer reinforcement. *Advanced Engineering Materials*, 6(9):754–761.

Zinkel, D.F. and J. Russell. 1989. *Naval Stores*. Pulp Chemicals Association, New York.

Zuluaga, R., J.L. Putaux, A. Restrepo, I. Mondragon, and P. Ganan. 2007. Cellulose microfibrils from banana farming residues: Isolation and characterization. *Cellulose*, 14:585–592.

7

Chinese Era

Xiaozhi (Jeff) Cao, Xiufang Sun, and Ivan Eastin

CONTENTS

China has important implications to international forest products trade and markets given its sheer size of exports and domestic demand backed by a fast-growing middle class. Due to historical and social factors, the Chinese forest sector differs from those in the Western world in the way how it is organized and governed. Understanding the Chinese forest sector's "unique" socioeconomical background and recent developments will provide useful insights for forest products marketing researchers and professionals. In this chapter, we provide an overview of the Chinese forest sector and discuss implications from four major perspectives: China as a raw material supplier, China as a global supply chain partner, China as a consumer, and China as a global investor.

7.1 China in Transition

Since Deng Xiaoping, the late leader of the Communist Party of China, launched his "Reform and Opening" program in 1978, China's economy has grown between 4% (in 1989) and 15% (in 1984) per year, and by 2010, China had replaced Japan to become the world's second largest economy in terms of purchasing power parity (PPP) and nominal GDP* (The World Bank 2012a).

China's economic reforms began in the agriculture sector with the introduction of the household responsibility system in the early 1980s, where individual households and farmers were allowed to contract land from people's communes and were free to choose which crops to grow, a process known as "decollectivization." Following initial success

---

* PPP and nominal GDP as both measures for the size of an economy have a major distinction in that PPP adjusts for currency rate difference while nominal GDP does not. Therefore, in international market research, the two measures are often used together to compare economic power across countries or regions.

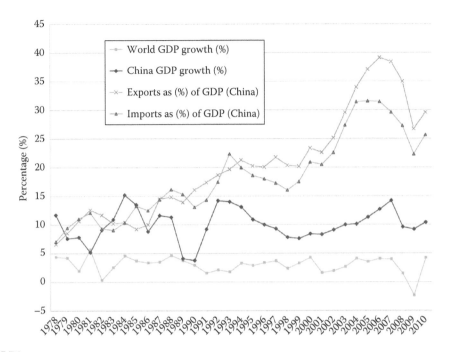

FIGURE 7.1
China's GDP and international trade growth. (From The World Bank, Databank: China, The World Bank, Washington, DC, 2012a, http://data.worldbank.org/country/china, accessed May 3, 2012.)

in the agriculture sector, economic reforms were gradually extended to other sectors of the national economy, including the manufacturing and service industries. In the meantime, policies were implemented that were designed to encourage investment in, and expansion of, foreign direct investment (FDI), international trade, and privatization and corporatization of state-owned companies. International trade, as a major indicator of China's openness, has experienced rapid expansion together with its economic growth. Exports as a percentage of GDP increased from less than 7% in 1978 to 30% in 2010, while imports' share of GDP also increased from 7% to over 25% over the same time period. China's accession into the World Trade Organization (WTO) in 2001 helped the country gain greater access to international markets. Both shares of imports and exports reached their peak between 2006 and 2007 before they declined due to combined impacts of the global economic downturn and the Chinese government's efforts to boost domestic demand (Figure 7.1).

As part of the broader reforms aimed at the decollectivization of China's rural landscape and the establishment of a market-driven economy, major reforms in the forest sector were initiated in the early 1980s. These changes reflected the government's desire to shift its focus from direct control over the forest industry through state ownership to encourage market-based competition and diversity in the ownership of the industry (Zhang 2008). The implementation of these reforms led to large increases in productivity. As a result, the forest sector has experienced unprecedented high rates of growth as seen in the boom of new factories and retail outlets, as well as the number of employees. In 2009, the forest industry's total output grew by over 23% to reach Chinese yuan (CNY) 1.8 trillion (US$265 billion) or about 6% of national GDP (SFA 2011a) compared to unofficial estimates of about 2% in 1978. The forestry sector has become a major employer in the national economy and is expected to provide 57 million jobs by 2012, 30 million of which will be created within the afforestation and the forest management sector. This has important implications to rural development and will also help absorb China's huge surplus of rural labor (SFA 2009a). China became the world's largest importer of wood products in 2010. In 2011, China was the leading importer of a broad range of primary wood products including logs and lumber, and it was the largest exporter of finished wood products such as wood flooring and wooden furniture (GTA 2012). The gradual integration of the Chinese forest products industry into the global supply chain has encouraged Chinese manufacturers to begin to follow international standards and pay increasing attention to the "responsible," "legal," and "sustainable" aspects of business development at the international scale. Privatization and FDI are driving major restructuring and consolidation within the forest industry, particularly in the processing and retail sectors. Meanwhile, China's outbound investments are rapidly increasing and driving the multinationalization of Chinese firms through mergers and acquisitions (China Daily 2011a).

Amid fast economic growth, social and cultural aspects of development have also been strengthened as part of industry's "soft power." Social interest groups (e.g., NGOs, trade associations, and community groups) are more active and involved in policy development. Multistakeholder consultation processes are increasingly being adopted by the government in the development of policy and standards during the course of forest tenure reform. Wood culture heritage preservation and cultural infrastructure development, particularly in rural areas, have received increasing support by the government, industry, and media (IWCS 2012).

The Chinese government is more open to engaging in international dialogues and partnerships designed to address significant ecological and environmental challenges, including climate change and illegal logging. China has signed more than 30 international environmental treaties, including, Convention on Biological Diversity, Convention on International Trade in Endangered Species of Wild Fauna and Flora (CITES), United Nations Framework Convention on Climate Change (UNFCCC), United Nations Convention to Combat Desertification (UNCCD), and International Tropical Timber Agreement (ITTA) (MOEP 2013). China has established cooperation relationships with a third of the countries of the world as well as with dozens of international organizations such as the IUCN. China has also signed 37 national-level cooperation agreements in the forest sector and 10 intergovernmental conventions with 8 countries (SFA 2011a).

However, the Chinese forest sector, which is often referred to as one of the "last castles of planned economy" in China, is still in the transition to become more market oriented. As a result, there are significant differences in the ways in which the Chinese forest industry is organized and governed relative to major industrialized countries. For example, the Chinese government still retains significant authority and responsibility over the forest products industry, international trade, and forest resource management through state-owned forest enterprises and policy directives such as annual harvest quotas. A wide variety of policies ranging from direct export subsidies to firms to maintaining an undervalued currency, and lax environmental regulations and enforcement, have long been considered to provide Chinese forest products companies with a low-cost competitive advantage in international trade. This transitional status also affects the way that nongovernmental organizations (NGOs) operate in China. Although NGOs are taking an increasingly active role in China in supporting policy development, building industry's capacity to meet international standards, and educating consumers to purchase "green" and "legal" forest products, the overall impact of NGOs (particularly foreign NGOs) remains limited, largely due to China's lack of explicit policies regarding the registration and operation of NGOs within the country. As a result, although many foreign NGOs have found alternative ways to register and operate in China (e.g., as a foreign consulting or training firm), their ability to work and their fund-raising capacity have been greatly restricted by Chinese laws for foreign businesses.

Overall, the Chinese forest sector has experienced significant developments in all aspects over the past three decades, thanks to Deng's "Reform and Opening" program. The forest sector will be expected to play a more important role as China moves forward toward a low-carbon economy (SFA 2011a). As the industry continues along the path of increasing internationalization, with the expansion of economic, social, and political reforms that this process brings, there remain many challenges, uncertainties, and opportunities ahead for China.

7.2 China as Raw Material Supplier

7.2.1 China's Forest Resources

China has 206.8 million hectares (ha) of forests, which is roughly the size of Mexico or 50 times the size of Switzerland. China is the fifth largest country in terms of forested area after Russia, Brazil, Canada, and the United States; the sixth largest in terms of growing forest stock (14,648 million cubic meters) after Brazil, Russia, the United States, the Democratic Republic of Congo, and Canada; and the ninth largest in terms of living forest carbon stock (6,203 million tons) after Brazil, Russia, Democratic Republic of the Congo, United States, Canada, Indonesia, and Peru. Despite this prominent ranking, China continues to run a forest and timber deficit because of its huge population and rapid economic and infrastructural development. Forests cover about 22% of China's total land area, which is lower than the world average (31%), and China's per capita forest area is below 0.15 ha per person, about one-fourth of the world average (0.62 ha per person) (FAO 2010a).

Since the early 1970s, the State Forestry Administration (SFA) of China has been conducting a national forest resources inventory survey every 5 years to monitor forest developments and key trends at the national and provincial levels (Table 7.1). According to the latest (seventh) survey results released in 2009 (SFA 2009), China's natural forests cover

TABLE 7.1

Inventory Data for China National Forest Resources

Inventory Report	Year	Forest Area (100 Million Hectares)	Forest Stock (100 Million Cubic Meters)	Forest Coverage (%)
1st report	1973–1976	1.22	87	12.7
2nd report	1977–1981	1.15	90	12
3rd report	1984–1988	1.25	91	12.98
4th report	1989–1993	1.34	101	13.92
5th report	1994–1998	1.59	113	16.55
6th report	1999–2003	1.75	125	18.21
7th report	2004–2008	1.95	137	20.36

Source: SFA 2009b forest resources statistics, 1–7 inventory reports.

119.7 million ha with a stocking volume of 11,402 million cubic meters, while plantation forests cover an additional 61.7 million hectares with a stocking volume of 1,961 million cubic meters. China's forest resources can be divided into two general categories: public welfare forests and commercial forests. Public welfare forests can be further divided into protective forests (including water, soil conservation, and fire protection forests) and special-purpose forests (including scientific, education, and environmental protection forests). Commercial forests include timber forests, firewood forests, and economic forests (or nontimber forests). Total forest area is almost evenly distributed between public welfare forests and commercial forests, with a ratio of 52%–48%, respectively (SFA 2009).

The majority of China's forests are located in five regions, accounting for 85% of total forestland. These five regions are the northeast Inner Mongolia forest region (including Heilongjiang, Inner Mongolia, and Jilin provinces), the southwest mountainous forest region (including parts of Yunnan province, Sichuan province, and the Tibet Autonomous Region), the southeast low mountain and hilly forest region (including the provinces of Jiangxi, Fujian, Zhejiang, Anhui, Hubei, Hunan, Guangdong, Guangxi, Guizhou, Sichuan, Chongqing, and Shaanxi), the northwest mountainous forest region (including parts of Xinjiang, Gansu, Ningxia, and Shaanxi province), and the southern tropical forest region (including parts of Yunnan, Guangdong, Guangxi, and Hainan provinces and the Tibet Autonomous Region) (Figure 7.2). These forestlands are owned by either the state or local

FIGURE 7.2
China's regional forest distribution. (From SFA 2009b. The seventh national forest resources inventory report. Beijing, China: The State Forestry Administration.)

communities (villages), while trees can be owned by private enterprises and individuals, in addition to the state-owned forest farms, forestry bureaus, and collective farms (FAO 2010a). Roughly 40% of China's forests are state owned, while the remaining 60% are largely collectively owned by local communities (e.g., villages) (SFA 2009; Xu et al. 2010). After decades of forest tenure reforms (see Section 7.2.3), the area of privately owned forests, as well as the total forestland area, has increased significantly. Between 2003 and 2008, the area of privately owned forests almost doubled, from 34 to 62 million hectares, second only to that of state-owned forests (77 million hectares), while the area of collective forests declined from more than 68 million hectares to less than 58 million hectares over the same period of time (FAO 2010a).

The major timber species (groups) that grow in China include Chinese oak (*Quercus* spp.), fir (*Abies* spp.), spruce (*Picea* spp.), larch (*Larix* spp.), birch (*Betula* spp.), China fir (*Cunninghamia* spp.), Masson pine (*Pinus massoniana*), Yunnan pine (*Pinus yunnanensis*), poplar (*Populus* spp.), and Fujian cypress (*Fokienia hodginsii*). These species cover 86.2 million hectares in area and represent 7,603 million cubic meters of timber volume (SFA 2011a).

7.2.2 Forest Governance and Domestic Timber Production

China's forest governance system has its roots in former Soviet Union traditions that were transferred largely unchanged to China after 1949. At the national level, the State Council is the chief administrative authority in China. The SFA, which reports to the State Council, is responsible for developing forestry policies, including setting harvest quotas and supervising law enforcement. Forestry bureaus at the provincial, district/county, and township levels are responsible for developing local timber harvest quotas and managing all license-based forest harvesting, transportation, management, and processing activities. Figure 7.3 outlines this governance structure in a simplified format.

There are more than 60 national laws, regulations, ministerial rules, and technical guidelines and standards that are available for forest management, landscape, afforestation, cultivation, water resources, labor protection, timber harvesting, processing, distribution, and

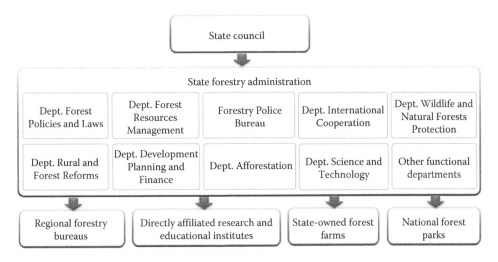

FIGURE 7.3
Forestry governance structure in China. (From State Forestry Administration. http://www.forestry.gov.en.)

trade of forest products. Two major national forestry laws are (1) *the Forest Law*, which was initially adopted in 1984 and revised in 1998 by the National People's Congress, and (2) *the Implementation Regulation of the Forest Law*, which was adopted in 2000 by the State Council. Other regulations, including the Regulations for Forestry Administrative Punishment Procedure, Guidelines for Forestry Administrative and Law Enforcement Supervision, and Regulations for Forestry Administrative Punishment and Hearing were established to provide a legal basis for protecting the rights and interests of forest communities and other parties (FAO 2010b).

The annual harvest quota system (HQS) is a major forest policy that was adopted after China's reform and opening and the HQS directly impacts commercial timber production in China. Adopted in 1987, HQS requires logging operations to apply for harvest licenses in order to cut trees or bamboo except when they are grown on rural residents' private plots or around houses and yards. HQS is based on data collected from state-owned forest management units and township-based collectively or privately owned forest management units. Forestry bureaus at the provincial level pool and audit these data before reporting the information to SFA. Every 5 years, SFA develops a national harvest quota based on the updated provincial-level forest statistics. After the national quota has been approved by the State Council, SFA allocates timber harvest quotas to the provinces and forestry bureaus of major timber-producing regions, which are then further divided and allocated at the county and township levels. Local forestry administrators are responsible for issuing logging licenses according to the quota that was allocated to their region. According to *the Forestry Law*, every shipment of timber must be accompanied by three permits—a timber harvest permit, a quarantine permit (certificate), and a transportation permit—as timber is taken out of the forests and before it enters the processing sector. These three permits are key documents necessary for establishing the chain of custody (CoC) for legal timber production in China. Law enforcement is carried out by forestry administrations and forestry police bureaus at the national and regional levels. Timber inspection stations are located in forested areas, ports, processing zones, and timber markets in order to enforce these regulations. In addition, SFA operates 13 forestry supervising commission offices in forest-rich regions in order to supervise the local utilization of forest resources.

Domestic timber production has been suppressed by a "logging ban" that was adopted in 1998 when the central government significantly reduced timber harvests in natural forest areas and greatly increased the area of forest designated as "protection" forests. The ban was implemented after China was hit by a series of devastating floods that cost thousands of lives and forced millions to leave their homes in the late 1990s. Between the periods of the fifth National Forest Inventory survey (1994–1998) and the sixth National Forest Inventory survey (1999–2003), the area of timber-producing natural forests was reduced by nearly 20 million hectares, or about 26%.

The control on domestic timber production was gradually eased after 2003 and reforms were made to the HQS to encourage timber production from plantations. Primary reform measures included making a distinction between ecological/public welfare forests and plantation/commercial forests. While maintaining the overall scheme of logging quota allocation, the new HQS allows more flexibility in allocating logging quotas within the plantation/commercial forests.

The current logging quota stipulated by the twelfth 5-Year Annual Logging Quota Plan, which governs the period of 2011–2015, has several points worth noting:

- The 5-year overall timber production quota is 258 million cubic meters, representing over 4% increase over the preceding 5-year quota.

- Guangxi province reports the biggest increase in harvest quota, making it the largest timber-producing province in China, followed by Yunnan and Fujian provinces. Guangxi's timber production is expected to increase by 47% to reach 36.8 million cubic meters during the 5-year period, or 14% of the total national timber harvest (Guangxi Forestry Bureau 2011).
- Plantation timber supplies, led by poplar (*Populus*) and eucalyptus (*Eucalyptus*) species, will contribute significantly to the domestic timber supply during this period (Flynn 2008).

The post-1998 HQS and policy changes clearly signal the Chinese government's intention to protect natural forests and to increase the role that plantations play in helping to achieve self-sufficiency in timber supply over the long term. But these measures have also triggered some concerns in the short term. The logging ban has led to a surge in the volume of timber imported by China due to the shortage of domestic timber supplies. It is estimated that imports account for at least 70% of China's total timber consumption, and the supply gap is projected to increase by at least two-thirds by 2015, at which time domestic plantations will have become mature enough to provide significant volumes of timber (Flynn 2008). China's heavy dependence on overseas timber, and the lack of a reliable and consistent supply of domestic timber, has become a major source of industry insecurity, contributed to a surge in timber commodity prices globally and driven by an increasing number of Chinese firms to invest in offshore forest resources in Russia, Africa, and Southeast Asia. In the meantime, environmental groups have become increasingly concerned that China's increasing demand for timber may further fuel illegal logging activities in source countries (Lague 2004). Domestically, illegal logging activities as well as social conflicts associated with the logging ban and forest tenure reform have also been reported (Feng 2009). For example, employment in China's forest industries dropped by 30% (or nearly 1 million jobs) in 1998 following the implementation of the logging ban (ILO 2003), despite the government's efforts to create job opportunities through afforestation/reforestation activities, developing community forest enterprises for nontimber forest products and encouraging forest tourism as alternative ways to support economic activity within local forest-dependent communities. It is widely supported in economic theory that environmental regulations have a positive effect on overall employment in the long run since they help generate new jobs elsewhere within the economy, although local job losses can be significant in the short term (Goodstein 1994). Therefore, national forest policy will need to address the local problem of economic dislocation to take into account both long-term economic gains and short-term economic losses (Deng 2003).

7.2.3 Afforestation and Reforestation Programs

Afforestation and reforestation are at the center of post-1998 forestry policies in China. In 1998, the Chinese government introduced a new sustainable forest management and environmental protection program that included six major forestry initiatives at a cost of US$85 billion through 2010. This program included the following components:

1. The Natural Forest Protection Program (NFPP) (often referred to as the logging ban)
2. The Land Conversion from Farmland back to Grassland and Forestland Program (the so-called Grain for Green Program)
3. The Program for Sandstorm Source Control in the Beijing–Tianjin region

4. The Shelterbelt Forest Development Program in the Three-North, Yangtze River Valley, and other Regions
5. The Program of Establishment of Fast-Growing and High-Yielding Timber Plantations in Key Areas
6. The Wildlife Protection and Nature Reserve Development Program

These forestry initiatives covered almost all of China's provinces and targeted nearly 80 million hectares of land for afforestation. In 2000, China formally launched the NFPP following a 2-year trial. NFPP was implemented between 2000 and 2010 in 734 counties and 167 forest enterprises in 17 provinces, including along the upper reaches of the Yangtze and Yellow Rivers and in the Northeast and Inner Mongolia regions, which have traditionally been major state-owned forest regions. The program requested a complete stop of all logging activities in natural forests along the middle and upper reaches of the Yangtze and Yellow Rivers, reduced logging in the natural forest in northeast China, and led to the transformation of forest logging enterprises into forest protection and tending companies.

During the first phase of NFPP (2000–2010), the Chinese government invested 111.9 billion CNY (about US$17 billion). In 2010, the Chinese government announced its decision to extend the NFPP to a second phase extending from 2011 to 2020. The second phase of the program will incorporate an additional 11 counties into the program with a total planned investment of 244 billion CNY (some US$37 billion). The first phase of NFPP resulted in a net increase of 10 million hectares of forest area and 725 million cubic meters of forest stocks in the project regions that resulted in a 3.7% increase in forest cover. It is expected that by the end of the second phase (in 2020), the NFPP will increase forest area by an additional 4.2 million hectares and increase the forest stock by 1.1 billion cubic meters (Zhao 2012).

The Land Conversion Program was implemented in 1999 and covered Beijing, Tianjin, Hebei, Shanxi, Inner Mongolia, Liaoning, Jilin, Heilongjiang, Anhui, Jiangxi, Henan, Hubei, Hunan, Guangxi, Hainan, Chongqing, Sichuan, Guizhou, Yunnan, Tibet Autonomous Region, Shaanxi, Gansu, Qinghai, Ningxia, and Xinjiang, a total of 25 provinces (autonomous regions, municipalities directly under the central government), and the Xinjiang Production and Construction Corps, involving 1897 counties (including municipalities, districts, and flags). The program regions were grouped into 10 categories, including high mountains and deep valleys in the southwest; mountainous and hilly areas in Sichuan, Chongqing, Hubei, and Hunan; low mountains and hilly areas in the middle and lower reaches of the Yangtze River; the Yunnan–Guizhou plateau; the hilly mountains in Hainan and Guangxi; the alpine steppes and meadows in the origins of Yangtze River and Yellow River; the arid desertification region in Xinjiang; the loess hills and valleys; the arid and semiarid areas in north China; and the mountains and sandy lands in northeast China.

The objectives and tasks of the program for 2010 included afforestation of 14.7 million hectares of reclaimed land, afforestation of 17.3 million hectares of land in barren mountains and barren land (both components included the pilot sites of land conversion from 1999 to 2000) suitable for growing trees, conversion of steep farmland and severe sandy farmland into forestland, and increase in the coverage of forest and grasses in the program regions by 4.5%. The ecological status of these regions and forestlands will be improved by this program (SFA 2010).

To implement the program, the government will provide subsidies for food and living allowances to all farmers who participate in land conversion activities. The subsidies provided to farmers participating in the program will include 2250 kg of food per year (unprocessed raw grain) for each ha of converted farmland in the Yangtze River basin and southern China and 1500 kg of food per year (unprocessed raw grain) for each ha of converted land in the Yellow River basin and northern China. Since 2004, the food subsidy has been changed to a cash subsidy (in principle) with the central government providing a subsidy calculated at 1.4 CNY for each kg of food (unprocessed raw grain) to the various provinces (autonomous regions and municipalities directly under the central government) on a contract basis. Specific subsidy criteria and methods of providing the subsidy were determined by the provincial governments according to their local conditions. The annual subsidy for living allowances was also established at 300 CNY/ha for land conversion. The duration for paying the food and living allowance subsidies was 5 years for land converted to grassland between 1999 and 2001, 2 years for land converted to grassland after 2002, 5 years for land converted to fruit trees, and 8 years for land converted to ecological forest. In 2007, the central government decided to extend these subsidies. Farmers will continue to get 1575 CNY annually for every ha of land converted from farmland to forest or grassland in the Yangtze River basin and southern China, and 1050 CNY for each ha of converted land in the Yellow River basin and northern China. The annual living allowance will remain at 300 CNY for each ha of converted land and the duration of these subsidies will remain the same (SFA 2010).

Subsidies for the costs of seeds, planting stock, and planting were also provided by the government to farmers who conducted land conversion. The subsidization criteria were set at 750 CNY/ha for the conversion of farmland and for planting trees in barren mountains and wastelands suitable for growing trees.

Through the Conversion of Farmland to Forest Program, a total of 28.4 million hectares was converted to forests, including 9.3 million hectares of forest from converted farmland and 19.2 million hectares of forest through afforestation of barren land and mountains and on mountains closed off for reforestation. The SFA reports that the program benefited some 124 million farmers (Zhao 2012).

The central government has spent 233.2 billion CNY (more than US$31 billion) on the 27.7 million hectares of new forests planted between 1999 and 2009. Given the success of this program, the SFA announced that the central government will earmark an additional 200 billion CNY (US$33 billion) toward afforestation programs through the end of 2021. These afforestation programs are also considered to be part of China's commitment to address climate change (Sun and Canby 2011).

Increasing the area of forest plantations has been an important part of SFA's overall strategy to increase domestic timber production and reduce China's reliance on imported timber products. Plantation forests now account for 38% of the total forest area in China, and the volume of timber harvested from plantation forests accounts for nearly 40% of the total harvest (SFA 2009b), although plantation quality and production volumes may well be lower than the official estimates. Many analysts suspect the ability of the Chinese plantations to supply domestic demand and therefore meet the government's ambitious goals of self-sufficiency, without further motivating small forest communities and low-income forest farmers to improve forest productivity (Xu 2011). There is also a growing consensus among Chinese government officials and researchers that the current approach to forest development has its limits. Nevertheless, with the rapid development of plantations and the gradual relaxation of the logging quota, China now has an opportunity to boost its domestic timber production in both public and collective forests (Sun and Canby 2011).

7.2.4 Forest Tenure Reforms

Under the broader trend of reforms aimed at the establishment of a market-driven economy, forestland ownership reform has been at the top of the agenda of Chinese forestry policy makers since the 1980s (SFA 2011b). Different from forestland ownership reforms as seen in many countries, China's forest tenure reform started from collective forests owned by local communities that account for 58% of China's total forestland area, instead of from state-owned forests. Also, Chinese reform is targeted at establishing private household ownership as the means of incentivizing farmers (Xu et al. 2010).

China's collective forest tenure reform has impacted approximately 100 million hectares of forests and 400 million people and therefore is considered to be one of the world's largest forestry reforms in modern times in terms of both the area and population impacted (Xu et al. 2010). Collective forest tenure reform was officially launched in July 2008, with the release of the "Guidelines on Fully Promoting Collective Forest Tenure System Reform." The reforms of the collective forest tenure system are focused on devolving land-use rights and forest ownership to individual households. In order to provide incentives to farmers to plant and sustainably manage trees, the government provides use rights of collective forest land to farmers so that they have full ownership of the trees grown on their contracted forestland. These reform measures served largely as an equalizer of opportunity and welfare between farmers living in heavily afforested areas and standard agricultural areas. To date, collective forest tenure reforms have been carried out across the country in 30 provinces (or cities/prefectures) involving 147 million farmer households and 510 million people—which represents 70% of the rural population. By the end of 2010, the total area of forestland with clearly defined forest tenure reached 162 million hectares accounting for 89% of the total area of collective forestland. More than 63 million certificates have been issued and 300 million farmers have benefited from the program (SFA 2011b). According to a recent study, the forest tenure reform program has yielded many benefits, including increased farmer's income and the reforestation area in many provinces, and helped to reduce fire incidents (Xu et al. 2010).

Tenure reform of state-owned forests is still in an experimental phase. Most pilot projects are being implemented in northeast China, which is home to over 90% of the state-owned forests and the largest state-owned forest enterprises (e.g., Jilin Forest Industry Group and Longjiang Forest Industry Group). Though there is no clear indication when or if the central government will officially introduce market-oriented reforms to state-owned forests and state-owned forest enterprises, many in the industry believe this will take place unavoidably as the forest reforms go further. Figure 7.4 provides a snapshot of daily life of local residents of the last wooden village in Changbai Mountain, which is located on the border of China and North Korea. Local forests are managed mainly by Baihe Forestry Bureau, which is part of state-owned Jilin Yanbian Forest Industry Group.

Despite the fact that many benefits have been achieved during the process of implementing forest tenure reforms, some problems have also been exposed. For example, there has been a lack of multistakeholder participation and transparency in the lease procedures. In some cases, village representative committees (or village assemblies) lease collective forestland to private companies without fully consulting with local villagers (He 2008). Once collectively owned forestlands are put under private management, conflicts between local communities (farmers) and private contractors seem to be unavoidable. Negative reports about private companies' abuse of forest rights and unfair rents have been well documented. For instance, Greenpeace China recently released a report (Ma and Liu 2009), accusing a large private forest company of misconduct and creating conflict with local

FIGURE 7.4
Villagers in Changbai Mountain, northeast China's Jilin province. (Photo courtesy of International Wood Culture Society (IWCS), Wood culture symposium and tour at Changbai Mountain, China, March 2012, http://www.iwcs.com/?p=blog&sp=content&bid=23, accessed May 3, 2012.)

communities during the process of land rent and timber purchasing in Hainan, Guangxi, Guangdong, and Yunnan provinces. As forest tenure reforms continue, further technical, financial, and policy support will be needed to help farmers to improve the efficiency of forest management, obtain legal harvest documents, establish processing facilities, and develop market channels for timber and nontimber forest products and eventually to make community-based forest operations more sustainable.

7.2.5 Forest Certification

Forest certification is typically conducted by independent third-party organizations to verify that forests are well managed and a CoC system is in place for tracking wood fiber materials in compliance with certain standards. Through labeling of finished products, forest certification can serve as a marketing tool to help companies gain competitive advantage in environmentally sensitive markets, therefore providing incentives for forest companies to adopt good and responsible forest management and sourcing practices.

Three forest certification standards are currently operational in China: the Forest Stewardship Council (FSC), the Programme for the Endorsement of Forest Certification (PEFC), and the China Forest Certification Council (CFCC) standards, and all of them are growing quickly. While each scheme has followed quite different development paths, they share many similarities. For example, each scheme takes a three-tier approach to assessment: principles, standards, and criteria.

The Chinese national scheme evolves from the Chinese industry standards for forest management and CoC that were developed by the China Forest Certification Council (CFCC), a standards development organization under the SFA. The CFCC scheme also draws heavily on the internationally developed Montréal Process Criteria and Indicators, just as the international schemes FSC and PEFC.

In China, standards development and accreditation is centrally controlled by the government. The Chinese State Administration of Quality Supervision and Inspection and Quarantine (AQSIQ) is the top administrative agency overseeing national standards development and certification in China. Under AQSIQ, China National Certification and Accreditation Administration (CNCA) and the Standardization Administration of the People's Republic of China (SAC) are two key functional departments responsible for certification, accreditation, and standards development and approval, respectively. Industry-specific ministries such as the SFA and the Ministry of Agriculture also play an important role in developing national standards for responsible industries. On the subnational level, China Inspection and Quarantine (CIQ) bureaus in each province and city are responsible for supervising local certification activities and regional standards development.

It is only recently that international voluntary standards such as FSC and the Marine Stewardship Council (MSC) have begun to play a more active role in China, as supported by international NGOs (Cao et al. 2012). Since the late 1990s, forest certification has grown rapidly in terms of the area of forests covered and the number of participating companies, largely driven by factors in the developed export markets (e.g., the United States, Europe, and Japan) (Yuan and Eastin 2007). It is estimated that as of 2011, approximately 2% of the forest area in China had been certified as being sustainably managed, twice the area that was certified in 2010. Similarly, the number of companies with CoC certification increased 10-fold between 2006 and 2011 and now exceeds 2000 companies. In terms of market share, FSC is the leading certification program, thanks in large part to the capacity building efforts of international NGOs such as World Wildlife Fund's Global Forest and Trade Network (WWF-GFTN) program as well as global retailers' support. The other two certification programs in China—PEFC and CFCC—which are supported by different interest groups, are becoming increasingly popular due to perceived advantages that include low cost, a larger and more reliable supply of raw materials, and government support (the latter of which is a key success factor for doing business in China). The newly developed CFCC program is expected to be fully implemented in 2012, and the Chinese government attaches great importance to the future success of CFCC as supplement to the annual HQS (Wang 2010).

At the same time, forest certification in China is facing many challenges: lack of market incentives and unreliable supplies of raw materials have been identified as major bottlenecks to greater acceptance of forest certification. As a result of unreliable supplies and the higher cost of certified wood, many certified companies don't produce any certified products (Cao et al. 2011). Poor law enforcement and corruption are reported in some areas, causing difficulty for forest farmers and processors to obtain the legal documentation necessary for meeting certification requirements (Anonymous 2011). Also, while the Chinese government encourages Chinese exports to meet international standards, it sets restrictions for international standards and foreign certification bodies to certify domestic-selling companies in China. According to a recent regulation issued by CNCA (2012), foreign standards will be required to go through a "localization" process to adapt their standards to the Chinese context, thereby essentially becoming a domestic standard. In many cases, foreign standards can also enter domestic Chinese market by endorsing local standards as observed in the organic and agriculture fields. This regulation reflects the Chinese government's intention to become a "game changer" by setting its own standards to compete with international schemes as China gains economic power (Long et al. 2010). This also reflects the government's intention to help more Chinese CBs obtain international accreditation to operate as local subcontractors. In this regard, PEFC's future endorsement of the CFCC certification program, which is widely expected to occur soon, will provide further synergies for both certification programs to grow within China and internationally. FSC, along with several other foreign standards organizations (including Fair Trade Standards (FLO)), has been officially approved by CNCA to operate in China, but they have been restricted to certifying only exports at this stage.

For Chinese certified exports to gain more popularity and consumer trust in international markets, the industry and authorities will need to take active measures to address credibility and transparency issues. Low transparency in public reporting, the exclusion of public organizations from the standards setting process, conflicts of interest, gaps between domestic and international standards, and misconducts by auditors are often reported and have led to major legislative actions in overseas markets (*Lincoln Journal Star* 2010).

7.3 China as Global Supply Chain Partner

China has become a major manufacturing hub for forest products, with employment in the Chinese primary and secondary wood–processing sectors expected to reach 11 million people in 2012 (SFA 2009). Between 1995 and 2011, China's total wood products exports (including wooden furniture) increased by an average of 16% per year, from US$1.8 to US$22.7 billion (GTA 2012). Meanwhile, China's wood products imports (including wooden furniture) grew by nearly 15% annually from $1.6 to $16.3 billion over the same period (Figure 7.5). It is estimated that wood imports have accounted for at least 70% of China's annual timber consumption since the early 2000s (Flynn 2008). In 2010, China's timber supply deficit was estimated to be 160–180 million cubic meters, and the gap is projected to increase to 300 million cubic meters by 2015, when Chinese plantations and natural forests are expected to reach full production (Flynn 2008). Russia is by far the largest timber supplier to China. In 2011, China imported over 20 million cubic meters of logs and lumber from Russia (valued at US$3.5 billion), over 60% of which was logs (GTA 2012). Increases in the Russian log export tax since 2008 have led many Chinese wood companies to set up sawmills just across the border in Russia. As a result, Russia's lumber exports to China have surged from less than 2 million cubic meters in 2008 to over 6 million cubic meters in 2011 (GTA 2012). During the same period of time, log imports from New Zealand, United States, and Canada increased from 2.5 million cubic meters to 15.6 million cubic meters as Chinese companies diversified their log suppliers (GTA 2012). Following Russia's accession into the WTO in December 2011, industry analysts expect the Russian government to lower log export tariffs while simultaneously imposing a log export quota. As a result, the general trend of China's increasing Russian lumber imports and diversification of timber suppliers is unlikely to change (Ekstrom 2012).

7.3.1 Key Factors Impacting Industry's Competitiveness

During the industry's early stage of development, generous government subsidies (including direct export subsidies such as export tax rebates and indirect export subsidies such as low-cost land rent and low-cost loans for the purchase of buildings and equipment as well as an undervalued currency and lax environmental and worker safety regulations), combined with abundant low-cost labor, provided Chinese manufacturers with a significant competitive advantage in international markets based on a strategy of low-cost

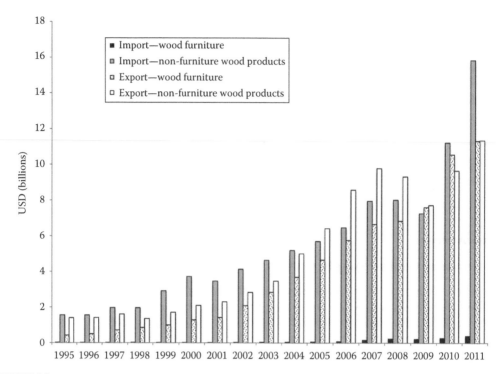

FIGURE 7.5
China's international trade of wood products, 1995–2011. (From GTA, *Global Trade Atlas,* Global Trade Atlas, Columbia, SC, 2012.)

leadership. However, over time and as the Chinese economy grew, the cost advantage of Chinese wood manufacturers began to decline. Beginning in the early 2000s, the Chinese government began removing and reducing export subsidies in order to control the growth of "high-pollution" and "resource-intensive" industries, including the wood manufacturing industry.

The CNY was pegged to the US dollar (USD) between 1994 and 2005 before China's central bank adopted a more flexible currency policy in July 2005 and allowed the CNY to appreciate. Between July 2005 and April 2012, the value of the yuan increased by over 30% against the US dollar, rising from roughly 1 USD = 8.3 CNY to 1 USD to 6.3 CNY. The impact of China's changing currency policy on the Chinese wood manufacturing sector has been diverse. An appreciating CNY has given Chinese buyers more purchasing power to import raw materials and invest overseas. But at the same time, it has made Chinese exports less competitive internationally due to increased prices while, at the same time, it is more difficult for domestic companies to compete with imported products that become relatively less expensive for Chinese consumers.

Labor and raw material costs account for a significant proportion of total production costs for manufactured wood products. Rising costs of labor and raw materials have direct impacts on the competitiveness of Chinese-manufactured wood products. Take the furniture sector as an example. Industry sources estimate that the labor cost of furniture manufacturers has increased by 20% since mid-2009. Many workers were laid off during the global recession in 2008 and early 2009, and when the market began to recover in the second half of 2009, many wood manufacturers in the east and coastal regions of China found they could not rehire enough workers, especially skilled workers. Increasingly, wood furniture manufacturing companies have been migrating into the Chinese interior areas like Sichuan province as well as into lower-cost countries such as Vietnam, Laos, and Cameroon to access lower-cost labor and natural resources. Rough estimates by furniture manufacturers suggest that raw materials account for around 50% of their total production cost and wood flooring manufacturers tend to have a higher raw material cost ratio. One large wooden flooring manufacturer (A&W Wood Products, Ltd., formerly known as Shanghai Anxin Flooring) reported that the cost of wood raw materials represents approximately 70% of their total production costs. The impact of rising costs is particularly significant to small- and medium-sized enterprises (SMEs) that account for about 90% of the industry's total population (Luo et al. 2009). In addition, forest products SMEs in China are also constrained by increasingly stringent environmental and legal timber trade regulations both domestically and internationally. Scholars and industry managers have called on the government and private sectors to provide more support to SMEs to help them build up their capacity to meet international standards and trade regulations (Zeng 2011).

FDI has contributed significantly to the industry's capacity expansion and export growth since the early 1990s. For example, between 2006 and 2007, Shanghai-based A&W Wood Products, Ltd. (A&W), obtained over $50 million in investment from overseas private equity firms and investment bankers including the Carlyle Group and Pantheon Venture, Inc., to facilitate the acquisition of a decoration material company and the construction of a new manufacturing facility to produce engineered wood flooring (EWF). Carlyle Group also used its distribution network to help A&W to launch its branded products into the US market. By 2007, A&W had successfully set up 50 sales outlets in the United States with Carlyle's help. The wood-based panel (WBP) sector in China has experienced significant consolidation and privatization as foreign investors set up joint ventures and use their international distribution networks to improve market share. For many foreign investors, the fast-growing secondary wood processing sector in China is a more attractive investment than timberland, given the quick turnover of the investment and the difficulty of securing a controlling stake in integrated state-owned enterprises where the Chinese government retains a tight control over forest resources.

At the same time, driven by the need for capacity expansion, many large Chinese firms are going public through initial public offerings (IPOs) and reverse mergers in international markets. The US market is the most popular choice for Chinese companies to raise capital and increase their company profile. However, following a string of recent accounting scandals that led to strengthened regulations and investigations of overseas-listed Chinese firms, the number of Chinese firms going public on the US market has dropped sharply since 2011 (Marvin and Brown 2011). According to the Bloomberg China-US 55 index of the most-traded Chinese stocks in the US market, Chinese stocks listed in New York fell by about 8% in 2011 after steep rises in 2009 and 2010 (Bloomberg 2012). The problem, although not unique to Chinese companies, suggests the urgent need for Chinese companies to improve accounting and corporate governance systems as they expand into global financial markets.

Meanwhile, Chinese enterprises are beginning to exploit new comparative advantages and actively participate in offshore development of forest resources, wood products processing enterprises, and wood-processing machinery repairing and gradually develop into multinational companies. According to the SFA, at the end of 2004, more than 40 overseas forestry projects had been approved. The total value of overseas investment was estimated to be $US650 million and involved the employment of over 3000 people, mainly in

Russia, Brazil, Gabon, Papua New Guinea (PNG), Suriname, Cambodia, Myanmar, Laos, and Malaysia. China's outbound investment will be discussed further in Section 7.6.

Overall, China's forest products industry has become more efficient and profitable over the past decade thanks in large part to continuous improvements in technology and innovations in product design and business strategies (Cao and Hansen 2006). The average productivity of the Chinese timber processing sector has more than doubled from CNY 11,800 per worker in 2000 to CNY 23,700 per worker in 2009, while the average profitability of the timber processing sector jumped by nearly 70% over the same time period, rising from 2.9% to 4.8% (Cao 2011). The rest of this section will provide a closer look at the major processing sectors of the wood industry.

7.3.2 Industry Profile of Major Wood Products Sectors

China is the leading producer and exporter of manufactured wood products, including WBPs, wooden furniture, plywood, wooden flooring, musical instruments, and a variety of other wood building products and handicrafts (Figure 7.6). The primary wood–processing sector in China is highly fragmented with many small-sized logging companies, sawmills, and plywood mills located in over 20 provinces (IBISWorld 2011a). By comparison, the secondary wood–processing sector is relatively concentrated within three economic development zones: the Pearl River Delta, the Yangtze River Delta, and the Bohai Sea Ring Area. These economic development zones are located in close vicinity to the Guangzhou, Shanghai, and Beijing metropolitan areas, respectively.

7.3.2.1 Logs and Sawnwood

Korean pine (*Pinus koraiensis*), Scots pine (*Pinus sylvestris*), white pine (dragon spruce, *Picea asperata*, and Faber's fir, *Abies fabri*), and dahurian larch (*Larix gmelinii*) harvested from the state-owned forests in the northern region of China (including the provinces of Hebei,

FIGURE 7.6
Musical instrument factory in eastern China.

Shandong, Heilongjiang, and Inner Mongolia) and Chinese fir (*Cunninghamia* spp.), Masson pine (*P. massoniana*), and Yunnan pine (*P. yunnanensis*) harvested from the collectively-owned forests in southern China (including the provinces of Jiangsu, Zhejiang, Guangdong, and Fujian) are the primary timber species harvested for wood production (SFA 2011a).

It is estimated that there are about 2,800 logging companies and over 200,000 sawmills in China (IBISWorld 2011a). The largest logging companies, such as Longjiang Forest Industry Group, China Inner Mongolia Forest Industry Group, Greater Higgnan Mountains Forestry Group Co., Ltd., and Jilin Forest Industry Group, tend to be state-owned enterprises or have the government as a majority shareholder. These state-owned forest product companies account for about 15% of the total output in the forest products sector (IBISWorld 2011a). In contrast, it is estimated that at least 90% of China's 200,000 sawmills are privately owned and most of them are small, inefficient mills using outdated production technology. In 2010, China's total sawnwood production reached 37.7 million cubic meters (about 40% of which was from softwood species), doubling the production volume of 2005 (FAO 2012).

7.3.2.2 Wood-Based Panels

China is the world's largest manufacturer of WBPs in terms of total production volume. Between 2000 and 2010, China's WBP production increased from 19 to 103 million cubic meters, representing an average annual growth rate of 16% (FAO 2012). Plywood, fiberboard, and particleboard combined accounted for over 90% of total WBP production in China over the last 10 years.

WBPs utilize a variety of wood materials, including wood by-products derived from other wood-manufacturing processes. Therefore, the production of WBPs is an efficient and cost-effective way to use waste wood material and conserve timber resources. In the face of raw material supply constraints (particularly a shortage of large-diameter trees and tropical timber imports), central government policy directives have been developed to favor the production of particleboard and medium-density fiberboard (MDF)/high-density fiberboard (HDF) over plywood. As a result, the relative proportion of MDF/HDF, particleboard, and plywood production has changed from 13:19:68 in 2000 to 40:13:47 in 2010 (FAO 2012). This change has forced plywood manufacturers to shift from producing commodity plywood products toward the production of higher value-added plywood products, including thick plywood, film-overlaid plywood, container decking, laminated veneer lumber (LVL), fire-retardant plywood, and exterior plywood. The particleboard sector is experiencing a similar value-added trend toward the production of thinner and more even-density products that can be used for furniture production and interior decoration. Domestic oriented strand board (OSB) production is growing steadily though slowly due to the absence of industry standards, effective quality control procedures, and rising competition from substitute products (such as thick plywood).

The rapid development of the Chinese woodworking machinery industry has facilitated the transition in the WBP's sector by providing new manufacturing technologies that facilitate the production of specialty products at competitive costs. However, compared to their US and European counterparts, the average productivity and capacity per production line within the Chinese panel industry remains low. This can be attributed to the fact that most panel manufacturers continue to use domestically produced multiopening presses designed to produce commodity-type products rather than using imported continuous press technologies. An exception to this trend includes a few leading panel manufacturers such as Dare Wood-Based Panel Group Co., Ltd., which is China's largest WBP manufacturer with headquarters in Fujian province. The company has eight manufacturing facilities across the country, including seven fiberboard production lines and one particle-board production line, with a total annual production capacity of over 2.5 million cubic meters (Dare 2012).

7.3.2.3 Wood Flooring

Wood flooring is a rapidly growing manufacturing sector in China. According to SFA statistics, between 2000 and 2010, total wood flooring production in China increased from 108 to 479 million square meters. The wood flooring sector consists of three main types of wood flooring: laminated wood flooring (LWF), solid wood flooring (SWF), and EWF. The strict domestic logging ban, in conjunction with diminishing supplies of large-diameter tropical timber, has provided an impetus for the rapid development of EWF products (such as LWF and EWF). In 2000, LWF and SWF production accounted for almost all wood floor-ing production in China. However, since 2000, the production of EWF has increased rap-idly from 10 million square meters to almost 118 million square meters in 2009. Under this trend, the relative proportion of LWF, SWF, and EWF production changed from 48:42:10 in 2000 to 39:25:36 in 2009. Solid bamboo flooring production also increased rapidly from 3 million square meters in 2000 to over 39 million square meters in 2010.

The wood flooring manufacturing sector in China is highly fragmented. According to the China Forest Products Industry Association's Flooring Committee, there are approxi-mately 4000 SWF manufacturers in China, producing between 70 and 80 million square meters of SWF annually. While there a number of SWF manufacturing clusters located in Zhejiang, Jiangsu, and Shandong provinces in east China, Nanxun, a small town with a population of just 200,000 located in Zhejiang province, is the major manufacturing base for SWF. Major manufacturers of wood flooring in China include China Flooring Holding Co., Ltd. (or Nature Flooring), Power Dekor Group, and the state-owned Jilin Forest Industry Group. However, given the fragmented nature of the wood flooring sector, these three companies accounted for less than 10% of total wood flooring production in China. Recent raw material cost increases, combined with the cooling of the housing industry, have forced many small and medium-sized flooring companies to close, and it is expected that the flooring sector will experience further consolidation in the future favoring large, integrated companies who can exert direct control over their raw material supplies and establish direct market channels.

7.3.2.4 Treated Wood

Over the past decade, China's rapid urbanization and booming housing industry have created a tremendous demand for treated wood in outdoor applications, includ-ing parks, walkways, and balcony decking for high-rise apartments (Cao et al. 2007). The treated wood market in China is dominated by Scots pine (*P. sylvestris*) and south-ern yellow pine (from the southern United States). Between 2005 and 2011, China's total imports of southern yellow pine logs from the United States grew from 2,546 to 111,070 m^3, while imports of untreated southern yellow pine lumber grew from to 6,938 to 89,033 m^3 (GTA 2012). In contrast, total imports of treated softwood lumber (primar-ily southern yellow pine) from the United States only increased from 2265 m^3 in 2007 to 4009 m^3 in 2011. Recent declines in US timber prices due to the impact of the housing crisis, plus an appreciating Chinese yuan relative to the US dollar, have continued to help Chinese buyers source untreated wood raw materials from overseas for use within the wood treating industry in China. Unfortunately, the quality and performance of treated wood produced in China is poor. The lack of strict product standards, which are essential for ensuring product quality and performance, are lacking in China and Chinese treated wood manufacturers are largely left to regulate themselves. This lack of regulatory oversight has created a race to the bottom in terms of quality control as manufacturers cut corners to reduce costs and undercut competitors' prices. The desire to low-cost treated lumber has not only eroded product quality and durability, but the environmental standards and treating procedures at a number of local treating plants are dismal as well (Cao et al. 2012).

7.3.2.5 Wooden Furniture

China is the largest manufacturer and exporter of wooden furniture in the world. In 2011, Chinese exports of wooden furniture exceeded US$10 billion, accounting for 30% of the world's total wooden furniture exports (GTA 2012). The United States is the biggest overseas market for Chinese wooden furniture, importing roughly one-third of China's total furniture exports over the past decade. Since the US government imposed antidumping tariffs on Chinese wooden bedroom furniture exports in December 2004, Chinese exporters have increasingly diversified their export markets to include emerging economies in the Middle East, and more recently, Russia and India, as well as increasing their sales within the domestic market. At the same time, many Chinese and foreign-invested companies have established furniture manufacturing facilities in Vietnam and other low-cost countries to avoid the US antidumping tariffs. Taisheng Furniture (commonly known as Lacquer Craft in the United States) is perhaps the largest wood furniture exporter to the United States with annual export sales exceeding $100 million at its peak. This Taiwanese-owned company has acquired several factories and retail brands in the US market. Similarly, Markor Furniture Group, one of the largest softwood furniture producers and exporters in China, has formed a strategic partnership with Ethan Allan (United States) designed to expand the market for their furniture products in both countries.

The wooden furniture sector in China is highly privatized compared to companies operating in the forest management and primary processing sectors. There are about 4000 wooden furniture companies operating in China, and most of them are small- to medium-sized operations with annual sales of less than US$36 million (CNY ¥300 million). Privately owned or foreign-invested companies represented over 80% of the sector's total production in 2011. Large state-owned integrated forest products companies such as Jilin Forest Industry Group also produce wooden furniture but have not yet gained a strong foothold in the marketplace due in part to intensive price competition (IBISWorld 2011b).

In 2011, China produced more than CNY314 billion (US$48 billion) worth of wooden furniture, which accounted for 60% of China's total furniture shipments by value or 30% by piece (CNFA 2011). Wooden furniture exports exceeded US$11 billion, about one-third of the furniture industry's total furniture exports (CNFA 2012). Hardwood lumber and veneer and WBPs are the primary wood raw materials used to produce wood furniture, and Malaysia, Indonesia, Russia, and the United States are the top hardwood product suppliers to China. Relative to hardwoods, softwoods are used to a much lesser extent in furniture production in China, although the volume of softwood use is growing, thanks in large part to outsourcing by large foreign retailers, including IKEA (Sweden) and Walmart (United States). The Pearl River Delta is China's most important manufacturing and export region for the wood furniture sector (CNFA 2011).

7.3.3 Distribution Channels

Chinese companies often have higher distribution costs relative to those in developed countries due to low distribution efficiency and the industry's fragmented nature. It is reported that distribution costs account for 18% of China's GDP compared to 11% in Japan and 7% in the EU (Li 2011).

According to industry estimates, China has about 2500 large wholesale timber markets that sell a wide variety of primary and secondary processed wood products. About 90% of these markets are traditional "booth-rental" style markets comprising a large number of small vendors who lack business plans, have a limited sales reach, and provide little after-sales service (Cao et al. 2006). It is estimated that wholesale timber markets that sell primary wood products (e.g., logs, lumber, veneer, and plywood) account for two-thirds of total sales of logs and wooden building and decoration materials in China. Around CNY 200 billion (roughly US$25 billion) in wooden building materials are sold in these markets annually. Major timber markets are generally located near manufacturing areas in the larger cities (e.g., Shanghai, Guangzhou, and Beijing). While wholesale markets do not offer the product warranties and after-sales service that home centers or major distributors may offer, they will continue to be a major supplier of building materials products due to their price competitiveness. Many wooden building material manufacturers also supply products directly to housing developers and builders on a project basis.

For many overseas suppliers of logs and lumber who don't have a local presence, this market channel typically involves distributors who carry inventory. Imported logs and lumber go through both water and land routes into China. Major land ports include Manzhouli and Suifenhe along the China–Russia border in Heilongjiang province (Figure 7.7). Major water ports include the port of Taicang and the port of Zhangjiagang in Jiangsu province and the port of Putian in Fujian province. The port of Taicang, China's largest water port and the major distribution center for imported logs and

FIGURE 7.7
Russian log importing gateway in northeast China.

lumber, is located about 60 miles from Shanghai. In 2011, the port of Taicang handled over 4 million cubic meters of timber shipments, mostly from Russia, New Zealand, Canada, and the United States.

7.4 China as a Consumer

China's domestic consumption of wood commodity and consumer products has been largely driven by its rapid urbanization and the recent privatization of the housing sector. Over the longer term, an aging society, as well as a fast-growing, brand-conscious middle class, will be important trends to monitor. These two social changes are expected to have a huge influence on the role of domestic consumption within the Chinese economy.

7.4.1 China's Housing Reforms

Up until the late 1990s, Chinese citizens lived in small, modest-quality publicly owned apartments that were provided by either the government or their employer. Residents didn't have full property rights, including the right to resell their apartments on the open market. Therefore, residents had little or no incentive to invest in remodeling and improvements in their housing units. Faced with the need to boost the economy and reduce the economic cost of providing and maintaining public housing, in 1997, the central government began to phase in privately owned housing. Aided by one-time government subsidies designed to help individuals buy their own homes, the privatization policy has led to a huge consumer demand for larger and better quality homes. In urban areas, aging 3–4 story walkup apartments are rapidly being razed and replaced with high-rise condominiums and apartments. Between 1999 and 2009, China's housing industry experienced a "golden decade" of development and added more than 550 million square meters of new residential floor space in urban areas annually. By the end of 2009, per capita residential floor space in urban areas exceeded 31 square meters compared to less than 10 square meters m^2 per capita in early 1990s. Meanwhile, the rapid urbanization process saw the gap between the urban and rural populations decline from 507 million in 1995 to just 44 million in 2011 (CDC 2012).

Beginning in 2010, the Chinese government adopted a series of policies aimed at cooling the "overheated" housing industry. Measures adopted to control investment within the housing sector included requiring relatively large down payments to qualify for housing mortgages and limiting the number of homes that a person can purchase to discourage speculation and flipping of homes. These policies have begun to have an effect in the large cities, and new home prices have begun to fall in the 10 biggest cities since September 2011. By November 2011, the average price of a new home had fallen by 0.2%–0.4% in the 100 biggest cities (*China Daily* 2011b). Many industry experts believe that the post-"golden decade" housing sector will see a continuous decline in home prices across China for some time. In the meantime, the government continues to encourage developers to provide housing for low-income urban households, and a large volume of these housing projects have begun in 2010. The most recent 5-year plan calls for the construction of 36 million affordable housing units in urban cities. This focus on affordable housing will continue to support the housing sector and has led some industry observers to predict that the strong domestic demand for furniture and wooden building products will continue for the next 5–10 years, driven by China's continuing urban development as well as a second round of interior renovations as the increasingly affluent middle class begins to replace furniture and refurbish condominiums and homes purchased during the "golden decade."

7.4.2 Demographic Trends

China is experiencing a demographic transition of historic proportions, caused by a combination of a declining birth rate, a smaller average family size, a rapidly rising senior population, an influx of migrant workers to urban areas, and a rapid expansion of the middle class. These changing demographic trends will significantly impact the economy and the mix of goods and services needed within China.

China's population reached 1.35 billion in 2011, more than doubling the level of 1949. In 1979, the Chinese government adopted its one-child policy that lowered the birth rate from 3.3% in 1970 to below 1.2% in 2011. Correspondingly, the average family size has also declined from 4 in 1990 to 3.1 in 2011 (CDC 2012). Analysts expect household size to continue to decline as the number of 2–4 person households increases and the number of multigeneration households declines.

Although the Chinese population is expected to continue growing before peaking at 1.45 billion in 2030, China's labor market will see the ratio of new entrants to retirees decline starting from mid-2010s as the number of retirees exceeds the number of new workers entering the labor force (Figure 7.8). China's working-age population (15–64) is 990 million in 2012 and will increase by less than 0.5% per year over the next 14 years before peaking at just over 1 billion in 2026. By 2050, the Chinese labor force will drop to 850 million (The World Bank 2012b). With continued development of the interior region of China and increasing urbanization in second- and third-tiered cities, a reverse flow of migrant workers has occurred in the major manufacturing areas of eastern China (e.g., the Pearl River

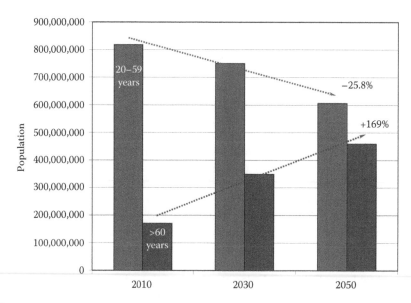

FIGURE 7.8
The number of workers in China will decline while the number of retirees will grow rapidly. (From United Nations, Child dependency ratio, Database maintained by the United Nations Department of Economic and Social Affairs, http://esa.un.org/wpp/JS-Charts/aging-child-dep-ratio_0.htm, accessed May 8, 2012, 2011.)

Delta in Guangdong). In these areas, factories are reporting that they are experiencing a labor shortage in the range of 2 million workers. This trend has provided further impetus for companies to migrate into the interior region of China.

An aging society will have a profound impact on the Chinese economy and society over the long term (The World Bank 2012b). Due to the "one-child" policy, China has one of the most rapidly aging societies in the world, and it is likely that China will become an aging society before it becomes a rich society. The proportion of those aged 65 and older increased from 5.6% of the population in 1990 to 8.9% in 2011, in contrast to a substantial drop in the 0–14 age group that declined from 27.7% in 1990 to 16.6% in 2011 (CDC 2012). The younger generation, particularly the so-called little emperors (children who are doted upon by their parents and grandparents as a direct result of the "one-child" policy), has become very brand conscious, creating a market for high-priced, international products such as IKEA and Nike. However, China's growing number of seniors could moderate the growth of this high-end of the consumer market as younger generations face increasing economic pressure to care for their elderly parents; a critical component of the traditional Confucian value of filial piety in China.

During China's transition to a consumption-driven economy, the Chinese domestic market will offer good opportunities for environmental products and technologies. China's middle class is estimated to be around 300 million (around a quarter of the population), which is roughly equivalent to the population of the United States. By 2020, the Chinese middle class is expected to reach 700 million (around 50% of the total population) (Credit Suisse 2011). The scale of this rapid expansion of the middle class is unprecedented in human history and will have dramatic implications for the marketing of wood products in China. Wealthy Chinese consumers traditionally favor dark-colored wood products and are willing to buy premium products such as rosewood furniture and foreign brands that enhance their social status (Kaplinsky et al. 2010). With growing concerns of wooden furniture and flooring that emit toxic off-gassing and contribute to indoor air pollution, Chinese consumers are increasingly favoring "green products." However, relative to Western consumers, the definition of "green products" to Chinese consumers may refer more to "trustworthy," "fashion," "high quality," "safe," and "healthy," and less to low pollution, sustainability, or nature conservation (Mol 2012).

7.5 "Green" Drivers in China

Policy directives, export market demand, and emerging Chinese consumer's demand for "green" and "healthy" products have provided strong incentives for Chinese forest products companies to adopt environmental strategies. Chinese manufacturers have begun to develop new technologies that reduce formaldehyde emissions and other health risks in their products, to reduce energy consumption and carbon dioxide emissions within their manufacturing operations, and to raise their corporate reputations with the public. Unlike businesses in developed economies that are widely thought of as responding to the threat of a consumer boycott and environmental groups' campaigns to ensure that businesses abide by stakeholder's wishes (Coddington 1993; Kochan and Rubinstein 2000), Chinese companies consider government (particularly local government) as one of their most important stakeholders. Government attitudes, as well as a company's relationship with the government (particularly with the local government), have a strong influence on business decision making. According to a Fortune China's corporate social responsibility survey, government policies and guidelines were identified as the most powerful drivers of industry's corporate social responsibility decisions (Wickerham and Zadek 2009).

China's government procurement sector, which currently accounts for roughly 20% of total domestic consumption, will provide good opportunities for green products. Both the Ministry of Finance and the Ministry of Environmental Protection released green procurement policies in 2006 designed to encourage the use of environmentally friendly building products certified to China Environmental Labeling Standards (CELS)—the country's largest eco-labeling program that was established in 1994. The CELS Program consists of more than 70 industry standards, including standards relating to wood and wood panel products, paint, and adhesives that are relevant to furniture and wood products manufacturers. So far, over 1600 firms have been certified under this program. It is estimated that the annual sales of eco-labeled products in China exceeds CNY 100 billion yuan (approximately US$15 billion) (Wu 2010). While the Chinese government procurement sector is based on competitive tendering procedures, it remains largely closed to overseas tenders. Though China has been a member of the WTO since 2001, the Chinese government has been hesitant to sign the WTO's government procurement agreement due to the relative standards set by Western countries and the potential impact of this on domestic standards (*China Daily* 2012).

In the building sector, China's green building program, though growing slowly, will help further expand domestic demand for green building products and energy-efficient technologies (Eastin et al. 2011). China's first green building standard "Evaluation Standard for Green Building" (GB/T 50378-2006), also referred to as the three-star system, was introduced in 2006 and is administered by the China Green Building Council. This standard complements Building Research Establishment Environmental Assessment Method (BREEAM) and Leadership in Energy and Environmental Design (LEED), which presently are used in China for office buildings for multinational firms or upscale apartments (Eastin et al. 2011). Although the three-star system green building standard does not specify the use of wood products, the Chinese government has shown a growing interest in wooden building materials and wood-frame construction technologies for low-rise public buildings in earthquake zones and as a strategy to increase the energy efficiency of these buildings. In 2010, the Chinese Ministry of Housing signed a memorandum of mutual understanding with the government of British Columbia, Canada, which includes a cooperative project to build a six-story wood-frame apartment building using materials and technologies provided by forest products companies in British Columbia (Canada Wood 2011). This project will provide Chinese government officials, builders, architects, and developers with the opportunity to evaluate the energy and carbon savings, seismic safety, speed of construction, and cost comparability of wood-frame construction relative to concrete and steel construction technologies. It has been reported that the government is likely to apply mandatory green building standards to all public building construction and residential building renovation projects that favor the use of energy-saving technologies, equipment, and materials. In large cities such as Beijing, the local government is expected to enforce GB requirements for all new high-end residential projects with a focus on energy-saving technologies. Leading developers such as Vanke Group have made the development of GB technologies and products a corporate long-term strategy and have set targets to gradually increase the proportion of GB projects undertaken every year. In January 2012, Vanke Group became a member of WWF's GFTN network in China and announced a new sourcing policy in favor of the use of certified wood products in its housing projects.

Industry demands for certified wood products are still largely driven by export market demand and emerging international trade regulations such as the amended Lacey Act and the recently adopted EU timber regulations (Cao et al. 2011). Seeking to improve environmental performance and reputation, an increasing number of global retailers are working with NGOs to pressure manufacturers to adopt better environmental management practices and obtain forest and/or CoC certification in a stepwise approach. Kingfisher (B&Q's mother company) is a good case in point. According to its CSR report (2007/2008), Kingfisher has set a corporate goal of having 75% of the timber volume sold in its stores be proven as coming from well-managed forests or from recycled sources by 2010/2011. These goals can be achieved in three ways, as specified by Kingfisher's "Steps to Responsible Growth" program: using known sources, using legal sources, and using forest certification. So far, 71% of the timber products that Kingfisher sourced globally have met the first step of requirement. Similar policies have also been adopted by other global retailers, including Walmart, Unilever, IKEA, Home Depot, and Staples.

Major international financing institutions such as the World Bank and the Asia Development Bank have put in place "green credit" policies (e.g., using certification programs to qualify for low interest loans) to encourage leading manufacturers to improve their environmental and social performance. China Flooring Holding Co., Ltd. (also known as Nature Flooring) is one of the largest flooring manufacturers in China. With an ambitious plan to grow into a world-class flooring manufacturer, the firm has committed to a US$120 million investment between 2007 and 2010 to expand its production facilities, develop plantation forests in China, and establish an integrated supply chain that is certified via the FSC program. Much of the funding for this expansion is being provided by the Morgan Stanley Asian Fund and the International Finance Corporation (IFC)-World Bank's member group. As part of the financing agreement, Nature Flooring has promised to increase its use of certified wood in its manufacturing process, with a long-term goal of reaching 100% use of certified wood. The firm's progress on these targets will be regularly assessed by the China Forest Trade Network (CFTN), the national body of the WWF's GFTN (IFC 2009). Similarly, the Chinese government also offers financial incentives designed to encourage banks to shift their financing focus from high-pollution and/or energy-consuming projects to "clean" projects that favor the use of energy-saving and emissions reduction products or technologies. The policy was jointly developed by the State Environmental Protection Agency (SEPA), the People's Bank of China (PBOC), and the China Banking Regulatory Commission (CBRC) in 2007.

Nongovernmental–nonprofit organizations are also playing an important role in the creation and promotion of green programs. The Nature Conservancy (TNC), WWF's GFTN, The Forest Trust (TFT), the Rainforest Alliance (RA), and Greenpeace are among the most active international NGOs in China. WWF-GFTN and TFT are both membership based organizations that focus on building market linkages and supply chain development for member companies. RA and TNC are nonmembership-based organizations. In terms of focus, TNC and WWF have much broader involvement in environmental issues, covering not only forestry but also watershed, wildlife, and other environmental and social issues. RA and TFT, as their names suggest, are mainly focused on the forest sector but with relatively deeper connections with the industry through certification/legality verification programs. Despite having different approaches and foci, like-minded NGOs are working closely with the Chinese government and industry in a broad range of environmental programs, including marketing and policy campaigns, transfer of sustainable forest management expertise, auditor training, "green" supply chain development, and certification/verification services.

7.6 China's Go Global Strategy and ODI

The Go Global Strategy (also referred to as the Going Out Policy) was an effort initiated in 1999 by the Chinese government to encourage Chinese companies to invest overseas in developing resources deemed as being in short supply within the country. The policy was made official in the tenth 5-year plan (2001–2005) and started the trend of multinationalization of Chinese companies. Outbound investment projects are administered through the Ministry of Commerce (MOFCOM) and the State Administration of Foreign Exchange. MOFCOM, together with the Ministry of Foreign Affairs, develops guidelines for industries looking to make external investments by country. Meanwhile, the National Development and Reform Commission (NDRC) and the Export–Import Bank of China provide a support system of loans for outbound investments and offer preferred rates for key projects recommended by the government (The Bank of Tokyo-Mitsubishi UFJ Ltd 2011).

The MOFCOM China Companies Overseas Investment Database contains the approved list of Chinese companies who were planning to make overseas green land investments between 1983 and 2010. The database shows that 15,895 Chinese companies declared their intention to invest overseas. Besides green land investment, there have been a large number of overseas merger and acquisition activities related to Chinese companies in recent years. In 2010, China's outbound direct investment (ODI) in the nonfinancial sector jumped by 36% to $59 billion, moving China to ninth place in the global investment rankings. In contrast, global FDI shrank by 40% in 2009 as a result of the global financial crisis. While this trend in overseas investment is predicted to continue, Chinese overseas investment will be increasingly focused on M&A, which are expected to be led by investments by state-owned enterprises. In 2010, China's overseas investment through M&A was US$23.8 billion, or 40.3% of the total, compared with US$19.2 billion (34%) in 2009. Most of these M&A projects were in the mining, manufacturing, and power supply sectors (CDC 2012).

China's overseas investment has been directed toward the Asia-Pacific region and Oceania, although it is expected that investment in the United States, EU, and Latin America will experience rapid growth. Chinese investment in Africa has also risen substantially, and the MOFCOM database shows that 36 forestry investment projects have been implemented in 14 African countries by 2010, covering forest management, logging, WBP, furniture, wooden flooring, and paper production. The main investment destinations for Chinese forestry companies in Africa are Gabon, Nigeria, the Republic of Congo, Ghana, and Zambia, with investment in Gabon accounting for more than 1/3 of the total investment. Gabon is an important timber supplier for China, and its rich forest resources and comparatively stable political environment are attractive to Chinese forestry companies. There are many big Chinese forestry companies operating in Gabon (e.g., Wenzhou Timber Group, Jiangsu Shengyang, and Sunly Wood), and these companies export a large volume of African logs to China (Huang and Wilkes 2011b). With Gabon's introduction of a log export ban in 2011, many Chinese companies are expected to establish wood-processing facilities in Gabon and sawnwood exports to China are expected to increase quickly.

With the Chinese government encouraging companies to develop overseas operations, MOFCOM and SFA jointly issued two guidelines, the Guidelines for Sustainable Overseas Silviculture by Chinese Enterprises (2007) and the Guidelines for Overseas Sustainable Forest Management and Utilization by Chinese Enterprises (2009), in order to regulate the activities of Chinese companies in the international forestry sector.

The 2007 guidelines laid out the fundamental principles that Chinese companies need to observe in terms of sustainable silviculture and described the basic requirements necessary for achieving sustainable silviculture. The 2007 guidelines regulate and guide the entire range of overseas activities for Chinese enterprises engaged in silviculture, ranging from evaluating the Chinese enterprises' activities related to silviculture, to guiding the Chinese enterprises in providing nontimber products as well as other services, to enabling Chinese enterprises to protect and develop global forest resources in a rational, efficient, and sustainable way. Pilot projects on implementation of the 2007 guidelines have been started in China's neighboring countries, including Cambodia, Myanmar, and Laos. Chinese enterprises in these countries are being encouraged to follow the guidelines in afforestation projects and to promote alternatives to the cultivation of narcotics.

The 2009 guidelines were developed with an aim to further standardize the management of forest resources as well as the wood-processing and utilization activities of Chinese enterprises in foreign countries. It was designed to enhance self-regulation within the industries, to promote legal and sustainable utilization of global forest resources as well as related trade activities. The 2009 guidelines cover all business and management aspects related to Chinese enterprises operating overseas including forest resource management, timber harvesting and wood processing, ecological and environmental protection, and local community development.

The 2007 and 2009 guidelines, though voluntary, indicate the Chinese government's changing emphasis focus us from achieving economic development at any cost to assuming more social and environmental responsibilities as China has risen to become the world's second largest economy (Huang and Wilkes 2011a).

7.7 Conclusion

Market-oriented forestry reforms and globalization have helped shape a modern and internationalized Chinese forest products industry over the past three decades. The Chinese government has promoted these developments as industry's most influential stakeholder and will continue to play a key role in the future. Under the national twelfth 5-year plan (2011–2015), boosting domestic consumption will be at the top of the Chinese government's working agenda. In addition, the SFA will continue to implement market-oriented reforms and focus more attention on the forest industry's sustainable development, while supporting the forest industry (particularly state-owned enterprises) expands in both domestic and international markets through the development and adoption of new technology/product development, increased product branding, and increased overseas investment. Policies for promoting ecotourism, forest carbon markets, and forest certification have also been advocated (SFA 2009).

At the same time, however, due to historical and social factors, the forest sector faces daunting ecological and environmental challenges. Raw material shortages and rising cost pressure will continue to threaten the industry's international competitiveness, particularly with respect to SMEs that comprise more than 90% of forest products companies. Globally, economic uncertainties (e.g., the European debt crisis) continue to loom and pose a significant risk to Chinese exporters. The future outlook for the domestic Chinese market is somewhat mixed. On the positive side, China's continued need for urbanization in second- and third-tier cities, combined with central government stimulus efforts to increase domestic consumption, will provide strong incentives for China's economy to grow and this should increase domestic demand for wood products. In contrast, China's economy has started to show signs of slowing due to tightening credit by the central government in the face of a perceived housing bubble, high inflation, and overheated investment in capacity expansion. Furthermore, the lack of, and poor enforcement of, existing product quality standards and environmental standards has created a "race to the bottom" in the domestic market as companies do whatever it takes to undercut their competitors' prices and gain a competitive advantage in the marketplace. The desire to produce at the lowest possible cost has eroded product quality and undermined the image of Chinese exports in international markets.

There is a clear need to focus industry's economic goals toward sustainable growth as China moves into a low-carbon economy. A "paradigm shift" within the Chinese forest products industry can be expected, the success of which will depend on combined efforts from the public and private sectors requiring continued forestry institutional reforms toward a market economy, improved transparency of policy and standards, as well as concerted efforts by the industry, environmental and social groups, media, and consumers.

References

Anonymous. 2011. Corruption charge over timber transportation in Ye County. http://www.ycsq.cc/tid-48420/ (accessed April 16, 2012) (in Chinese).

The Bank of Tokyo-Mitsubishi UFJ Ltd. November, 2006. Outward investment by China gathering stream under the go global strategy, *Economic Review* 1(17).

Bloomberg. 2012. China-US 55 Index. Bloomberg: http://www.bloomberg.com/quote/CH55BN:IND (accessed April 16, 2012).

Canada Wood. 2011. China program introduction: 2005–2011 projects. http://www.canadawood.cn/downloads/pdf/international-series/chineseprojects-June22.pdf (accessed April 16, 2012).

Cao, X. 2011. Does it pay to be green? An integrated view of environmental marketing with evidence from the forest products industry in China. PhD dissertation. University of Washington, Seattle, WA. 170 pp.

Cao, X., Braden, R., and I. Eastin. 2006. Distribution systems for value-added wood products in China. CINTRAFOR Working Paper 102. University of Washington. Seattle, WA. 76p.

Cao, X., Braden, R., Eastin, I., and J. Morrell. 2007. China treated lumber market study. CINTRAFOR Working Paper 107. University of Washington, Seattle, WA. 54pp.

Cao, X. and Hansen, E. 2006. Innovations in China's furniture industry. *Forest Products Journal* 56 (11/12), 33–42.

Cao, X., Morrell, J., and E. Hansen. 2012. The Chinese treated wood market: Current status and future perspectives. *Forest Products Journal*. 61(8): 644–648.

Cao, X., Seol, M., and I. Eastin. 2011. An overview of forest certification in China: Benefits and constraints. *3rd International Scientific Conference on Hardwood Processing*. pp. 169–176. Blacksburg, VA: Virginia Tech.

CDC. 2012. The China Data Center (CDC), Ann Arbor, MI. http://chinadatacenter.org/ (accessed April 16, 2012).

China Daily. 2011a. ODI set to overtake FDI within three years. May 6, 2011. http://www.china.org.cn/business/2011-05/06/content_22505545.htm (accessed April 16, 2012).

China Daily. 2011b. Property prices in big cities heading south. http://www.chinadaily.com.cn/bizchina/2011-11/19/content_14124429.htm (accessed April 10, 2012).

China Daily. 2012. China unlikely to join WTO agreement. http://news.xinhuanet.com/english/china/2012-03/21/c_131489311.htm (accessed on April 10, 2012).

CNCA. 2012. Guidelines for certification organizations to apply for sub-contracting businesses. Beijing, China: China National Certification and Accreditation Administration. http://www.cnca.gov.cn/cnca/zxbs/xzxk/slrzjgsp/kdrzyw/09/473784.shtml (accessed April 10, 2012).

CNFA. 2011. China national furniture industry report 2009–2010. Beijing, China: China National Furniture Association.

CNFA. 2012. Statistical Data. China National Furniture Association: http://www.cnfa.com.cn/cnfa2009/tjsj/ (accessed April 16, 2012).

Coddington, W. 1993. *Environmental Marketing: Positive Strategies for Reaching the Green Consumers.* New York: McGraw-Hill, Inc.

Credit Suisse. 2011. China Consumer Survey 2011. January 2011. Credit Suisse Research Institute. Switzerland, 235 p.

Dare. 2012. Dare wood based panel group. Retrieved February 15, 2012, from company profile: http://www.dareglobalwood.com/en/about.asp

Deng, W. 2003. *Challenging the National Logging Ban in China: An Experience of Community-Based Natural Forest Management in Sichuan Province.* Montreal, Quebec, Canada: World Forestry Congress.

Eastin, I., Sasataini, D., Ganguly, I. Cao, X., and M. Seol. 2011. The impact of green building programs on the Japanese and Chinese residential construction industries and the market for imported wooden building materials. CINTRAFOR Working Paper 121, University of Washington, Seattle, WA. 74 p.

Ekstrom, H. 2012. *Reduced Log Export Tariffs in Russia Unlikely to Boost the Country's Log Export Volumes Back up to Historic Levels.* Seattle, WA: Wood Resources International LLC.

FAO. 2010a. *The Global Forest Resources Assessment.* Rome, Italy: FAO.

FAO. 2010b. *Forestry Policies, Legislation and Institutions in Asia and the Pacific: Trends and Emerging Needs for 2020.* Bangkok, Thailand: FAO Regional Office for Asia and the Pacific.

FAO. 2012. Forest trade flows. FAOSTAT. http://faostat.fao.org/site/626/default.aspx#ancor (accessed May 3, 2012).

Feng, Y. 2009. Collective forest tenure reforms and forest crisis. China.com.cn http://www.china.com.cn/news/zhuanti/hblps/2009-05/08/content_17745715.htm (accessed April 9, 2012).

Flynn, R. 2008. *China Timber Supply Outlook 2008–2012.* Boston, MA: RISI, Inc.

Goodstein, E.B. 1994. *Jobs and the Environment: The Myth of a National Trade-Off.* Washington, DC: Economic Policy Institute.

GTA. 2012. *Global Trade Atlas.* Columbia, SC: Global Trade Atlas.

He, D. 2008. Issues with China's forest tenure reform. *Theory Frontier* (理论前沿), 19–21.

Huang, W. and A. Wilkes. 2011a. Analysis of China's overseas investment policies. Working Paper 79, CIFOR, Bogor, Indonesia.

Huang, W. and A. Wilkes. 2011b. Analysis of approvals for Chinese companies to invest in Africa's mining, agriculture and forestry sectors. Working Paper 81, CIFOR, Bogor, Indonesia.

IBISWorld. 2011a. Logging in China (IBISWorld Industry Report 0221). Santa Monica, CA: IBISWorld Inc.

IBISWorld. 2011b. *Wood Furniture Manufacturing in China.* Santa Monica, CA: IBISWorld, Inc.

IWCS. 2012. Wood culture symposium and tour at Changbai Mountain, China. March 2012. http://www.iwcs.com/?p=blog&sp=content&bid=23 (accessed May 3, 2012).

Kaplinsky, R., Terheggen, A., and J. Tejaja. 2010. What happens when the market shifts to China? The Carbon timber and Thai Cassava value chains. In O. Cattaneo, G. Gereffi, and C. Stariz (eds.), *Global Value Chains in a Postcrisis World: A Development Perspective.* p. 408. Washington, DC: The World Bank.

Kochan, T.A. and S.A. Rubinstein. 2000. Toward a stakeholder theory of the firm: The Saturn pattern. *Organization Science*, 11(4): 367–386.

Lague, D. 2004. Asia's forests head to China–Beijing's logging ban sparks fears of deforestation abroad. Illegal Logging Info: http://www.illegal-logging.info/item_single.php?it_id=124&it=news (accessed April 16, 2012).

Li, H. 2011. Speech at China wood panel distribution forum. November 2011, Beijing, China. http://home.focus.cn/news/2011-11-29/240737.html (accessed April 10, 2012).

Lincoln Journal Star. June 28, 2010. Organic certification agency defends itself. http://journalstar.com/business/local/article_67b104fe-82d0-11df-9f33-001cc4c03286.html (accessed April 16, 2012).

Long, G., Zadek, S., and J. Wickerham. 2010. *Advancing the Sustainability Practices of China's Transnational Corporations*. Winnipeg, MB: International Institute for Sustainable Development.

Luo, X., Li, R., Lin, L., Gao, X., Pan, G., Xia, E. et al. 2009. *Challenges and Opportunities for China's Small and Medium Forest Enterprises*. Rome, Italy: FAO.

Ma, L. and B. Liu. 2009. *Witness APP 30 Years of Deforestation* (见证金光集团APP毁林三十年). Beijing, China: Greenpeace.

Marvin, D. and K. Brown. November 16, 2011. Embattled sino-forest fights back against muddy waters. Retrieved February 18, 2012, from *Wall Street Journal*: http://online.wsj.com/article/SB10001424052970203503204577039930338094506.html (accessed April 16, 2012).

Ministry of Environmental Protection of China (MOEP). 2012. International environmental conventions to which China is a signatory. http://gjs.mep.gov.cn/gjhjhz/index.htm (accessed April 16, 2012).

Mol, A. 2012. China's middle class as environmental frontier. Al Jazeera. http://www.aljazeera.com/indepth/opinion/2012/01/201217122143498595.html (accessed April 16, 2012).

SFA. 2009a. China forestry revitalization plan 2010–2012. China eucalyptus research center of state forestry administration: http://www.chinaeuc.com/show.asp?id=358 (accessed April 16, 2012).

SFA. 2009b. The seventh national forest resources inventory report. Beijing, China: The State Forestry Administration.

SFA. 2010. Land conversion from farmland back to forestland. http://english.forestry.gov.cn/web/article.do?cid=200911161236085325 (accessed October 4, 2011).

SFA. 2011a. 2010 China forestry development report. Beijing, China: State Forestry Administration.

SFA. 2011b. *The Reform of the Collective Forest Tenure System in China*. Beijing, China: The Department of Rural Forestry Reform and Development of SFA.

SFA. 2011c. China's Forest Restoration (in Chinese).

Sun, X. and K. Canby. 2011. China: Overview of forest governance, markets and trade. June 2011. European Forest Institute & Forest Trends. 52p. http://forestindustries.eu/sites/default/files/userfiles/1file/baseline_study_china_report_en.pdf (accessed April 16, 2012).

Wang, W. 2010. Status quo of forest certification in China. Conference presentation at Forests, Markets, Policy, and Practice, September 7–8, 2010, Beijing, China. http://www.cfcn.cn/cmc3/download/%E5%8F%91%E8%A8%80ppt/7-1-5-Wang%20Wei-%E4%B8%AD%E8%8B%B1.pdf (accessed April 16, 2012).

Wickerham, J. and S. Zadek. 2009. China's corporate social responsibility change makers. Fortune China, March 2009: cover story.

The World Bank. 2012a. *Databank: China*. Washington, DC: The World Bank. http://data.worldbank.org/country/china (accessed May 3, 2012).

The World Bank. 2012b. *China 2030: Building a Modern, Harmonious, and Creative High-Income Society*. Conference Edition. Washington, DC: The World Bank. 468 pp.

Xu, J. 2011. China's new forests aren't as green as they seem. *Nature*. http://www.nature.com/news/2011/110921/full/477371a.html (accessed April 16, 2012).

Xu, J., White, A., and U. Lele. 2010. *China's Forest Tenure Reforms*. Washington, DC: Rights and Resources Initiative.

Yuan, Y. and I. Eastin. 2007. Forest certification and its influence on the forest products industry in China. CINTRAFOR Working Paper 110. University of Washington, Seattle, WA. 69pp.

Zeng, Y. 2011. Conference agenda and outputs. Forests, Markets, Policy and Practice (FMP&P) 2011: International conference and workshops: http://www.cfcn.cn/cmc4/download/发言材料/Zeng%20Yinchu-EN.pdf (accessed April 16, 2012).

Zhang, L. 2008. Reform of the forest sector in China. In *Re-Inventing Forestry Agencies: Experiences of Institutional Restructuring in Asia and the Pacific*. pp. 215–234. Bangkok, Thailand: FAO Regional Officer for Asia and the Pacific.

Zhao, S. 2012. China's Green Growth-Forestry Development in China since the Sixteenth National Congress of the Communist Party of China. China Foresty Press. Beijing. September 7, 2012, p. 680.

8

Russia in the Global Forest Sector

Eduard L. Akim, Nikolay Burdin, Anatoly Petrov, and Leonid Akim

CONTENTS

8.1 Introduction

The forest sector of the Russian Federation includes forestry, logging, sawmilling, plywood, panel, and pulp and paper industries. Over 60,000 large, medium, and small enterprises located throughout the country are engaged in extensive forest regeneration, increasing their productivity, protection from fire and insects, logging, and all types of wood processing. All sectors of the forest industry are technologically interrelated on the basis of production and utilization of the natural resource—raw wood material. The products of the forest industry are widely used in construction, agriculture, the printing industry, trade, and medicine.

The role of the forest sector in the Russian economy is modest—in 2011, its share of Russian GDP was just over 1.3% with a share of total industrial production of 3.7%. The sector employed just above 1% of the population. At the same time, the forest sector is one of the most stable exporters in the Russian Federation bringing in an additional 2.4% of GDP in export revenues. Although the forest sector's contribution to GDP and exports are small compared to Russia's main export products of oil and gas, there is potential for future development, being based on a natural resource that, in contrast to oil, gas, coal, ore, and other minerals, is renewable.

Despite the world's richest forest resources, Russia's contribution to the global forest sector is low according to the 2011 data from the Food and Agriculture Organization (FAO) of the United Nations (2012): 5.3% of global harvest, 5.0% of lumber production, 3.0%–3.3% of wood-based panel production, and 2.0% for paper and paperboard production. Russia also lags behind developed countries with regard to per capita consumption of lumber, plywood, particleboard and fiberboard, paper, and paperboard.

In this chapter, we begin by providing an overview of Russia's forestry sector, touching on domestic harvests and certification. Next, we discuss Russia's role as a producer of forest products and its role in global trade. Each major product sector, roundwood, lumber, plywood, composite panels, and pulp and paper are covered. Emerging wood-based biofuels and chemical sectors are discussed as well as the small but potentially important furniture sector. We then discuss domestic markets for wood products and policy issues that impact the forest sector as a whole. Finally, we draw conclusions on where the future may lie for Russia's forest products industry.

8.2 Russia's Forests and Domestic Harvest

Russia is a federation that, since March 1, 2008, consists of 83 federal subjects, also known as the constituent entities of the Russian Federation. In 1993, when the Constitution of Russia was adopted, there were 89 federal subjects listed. By 2008, the number of federal subjects had been decreased to 83 because of several mergers. The federal subjects are of equal federal rights in the sense that they have equal representation—two delegates each—in the Federation Council (upper house of the Federal Assembly). They do, however, differ in the degree of autonomy they enjoy. Federal subjects should not be confused with Federal districts of Russia that are much larger, and each encompasses many federal subjects. There are 83 federal subjects in the Russian Federation, while there are only 8 federal districts.

Due to geographic, logistic, and transportation challenges, only about 45% of all forest areas in the Russian Federation are available for harvesting with the rest situated in inaccessible regions. The primary regions with available forests are in the North European and Ural regions and along Trans-Siberian Railway in Siberia (Petrov 2005, 2008; Russian Federation Ministry of Industry and Trade and Ministry of Agriculture 2008; Figure 8.1).

Nearly half (46.6%) of all land area in the Russian Federation is covered with forest (Figure 8.2). Recent forest inventory data estimate that the country has 890 million ha of forests, of which about 19.6% are situated in European and Ural areas of Russia (FAO 2012). Most of these forests are boreal forest, composed primarily of larch (*Larix* spp.) (35.8%), pine (*Pinus* spp.) (15.6%), and birch (*Betula* spp.) (15.0%).

According to government inventory data (Table 8.1), the total forest growing stock in Russia is more than 83 billion m³. The annual increment available for harvest is estimated to be 990 million m³ and the annual allowable cut is 635 million m³. In 2010, total harvest volume was 173.6 million m³, predominantly coniferous species (FAO 2012; Russian

FIGURE 8.1
Map of Russia.

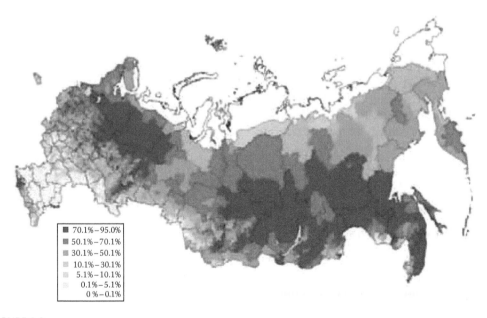

FIGURE 8.2
Forest cover of the Russian Federation, 75.4% of total forest area, of which 67.4% is covered by forest vegetation, and 8.0% is not covered by forest vegetation. Non-forested land accounts for 24.6%. (From Food and Agricultural Organization of the United Nations (FAO), *The Russian Federation Forest Sector Outlook Study to 2030*, Rome, Italy, 2012.)

TABLE 8.1

General Indices of Forest Resources of the Russian Federation

	Unit	Russian Federation, Total	European Part	%	Asian Part	%
Forest land area	Million ha	890.8	174.2	19.6	716.6	80.4
Stocked area	Million ha	796.2	179.3	22.5	625.9	77.5
Growing stock, total	Billion m³	83.3	22.9	27.5	60.4	72.5
Mature and overmature of which	Billion m³	44.3	12.7	28.7	31.6	71.3
Coniferous	Billion m³	34.2	8.0	23.4	26.2	76.6
Non-coniferous	Billion m³	9.5	3.3	34.7	6.2	65.3
Total average annual increment	Million m³	993.8	377.2	38.0	616.6	62.0
Annual allowable cut	Million m³	635.0	313.2	49.3	321.8	50.7

Federation Ministry of Industry and Trade and Ministry of Agriculture 2008). Illegal logging is estimated to account for only about 1% of wood harvesting. However, these official data significantly differ from the estimates from a number of sources including the World Wildlife Fund (WWF) and the World Bank (Nilsson 2006; Petrov 2011; Shmatkov 2009, 2011; World Bank 2007, 2011). According to these sources, illegally logged wood accounts for about 20% of the annual harvest with associated annual financial losses of 20 billion ($US 706 million) rubles in 2011, however, these number are somewhat controversial.

The other significant harvesting challenge is due to the extremely low average output of forest products from existing forestland. For example, the index of wood cut per 1 ha of forested area was 0.22 m³ in Russia in 2000, whereas in the countries with developed forest industries, it was 2.5–3.5 m³/ha. In 2010, output per hectare was even lower than in previous decades. This can be partially explained, by the fact that forests in the Russian Federation are predominantly state-owned and logging entities are not timber landowners, limiting incentives to practice silvicultural methods to achieve high yields. Enterprises and organizations engaged in wood harvesting carry out harvesting and silvicultural activities mainly through long-term leasing or through auctions that seem to further contribute to the disconnect between yield optimization and short-term management.

8.3 Forest Certification

Although problems exist regarding harvesting and silvicultural activities, at present there are two systems of forest certification active in Russia: the Forest Stewardship Council (FSC) and the Russian National Council of Forest Certification accredited by the Program for the Endorsement of Forest Certification (PEFC). At the end of 2011, the total forest area of certified forests was 30 million ha (less than 4% of total Russian forest land) with a total of 300 certificates, including 70 where forest management was certified. The balance of certificates is for chain-of-custody held by other members of the forest products supply chain. Certified forests are located in 17 subjects of the Russian Federation and occupy an area that is 26% of all forest lands leased for logging (FAO 2012).

Further expansion of certified forest areas is expected to continue in the future due to several factors. The main factor is an export orientation of the forest industry into ecologically sensitive markets, especially the European Union (EU). There is also an increasing influence of Russian and foreign nongovernmental organizations (NGOs) encouraging both state agencies and private businesses to certify forest lands. Finally, forest certification is regarded as a tool in preventing illegal logging and corruption. As such, the National Forest Plan and forest plans of several subjects of the Russian Federation are promoting certification. With all these factors in play, the increase of FSC-certified forest areas is forecast to continue at an annual rate of 3 million ha through 2015.

The further expansion of certified areas strongly depends on demand for certified wood from China, which is the largest importer of Russian roundwood and lumber. As a majority of Russian forest products sold to China finds its way in the form of finished products to US and EU markets, the pressure for chain-of-custody certification will increase.

8.4 Russia as a Producer and Global Supplier of Forest Products

8.4.1 Factors Impacting Competitiveness

The main competitive advantage of Russia's forest industry is a huge commercial forest resource of highly-demanded coniferous species (80% of total stock). Other advantages are a well-educated and low-wage workforce, potential to adopt scientific methods to achieve growth and harvest potential, an alignment of research programs with the European and FAO research.

The integration of the Russian forest sector, particularly the pulp and paper industry into global trade, is reflected by continuous growth of export and import volumes of forest products and in the investment of major global companies such as International Paper, SCA, and Mondi in the Russian forest sector.

Currently, the transition to a free market economy is complete in the forest industry with over 97% of enterprises having been privatized or converted to joint stock companies. Prices fluctuate in the domestic market based on supply and demand, external trade in forest and paper products is liberalized, and the former rigid system of management of forest enterprises at federal and regional levels no longer exists. Market forces are hoped to encourage modernization of equipment and promote further development in forest management and throughout the value chain for forest products. However, in reality, many forest enterprises remain financially unstable, have outdated equipment, and are facing insolvency.

At present time, the structure of forest industry production and exports is focused on roundwood and lumber exports and not on value-added products such as pulp and paper or finished wood products such as furniture, flooring, molding, and millwork. In the last few years, the country has imposed high tariff barriers and has started to limit roundwood exports in an attempt to reorient exports toward higher value products. With Russia's membership in the World Trade Organization (WTO), it is unclear if these tariffs will remain as they are strongly opposed by China and some other countries that import roundwood from Russia.

Another factor limiting Russia's potential to be a stable and consistent producer of forest products is the uneven geographic distribution of the forest industry. The concentration of wood processing in the North European part of the country has resulted in overexploitation of forests in this region, while Siberian forests, the largest nontropical forest region in the world, remain underutilized. Despite multiple announced government programs, all expansion into Siberia has been limited by minor road development and a resulting low population in these vast territories.

In summary, the country faces low levels of forest management, and an unsatisfactory forest industry structure resulting in insufficient output of value-added competitive products. As such, raw materials and commodities are the predominant Russian forest products exports. However, the forest sector of the Russian Federation, with its huge forest resources, has potential for contributing significantly to the development of the Russian economy and of becoming a sophisticated player in the global forest sector.

8.5 Profile of Major Forest Industry Sectors

8.5.1 Roundwood

In the period of 1990–2007, roundwood has represented the major export category of forest-based products from Russia at about 55% (Table 8.2). However, as with the situation with illegal logging, official statistics underestimate the volume of roundwood exports. In particular, significant volumes are being exported to China without official reporting or payment of export taxes (Shmatkov 2011).

In 2007, as a means to reduce raw material exports and to encourage value-added processing, the Russian government signed into law Resolution 75 that gradually established new levels of export taxes on roundwood over the period of 2007–2011. The export tariff on saw logs was scheduled to increase from €4 ($6) per m^3 in 2006 to €50 ($77) per m^3 in 2009 with these tariffs to be applied to birch pulpwood by 2011. Even though full implementation of these new tariffs was delayed several times, they fostered a reduction of roundwood exports by 2.3 times in 2011 compared to 2007; this reduction projected to continue (Akim and Kovalenko 2009; UNECE/FAO-United Nations 2012a).

8.5.2 Lumber

Lumber production is one of Russia's oldest and traditional industries. By volume, the Russian Federation is the biggest European exporter. The main importers of Russian lumber are China, Uzbekistan, Egypt, Iran, Azerbaijan, Japan, Germany, Afghanistan, Finland, Estonia, France, North Korea, and the Netherlands. In total, 73 countries purchased lumber from Russia in 2011.

According to official government statistics, lumber production in 2011 was 20.1 million m^3; however, these data are significantly lower than reality. Reliable data are not available because of a large number of small privately owned sawmills which do not submit any production information to governmental agencies. Very often, these sawmills do not even measure their production volumes. Typically, the small companies produce low-quality materials using very simple and outdated equipment and sell their production at the local market, primarily for residential construction. In many instances, their raw materials are illegally logged. The long-term forecast of lumber production, according to the FAO (2012), is expected to grow to 42.0 million m^3 by 2015 and exceed 66.2 million m^3 by 2030.

TABLE 8.2

Structure of Russian Exports of Forest-Based Products in 1990–2010

	1990	1998	1999	2000	2001	2002	2005	2006	2007	2008	2009	2010
Roundwood, million m³	15.0	20.7	28.3	32.0	32.9	37.8	48.3	51.1	49.3	37.1	22.3	21.2
Sawnwood, million m³	7.1	4.8	6.4	7.8	7.7	9.0	14.8	15.9	17.3	15.3	16.2	17.7
In terms[a] of roundwood, million m³	11.4	7.7	10.2	12.5	12.3	14.4	23.7	25.4	27.7	24.5	25.9	28.3
Market pulp, million MT	0.628	1.029	1.390	1.648	1.785	1.885	1.946	1.918	1.899	2.035	1.715	1.705
Paper and paperboard, million MT	0.967	1.705	1.998	2.253	2.347	2.458	2.737	2.701	2.512	2.634	2.717	2.600
Pulp, paper, and paperboard, million MT	1.595	2.734	3.398	3.901	4.132	4.328	4.683	4.619	4.411	4.669	4.432	4.305
In terms[b] of roundwood, million m³	5.7	9.3	11.5	13.2	14.0	14.7	15.9	15.7	15.0	15.8	15.0	14.6
Total exports of forest and paper products in terms of roundwood, million m³	32.1	37.7	50.0	57.7	59.2	66.9	87.9	92.2	92.0	77.4	63.2	64.1
Percentage of pulp, paper, and paperboard in terms of roundwood	17.7	24.7	23.0	22.9	23.6	21.9	18.1	17.0	16.3	20.4	23.7	22.8
Percentage of roundwood exports	47	55	57	55	56	56	57	56	54	48	35	33

Source: Food and Agricultural Organization of the United Nations (FAO), *The Russian Federation Forest Sector Outlook Study to 2030*, Rome, Italy, 2012.

[a] The factor 1.6 is used—source: FAO.

[b] The factor 3.39 is used—source: FAO.

8.5.3 Plywood

Plywood is the fastest growing sector of forest-based industry with a 47% increase from 2.1 million m³ in 2009, a postrecession year, to an estimated 3.1 million m³ in 2012 (Russian State Statistics Service 2012; Table 8.3). Most of the plywood production is found in the northwestern region of the European part of the Russian Federation. Plywood is widely used domestically in the construction industry and to a lesser degree in the small furniture sector as well as being a significant export product. The most important product is currently birch plywood that is valued for its surface hardness and is exported to the

TABLE 8.3

Production Volume of Major Sectors (2007–2012)

	2007	2008	2009	2010	2011	2012 (Forecast)
Roundwood, million m³	207.0	181.4	161.2	173.6	180.0	191.0
Lumber, million m³	24.3	21.6	19.0	19.0	20.1	20.6
Plywood, million m³	2.777	2.592	2.107	2.688	3.003	3.100
Particleboard, million m³	5.500	5.750	4.562	5.484	6.600	6.606
Fiberboard, million m²	497	479	386	395	442	458
Market pulp, million MT	2.421	2.286	2.176	2.100	2.200	2.215
Paper and paperboard, million MT	7.591	7.700	7.373	7.363	7.406	7.610

Source: Data from Russian Federation Statistical Bureau. (Russian State Statistics Service, 2012).

Note: The numbers are given according to ROSSTAT (Russian Federation Statistical Bureau) data.

United States, Germany, Egypt, Turkey, Italy, the United Kingdom, and many other countries. Total volume of export in the period of 2005–2010 was stable at 1.3–1.5 million m³/year. The biggest opportunity for further development in the plywood sector is production of large size sheets that are used in the home-construction industry.

8.5.4 Particleboard and Fiberboard

Driven primarily by furniture production (90%–95% of consumption) and by demand for packaging materials, particleboard production is growing faster than the forest industry as a whole (Russian State Statistics Services 2012; Table 8.3). In 2010, production of particleboard was 5.48 million m³ and fiberboard, 1.67 million m³. According to forecasts, production of particleboard will reach 11.7 million m³ by the year 2030 and production of fiberboard, 4.2 million m³.

The main problem in particleboard production is the outdated technology used by many producers and a slow rate of implementation of new technologies, particularly of oriented strand board (OSB), which is planned to be produced in the Russian Federation in 2013. The situation is similar with the production of medium-density fiberboard (MDF). New technologies, particularly dry process methods and continuous methods, have recently been recently adopted with their production volume increasing to 1.2 million m³ in 2011. MDF is replacing particleboard in furniture production that will lead to growth in production and consumption of MDF in the medium and long term.

Currently, because of low volumes of OSB and MDF domestic production, these products are imported; in 2010 the combined import was 0.78 million m³. As these products begin to be manufactured in higher volumes, imports are expected to decrease in the next 5–10 years (FAO 2012; UNECE 2003; UNECE/FAO-United Nations 2012).

8.5.5 Pulp and Paper

Despite the drawbacks of a planned economy, the USSR consistently was a top global producer of paper and paperboard and in 1990 accounted for 5.2% of global output of these products. The following decades of social and economic turmoil greatly affected

TABLE 8.4

Output of Chemical Pulp, Paper, and Paperboard in the Russian Federation for Various Years (1988, 1996, 1998, 1999, 2010, 2011) (Thousand Metric Tons)

	1988	1996	1998	1999	2010	2011
Chemical pulp	8331	3028	3205	4225	7285	7361
Paper	5465	2274	2325	2966	4674	4672
Paperboard	3167	962	1102	1569	2689	2734

the Russian economy and by the mid-1990s, the Russian global share of pulp and paper production dropped to 1.1% (Akim 2007, 2009; Akim and Kovalenko 2009; UNECE/FAO-United Nations 2012).

Table 8.4 shows the changes in output of chemical pulp, paper, and paperboard for various years between 1988 and 2011. Owing to relative economic and political stability established in the country since a major currency revaluation in 1998 and an expansionary macroeconomic policy since 1999, Russia has experienced a continuous increase in output of pulp, paper, and paperboard. Production has doubled since 1996, although output has not reached record levels seen in pretransition periods of 1988–1989 (in the late Soviet era). In 2011, the country constituted about 2% of world paper and paperboard production.

In 2007–2008, the Russian pulp and paper sector continued to expand production of pulp, paper, and paperboard, particularly the output of paperboard for packaging. However, the economic crisis of 2009–2010 negatively affected the industry. In 2009, Russia's output of market pulp decreased by 4.8%, and the output of paper and paperboard decreased by 4.2%. The only exception was newsprint production that increased by 1.0%. During 2010–2011, the situation stabilized and output started to grow again reaching 2.2 million metric tons (MT) of market pulp and 7.4 million MT of paper and paperboard (Figure 8.3; Table 8.3).

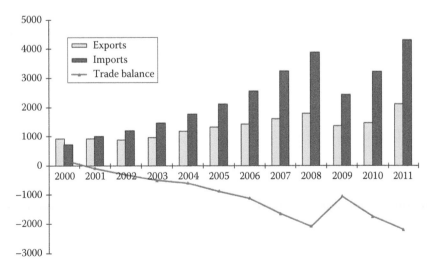

FIGURE 8.3
Russian exports, imports, and trade balance of paper and paperboard in 2000–2011 (million USD).

TABLE 8.5

Major Russian Pulp, Paper, and Board Producers 2009 (2007)

	Total Output, × 1000 MT	Market Pulp, × 1000 MT	Paper, × 1000 MT	Board, × 1000 MT
Ilim Group	2356 (2400)	1477 (1590)	206 (260)	672 (550)
Kotlas	905 (945)	257 (354)	206 (260)	442 (331)
Bratsk	744 (734)	513 (515)		230 (219)
Ust-Ilimsk	707 (721)	707 (721)		
Svetogorsk	444 (411)		36 (333)	82 (78)
St. Petersburg KPK	229 (247)			229 (247)
Arkhangelsk	769 (771)	221 (211)	78 (82)	471 (478)
Syktyvkar	849 (836)		641 (622)	208 (214)
Kondopoga	760 (733)		760 (733)	
Volga	589 (579)		589 (579)	
Solikamskbumprom	451 (418)		451 (418)	
Segezha	237 (283)		231 (232)	7 (53)

Sources: State Customs Committee, Pulp Paper Board Magazine, PPB-Express, 2012.

A majority of Russian pulp and paper production is oriented toward exports that account for about 80% of market pulp production and around 30% of paper and paperboard production. Major export destinations for these products are China (market pulp and kraft linerboard), Ireland (market pulp and kraft linerboard), India (newsprint), and Turkey (newsprint). The abundance of forest resources containing northern softwood fiber, essential for reinforcing paper and board sheets, assures a strong long-term position for Russia in exporting to neighboring developing markets, particularly China and India. Overall, this upward trend of production and exports is expected to continue as global economic conditions improve (Russian State Statistics Service 2012).

A fairly recent phenomenon is having an impact on the sector's structure. The industry is experiencing multiple mergers and acquisitions resulting in a high degree of consolidation. The biggest Russian company, Ilim Group, formed by an alliance between International Paper and Ilim Pulp Enterprise in 2006, produces 68% of the country's market pulp and 37% of all paperboard (Table 8.5).

8.6 Wood-Derived Fuel

Recent demand in Europe for wood-derived fuels, such as wood pellets, has created potential for development of new production facilities oriented to this market segment. Bioenergy is one way to increase wood utilization, as it often uses low-value wood or wood residues from sawmilling and other production processes. Potential volume of wood-based raw materials from Russia for bio-based production is estimated to be 300 million m^3 annually. Part of this volume can come from utilization of low-quality species, such as aspen (*Populus* spp.) (up to 100 million m^3), and the balance coming from post harvest residuals and industrial waste. Presently, there is rapid development in this biofuel area. In 2011, the forest sector produced 1 million MT of wood-based fuel, mainly pellets. New production facilities are being colocated at several paper mills and sawmills. Several companies are considering additional stand-alone investments in this sector with announced plans for 1.3 million MT of new pellet production.

Even current or future competitive prices of fossil fuels such as oil, natural gas, and coal do not seem to have any major effects on the trends toward stronger environmental pressure that continue to grow in Europe and other countries (FAO 2012; Jansen 2012; UNECE 2012a; UNECE/FAO-United Nations 2012).

8.6.1 Potential for Wood-Derived Chemicals

One of the major challenges of Russia as the most forested country in the world is sustainable management and multipurpose utilization of forest resources. Russian forests are 58% larch, mainly Siberian larch (*Larix sibirica*) and dahurian larch (*Larix gmelinii*), and pine (*Pinus* spp.). Siberian and dahurian larches have strong and dense wood with unique chemical compositions. Unlike species of larch that grow in North America and Western Europe, Siberian and dahurian larch wood contains 7%–30% of a water-soluble polysaccharide arabinogalactan. As such, Siberian and dahurian larch wood cannot be pulped using conventional technology. When larch wood is used only for house construction and furniture manufacturing, half of the wood becomes waste. Until there is a scientifically and economically sound technology to pulp larch wood, larch harvesting is not economically feasible. This hinders development of the forest industry in Siberia and the Far East and contributes to population outflow from these depressed areas.

One of the most important challenges (and opportunities) of the Russian forest industry is to create and implement innovative integrated technologies for chemical processing of larch wood in addition to expansion in manufacturing finished goods. Success could help keep existing idled Siberian facilities to be reactivated such as pulp mills at Bratsk and Ust-Ilimsk and to potentially foster the construction of new green field facilities.

One of the most efficient areas of larch utilization may be its integrated converting into new products demanded by the world and domestic markets. (Akim 2007, 2009; Akim and Kovalenko 2009; FAO 2012; Herbert 2012; UNECE 2012b; UNECE/FAO-United Nations 2012). Accordingly, in response to this issue, there is a national plan to create technologies to make pulp out of larch wood for further chemical processing, production of tissue products, and packaging paper and board. Larch wood waste would be used to obtain various chemical products such as arabinogalactan, dihydroquercetin, and microcrystalline cellulose. Conceptually, new products would enter global markets, and competitive advantage would be developed for Russia's most prevalent wood resource. Potential larch harvesting capacity in Russia is 105 million m^3.

8.7 Furniture

A significant concentration of furniture manufacturing is located in the European part of Russia, predominantly in the Central region (40%) and Volga region (26%). Only 9% of manufacturing is located in Siberia and the Far East. Consumption is primarily concentrated in major urban centers corresponding to population, income and purchasing power—high in Moscow and several other urban regions such as St. Petersburg and Tatarstan and much lower in small towns and in rural areas (Makeev 2009).

TABLE 8.6

Furniture Production in the Russian Federation

	2000	2001	2002	2003	2004	2005	2006	2007	2008	2009	2010	2011
Furniture production, million USD	761.1	888.4	1006.1	1283.3	1633.7	1448.3	1819.5	2618.6	3285.3	2171.3	2812.2	3092.8

Source: Data from Russian Federation Statistical Bureau. (Russian State Statistics Service, 2012.)

After a dramatic decrease in furniture production through the 1990s, the industry experienced relatively stable growth from 2001 to 2008, fueled by increased domestic purchasing power and consumption. Two additional factors for growth were high import customs tariffs and relatively low capital requirements for modernization or start-up production. According to official government statistics (Russian State Statistics Service 2012), in 2008 furniture production reached $US 3.3 billion (Table 8.6). The economic crisis of 2009–2010 impacted the industry severely with manufacturing value in 2009 dropping to 66% of 2008 levels. A slow recovery began in 2011 and continued through 2012.

Throughout the past two decades, the Russian furniture industry has experienced strong competition from imported products. Even high custom tariffs, making imports of inexpensive furniture prohibitively expensive, could not prevent domination of imports in domestic markets. Throughout the past 10 years, the share of Russian-made furniture in domestic consumption never exceeded 40% although this ratio is slowly increasing. This has created a persistent negative trade balance; furniture exports represented only 6.8% of total production in 2011 (Mebelny Business 2012).

8.8 Russia as a Consumer: Internal Consumption and Demand Trends

Although domestic consumption of wood products is driven by new home construction, consumption is limited by preferences for brick and concrete construction, even for individual homes. Home construction based on heavy use of lumber or wood-frame types is still in early stages of acceptance.

The development of operational internal markets in Russia has also been slow in the last few years primarily due to impacts from the global financial crisis. Also, in general, the internal market is characterized by an income gap between populations in major urban centers, particularly in Moscow, and rest of the country. Low income levels in small and medium towns and in the countryside significantly limit consumption of goods, including forest products. At the same time, higher earners are more oriented toward high-end imported products (Figure 8.4).

Demand dynamics have driven imports of value-added products, such as modern construction materials (OSB, MDF, and LVL), coated paper, and board for the printing industry and packaging. The demand for value-added, further processed products is also limited by income effects as well as slow growth in derived-demand industries, such as home construction, publishing, and furniture production.

8.9 Major Policy Issues

Two decades of political, institutional, and economic reforms in the Russian Federation demonstrate that the forest sector, including forest products manufacturing industries and the forestry sector, are slowly making the transition to a market economy framework and to sustainable forest management standards. The speed of this transition is impeded, however, by frequent changes in regulations governing forest ownership.

During last 20 years, the federal forest legislation has changed several times. The "Basic Forestry Legislation," "Osnovy lesnogo zakonodatelstva," 1993, decentralized forest management and transferred the management functions related to forest land to local government authorities in administrative districts. The 1997 "Forest Code" subsequently transferred management functions related to forest land to the government authorities of the 83 federal subjects of the Russian Federation, leaving legislative and supervisory functions to federal government "Federal Law, 1997." A few years later, "Federal Law no. 122," 2004, centralized forest management and transferred forest management functions to federal executive bodies. Then, Acting Forest Code, "Federal Law," 2006, once again decentralized forest management and transferred managing state supervisory functions related to forest land to government authorities of the subjects of the Russian Federation (Forest Code of the Russian Federation (Lesnoi Kodeks Rossiiskoi Federatsii) 2006).

All these legislative transitions were accompanied by drastic changes in the institutional organization of forest administration related to power sharing between federal, regional, and municipal forest authorities. The constant legislative changes were accompanied by frequent transfers of the Federal Forestry Agency, federal executive body responsible for the sphere of forest relations, from one ministry to another. From 2000 to 2012, the position of the Federal Forestry Agency changed four times: until 2008 when it fell under the Ministry of Natural Resources; from 2008 to 2010, it was subordinated to the Ministry of Agriculture; from 2010 to May 2012, it answered directly to the Government of the Russian Federation; and as of May 2012, it reports to Ministry of Natural Resources and Ecology. One thing, however, did not change—the ownership of the forests remained with the state, which interestingly is said to be supported by a great majority of the population. The current system of forest management in the Russian Federation is shown in Figures 8.5 and 8.6.

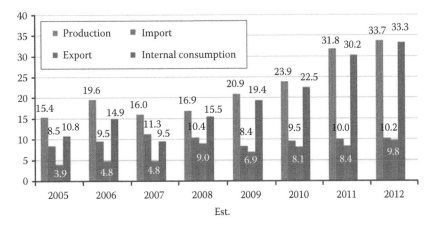

FIGURE 8.4
Dynamics of internal consumption of forest industry products, billion USD.

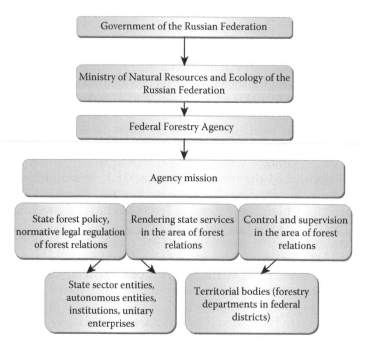

FIGURE 8.5
Institutional organization of forest management in the Russian Federation. (From Food and Agricultural Organization of the United Nations (FAO), *The Russian Federation Forest Sector Outlook Study to 2030*, Rome, Italy, 2012.)

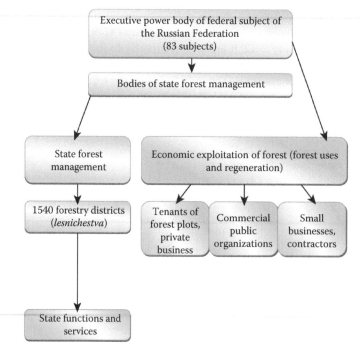

FIGURE 8.6
Institutional organization of forest management on the regional level. (From Food and Agricultural Organization of the United Nations (FAO), *The Russian Federation Forest Sector Outlook Study to 2030*, Rome, Italy, 2012.)

8.10 Conclusions

In the 22 years since the demise of the Soviet Union and its state-owned economy, the forest industry of the Russian Federation has lived through sharp decline and painful reconstruction. The rise from the abyss was slow and many industry sectors never reached production volumes achieved in the 1980s. The current system underpinning the forest sector in Russia was mostly designed, constructed, and developed in the middle of the past century on the basis of the political decisions, technology, and knowledge of that time. The system includes capital assets, institutions, policies, science, and education in the forest sector. As of now, the assets are mostly physically obsolete, and the administrative organizations and policies are inconsistent and in need of profound, fundamental reconstruction. The difficulties in the development of forest management and forest-related industries caused by these changes are further augmented by an absence of coherent national forest policy responsibility for political decision making and establishing the national and regional priorities in long-term planning to develop the forest sector. Stabilization of forest legislature will arrive only with development and adoption of such national policy.

On the positive side, structural transformations in the overall economy are complete and private property and market forces govern most of the forest industry. The sector has also rebuilt ties to the world economy with Russian exports of forest products playing important roles for wood needs in China, Finland, and in many other countries. Foreign investment and modernization of production facilities will increase over time.

The further growth of the Russian forest industry strongly depends on the development of internal markets and on conditions and development of world economy. In any case, one cannot ignore the biggest forest reserves on the planet possessed by Russia, its significant industrial and scientific potential, and growing internal consumption. These factors will keep developments in Russia's forest sector on the radar screens of investors, market analysts, and competitors.

References

Akim, E. 2007. Outlook for Russian fiber in a global context. *RISI European Pulp and Paper Outlook Conference*, Barcelona, Catalonia, March 25–27, 2007.

Akim, E. 2009. Private sector perspective on the use of international forest products information. *Proceedings of the Workshop on Forest Products Statistics for CIS-Region National Statistical Correspondents*. Food and Agriculture Organization of the United Nations, United Nations Economic Commission for Europe, Geneva, Switzerland.

Akim, E. and Kovalenko, M. 2009. Pulp and paper industry-opportunities and challenges in Russia. *O Papel*. LXX:21–29.

FAO. 2012. *The Russian Federation Forest Sector Outlook Study to 2030*. Rome, Italy, 2012.

Forest Code of the Russian Federation (Lesnoi Kodeks Rossiiskoi Federatsii). 2006. Available online at http://www.rg.ru/2006/12/08/lesnoy-kodeks-dok.html (Accessed February 18, 2013).

Herbert, P. 2012. Trends in global pulp and paper markets and larch project. Globalization of pulp and paper market. In: *Innovative Technology in the Russian Forest Sector. The Way to the Green Economy*. United Nations Economic Commission for Europe, New York.

Jansen, H. 2012. Results of the UNECE biomass project: Improved trade logistics for the sustainable use of biomass in Northwest Russia and development of sustainable biomass trade and export opportunities for selected regions of the Russian Federation. In: *Innovative Technology in the Russian Forest Sector. The Way to the Green Economy*. United Nations Economic Commission for Europe, Geneva, Switzerland.

Makeev, V. 2009. Russia: Understanding its furniture market and business culture. *Furniture and Furnishing Export International*. September and October 2009.

Mebelny, B. 2012. *Recovery of Furniture Business* (in Russian). 107(2012):35–38.

Nilsson, S. 2006. Russia, China and the rest of the world. *47th Session of FAO ACPWP*, Rome, Italy, June 6, 2006.

Petrov, A.P. 2005. Forest management under long-term forest lease: shared responsibility of the government and private business. *The Russian-Swedish Seminar Proceedings*. October 26–27, 2005. p. 105.

Petrov, A.P. 2008. The long-term strategy of forest sector development in Russian Federation: from products market to forest resources. *The Russian-Swedish Seminar Proceedings*. November 27–28, 2008. p. 130.

Petrov, A.P. 2011. *Improvement of Forest Law Enforcement and Governance in the Russian Federation*. World Bank, Moscow, Russia.

PPB-Express. 2012. PPB exports, PPB imports, pulp paper board (PPB) magazine (of Russia). 2011. Available at: www.cbk.ru/eng/cbk_mag.php (Accessed January 12, 2013).

Russian Federation Ministry of Industry and Trade and Ministry of Agriculture. 2008. Strategy of Russian forest complex development until 2020, Moscow, Russian Federation.

Russian State Statistics Service. 2012. Russian Industry–2012. Available at http://www.gks.ru/bgd/regl/b12_48/Main.htm (Accessed February 15, 2013).

Shmatkov, N. 2009. The Lacey Act as instrument to prevent illegal logging. *Sustainable Forestry*. 21:8–11.

Shmatkov, N. 2011. European market against the illegal logging. *Sustainable Forestry*. 26:15–26.

UNECE/FAO-United Nations. 2012. *Forest Products Annual Market Review 2011–2012*. Geneva, Switzerland.

United Nations Economic Commission for Europe (UNECE). 2003. *Russian Federation Forest Sector Outlook Study*. Geneva Timber and Forest Discussion Paper 27. United Nations, Geneva, Switzerland.

United Nations Economic Commission for Europe (UNECE). 2012b. *The Development of Regional Biomass Action Plans of the Russian Federation*. United Nations, Geneva, Switzerland.

United Nations Economic Commission for Europe (UNECE). 2012a. *Innovative Technology in the Russian Forest Sector. The Way to the Green Economy*. United Nations, Geneva, Switzerland.

World Bank. 2007. Systems to watch the wood origin as an instrument preventing illegal logging. Alex. Moscow. 2007: 136 (in Russian).

World Bank. 2011. State and scale of illegal forest use in Russia. Alyans. Moscow. 2011:176 (in Russian).

Section III

The Forest Sector within an Environmental Paradigm

9

Environmental Activism and the Global Forest Sector

Jacki Schirmer

CONTENTS

9.1 Introduction

Modern environmental activism has had a profound influence on many industries globally in recent decades. In the global forest sector, campaigns by environmental activists have influenced timber supply and have also shaped the international markets into which forest products are sold. Within the last two decades, three key changes have happened in the global environmental activism field. Firstly, environmental activism has evolved from having a reliance on direct protest against and lobbying of various stakeholders to implementing sophisticated, multifaceted campaigns that target the entire forest products supply chain. Secondly, the number and type of concerns raised by environmental non-governmental organizations (ENGOs) broadened during this period from a predominant focus on natural forests to include tree plantations, biofuels, and carbon sequestration. Thirdly, environmental activism has increasingly focused on social justice issues in addition to environmental concerns.

Despite its expanding domain and growing influence, environmental activism is sometimes dismissed as an unnecessary action that lacks legitimacy. In the Australian plantation forestry sector, for example, Schirmer (2002) found that forest managers dismissed activists by calling them a "minority" who were uninformed and lacked legitimacy. In the United States, loggers dismissed environmentalists' concerns as being "idiocy" and based on "doomsday science" (Satterfield 1997). In British Columbia (BC), Canada, MacMillan Bloedel, a large forest company, rejected environmental activists' views as illegitimate in the 1990s (Zietsma et al. 2002). Such reactions are common when companies find their beliefs and existing processes challenged (Zietsma et al. 2002) and especially when activists pursue acts of civil disobedience to gain attention, such as blockades of forestry roads (Patterson and Allen 1997). Company reluctance to consider or respond to activists' views, due to fear that doing so might legitimate these views, reduces opportunities for learning and change, resulting in forest companies remaining defensive of—and hence "trapped" in—their original positions (Zietsma et al. 2002).

Whether environmental activism is viewed as a welcome intervention to achieve change, or as an unnecessary hindrance to the forest sector, it is ubiquitous across the forest products sector globally. By ignoring the increasingly influential role of environmental activism in the sector, forest products companies not only risk losing important opportunities to learn and develop but also risk disrupted operations, decline in sales or restricted access to funding from financial institutions (Gritten 2009). This chapter explores trends in environmental activism in the global forest sector. It does not attempt to evaluate the validity of the many claims made by environmental activists, governments, forestry companies, or any of the other stakeholders who interact as part of environmental campaigns. The chapter is organized as follows: First, I document the history of and emerging trends in environmental activism with a particular focus on environmental campaigns in the forest sector. Then, I examine the factors that influence the success of ENGO campaigns. I conclude with a brief discussion of the likely future for environmental activism in the global forest sector.

9.2 Environmental Activism: Growth of a Global Movement

The modern environmental "movement" has no single origin; it emerged "in different places at different times, and usually for different reasons" (McCormick 1989, 1). It is generally considered to have coalesced into an internationally recognizable force in the 1960s. During this decade, new environmental groups with a new and radically different approach for achieving change on environmental issues were formed. Unlike previous conservation groups that sought the patronage of government or businesses, these new groups adopted confrontational methods including media campaigns, street protests, blockades, and other acts of civil disobedience. The prominent activist group, Greenpeace, exemplified this type of protest by repeatedly sailing protest boats into waters near nuclear test zones in the 1970s and 1980s (McCormick 1989).

Despite its relatively recent manifestation, this modern form of environmental protest can be traced back to the impacts of unabated and exploitative use of natural resources during the last two centuries. In particular, the large-scale extraction of resources associated with the colonial era (from the seventeenth to the nineteenth century) and the simultaneous emergence of modern scientific methods for understanding the impact of humans on the environment led to the emergence of this modern form of environmental protest (Grove 1995).

Several developments triggered the emergence of new forms of environmental activism globally from the 1960s. The publication of *Silent Spring* by Rachel Carson in the United States in 1962 and environmental disasters such as the collapse of a mine pit heap in South Wales in 1966; oil spills on the coastline of Britain and the United States in the 1960s and 1970s; mercury poisoning in Minamata, Japan, in the 1960s–1970s; and the gas leakage from a Union Carbide plant in Bhopal, India, in 1984 are some of the milestone developments (McCormick 1989) that triggered a rapid increase in public awareness of the potential environmental implications of modern industrial practices. Concern over the implications of use of nuclear power and weaponry was also a strong factor that increased public concern about the environment and triggered the formation of many environmental groups (Athanasiou 1998). The simultaneous emergence of television and modern media helped spread these concerns among global audiences and provided a platform for prominent environmentalists such as Paul Ehrlich, Barry Commoner, Garrett Hardin, and others to influence the general public (McCormick 1989).

These various factors led to the evolution of concerns that were previously confined to a relatively small number of conservation professionals and interested citizens "into a fervent mass movement that swept the industrialized world" (McCormick 1989, 47). Hundreds of environmental groups were formed in Britain, the United States, and parts of Western Europe, seeking to take action to address the environmental consequences of modern industrial practices. These groups rapidly moved beyond their initial grassroots focus on local environmental issues to become "a major force in the politics of the environment" (Porter et al. 2000, 61) at both regional and international levels.

Environmental activism emerged primarily in the developed world but also grew rapidly in less developed countries from the 1960s, partly through active campaigns led by groups based in developed countries and partly through emergence of local grassroots movements. For example, by the 1980s, thousands of citizen groups in India were taking action on environmental issues, while the Green Belt Movement in Kenya was helping local communities combat desertification through tree planting (McCormick 1989). It must also be noted that underlying motivations and objectives of environmental activism differ between the developed and less developed countries. In less developed countries, environmental action is often oriented as much to improving the livelihoods of the poor through focusing on land rights, community development, and empowerment, as to achieving specific environmental outcomes. The environmental movement in developed countries, however, has typically focused more on preservation of wildlife and other such areas not directly concerned with human well-being (Taylor et al. 1993; Athanasiou 1998; Porter et al. 2000).

The term "environmental activism" in the forest sector typically conjures up images of protestors blockading forest roads, staging "tree sits," or chaining themselves to harvesting machinery. While this protest-based action was an initial focus of a great number of ENGOs, they rapidly adopted the use of sophisticated media campaigns, political lobbying, engagement with the government and business, and the development and implementation of market-based mechanisms,* to achieve their objectives. ENGOs also developed partnerships or otherwise expanded their reach to tackle issues globally. Increasing pressure on ENGOs to recognize issues of human rights has resulted in the refocusing of environmental discourses to include notions of justice, rights, and equity in addition to concerns about environmental sustainability (Sarkki and Heikkinen 2010). Overall, ENGO actions now span a much broader range of actions than the stereotypes of direct action that environmental activism is best known for. In the forest sector, their actions go well beyond the better-known acts of civil disobedience to include activities such as lobbying, participation on forest management committees, and collaboration between environmental activists and forest products companies.

Environmental activism is, almost by definition, spearheaded by ENGOs and is defined by the emergence of nongovernment organizations dedicated to acting on environmental issues. It is important to consider the different forms these organizations take before examining their activities in the forest sector. An ENGO is defined as a group that is established independently of governments or corporations and that operates as a nonprofit organization to promote environmental issues (Oberthur et al. 2002; Hayter 2003).

9.3 Growth of Environmental Activism in the Global Forest Sector

Modern environmental activism in the forest sector mirrors the evolution of environmental activism as a whole. It evolved in the 1960s and 1970s from small, localized campaigns to involve campaigns that rapidly gained wider national and international attention, with some of the better-known campaigns occurring in developed countries such as the United States, Canada, Finland, and Australia (Dargavel 1995; Wilson 1998; Gritten et al. 2012). As the forest products industry globalized, environmental activism globalized as well. Examples of globalized environmental activism are represented in two of the best-known environmental conflicts over forestry worldwide: disputes over logging in the province of BC in Canada (described in Box 9.1) and the state of Tasmania in Australia (Box 9.2). In both cases, concerns over logging of forests were addressed not only by protests in the forests themselves but by far-reaching campaigns targeting global markets for wood products originating from these forests. These examples, together with those in Boxes 9.3 and 9.4, are drawn on throughout this chapter to highlight key characteristics of environmental activism in the global forest sector. Activism in the forest sector has not been confined to developed countries. By the end of the 1970s, ENGO protests against forest management and the building of wood and paper-processing facilities were occurring in countries as diverse as Kenya and Indonesia (Gritten et al. 2012). The Chipko movement in India, described in Box 9.3, is perhaps the best-known and most often-cited example of grassroots environmental activism in the developing world involving the forestry sector. More recently, prominent environmental campaigns aiming to change forest practices have occurred in the Karelia region of Russia. Similarly, protests against deforestation, unsustainable

---

* Market-based mechanisms are described in more detail subsequently in this chapter, but broadly refer to actions that aim to influence the markets for wood and paper products, either by creating demand for products considered to be sustainably produced or by reducing demand for products considered to be unsustainably produced.

BOX 9.1 ENVIRONMENTAL ACTIVISM OVER THE COASTAL FORESTS OF BRITISH COLUMBIA, CANADA

Following decades of ongoing protest over logging of forests in various locations in BC, Canada, including the Clayoquot Sound on Vancouver Island, from the mid-1990s a coalition of ENGOs, including Greenpeace, Forest Ethics, the Sierra Club, and the Rainforest Action Network (RAN), among many others, launched ongoing campaigns aimed at changing logging practices in the central and northern coastal forests of BC. Environmental activists used a range of methods, including marketing campaigns that rebranded these forests in Canada and internationally as the "Great Bear Rainforest"; protests at logging sites; and targeting American and European destination markets for the wood products harvested from these forests. For example, protestors picketed Home Depot and IKEA stores to place pressure on these companies to stop purchasing products made from wood harvested in the region.

Following this initial phase of protest and influence building, during which ENGOs refused to participate in negotiations to end the conflict, a shift to more collaborative approaches occurred in 1999. Following initial meetings between the forest industry, government, and ENGOs, the "Joint Solutions Project" was initiated, involving these groups as well as representatives of First Nations. Environmental activists agreed to suspend market campaigns, and logging was deferred in 30 watersheds while discussions took place. In 2000, they agreed to defer logging while continuing to negotiate on the outcomes for forest management. A Coast Information Team (CIT) was initiated in which partisan and nonpartisan experts jointly evaluated evidence and made recommendations to the negotiating parties. The group included technical scientists and experts from both the forest sector and ENGOs and channeled debates over science into various working groups that were charged with assimilating the dueling priorities, models, and databases produced by partisan and nonpartisan interests (Clapp and Mortenson 2011, 908–909) and developing options and opinions that were then presented to the main negotiators.

This process facilitated the development of a shared understanding that formed the basis for generation of agreed approaches to forest management in the multi-stakeholder negotiations. In 2003, the groups reached consensus and recommended that around one-third of the region be placed into some form of conservation reserve and that logging practices be changed in much of the rest; this was followed by negotiations between the First Nations and the BC government, with the new plan of management adopted in 2006, and further rules and implementation processes continuing after this.

Sources: Hayter, R., *Ann. Assoc. Am. Geogr.*, 93(3), 706, 2003; Rossiter, D., Cult. Geogr., 11(2), 139, 2004; Affolderbach, J., *Econ. Geogr.*, 87(2), 181, 2011; Clapp, R.A. and Mortenson, C., Soc. *Nat. Resour.*, 24(9), 902, 2011.

logging, the establishment of tree plantations, and the construction and operation of pulp mills have taken place in regions as diverse as Indonesia, China, and Uruguay (Table 9.1). In many campaigns involving developing countries, international ENGOs invest in capacity building of local ENGOs. For example, Lei (2009) has documented how international ENGOs, such as Greenpeace and World Wildlife Fund (WWF), have

BOX 9.2   ENVIRONMENTAL ACTIVISM OVER HARVEST
OF NATIVE FORESTS IN TASMANIA, AUSTRALIA

The state of Tasmania is the site of some of Australia's most famous environmental protests, some of which—such as campaigns against the damming of rivers for hydropower—were catalysts for development of the environmental movement nationwide. Environmental activists raised concerns over logging of natural (native) forests from the 1970s, as the practice of producing woodchips for export expanded rapidly. During the following decades, coalitions of local and national ENGOs protested both the harvest of native forests and the building of large new processing plants. Examples of high-profile disputes include protests against the establishment of a proposed pulp mill in Wesley Vale in the late 1980s (the protests were successful in stopping the mill being established), campaigns to have areas of the Lemonthyme and Southern Forests recognized as World Heritage areas and reserved from timber harvest, and a series of campaigns to halt logging in other areas of forest argued by ENGOs to have important conservation values (particularly campaigns related to forests in the Styx Valley and Tarkine regions).

In 2003, the forest company Gunns Ltd proposed building a pulp mill in the Tamar Valley. This triggered an ongoing campaign by a range of groups, including many ENGOs, against the pulp mill, which continued at the time this chapter was written. Direct protest, legal challenges, lobbying, and market campaigns have all been used to protest the mill.

In 2010, representatives of unions, industry, and ENGOs negotiated for several months, reaching a "Statement of Principles" that provided a basis for moving forward with seeking resolution of forest conflict. Following this, the negotiating parties invited the Tasmanian and Australian governments into the negotiation process. On July 24, 2011, an agreement was made in which the governments committed to support an independent assessment process to identify areas of forest to be reserved from harvest. In late 2012, a "peace agreement'" was signed, and in early 2013 passed into legislation.

Sources: Dargavel, J., *Fashioning Australia's Forests.* Oxford University Press, South Melbourne, Victoria, Australia, 1995; Gale, F., *Aust. J. Publ. Admin.*, 67(3), 261, 2008; Affolderbach, J., *Econ Geogr.*, 87(2), 181, 2011; Schirmer, J. et al., *Socioeconomic Impacts of Forest Industry Change: A Baseline Study of the Tasmanian Forest Industry*, CRC for Forestry, Canberra, Hobart, Australia, 2011; author's personal communication and participation in the IVG.

supported local Chinese ENGOs to monitor and campaign against logging operations. Greenpeace and WWF established Chinese offices through which they worked with a Yunnan-based ENGO, Green Watershed, to campaign against what they argued were illegal logging operations by Asia Pulp and Paper (APP) in the Yunnan province. These campaigns gained media attention within China and triggered investigation of APP's practices by the Chinese government.

In addition to the environmental campaigns described in Boxes 9.1 through 9.4, Table 9.1 outlines several other high-profile environmental campaigns in the forest sector. This illustrative list represents only a small fraction of the hundreds of conflicts documented

BOX 9.3   THE CHIPKO MOVEMENT

Rapid economic development in post-independence India included the granting of logging licenses to a number of companies, covering large areas of Himalayan forests. The logging, often involving widespread deforestation, had multiple negative consequences: local people, particularly women, often lost access to resources critical for their livelihood including the right to gather firewood and nontimber forest products. In addition, the deforestation resulted in landslides and erosion.

In 1973, members of some Himalayan communities, frustrated at the removal of their rights to access and use local forests, formed what was later termed the Chipko movement. This movement was characterized by a unique form of nonviolent activism: in the first of the protests that characterized the movement, a group of female villagers hugged trees to prevent them from being felled (the word "chipko" means to "hug" or "cling"). These protests acted to prevent logging in many cases, and forced changed in how the government managed access to forests.

The Chipko movement is itself contested, with a diverse literature variously representing the movement as a peasant movement focused on livelihood, an ecological movement, and an eco-feminist movement; in reality, the movement evolved in all of these directions, with some activists promoting employment of local people, and others on stopping logging irrespective of who benefited from it.

The movement rapidly spread from one village to encompass multiple villages in the Himalayas through the 1970s, and spread further throughout India in the 1980s. It became a catalyst for civil society movements that subsequently spread throughout India and in the developing world more broadly, and inspired environmental groups worldwide to use similar nonviolent protest methods. It was successful in achieving change in how rights to use forests were granted in some parts of India.

Sources: Guha, R., *The Unquiet Woods: Ecological Change and Peasant Resistance in the Himalaya*, Oxford University Press, New Delhi, India, 1989; Taylor, B. et al. Grassroots resistance: The emergence of popular environmental movements in less affluent countries, in Kamieniecki, S. (ed.) *Environmental Politics in the International Arena: Movements, Parties, Organizations and Policy*, State University of New York Press, Albany, NY, 1993, pp. 69–89, 1993; Linkenbach, A., *Forest Futures: Global Representations and Ground Realities in the Himalayas*, Permanent Black, Uttarakhand, India, 2007.

worldwide that involve the global forest sector. Gritten and Mola-Yudego (2011) have identified over 300 instances of conflict in a recent review of academic and ENGO literature. Further, Mola-Yudego et al. (2012) and Mola-Yudego and Gritten (2010) found that forest conflicts are more common in Europe, North America, South America, and Southeast Asia than in African and central Asian countries. Overt conflicts only represent a fraction of environmental activism because, for a variety of cultural, political, or financial reasons, environmental activism does not always involve an overt conflict and is not always brought to public attention.

Around the world, hundreds of ENGOs are active on forestry and forest products-related issues. These include well-known and prominent ENGOs such as Greenpeace, the WWF, the RAN, and Global Forest Watch, to name a few. While providing an exhaustive

**BOX 9.4   ENVIRONMENTAL ACTIVISM FOCUSED ON A
FOREST COMPANY: THE CASE OF APRIL IN INDONESIA**

In some cases, environmental activism in the forest products sector involves multiple
interacting issues. The case of ENGO campaigns against the Indonesian operations
of Asia Pacific Resources International Holdings Ltd (APRIL), a privately owned pulp
and paper company, is one example. APRIL has 330,000 hectares of government-
awarded concessions in Indonesia. It operates a pulp and paper mill in Sumatra that
uses wood harvested from natural forests and from plantations is has established
since 1993.

More than 50 ENGOs, many operating internationally as well as some based
locally, have argued that APRIL is engaged in inappropriate practices such as illegal
logging, establishing plantations on land that Indigenous people and other members
of local communities claim rights to, and deforestation.

ENGOs have campaigned using multiple methods. In particular, they have used
market campaigns in which they lobby financiers of APRIL and purchasers of its
products to sever ties with the company unless it changes its practices. These campaigns have resulted in some changes to company practices, according to Gritten
and Kant (2007), although the company itself argues that actions taken that appear to
address some ENGO concerns have not been driven by ENGO pressure.

ENGOs have also invested in producing reports that they argue verify their claims
and concerns, while APRIL has responded by funding research and audits of its
practices to support its claims that it is operating in a legal and sustainable manner.

Sources: Gritten, D. and Kant, P., *Int. For. Rev.*, 9(4), 819, 2007; Gritten, D. and
Saastamoinen, O., *Soc. Nat. Resourc.*, 24(1), 49, 2010.

list of all ENGOs and the issues they are pursuing is a daunting task, Table 9.2 provides a
brief overview of some of the prominent international ENGOs engaged in activism in the
global forest sector and some of the issues and campaigns they were actively engaged in
during 2011 and 2012.

As can be seen, international ENGOs vary considerably in their focus and activities:
some focus on information provision and networking (such as forests.org and the World
Resources Institute [WRI]), others on environmental education and working with businesses and consumers to change practices (such as the Rainforest Alliance and WWF),
and still others on direct protest and investigations of forest practices (such as the
Environmental Investigation Agency [EIA] and Greenpeace International). Some have a
strong focus on social justice issues, such as land rights and reducing poverty (e.g., World
Rainforest Movement [WRM] and Forest Peoples Programme [FPP]); others focus less on
these issues. Some focus solely on issues related to forests, but most campaign on both
forest-related issues and a broader set of environmental issues.

Sometimes, ENGOs also differ in how they approach forest products issues: for example, Global Witness has criticized WWF's Global Forest and Trade Network (GFTN) program (discussed later), arguing that WWF is allowing unsustainable timber harvest to
occur while giving it a stamp of approval (Global Witness 2011). However, collaboration
among ENGOs is probably more common than conflict: It is common for local branches
of these international groups to form alliances to work on specific forest-related issues.

TABLE 9.1

Examples of Environmental Activism Involving the Global Forest Sector

Location, Forest Company/ Agencies Involved, and Time Frame	Issues Involved and Dates
Tasmania, Australia Gunns, Forestry Tasmania 1960s onward	This conflict includes ongoing protests against logging of publicly owned native forests in multiple locations in Tasmania, as well as campaigns protesting proposals to establish a pulp mills in the 1980s and in the late 2000s. In addition, some conflict has occurred over the establishment and management of tree plantations.
APRIL, Sumatra, Indonesia, and APP, Indonesia 1990s onward	These conflicts are focused on APRIL and APP, pulp and paper companies operating in Indonesia. Multiple ENGOs have protested clearing of natural forests for the establishment of plantations by these companies, as well as the other sources of timber used in the mills. The conflicts include social issues focused around indigenous rights and rights of local communities to forest access, in addition to concerns about environmental impacts. Several other conflicts have also occurred in Indonesia, often involving the same ENGOs acting in the APRIL and APP conflicts.
Veracel pulp mill, Brazil 1990s onward	Conflict over the establishment of plantations and their processing by the Veracel pulp mill in Brazil.
APP, China 2000s onward	ENGOs including the World Rainforest Movement have criticized a number of operations of APP in southern China, including establishment of plantations and of pulp and paper mills, with concerns about both the plantations established in China and claims that the mills use illegally logged timber.
The Flow Country, Scotland 1970s–1990s	Conflict over the "Flow Country" involved concerns over the establishment of tree plantations on bog areas that forms critical habitat for migrating bird species. The conflict, which was the culmination of many years of concern over afforestation in other areas, led to major changes in management of afforested land throughout Great Britain. This conflict ended in the 1990s when afforestation ceased.
Pan African Paper Mill, Kenya 1970s–2000s	ENGOs campaigned about issues of air and water pollution by the Pan African Paper Mill, as well as concerns about deforestation believed to result from the mill's operations.
Pacific Northwest "spotted owl" conflict, United States 1970s–1990s	ENGO campaigns against harvesting of publicly owned natural forests in the Pacific Northwest of the United States resulted in a drastic reduction of logging in 1991.
Uruguay Metsä Botnia and ENCE pulp and paper mills 2000s	Environmental activism has played a part in complex conflicts that also involve other dimensions. The conflicts are over construction of pulp mills on the River Plate, with environmental and labor rights concerns among those raised about the mills. This is also an example of border conflict, with the conflict involving Argentinian ENGOs and residents who believed the pulp and paper mills would pollute waters in the river that forms a border between the two countries. The Argentinian and Uruguay governments brought the conflict to the International Court of Justice to seek resolution and in 2010 established joint management of river activities.
Finland 1990s (with conflict in many cases having earlier origins)	Conflicts over use of forests have occurred in many locations, often focused on concerns over harvesting of publicly managed forests. High-profile instances include conflict over harvesting of state-managed forest in the Inari region, which includes issues around indigenous land rights as well as concerns about environmental sustainability, and more recent conflict in the Muonio region, again over harvest of forests by the state forest agency Metsähallitus.

(continued)

TABLE 9.1 (continued)

Examples of Environmental Activism Involving the Global Forest Sector

Location, Forest Company/ Agencies Involved, and Time Frame	Issues Involved and Dates
Malaysia 1980s onward	Indigenous people protested logging of forests in Sarawak, through actions including blockades of logging roads. The campaigns have involved partnerships between the indigenous people and a range of ENGOs, with the campaigns gaining wide international attention.
Karelia, Russia, 1990s onward	ENGOs campaigned against harvesting of areas of Karelia's forests they argued had high conservation value, with a strong focus on market campaigns targeting international purchasers of wood from the region.

Sources: WRM/SAM (World Rainforest Movement/Sahabat Alam Malaysia), *The Battle for Sarawak's Forests,* World Rainforest Movement and Sahabat Alam Malaysia, Penang, Malaysia, 1990; Bendix, J. and Liebler, C.M., *Ann. Assoc. Am. Geogr.*, 89(4), 658, 1999; Warren, C., *Scott. Geogr. J.*, 116(4), 315, 2000; Gritten, D. and Kant, P., *Int. Forest. Rev.*, 9(4), 819, 2007; Kortelainen, J., *Environ. Plann. A.*, 40(6), 1294, 2008; Gritten, D. and Mola-Yudego, B., *Int. J. Comm.*, 4(2), 729, 2010; Gritten, D. and Saastamoinen, O., *Soc. Nat. Resour.*, 24(1), 49, 2010; Sarkki and Heikkinen, 2010; Affolderbach, J., *Econ. Geogr.*, 87(2), 181, 2011; Kröger, M. and Nylund, J.-E., *Forest. Pol. Econ.*, 14(1), 74, 2012.

Illustratively, Indonesia-based branches of Friends of the Earth (FOE) and WWF have partnered with a local ENGO, Jikalahari "Riau Forest Rescue Network" to monitor logging activities in the Riau province of Sumatra, Indonesia (EOTF 2012), in just one of countless examples of ENGO alliances working on specific forest campaigns around the world.

9.4 Issues and Campaigns: What Triggers Environmental Activism in the Global Forest Sector?

Mola-Yudego and Gritten (2010) highlighted the wide array of issues igniting conflict in the forest sector globally. These issues include clearing of forest for agricultural purposes, bioenergy plantations, establishment of nature reserves, deforestation, use of genetically modified material, illegal logging, indigenous rights, operations of forest companies including wood and paper processors, establishment and management of plantations, and conflicts between stakeholders about how to best manage forests.

While these issues are varied and have changed over time, some underlying themes can be observed in the concerns raised by environmental activists in the global forestry arena. In particular, concerns about the sustainability of the following practices or activities have commonly formed the bases of activism by ENGOs:

- Deforestation, in which forests are cleared for agriculture, establishment of tree plantations, or other purposes and the cleared trees are sold into wood products markets
- Harvest of natural forests, particularly where clear-felling harvest techniques are used

TABLE 9.2

Examples of ENGOs Working on Issues Related to the Global Forest Sector at the International Scale

ENGO Name	What Forest Sector Objectives Do the Group Have?	Examples of Forest-Related Activities and Campaigns Operating in 2011 and 2012
Environmental Defense Fund (EDF)	EDF aims to achieve change in forest practices through legal reform and market-based responses to environmental problems, with a strong focus on evidence-based response to environmental problems.	The Amazon Basin project aims to build capacity of indigenous peoples to take part in deforestation dialogues, access market opportunities such as Reducing Emissions from Deforestation and Forest Degradation (REDD), and ensure they benefit from efforts to reduce deforestation. (http://www.edf.org/)
Environmental Investigation Agency (EIA)	The EIA uses undercover investigations to gather evidence of unsustainable forest management, also involved in international negotiation processes.	In April 2012, EIA released a report on imports of illegally felled timber from the Peruvian Amazon into the United States. (http://www.eia-international.org/)
Forest Peoples Programme (FPP)	Focuses on supporting people whose livelihood depends on forests to secure their rights to the forests and ability to control their land and be heard in decision-making processes.	FPP has multiple projects in which they work with local communities in a range of countries. They also take part in international discussions and negotiations on issues related to forests and livelihoods. (http://www.forestpeoples.org/news)
forests.org (part of Ecological Internet)	forests.org aims to facilitate the sharing of information on deforestation and forest degradation and through this to facilitate interaction and action on these issues globally.	forests.org has regularly updated information on various ENGO campaigns to address deforestation and forest degradation globally and provides links to a range of organizations and information sources on these issues. (http://forests.org)
Friends of the Earth (FOE) (umbrella group for multiple national and local ENGOs)	FOE engages in a range of campaigns related to use of forests for wood products and campaign on both environmental and social justice issues. The specific issues vary for different member groups.	2012 campaigns include the "land grabs" campaign protesting communities losing use of land when it is leased or sold to corporations for uses including tree plantations for wood and biofuel and the "climate and biodiversity" campaign protesting actions such as REDD and other carbon offset actions. (http://www.foei.org)
Global Witness	Campaigns against unsustainable use of natural resources, with a strong focus on social justice issues.	Recent campaigns include a focus on stopping illegal logging in Madagascar and campaigning against WWF's GFTN based on concern over actions of network members (Global Witness 2011). (http://www.globalwitness.org/)
Greenpeace International (umbrella group for multiple Greenpeace organizations worldwide)	Greenpeace aims to reduce forest deforestation and degradation, through direct action, lobbying, working with indigenous communities, and investigations.	In early 2012, campaigns included the sailing of the Rainbow Warrior (with a flotilla of other boats) in the Amazon to protest deforestation and release of an investigation into logging practices by APP that argued APP was breaking Indonesian laws by processing illegally felled trees. (http://greenpeace.org/international/en/campaign/forests)

(continued)

TABLE 9.2 (continued)

Examples of ENGOs Working on Issues Related to the Global Forest Sector at the International Scale

ENGO Name	What Forest Sector Objectives Do the Group Have?	Examples of Forest-Related Activities and Campaigns Operating in 2011 and 2012
International Alliance of Indigenous and Tribal Peoples of the Tropical Forests (IAITPTF) (umbrella organization for multiple member groups)	The IAITPTF aims to promote recognition of the rights of indigenous and tribal peoples living in tropical forest regions to land, to participate in decision making, and to assist these peoples to network, share information, and act jointly.	The IAITPTF holds meetings bringing its members groups together as well as participating in international discussions and negotiations. (http://www.international-alliance.org/index.html)
Rainforest Action Network (RAN)	RAN uses direct action and campaigns targeting markets for paper products to protest harvesting of natural forests and establishment of plantations for paper production.	The "Rainforest-Free Paper Campaign" is seeking a moratorium on conversion of forests and peatlands to tree plantations for paper production in Indonesia, with a focus on market campaigns to leverage change. (http://ran.org)
Rainforest Alliance	The Rainforest Alliance aims to achieve tropical forest conservation and alleviate poverty through working with communities and businesses to ensure sustainable forest management can also be profitable. They have a particular focus on market-based certification, with associated environmental education campaigns.	The Rainforest Alliance has continued ongoing work implementing and promoting its forest certification programs and encouraging development of markets for certified wood. (http://www.rainforest-alliance.org/)
World Rainforest Movement (WRM)	Focuses on improving local people's rights through campaigns against expansion of tree plantations (particularly monocultures and genetically modified plantations established by corporations), deforestation, pulp mill developments, and promotion of social justice.	WRM facilitates communication and networking among a wide range of groups, as well as representing their views in international processes discussing forest-related issues. It does not work in partnership with governments or business. (http://www.wrm.org.uy/)
World Resources Institute (WRI)	The WRI aims to provide timely information on issues such as illegal logging, deforestation pressures, and other threats to forest sustainability, for a range of regions globally. They work with a range of institutions including governments and NGOs in these countries.	The Forest Landscapes Initiative "seeks to increase the ability of governments, businesses, and civil society to protect intact forest landscapes, manage working forests more effectively, and restore deforested lands" in Central Africa, Southeast Asia, Russia, and South America. (http://www.wri.org/project/global-forest-watch)

TABLE 9.2 (continued)

Examples of ENGOs Working on Issues Related to the Global Forest Sector at the International Scale

ENGO Name	What Forest Sector Objectives Do the Group Have?	Examples of Forest-Related Activities and Campaigns Operating in 2011 and 2012
World Wildlife Fund (WWF) (umbrella group for multiple WWF organizations worldwide)	WWF aims to achieving forest conservation and improve local livelihoods that depend on forests.	WWF has a long history of forest-related campaigning. In 2012, one of WWF's 13 global initiatives was to focus on promoting REDD as a way of promoting forest-based carbon and reducing climate change impacts. WWF leads the GFTN, a global network of forest sector businesses, communities, and NGOs in more than 30 countries aimed at creating markets for sustainable forest products. (http://www.wwf.org)

Source: Information in this table is drawn from the website of the organization unless otherwise specified.

- Establishment of tree plantations on (1) land cleared of natural forests for the purpose of plantation establishment or (2) agricultural land already cleared of trees and (3) establishment of genetically modified trees in plantations
- Harvest of natural forest or tree plantations for particular products, particularly woodchips and bioenergy production
- Establishment of pulp and paper mills or other wood-processing facilities

In the following discussion, I briefly describe how environmental activism has sought to address each of these themes.

9.4.1 Deforestation

The clearing of natural forests is a common focus of forest-related activism globally, as can be seen in Tables 9.1 and 9.2. Clearing here refers to the clearing of forests and conversion of land to non-forest uses such as agriculture. Numerous campaigns have been launched in a number of countries to protest deforestation including the Chipko movement described in Box 9.3. Other prominent deforestation campaigns involving grassroots and international ENGOs, often acting in collaboration with each other, include the following*:

- Sarawak, Malaysia, 1980s onward: Campaigns by indigenous groups, in collaboration with international ENGOs such as RAN, to end logging of the forests they have traditionally depended upon, using methods ranging from nonviolent action to "ecotage" involving destruction of forest machinery, and market campaigns (Taylor et al. 1993).
- Brazil, 1980s onward: Rubber tappers and indigenous groups, joined by a range of local, national, and international ENGOs, have protested the deforestation of the Amazon, seeking to preserve the forests, communities' way of life, and their livelihoods (Taylor et al. 1993).

---

* It should be noted that in some cases, claims that areas are being deforested are themselves contested, with governments or the forest products industry arguing that what ENGOs claim to be deforestation is sustainable harvest. No attempt is made here to evaluate the various claims and counterclaims.

- Cambodia, 1990s onward: Illegal deforestation was monitored by Global Witness (also see Table 9.2). Subsequently, the World Bank and other agencies put pressure on the Cambodian government to introduce independent forest monitoring. Between 1999 and 2004, Global Witness conducted this independent monitoring with financial support from the Australian and UK governments (Humphreys 2006).

9.4.2 Harvest of Natural Forests Not Involving Deforestation

The harvest of natural forests is commonly a target of environmental activism even when it does not involve deforestation (in other words, when the goal is to harvest and regrow the forest rather than to clear it). Anti-logging campaigns have occurred in many countries (see Table 9.1 and Boxes 9.1 and 9.2, for examples). Concerns expressed by ENGOs typically focus on the sustainability of the harvest techniques used, particularly whether they present a threat to wildlife or plant biodiversity or cause soil erosion and siltation of waterways. Other concerns include the use of clear-felling techniques (logging of all trees in a large single area followed by forest regeneration), the methods used to regenerate forests post logging, and the harvesting of old-growth forests. Sometimes, all these issues are raised simultaneously: Environmental activists in Tasmania (e.g., The Wilderness Society) have raised all these issues as part of ongoing campaigns aimed at reducing logging of natural forests (see Box 9.2).

9.4.3 Tree Plantations

Plantation forestry is also a source of concern for many environmental activists, particularly where a natural forest is cleared to provide land for plantation. Plantations established on cleared agricultural land are also the subject of environmental activism in many cases (see Carrere and Lohmann 1996; Schirmer 2007), though some ENGOs differ on this issue. In fact, some ENGOs promote plantations as an alternative to logging of natural forests, while others argue that the rapid expansion of plantations in recent decades has created new environmental problems (Cossalter and Pye-Smith 2003). Briefly, issues raised by ENGOs include concerns about (Cossalter and Pye-Smith 2003) the following:

- The impacts of establishing monoculture plantations of tree species exotic to the location in which they are planted, with concerns about risks of pest outbreaks, loss of biodiversity, and interbreeding with local tree species
- The use of genetically modified trees, with concerns about impacts on biodiversity, interbreeding, and pest outbreaks
- The practices used to establish and manage plantations, particularly whether they result in erosion, siltation of waterways, and the impacts of chemical application for weed and pest control
- Displacement of people whose livelihood previously relied on the land on which plantations are established, particularly in countries with insecure land tenure*
- Whether plantations generate local employment or similar amounts of employment compared to previous land uses

---

* Examples of insecure land tenure include where no documented ownership or use rights exist over an area of land or in which land ownership and access rights are contested.

9.4.4 Use of Trees for Biofuel

The 2000s saw the emergence of ENGO concerns about the harvest of natural forests for bioenergy and biofuels production. Proponents argue that bioenergy and biofuels present opportunities for renewable, sustainable energy sources that can substitute for use of fossil fuels. ENGOs contend that demand for low-priced, high-volume products will trigger unsustainable logging and potentially also forest management techniques that are targeted to maximizing production of wood volume for biofuel at the expense of ecological characteristics and integrity of the forest. Such concerns are often linked to debates about how best to respond to climate change and the role of forests in these processes, with some ENGOs arguing forests should be focused on as carbon sinks rather than harvested for biofuel (see Penn 2011 for a discussion of ENGO concerns).

9.4.5 Wood and Paper-Processing Facilities

The establishment and operation of wood-processing and paper-manufacturing facilities often become a subject of environmental activism. This is particularly true for pulp and paper mills, where concerns focus on both the volume of wood harvest required from nearby forests to meet their fiber demand and the nature of the operations of mills—for example, whether they use unsustainable volumes of water and/or electricity, and pollute waterways and air with effluent. For example, as described in Table 9.1, the construction of two pulp mills in Uruguay was opposed in part due to concerns about whether they would pollute the River Plate. Similarly, ENGOs have campaigned against the operations of Ta Ann, a producer of veneer products in Tasmania, Australia, although the concerns in this case are primarily related to their belief that the establishment of Ta Ann resulted in unsustainable volumes of harvesting from natural forests (for details, see Arndt 2012).

9.5 Effects of Globalization on ENGO Campaigns in the Forest Sector

The rapid globalization of environmental activism has had an impact on how environmental campaigns emerge in the forest sector. Rather than relying on "bottom-up" campaigns that are initiated by locally based ENGOs, international and national ENGOs increasingly generate local campaigns from the "top-down." One example of this "top-down" approach is on page X, which outlines how international ENGOs have built local ENGO capacity in China to protest logging operations (Lei 2009). In other cases, international ENGOs directly invest in building local campaigns around issues they have identified as being of interest at a national or international scale. For example, after the collapse of the Soviet Union, Greenpeace and WWF invested in building both international and local campaigns aimed at changing forest practices in forests in the Karelia region of Russia, which they argued should be considered "old-growth" forest with high biodiversity value (Kortelainen 2008).

    While becoming more global in reach over time, environmental activism has become more complex as well. Instead of involving a single group protesting in a forest or at a forest products company site, activism now typically involves a network of ENGOs, often working in alliance with other organizations outside the ENGO sector. These networks of groups often design an overarching campaign that incorporates multiple simultaneous actions such as public education, lobbying of governments, protests and demonstrations, and market-based actions.

A typical example of an overarching campaign is Tasmania, Australia, where environmental activists have campaigned against the logging of publicly owned native forests for four decades (Box 9.2). During this period, multiple campaigns have protested logging in different locations within Tasmania. In addition, there have been campaigns to prevent establishment of a pulp mills in Wesley Vale and Bell Bay (Sonnenfeld 2002; Gale 2008). In each of these campaigns, a diverse coalition of ENGOs has been involved and applied a diversity of actions as part of the overarching campaign. Affolderbach (2008) identified coalitions that included local ENGOs focused on Tasmania (e.g., the North West Environment Centre, Environment Tasmania, the Tasmanian Conservation Trust, and the Tasmanian National Parks Association), national ENGOs (e.g., The Wilderness Society, the Australian Conservation Foundation, and FOE Australia), and international ENGOs (e.g., WWF, Greenpeace, and RAN). Such multilayered, complex coalitions have provided the skills and resources needed for ENGOs to undertake wide-ranging actions from locally based protests to international lobbying of Japanese firms purchasing Tasmanian woodchips.

A North American example of complex coalitions of ENGOs is given by Affolderbach (2008) who lists 13 core ENGOs involved in a forest conflict in BC and identified how they each took responsibility for different actions that included direct protest, providing scientific reports and evidence, market campaigns, and government lobbying.

Multilayered, complex coalitions serve multiple purposes. First, they provide advantage arising out of differentiation among various ENGOs. Based on their focus, ENGOs are often placed on a spectrum ranging from "bright green" to "dark green." "Bright green" or "light green" ENGOs are considered moderate in their action. They typically engage in collaborative and moderate strategies to achieve change, such as direct negotiations and collaboration with forest companies and governments. They often focus on social issues as well as environmental issues (e.g., ensuring fair labor laws). In the forest sector, groups such as WWF and the Rainforest Alliance would typify the "lighter green" approach, as their work involves collaboration with industry and governments, and using market-based approaches. "Dark green" ENGOs, on the other hand, are more radical and engage in direct forest action such as blockades, protests, and site invasions and typically have a "deep ecology" philosophy that is highly biocentric, as it focuses on considering the needs of the environment independently of the needs of humans to use the environment (Hayter 2003; Hoffman 2009). Greenpeace International is an example of a "dark green" ENGO relative to, for example, WWF and the Rainforest Alliance. In general, more moderate ENGOs are more likely to work within existing institutions and decision-making processes, while more radical organizations are more likely to engage in confrontations via direct protest (Alcock 2008). While these differences can sometimes result in ENGOs choosing not to partner with each other, often they result in strong coalitions in which each engages in actions best suited to their philosophy, support base, and skills.

9.6 Strategies Used by Environmental Activists to Achieve Change

ENGOs are unique in their ability to use a diverse combination of strategies ranging from acts of civil disobedience to collaboration with forest products companies (Affolderbach 2011). The stereotypical image of an "environmental activist" masks the reality of the sophisticated, wide-ranging, and multipurpose forms that modern environmental activism takes.

The choice of the methods used in any particular environmental campaign depends on many factors. Firstly, it depends on the ENGO itself: As ENGOs have diversified, so have the methods they use, and an ENGO will select the strategies best suited to their particular goals and philosophical perspective (Alcock 2008). For example, Greenpeace favors direct protest, which is consistent with its "deep green" perspective and its philosophical perspective that more collaborative engagement with groups such as government or industry risks inappropriate compromise of the goals the group seeks to achieve. Secondly, choice of methods depends on the cultural context within which an ENGO works. For example, in a situation where a local community is hostile to ENGOs, less confrontational methods designed to gain local support, such as holding community gatherings and delivering education campaigns, may be used in preference to direct forest protests (Affolderbach 2011). The methods chosen also depend on whether the local culture or those in power are accepting of the use of methods such as protests and litigation (Lovelock 2005); for example, the 2012 murders of two Cambodian activists who had protested illegal logging in the media may reduce the use of media-based activism in that country (Plokhii 2012) Thirdly, governance frameworks also influence the choice of methods. If existing decision-making processes provide opportunities for ENGOs to be heard and bring about change, they will likely be used in preference to taking external actions such as protest. Cartwright (2003, 116) suggests that to be effective, ENGOs must choose methods compatible with the views and values of their direct support base, the general public, the groups they seek to change, and decision makers such as government policy makers.

Figure 9.1 illustrates the range of methods drawn on by environmental activists in campaigns involving the forest sector. These methods range from those that seek to deliberately provoke conflict and antagonism to others that seek to improve relations and ability to work directly with the forest sector. As such, they are represented on a spectrum from conflict to collaboration. Actions involving high conflict include direct protests such as blockading roads to forest harvest sites and picketing outside markets where forest products are sold and protestors chaining themselves to forest machinery. Collaborative methods include round-table stakeholder negotiating processes and situations where ENGOs are included in decision-making processes regarding forest management. Between these two extremes lie a wide range of methods that include litigation, media and education campaigns, and market persuasion.

In order to meet their overall goals, ENGOs often draw on a range of common methods. In the following discussion, I identify and describe some of the key strategies that ENGOs typically choose from to achieve their specific objectives.

FIGURE 9.1
Range of methods used by environmental activists in forest sector campaigns.

9.6.1 Creating a "Brand" for the Forest

Many ENGOs aim to increase public recognition of and attachment to the forest that is the focus of their campaigns (Affolderbach 2011). Recognition development requires creating an identity for the forest—making it a recognizable "place"—through naming it, mapping it, and describing it (Hayter 2003; Rossiter 2004). In BC, for example, the name "Great Bear Rainforest" was given to central and north coast forests by ENGOs in the 1990s to provide an identity to their campaigning in this region (Rossiter 2004). Similarly, in Tasmania, names such as "Valley of the Giants" and the "Tarkine" have been used to provide forests a readily accessible identity (Affolderbach 2011). These names are commonly accompanied by the creation of maps delineating the boundaries of the forest that a campaign focuses on, again emphasizing and strengthening public recognition of that forest area.

Forests are also given individual identities through what are sometimes termed "valley by valley" campaigns, in which campaigns shift to different forest localities sequentially within a region (Hayter 2003). Thus, in any given region, an overarching campaign will often be an umbrella for multiple distinct campaigns specifically linked to individual areas of forest. Tasmanian forests are a case in point: The long-running campaigns in that region described earlier have involved the naming of multiple different areas of forest as the logging of each is protested (Affolderbach 2011).

To accompany the naming of forests and focus on place-based campaigns, ENGOs also promote specific images of the forest, particularly those suggesting a pristine nature or in the case of conflicts involving debate about old-growth forests, reinforcing the concept of the forests as being ancient and irreplaceable. These terms are then contrasted with evocative language that associates forest harvest with the theme of destruction. For example, Rossiter (2004) documents how the use of terms such as "biological desert" or of forests being "stripped" was used in promotion materials disseminated by Greenpeace as part of protesting forest harvest in BC. These terms were subsequently used by the general public in their letters to newspapers and at other public forums, providing evidence of the influence of the terminology promoted by Greenpeace on public perception and communication of the impacts of logging.

9.6.2 Using Scientific Information

Scientific data play a critical role in many conflicts surrounding environmental issues. In the forest sector realm, environmental activists often claim to have scientific evidence of the unsustainability of logging practices. Forest products companies, as a response, typically draw on their own scientific evidence to refute activists' claims. As a result, any campaign over an environmental issue typically involves claims and counterclaims about the "scientific evidence" surrounding an issue, as well as calls for the use of "science" to evaluate the claims of different stakeholders (Satterfield 1997; Johnston and Soulsby 2006). ENGOs often fund research and employ scientists to generate scientific information on the issues they campaign on (Affolderbach 2011; Clapp and Mortenson 2011). The presentation of conflicting scientific evidence by different stakeholder groups (ENGOs, forest products companies, etc.) typically leads to the problem of "adversarial science," wherein each group claims validity of their findings while refuting others' (Satterfield 1997).

An example of adversarial science can be seen in environmental activism over harvest of natural forests in Australia, where a number of ENGOs have campaigned for cessation of harvesting of natural forests in the state of Victoria. To provide support for their case, two Australian ENGOs—the Australian Conservation Foundation and The Wilderness

Society—commissioned a report in 2010 that sought to demonstrate that it was technically feasible to cease logging of natural forests while still supporting the Victorian wood and paper products industries by shifting to using wood grown in plantations (NIEIR, 2010). In response, VicForests, the state-owned agency responsible for managing publicly owned natural forests for harvesting, and Australian Paper, one of the largest customers for wood harvested from natural forests, commissioned a report prepared by Pöyry Management Consulting (2011) that refuted many of the arguments made by NIEIR (2010) and argued that it would be difficult to transition to using plantation wood.

Sometimes, however, ENGOs and forest companies engage in processes that enable them to resolve their science-based adversarial stances. In the multistakeholder negotiations described in Box 9.1, forest industry and ENGOs were able to develop a shared and agreed understanding of scientific data in BC, Canada, that was crucial to achieving a joint agreement about how to manage those forests.

9.6.3 Influencing Governments

Environmental activism in the forest sector in the 1970s and 1980s was often directed at lobbying governments to seek changes in government regulation of forest management practices (Wilson et al. 2001). Putting pressure on governments can involve face-to-face lobbying, letter writing/e-mailing, and phone campaigns to pressure government officials and representatives to take action (Affolderbach 2008), as well as litigation (discussed further in Section 9.6.4).

To convince governments to change is, however, not always the most effective approach for ENGOs, because it can be a lengthy and daunting process. Because of these challenges, many ENGOs have, over the last two decades, shifted away from targeting governments and embraced market-based strategies that directly target forest-based companies. Directly targeting forest products companies can provide ENGOs with leverage to bring about desired changes more readily than lobbying governments (Alcock 2008; Gritten and Mola-Yudego 2010); as Walter (2003, 532) explains: "market based campaigns enable ENGOs to move forward with civil society-based initiatives that can influence logging companies' activities, without getting bogged down in state-based processes."

While ENGOs have increasingly shifted to using market-focused methods, government lobbying remains common (Alcock 2008), but is often less visible than other methods, as is evident from the following quote from a Tasmanian environmental activist:

> we are known for those [campaigns] up the tree and yet we probably spend more of our time and effort in parliament lobbying and dressed up in suits talking nicely to decision makers.... Affolderbach (2011, 195).

The extent of resources an ENGO devotes to lobbying governments depends on the strength of the governance regime in place. In a country with high levels of public corruption, ENGOs are likely to be more effective through targeting companies than lobbying government (see "Targeting the Supply Chain" on p. X) and consequently choose to invest their resources differently than those campaigning in countries with strong and stable governance (Lovelock 2005).

9.6.4 Targeting Forest Companies

The pressure that ENGOs put on forest companies generally involves direct protest and/or litigation. Direct protests, particularly at the site of company operations, can cause economic loss, as well as act to reduce consumer demand for a company's products by raising public awareness of the ENGOs' concerns about company practices. In Tasmania in 2007, the state forestry agency, Forestry Tasmania, sought compensation for the losses incurred as a result of blockading of a forest road by a protestor, arguing it had lost money due to lost business activity and the costs of removing the wooden tripod the protestor had used to blockade the road (*Sydney Morning Herald* 2007).

A second strategy is direct litigation against a forest company or a government agency managing public forests. Litigation takes place only when a company/agency is considered to be engaging in actions violating existing laws and regulations and hence is typically used in countries with strong governance systems. Crowfoot and Wondolleck (1990) have noted, for example, that litigation was commonly used in the United States in the 1980s. Despite the worldwide attention given to some of the successful cases of litigation brought by ENGOs, the use of litigation has remained limited because it requires there to be strong and relevant laws and regulations, can be a lengthy and expensive way to achieve change, and often depends on achieving changed interpretations of existing laws as they are applied to new situations. In addition, where the government strongly protects the forest industry, it may change or modify laws to prevent or reduce the possibility of a legal challenge, to reduce scope for objections to decisions about forestry, or simply to speed proposed forest sector developments. For example, in 2007, the Tasmanian state government replaced the existing planning procedures to simplify the approval process for a proposed pulp mill in the Tamar Valley. The new approval procedures were widely criticized as a "watered-down" process that lacked accountability and transparency (Gale 2008). In regions where governance is weak, ENGOs are likely to first work to strengthen legal frameworks before using litigation. In Indonesia, for example, ENGOs have worked to change the processes by which forest concessions are granted, aiming to improve the transparency of the process and the likelihood that only responsible forest managers are granted concessions (Gritten and Saastamoinen 2010).

Which companies do ENGOs target? This question has generated high interest among activism scholars; recent evidence suggests that ENGOs choose a "target" company based on their perception of the environmental impacts of that company, the size and prominence (visibility) of the company, and its likely influence on the rest of the forest sector (Hendry 2006; Mola-Yudego and Gritten 2010). Large, multinational companies are, therefore, more likely to be targeted by ENGOs.

9.6.5 Targeting the Supply Chain: Direct Protest Methods

In addition to direct protest and litigation, other methods used to influence forest company practices include pressuring their financiers/investors, customers, and end consumers—in other words, targeting the supply chain on which forest companies depend, particularly their markets. While ENGOs initially focused on achieving change in the "production space" of forestry, since the 1990s, a rapid shift has occurred to target what Kortelainen (2008) terms the "consumption space," meaning the markets into which forest products are sold (Wilson et al. 2001). Often, targeting the "consumption space" means targeting forest products retailers with strong brand name and high consumer recognition (such as IKEA and Home Depot). The use of market-based campaigns involving forest products retailers first emerged in the 1990s as a prominent strategy in BC, Canada, where ENGOs dissatisfied with the success of campaigns targeting government turned to lobbying prominent European and US retail businesses selling wood products made from timber harvested in BC forests (Hayter 2003; Walter 2003; Affolderbach 2011). Greenpeace and the RAN protested outside Home Depot, and IKEA stores in the United States and throughout Europe, performing media stunts that gained widespread public attention to their call for consumers to boycott the two stores unless they stopped purchasing wood from BC forests (Affolderbach 2011). These protests formed the basis for a widespread media campaigns and were accompanied by education campaigns targeted at the companies purchasing products made of BC timber, aimed at changing their wood-purchasing decisions. Consequently, several large forest products purchasers announced they would no longer purchase wood products harvested from the areas that ENGOs had targeted (Affolderbach 2011). Since the success of this campaign, market-based campaigns have become an integral part of environmental activism involving the forest sector globally. In some cases (e.g., the case described in Box 9.4), market-based campaigns form the central part of ENGOs strategies to achieve change. Market-based campaigns, in effect, target the entire supply chain of the forest sector, bringing about changes in forest management practices by pressuring both suppliers and buyers of forest companies. Large and powerful supply-chain partners can often more effectively change a company's practices than any direct attempt of an ENGO.

The success of market-based campaigns primarily relies on the presence of socially conscious consumers, also often known as "green consumers." Large forest products companies are more likely to respond to ENGO protests if they feel that their consumer base could be threatened. The methods used by ENGOs aim not only to attract the attention of these consumers but also to create more of them, through education campaigns targeted at the general public. Similar to direct protests and government lobbying, supply-chain-focused strategy also involves actions such as media campaigns in consumer countries, direct protests outside financiers and retail chains selling wood products, boycotts of wood products, and education campaigns aimed at influencing decision makers within upstream and downstream businesses. These actions ultimately aim at ensuring that the general public—the end consumers of forest products companies—joins in to create pressure on the entire supply chain.

The shift to a supply-chain focus has been a prime driver of the globalization of environmental activism in the forest sector. In practice, a supply-chain-focused approach entails "blanketing" the entire supply chain. In a globalized forest products industry, the supply chains of forest companies are often spread across the globe. A supply-chain-focused strategy is therefore most effective when multiple ENGOs that can act successfully in different locations form alliances (Gritten and Mola-Yudego 2010). Focusing on the supply chain has also resulted in a shift away from the traditional "place-based" focus of ENGO campaigns that only involved the region where a forest was located. For example, ENGO campaigns against establishment of a Gunns pulp mill in Tasmania have included media campaigns in Japan aimed at influencing purchasers of products from Gunns Ltd, and campaigns targeting multiple banks (based in Australia and other countries) proposed as finance partners for the mill (Affolderbach 2011; Manning 2011). Similarly, ENGOs concerned about forest practices in Indonesia have targeted global financial institutions such as Barclays and HSBC that provide finances for pulp and paper mills in Indonesia, as well as the companies that purchase and distribute paper products produced by these mills worldwide (Gritten and Kant 2007; Gritten and Saastamoinen 2010).

The shift to supply-chain-based campaigns has been so profound that in some cases it has effectively replaced grassroots, place-based campaigning. This is particularly true in countries where environmental activism has only recently emerged. For example, in Russia where environmental activism emerged largely after the collapse of Soviet Union, international ENGOs initiated campaigns against harvesting of old-growth forests firstly through influencing international markets and transnational forest products firms that sourced timber from Russia (Kortelainen 2008). After international ENGOs had invested in pressuring international markets and supply-chain partners, their secondary action involved fostering the creation of local campaigns.

The "discovery" of market-focused campaigns "has been likened to the discovery of dynamite" (Walter 2003, 533). It is not without its detractors, however; it has been criticized as representing a shift from "civil disobedience to cranky, but nevertheless obedient, consumerism" (Walter 2003, 533). This critique suggests that ENGOs are faced with a double-edged sword when they use market campaigns: their power in these campaigns relies on the presence of wood products markets—by targeting these markets, do ENGOs inadvertently become reliant on the maintenance of these wood products markets? This critique notwithstanding the growth of campaigns targeting the entire supply chain has been and continues to be rapid worldwide.

9.6.6 Targeting the Supply Chain: Forest Certification

The market campaigns previously described have a natural parallel: While providing the stick of protests and reduced consumer demand, ENGOs have provided the carrot of increasing demand for wood certified as being sustainably produced. Interest in market-based certification schemes that provide incentives for forest companies to use practices approved by ENGOs emerged in the 1990s (Cashore et al. 2003), and the Forest Stewardship Council, a group formed in 1993 by a coalition of stakeholders including prominent ENGOs (Cashore et al. 2007), has been particularly influential in changing the shape of the global forest sector.

While certification is examined in detail elsewhere in this book (see Chapters 2 and 10), it is discussed here for two reasons: First, it is an important market-based method by which ENGOs exert influence on the forest industry, despite ongoing debate about how best to achieve sustainability via certification (Alcock 2008). Second, the success of certification is dependent in many cases on the ability of ENGOs to influence markets through the protest campaigns previously described. Certification is predicated on the idea of providing access to markets that would otherwise not exist for forest products companies. This requires ENGOs to create markets for certified wood. There are many examples of ENGOs partnering a call for certification with direct market campaigns that influence demand by encouraging wood products retailers to boycott noncertified timber, particularly in European wood products markets (Cashore et al. 2007).

The creation of the green market has probably been most successful in Europe and North America, with less creation of distinct demand for certified wood in other regions. For example, Kortelainen (2008) identified that certification became a driver for Russian forest products companies only when they sought to gain access to European markets in which there was buyer demand for certified products.

Certification is also yet another factor supporting the globalization of ENGO influence, through creating voluntary governance frameworks that are promoted by transnational coalitions of ENGOs and implemented by forest products companies in order to satisfy the demands of "consumer countries" often located a considerable distance from the forests from which their wood products are sourced (Kortelainen 2012). ENGOs play a critical role not only in creating market demand but also in actively organizing networks of buyer groups to create demand for certified wood. In the largest and best-known example, by 2008, the WWF-led GFTN had members who traded almost 200 million cubic meters of wood (Auld et al. 2008). Originating in 1995 under a different name, the network promotes trade of legally produced, sustainable timber products, with members provided assistance by WWF to achieve sustainable and legal trade and in return being permitted to use the WWF "panda" brand on their products, providing a market advantage. These members provide substantial demand for timber certified under the Forest Stewardship Council and other certification schemes (Auld et al. 2008).

9.6.7 Framing Public Opinion

Public support is critical to the success of ENGO campaigns. Therefore, a critical part of any ENGO campaign is to frame public opinion through media and education campaigns and, in recent years, via social media such as Facebook and Twitter. Public opinion is often framed using high-profile "media stunts" such as blockades, "tree sits," invasions of processing sites, and rallies, among others. Education campaigns typically involve production of newsletters and documentaries, creating Internet sites, holding school and public talks, organizing forest site visits, and using other strategies that draw media attention and increase public awareness of and support for ENGOs (Affolderbach 2008, 2011). The following are some representative examples to illustrate the breadth of the methods used to draw media attention and influence public opinion:

- In the 1970s and 1980s, Irish farmers in County Leitrim physically blocked forest companies from entering agricultural properties, in order to stop them from establishing Sitka spruce plantations (Schirmer 2007).
- In 1987–1988, indigenous communities barricaded logging roads in Sarawak, Malaysia, for several months, to prevent logging of the forests they depended on and lived in (WRM/SAM 1990). Protests have continued since; in 2008, a blockade of roads by indigenous communities continued for over a month (ENS 2008).
- In 2011, protestors from Cascadia Earth First, a US-based ENGO, organized tree sits in Oregon, United States, by establishing small platforms high in trees that were due for harvesting. The removal of protestors was made deliberately difficult by connecting ropes and platforms in a way that, if removed by police or others, would result in injury to the protestors (Higgins 2011). Tree sits such as these are sometimes combined with use of social media to draw increased attention. For example, Miranda Gibson began a tree sit in southern Tasmanian in December 2011 to protest against a planned forest harvest. Her tree sit protest was ongoing at the time of writing in 2012 and used solar power panels to power a laptop computer connected to the Internet from which she participated in online blogs and media interviews and developed a strong Facebook following (see http://observertree.org/for the website of this protest). Her protest achieved a strong media following not only in Australia but also in Japan, which is a large market for wood products harvested in Tasmania.
- ENGOs increasingly turn to social media to interface with the public. It is common for ENGOs to create Facebook pages that act as a platform for specific forest campaigns, particularly those involving an alliance of multiple ENGOs. For example, the "SAF and the Anti-Biomass Incineration/Forest Protection Campaign" Facebook page is dedicated to a campaign against use of trees for bioenergy in the United States, involving multiple ENGOs led by "Save America's Forests."

Similarly, Twitter feeds are used by a number of organizations (such as the EIA and RAN) to update their followers about actions and plans.

- Protestors illegally entered the site of the Ta Ann "Southwood" veneer processing factory in Tasmania, Australia, in January 2012 and temporarily stopped production to protest against its use of wood harvested from natural forests (Arndt 2012).

9.6.8 Using Alliances

In order to bring about desired changes in company practices, ENGOs often form complex alliances and networks with one another (Hendry 2006). Alliances are essential for ENGOs to effectively and rapidly influence the widely distributed forest products supply chain (Gritten and Mola-Yudego 2010). An important part of any environmental campaign is investing in building and maintaining alliances and networks that require continued negotiation and discussion among alliance partners.

Alliance formation occurs at all levels—from local to international. At the local level, an ENGO may effectively partner with recreation groups, other ENGOs, and political organizations to leverage attention to a specific area of forest or forest management issue (see Fagan, 2006). At the international scale, the same ENGOs often partner with each other and with nonenvironmental NGOs across state boundaries to achieve change. These complex networks of ENGOs "criss-cross political boundaries in bewildering fashion at high speed" (Hayter 2003, 711). Alliance formation enables ENGOs to increase not only access to financial resources but also to leverage nonfinancial resources such as access to knowledge, decision makers, and critical social networks (Fagan 2006; Gritten 2009; Gritten and Mola-Yudego 2010). For example, Kröger and Nylund (2012, 77) describe the importance of alliance formation among various stakeholders to jointly act on concerns over establishing plantation forests and associated pulp processing in Brazil:

> … different actors reap synergy benefits by sharing and replicating knowledge of experts on other areas than their own, and networking allows creating new hybrid discourses of socio-environmentalism, which then bridges different groups into an alliance or at least positions them on a cooperative platform.

Alliance formation and coalition building is, however, a complex and challenging task. ENGOs seeking to partner with indigenous peoples and nonenvironmental NGOs, or even with ENGOs that have different agendas to their own, are sometimes accused of using these alliances to increase their bargaining power while not genuinely sharing the interests of their allies (Affolderbach 2011). As a result of such limiting challenges, coalitions do have varying success (Hayter 2003). When they are successful, however, they substantially increase the potential for the coalition as a whole to achieve its goals (see, e.g., Sarkki and Heikkinen 2010).

9.7 Choosing the Right Methods

This chapter has discussed both protest and collaboration-oriented approaches. Notwithstanding their relative merits and demerits, it is important to examine when and why ENGOs decide to engage in confrontational versus collaborative methods.

Confrontational methods such as direct protest typically involve acting outside existing decision-making processes. Some authors debate whether an ENGO campaign will be more effective if the ENGO stays outside the "status quo" to protest as an outsider or chooses to become "mainstreamed" and work with and within existing decision-making structures as an "insider" (see, e.g., Johnson and Prakash 2007; Hoffman 2009). In the forest products sector, examples of both these approaches can be found; for example, Greenpeace International tends to maintain their influence through ongoing direct action and conflict methods and does not typically engage in "insider" partnerships. On the other hand, the Rainforest Alliance works in partnerships with businesses to promote sustainable forest management and has been involved in forest certification programs since 1989. Others have combined both approaches or shifted between them over time. A recent example is The Wilderness Society in Australia, which in 2010 shifted from a focus on protest campaigns about harvesting of natural forests in Tasmania to engaging in direct negotiations with the forest industry and other groups (TWS 2010), while reducing its direct protest campaigns.

Will a campaign be more successful if pressure is applied through protests that affect a cascade of stakeholders such as the financiers and customers of a forest products company or if the ENGO works directly with a company? This question comes back to the power and influence of the ENGO relative to the organizations it seeks to influence and change; ENGOs with high influence may be successful in using collaborative methods (see, e.g., Johnson and Prakash 2007), while those with little influence may use "high-conflict" methods such as direct protest to increase their influence and hence their ability to achieve change (Affolderbach 2011). For this reason, many environmental campaigns begin with the use of direct protest at forest or processing sites accompanied by lobbying of government and media campaigns targeted to influence public opinion. As the bargaining power of ENGOs increases, they typically include methods such as direct stakeholder negotiation in their mix of methods, although some ENGOs such as Greenpeace continue to focus on a high-conflict approach despite having become highly influential. A one-time decision of an ENGO to participate in multistakeholder negotiations does not necessarily mean a permanent, irrevocable shift to more collaborative methods or vice versa. If "success" is defined as the durability of agreements made as a result of collaborative processes, then these processes have had varying outcomes in the forest sector. On the one hand are examples such as the Joint Solutions Project and subsequent stakeholder collaborations in BC, where direct negotiations between ENGOs and forest industry groups formed the basis of collaborative forest management and brought about lasting changes in the relations between ENGOs and the forest industry (Zietsma and Lawrence 2010). On the other hand, in the Australian government's Regional Forest Agreement (RFA) process, regionally based negotiation between the forest industry, government, ENGOs, and other stakeholders achieved some regionally based agreements regarding forest management but was followed by a return to confrontational protest by ENGOs within a relatively short period in many of these regions (Musselwhite and Herath 2005). Another mixed outcome has occurred in Finland, where some ENGOs involved in multistakeholder planning committees and working groups from the early 1990s—in which forest companies, government forest agencies, and ENGOs attempted to collaboratively address their differences—have withdrawn from these processes (Sarkki and Heikkinen 2010).

Gritten (2009) argues that the theory of stakeholder salience can help identify when and why different methods will achieve an ENGO's goals in different situations. According to stakeholder salience theory (Mitchell et al. 1997), the extent to which a stakeholder is influential depends on three things: the stakeholder's power (ability to impose its views on those it seeks to influence); legitimacy (its acceptance as a group taking appropriate actions and having legitimate views); and urgency (its ability to create a pressing need for attention). Following this argument, ENGOs will use different methods to increase their power, legitimacy, and/or urgency, depending on which of these factors better promise to serve their needs. For example, an ENGO that lacks legitimacy and acceptance in traditional decision-making processes will be more likely to use direct protest methods, such as blockades of forest sites, to build public support and hence increase its power and urgency (Affolderbach 2011).

This theory suggests that ENGOs can successfully use more collaborative methods only if they have sufficient legitimacy, urgency, and power. However, if they lack legitimacy, urgency, or power, then confrontational methods may be more effective for them. For example, in the BC forest conflict in the 1990s, ENGOs built their bargaining power through refusing to participate in proposed stakeholder negotiations that they criticized as being "talk and log" exercises (Wilson 1998). In this case, ENGOs had to first disrupt existing decision-making processes by using protest-based methods in order to create a new collaborative space in which they could have genuine influence (Zietsma and Lawrence 2010; Clapp and Mortenson 2011). In other processes, ENGOs agreed to collaborate only after an agreement was reached to temporarily halt forest harvesting: A key example is Australia's RFA process, in which areas of forest were set aside in "interim deferred areas" at the start of the process, until stakeholders reached agreement on which areas should be harvested (Musselwhite and Herath 2005).

If a shift to collaboration is made too early in the process, it is likely to fail and as ENGOs seek to further increase their influence and ability to achieve change, confrontation is more likely to occur. That said, the influence an ENGO develops through protest generally achieves little change unless it is subsequently followed by less confrontational and more collaborative processes. The question is therefore not whether to engage in collaborative activism, but at what point an ENGO can do this in a way that does not compromise its values and desired goals. A multistakeholder collaborative process is more likely to be successful when adequate time is provided for stakeholders to interact with each other, and when the process involves first agreeing on a set of broad principles, and then identifying how to implement these principles (McKnight and Zietsma 2007).

9.8 What Makes an ENGO Campaign Successful?

Different groups view and define success differently, and therefore collaborating ENGOs may sometimes have different—and potentially incompatible—goals. Anshelm and Hansson (2011) have identified critical differences in the views of prominent international ENGOs about the actions needed to successfully address climate change. For example, WWF believes in collaboration among government, ENGOs and businesses for achieving change, while Greenpeace and FOE are more oriented to achieving change via regulatory approaches and direct government intervention. Differences in these fundamental viewpoints will dictate how these different ENGOs define success: WWF will be more likely to view the implementation of voluntary market-based mechanisms as a successful step in achieving change, while the other organizations may not believe successful change has occurred until government regulation is put in place. Differences in these fundamental premises are also visible in some ENGO alliances. Gritten and Kant (2007) identified that in the case of environmental activism focused on APRIL in Indonesia (see Box 9.4), WWF was "willing to accept a step wise approach" in which the company gradually changed operations, while another partner, FOE Finland, would only consider the campaign successful if the company changed all the practices (p. 831).

The difference in ENGO viewpoints notwithstanding, it is critical to recognize the role of external factors as well as the role of the actions taken by ENGOs. In particular, an environmental campaign is more likely to succeed if it occurs at a time when the forest industry is already experiencing stress, such as a market downturn (Gritten and Mola Yudego 2010). For example, Hayter (2003) argued that ENGOs were successful in achieving change in the BC forest conflict in the 1980s partly because the industry was undergoing a substantial downturn, and this increased the relative power of the ENGOs to achieve change.

Access to funding is another critical factor in ENGO success. While a strong funding base generally enables ENGOs to take coordinated and well-organized action (Affolderbach 2011), it can also sometimes reduce their credibility. Accepting funding from corporations, for example, is sometimes viewed as co-optation (Gritten and Kant 2007; Hoffman 2009). Funding dynamics can be intricate and sometimes ENGOs are argued to engage in "switch and change" where they champion issues largely because they can obtain funding if they do so. Fagan (2006, 792) documented that many ENGOs in postconflict Bosnia–Herzegovina have not emerged "as a direct response to specific ecological issues, but in response to the availability of project funding to work on particular issues." In such cases, ENGO agendas are not "grass roots" but a reflection of the agenda of donor groups. Similarly, Mandondo (2003) documented how the focus of the Rural Development Forestry Network (RDFN) changed over time with changing funding sources. Interestingly, Fagan (2006) also noted that a coalition of ENGOs protesting a road being developed by a forest company in Bosnia–Herzegovina did not receive funding that many other ENGOs had received, since the actions of this genuinely "grassroots" coalition were viewed as a "political campaign," disqualifying the coalition from EU funding. In short, funding mechanism and sources may sometimes critically influence ENGO action and their ability to achieve change.

Overall, the success of an ENGO campaign depends on its ability to balance the demands of funders, supporters, and its credibility as a true champion of issues of public concern.

9.9 ENGOs and the Forest Products Industry: Future Outlook

Recent decades have seen a shift in the focus of environmental activism from the "production space" to the "consumption space." The shift to use of market-based methods, ranging from protests targeting markets to providing positive incentives for change through certification, is being achieved through the gradual creation of markets for the types of forest products considered acceptable by ENGOs. Additionally, unlike previous environmental campaigns that typically started as "grassroots" campaigns and then scaled up their scope to cover the entire supply chain, many campaigns now follow a "top-down" approach, wherein campaigns are initiated and enacted by international ENGOs, who bring the activism to the local level, build public support, and develop local ENGO capabilities. These changes have led to the increasing use of networks and alliances stretching across national boundaries to parallel the global reach of globalizing forest products companies. In the future, a focus on market-based methods is likely to continue and even expand.

The extent of growth in market-based measures, however, will be dependent on the extent of popularity of certified forest products outside the non-Western world. Given the rise of demand for wood products in countries that do not currently have a substantial culture of green consumerism, such as China, it is highly likely that market-based campaigns will reach their natural limits unless ENGOs successfully foster such a culture in these countries in coming years. Also, while many hope for a future in which environmental activism shifts to greater use of collaborative processes, this will only occur if ENGOs have adequate influence on these processes—otherwise confrontational environmental activism will continue. ENGOs engage in collaborative processes when they are confident they can achieve substantial change in practices; these prerequisites for ENGO participation in collaborative negotiations are unlikely to change.

Overall, increasing concerns about environmental degradation, increasing awareness of the role of forests in climate change mitigation, and globalization of forest industry supply chains, mean that environmental activism will continue to play an integral role in monitoring and influencing forest sector policies globally.

This chapter has focused on understanding the motivations, methods, and campaigns used by ENGOs. This raises the obvious question: How can forest products companies do a better job of working with ENGOs?

The first answer to this question is that forest products companies must stop doing what doesn't work. In particular, ignoring ENGOs, or dismissing their views as illegitimate, doesn't work. Instead, it typically serves to inflame conflict, and to facilitate growth in ENGO power and influence, as when rejected by a forest products company, ENGOs instead turn to campaigning in the media and within the company's supply chains to achieve their desired outcomes.

Similarly, engaging in confrontational action such as litigation can result in a forest products company "winning the battle, but losing the war." Even where companies that bring litigation against environmental activists win their case, the process of litigation often reinforces consumer support for ENGOs, particularly by providing free publicity for the ENGO cause. For example, in 2004, the Australian forest products company Gunns sued 20 environmental activists and organizations, claiming the activist's protests had damaged Gunns' reputation and business. The ensuing 5 years of legal action were used as a platform by the ENGOs involved to promote their ongoing activism and gain widespread media attention for their cause. While the company gained a small amount in settlements from some of the ENGOs sued, it paid far more in legal costs (and in a final settlement) than it received. More importantly, the cost to its reputation in the minds of consumers of its products was incalculable, an outcome predicted in advance by scholars such as Vanclay (2005). Similar lessons have been repeatedly documented in the corporate world, perhaps most famously in the "McLibel" case in which McDonalds sued critics they claimed had defamed the company. The ensuing media publicity gained widespread attention for the views of McDonalds' critics, attention afforded largely by the court case rather than the previous methods used by these activists (Vick and Campbell 2001).

We know what doesn't work. Methods that do work require a different approach by forest products companies, one in which they view ENGOs as raising legitimate and important questions that need to be addressed by the forest sector, either through changing practices, or better demonstrating the sustainability of their practices. This is, however, easier said than done: many companies fear that such a move will increase ENGOs' relative power and concomitantly reduce the ability of the forest products companies to have their voice and viewpoints heard.

While this fear is understandable, it is often counterproductive for several reasons. First, early acknowledgment of the legitimacy of ENGOs by forest products companies reduces the need for ENGOs to engage in highly visible protest actions that often result in significant costs to a forest products company's finances, market, and reputation. Second, early acknowledgment enables the company to avoid the "legitimacy trap" described earlier, in which the company becomes so focused on continually rejecting concerns raised by ENGOs that legitimate concerns are not acknowledged or acted on. Third acknowledgment can enable a more rapid shift to collaborative processes that seek real solutions to concerns raised.

Forest products companies have been partners in several collaborative processes discussed in this chapter, such as the Joint Solutions Project in BC, Canada (see Box 9.1), and Australia's RFAs. These collaborative processes are a way forward for forest products companies in some cases, but they are not easy and have had varying success. Success appears reliant on all the collaborating parties having a genuine desire to engage in dialogue, on having adequate time to conduct this dialogue, and on ensuring ongoing dialogue over time, rather than having a "once-off" collaborative process meant to provide a single point-in-time solution. Processes that allow for ongoing discussion are more likely to succeed as they enable changing views, value, and knowledge to be continually negotiated on an ongoing basis in a manner that avoids a return to protest-based activism.

Reaching this ideal point takes time: time to build trust in what have previously been confrontational relationships, time to enable ENGOs and forest products companies to understand each other's points of view and jointly evaluate scientific evidence, and time to identify creative solutions to the dilemmas posed by ENGO criticism of the sector. A good beginning for any forest products company is to ensure they are open to discussion with ENGOs, even where that discussion involves the expression of views they do not agree with. Being open can in time lead to trust and to collaboration in which ENGOs work with forest products companies to find solutions that enable maintenance of wood products businesses while addressing ENGO concerns.

References

Affolderbach, J. 2008. ENGOs and environmental bargains: A comparative analysis of forest conflicts in Tasmania and British Columbia. Thesis submitted in partial fulfilment of the requirements for the degree of Doctor of Philosophy. Simon Fraser University, Vancouver, British Columbia, Canada.

Affolderbach, J. 2011. Environmental bargains: Power struggles and decision making over British Columbia's and Tasmania's old-growth forests. *Economic Geography* 87(2): 181–206.

Alcock, F. 2008. Conflicts and coalitions within and across the ENGO community. *Global Environmental Politics* 8(4): 66–91.

Anshelm, J. and Hansson, A. 2011. Climate change and the convergence between ENGOs and business: On the loss of utopian energies. *Environmental Values* 20(1): 75–94.

Arndt, D. 2012. Protestors temporarily disrupt Ta Ann operations. *The Examiner*, January 26, 2012, Launceston. URL: http://www.examiner.com.au/news/local/news/environment/protesters-temporarily-disrupt-ta-ann-operations/2431355.aspx (Accessed April 21, 2012).

Athanasiou, T. 1998. *Divided Planet: The Ecology of Rich and Poor*. The University of Georgia Press, Athens, GA.

Auld, G., Gulbrandsen, L.H., and McDermott, C.L. 2008. Certification schemes and the impacts on forests and forestry. *Annual Review of Environmental Resources* 33: 187–211.

Bendix, J. and Liebler, C.M. 1999. Place, distance, and environmental news: Geographic variation in newspaper coverage of the Spotted Owl conflict. *Annals of the Association of American Geographers* 89(4): 658–676.

Carrere, R. and Lohmann, L. 1996. *Pulping the South: Industrial Tree Plantations and the World Paper Economy*. Zed Books, London, U.K.

Cartwright, J. 2003. Environmental groups, Ontario's lands for life process and the forest accord. *Environmental Politics* 12(2): 115–132.

Cashore, B., Auld, G., and Newsom, D. 2003. Forest certification (eco-labeling) programs and their policy-making authority: Explaining divergence among North American and European case studies. *Forest Policy and Economics* 5(3): 225–247.

Cashore, B., Egan, E., Auld, G., and Newsom, D. 2007. Revising theories of nonstate market-driven (NSMD) governance: Lessons from the finnish forest certification experience. *Global Environmental Politics* 7(1): 1–44.

Clapp, R.A. and Mortenson, C. 2011. Adversarial science: Conflict resolution and scientific review in British Columbia's Central Coast. *Society and Natural Resources* 24(9): 902–916.

Cossalter, C. and Pye-Smith, C. 2003. *Fast-Wood Forestry: Myths and Realities*. Center for International Forestry Research, Bogor, Indonesia.

Crowfoot, J.E. and Wondolleck, J.M. 1990. *Environmental Disputes: Community Involvement in Conflict Resolution*. Island Press, Washington, DC.

Dargavel, J. 1995. *Fashioning Australia's Forests*. Oxford University Press, South Melbourne, Victoria, Australia.

ENS (Environmental News Service). 2008. Malaysian Indigenous people face arrest at logging blockade. June 17, 2008. URL: http://www.ens-newswire.com/ens/jun2008/2008-06-17-02.html (Accessed April 21, 2012).

EOTF (Eyes on the Forests). 2012. APP: Default on environmental covenant. Investigative report, Eyes on the Forests. URL: http://www.eyesontheforest.or.id (Accessed April 21, 2012).

The Examiner. 2012. ANZ branches targeted by mill protestors. *The Examiner*, January 26, 2012, Launceston. URL: http://www.examiner.com.au/news/local/news/environment/anz-branches-targeted-by-mill-protesters/2433168.aspx (Accessed April 21, 2012).

Fagan, A. 2006. Neither "north" nor "south": The environment and civil society in post-conflict Bosnia-Herzegovina. *Environmental Politics* 15(5): 787–802.

Gale, F. 2008. Tasmania's Tamar Valley pulp mill: A comparison of planning processes using a good environmental government framework. *The Australian Journal of Public Administration* 67(3): 261–282.

Global Witness. 2011. Pandering to the loggers: Why WWF's global forest and trade network isn't working. URL: www.globalwitness.org/sites/default/.../Pandering_to_the_loggers.pdf (Accessed April 21, 2012).

Gritten, D. 2009. Facilitating resolution of forest conflicts through understanding the complexity of the relationship between forest industry and environmental groups. Dissertationes Forestales 91. Finnish Society of Forest Science, Finnish Forest Research Institute, University of Helsinki and University of Joensuu. Vantaa, Finland.

Gritten, D. and Kant, P. 2007. Assessing the impact of environmental campaigns against the activities of a pulp and paper company in Indonesia. *International Forestry Review* 9(4): 819–834.

Gritten, D. and Mola-Yudego, B. 2010. Blanket strategy: A response of environmental groups to the globalising forest industry. *International Journal of the Commons* 4(2): 729–757.

Gritten, D. and Mola-Yudego, B. 2011. Exploration of the relevance of geographical, environmental and socio-economic indicators on forest conflict types. *International Forestry Review* 13(1): 46–55.

Gritten, D., Mola-Yudego, B., and Delgado-Matas, C. 2012. Media coverage of forest conflicts: A reflection of the conflicts' intensity and impact? *Scandinavian Journal of Forest Research*. 27(2): 143–153.

Gritten, D. and Saastamoinen, O. 2010. The roles of legitimacy in environmental conflict: An Indonesian case study. *Society and Natural Resources* 24(1): 49–64.

Grove, R.H. 1995. *Green Imperialism: Colonial Expansion, Tropical Island Edens and the Origins of Environmentalism, 1600–1860*. Cambridge University Press, New York.

Guha, R. 1989. *The Unquiet Woods: Ecological Change and Peasant Resistance in the Himalaya*. Oxford University Press, New Delhi, India.

Hayter, R. 2003. The war in the woods: Post-fordist restructuring, globalization, and the contested remapping of British Columbia's forest economy. *Annals of the Association of American Geographers* 93(3): 706–729.

Hendry, J.R. 2006. Taking aim at business: What factors lead environmental non-governmental organizations to target particular firms? *Business Society* 45(1): 47–86.

Higgins, J. 2011. Protesters get ready to confront officials who consider their sit-in an illegal act: Activists hope to stop logging. *The World* (online media publication) URL: http://theworld-link.com/news/local/article_43b1114d-821c-5c65–9263–82d896193166.html#ixzz1s5COrEmk (Accessed April 15, 2012).

Hoffman, A.J. 2009. Shades of green. *Stanford Social Innovation Review* Spring: 40–49.

Humphreys, D. 2006. *Logjam: Deforestation and the Crisis of Global Governance*. Earthscan, London, U.K.

Johnson, E. and Prakash, A. 2007. NGO research program: A collective action perspective. *Policy Sciences* 40(3): 221–240.

Johnston, E. and Soulsby, C. 2006. The role of science in environmental policy: An examination of the local context. *Land Use Policy* 23(2): 161–169.

Kortelainen, J. 2008. Performing the green market—Creating space: Emergence of the green consumer in the Russian woodlands. *Environment and Planning A* 40(6): 1294–1311.

Kortelainen, J. 2012. Boundaries of transnational forest governance in Russia. *Scandinavian Journal of Forest Research* 27(2): 221–228.

Kröger, M. and Nylund, J.-E. 2012. The conflict over Veracel pulpwood plantations in Brazil—Application of ethical analysis. *Forest Policy and Economics* 14(1): 74–82.

Lei, X. 2009. Public initiatives and local practices in China's response to climate change. In *China and Global Climate Change: Proceedings of the Conference Held at Lingnan University, Hong Kong, June 18–19, 2009*. Lingnan University, Hong Kong, People's Republic of China, pp. 545–559.

Linkenbach, A. 2007. *Forest Futures: Global Representations and Ground Realities in the Himalayas*. Permanent Black, Uttarakhand, India.

Lovelock, B. 2005. Tea-sippers or arsonists? Environmental NGOs and their responses to protected area tourism: A study of the Royal Forest and Bird Protection Society of New Zealand. *Journal of Sustainable Tourism* 13(6): 529–545.

Mandondo, A. 2003. Snapshot views of International Community Forestry Networks: Rural Development Forestry Network Study, CIFOR/Ford Foundation (unpublished), March 2003. URL: http://www.cifor.org/publications/pdf_files/CF/RDFN.pdf (Accessed April 15, 2012).

Manning, P. 2011. The felling of Gunns. *The Age*, March 12, 2011. Article accessible online at http://www.theage.com.au/national/the-felling-of-gunns-20110311–1br9d.html (Accessed January 25, 2012).

McCormick, J. 1989. *Reclaiming Paradise: The Global Environmental Movement*. Indiana University Press, Bloomington, IN.

McKnight, B. and Zietsma, C. 2007. Local understandings: Boundary objects in high conflict settings. *Proceedings of the Organization Learning, Knowledge and Capabilities Conference*, June 14–17, 2007, London, U.K., pp. 688–698.

Mitchell, R., Agle, B., and Wood, D. 1997. Toward a theory of stakeholder identification and salience: Defining the principle of who and what really counts. *The Academy of Management Review* 22(4): 853–886.

Mola-Yudego, B. and Gritten, D. 2010. Determining forest conflict hotspots according to academic and environmental groups. *Forest Policy and Economics* 12(8): 575–580.

Mola-Yudego, B., Gritten, D., and Delgado-Matas. 2012. Quantitative investigation of forest conflicts using different data collection methods. *Scandinavian Journal of Forest Research*. 27(2): 130–142.

Musselwhite, G. and Herath, G. 2005. Australia's regional forest agreement process: Analysis of the potential and problems. *Forest Policy and Economics* 7(4): 579–588.

NIEIR (National Institute of Economic and Industry Research). 2010. Opportunities, issues and implications for a transition of the Victorian wood products industry from native forests to plantations. A report for the Australian Conservation Foundation and the Wilderness Society. NIEIR, Melbourne, Victoria, Australia.

Oberthür, S., Matthias, B., Sebastian, M., Stefanie, P., Richard, G.T., Jacob, W., and Alice, P. 2002. *Participation of Non-Governmental Organisations in International Environmental Governance.* Ecologic Institute, Berlin, Germany.

Patterson, J.D. and Allen, M.W. 1997. Accounting for your actions: How stakeholders respond to the strategic communication of environmental activist organizations. *Journal of Applied Communication Research* 25(4): 293–316.

Penn, B. 2011. A little lecture on the big burn: Bioenergy and the privatization of British Columbia's Crown forests. *The Forestry Chronicle* 87(5): 598–602.

Plokhii, O. 2012. Murders in the forest. Reuters online edition. URL: http://blogs.reuters.com/great-debate/2012/09/20/murders-in-the-forest/(Accessed September 22, 2012).

Porter, G., Welsh Brown, J., and Chasek, P.S. 2000. *Global Environmental Politics*, 3rd edn. Westview Press, Oxford, U.K.

Poyry Management Consulting. 2011. Review of issues affecting the transition of Victoria's hardwood processing industry from native forest to plantations. Report prepared for Victoria Forests and Australian Paper. Poyry Management Consulting, Melbourne, Victoria, Australia.

Rossiter, D. 2004. The nature of protest: Constructing the spaces of British Columbia's rainforests. *Cultural Geographies* 11(2): 139–164.

Sarkki, S. and Heikkinen, H.I. 2010. Social movements' pressure strategies during forest disputes in Finland. *The Journal of Natural Resources Policy Research* 2(3): 281–296.

Satterfield, T. 1997. "Voodoo science" and common sense: Ways of knowing old-growth forests. *Journal of Anthropological Research* 53(4): 443–459.

Schirmer, J. 2002. *Plantation Forestry Disputes: Case Studies on Concerns, Causes, Processes and Paths Towards Resolution.* Technical Report 42 (revised). Cooperative Research Centre for Sustainable Production Forestry, Canberra, Hobart, Australia.

Schirmer, J. 2007. Plantations and social conflict: Exploring the differences between small-scale and large-scale plantation forestry. *Small-Scale Forestry* 6(1): 19–33.

Schirmer, J., Dunn, C., Loxton, E., and Dare, M. 2011. *Socioeconomic Impacts of Forest Industry Change: A Baseline Study of the Tasmanian Forest Industry.* CRC for Forestry, Canberra, Hobart, Australia.

Sonnenfeld, D.A. 2002. Social movements and ecological modernization: The transformation of pulp and paper manufacturing. *Development and Change* 33(1): 1–27.

Sydney Morning Herald. 2007. Forestry Tas seeks compo from Weld Angel. October 3, 2007. URL: http://www.smh.com.au/news/National/Forestry-Tas-seeks-compo-from-Weld-Angel/2007/10/03/1191091164764.html (Accessed April 21, 2012).

Taylor, B., Hadsell, H., Lorentzen, L., and Scarce, R. 1993. Grass-roots resistance: The emergence of popular environmental movements in less affluent countries. In Kamieniecki, S. (ed.) *Environmental Politics in the International Arena: Movements, Parties, Organizations and Policy.* State University of New York Press, Albany, NY, pp. 69–89.

Vanclay, J.K. 2005. Gunns, greens and silk: Alternative approaches to resource conflicts. Paper presented to *"Burning Issues in Forestry", 22nd Biennial Conference of the Institute of Foresters of Australia, April 10–14, 2005, Mount Gambier, South Australia.* Institute of Foresters of Australia, Yarralumla, Canberra, Australia.

Vick, D.W. and Campbell, K. 2001. Public protests, private lawsuits, and the market: The investor response to the McLibel case. *Journal of Law and Society* 28(2): 204–241.

Walter, E. 2003. From civil disobedience to obedient consumerism? Influences of market-based activism and eco-certification on forest governance. *Osgoode Hall Law Journal* 42(2–3): 531–563.

Warren, C. 2000. Birds, bogs and forestry' revisited: The significance of the flow country controversy. *Scottish Geographical Journal* 116(4): 315–337.

The Wilderness Society (TWS). 2010. ENGOs welcome Federal Government action to protect Tasmanian Forests and rebuild forest industry. Media release, December 7, 2010. URL: http://www.wilderness.org.au/regions/tasmania/engos-welcome-federal-government-action-to-protect-tasmanian-forests-and-rebuild-forest-industry (Accessed April 15, 2012).

Wilson, B., Takahashi, T., and Vertinsky, I. 2001. The Canadian commercial forestry perspective on certification: National survey results. *The Forestry Chronicle* 77(2): 309–313.

Wilson, J. 1998. *Talk and Log: Wilderness Politics in British Columbia*. UBC Press, Vancouver, British Columbia, Canada.

WRM/SAM (World Rainforest Movement/Sahabat Alam Malaysia). 1990. *The Battle for Sarawak's Forests*. World Rainforest Movement and Sahabat Alam Malaysia, Penang, Malaysia.

Zietsma, C. and Lawrence, T.B. 2010. Institutional work in the transformation of an organizational field: The interplay of boundary work and practice work. *Administrative Science Quarterly* 55(2): 189–221.

Zietsma, C., Winn, M., Branzei, O., and Vertinsky, I. 2002. The war of the woods: Facilitators and impediments of organizational learning processes. *British Journal of Management* 13(S2): S61–S74.

10

Implementing Sustainability in the Global Forest Sector: Toward the Convergence of Public and Private Forest Policy

Timothy M. Smith, Sergio A. Molina Murillo, and Britta M. Anderson

CONTENTS

10.1 Introduction

Sustainable forest management (SFM) has remained at the forefront of forestry discussions globally for several decades and is an important concept both for the forestry and forest products sectors. While the United Nation's Food and Agriculture Organization (FAO) defines SFM as "the stewardship and use of forests and forest lands in a way, and at a rate, that maintains their biodiversity, productivity, regeneration capacity, vitality and their potential to fulfill, now and in the future, relevant ecological, economic and social functions, at local, national and global levels, and that does not cause damage to other ecosystems" (MCPFE 2001), countless derivations from this basic definition are reflected across the various regulatory and voluntary programs that are in operation globally. This variation occurs largely because each program adopts its own criteria and indicators to assess, verify, and incentivize SFM practice, based on its unique interests and geographic focus (Prabhu et al. 1999).

Because forests and societies are in constant flux, the criteria, indicators, and desired outcomes of SFM also remain moving targets—shifting as scientific knowledge improves, as technologies evolve, and as societal values change. That said, there appears to be growing international consensus that key elements of SFM include some notion of balancing the productive functions of forest resources (supply of timber and/or non-timber forest products), the protective functions of forests (carbon storage and sequestration, water quality, wildlife habitat, the preservation of biological diversity, etc.), and the development of socioeconomic and political institutions to carry out SFM plans and policies.

As multi-stakeholder views of managing forests and forest resources have become prominent over the years, SFM has also evolved and adapted to meet these divergent interests. Similarly, globalization of forest sector operations has necessitated creation of transnational SFM initiatives to ensure conformity to the same standards across the value chain. Thus, SFM boundaries encompass both national and transnational initiatives and programs. As a result, SFM initiatives are not governed exclusively by state-centered regulation, but rather private actors, nongovernmental organizations, and hybrid public–private initiatives join in to form a broad governance landscape surrounding SFM. Finally, these initiatives or programs can either be regulatory (mandated by state-based authority) or voluntary (privately adopted largely based on market or ethical grounds).

Thus, governance structure surrounding SFM is multilevel and complex. Indeed, previous scholarship has focused on understanding these multilevel governance arrangements but has largely remained limited to analyses of these institutions in isolation—investigating the emergence, effectiveness, and legitimacy of various initiatives (Hooghe and Marks 2001; Buchanan and Keohane 2006; McDermott et al. 2010; Blackman and Rivera 2011). In this chapter, we take a broader approach. We first start with providing a typology of various SFM-focused programs and policies that, we believe, capture the spectrum of SFM governance structures. We label them (1) private governance networks, (2) transnational regulatory policies, (3) transnational voluntary policies, (4) state-level regulatory policies, and (5) state-level voluntary policies. Within each category, we identify key examples of specific SFM policies/regulations and discuss them at some length. Second, we focus our discussion on the interaction among these various governance mechanisms. Finally, we conclude by suggesting that resources developed within private initiatives—particularly those addressing MRV (monitory, reporting, and verification) capacities—provide important capabilities and opportunities to state-based and transnational policy makers attempting to regulate and incentivize sustainable forestry and forest products.

10.2 Private Governance Networks

For understanding private initiatives in the sustainable forestry arena, we borrow from the rival private governance network (RPGN) concept (Smith and Fischlein 2010). RPGNs draw from resources generated within existing, external commercial and civil society domains and collectivize in an effort to gain legitimacy and authority over the rules of sustainability governance. In doing so, these networks compete based on their ability to access shared, or "relational," resources from network partners—commonly, in the form of reputational assets, tacit knowledge, or access to markets and financing. In the absence of government, private actors gain authority—the ability to decide, direct, make rules, and obtain performance from others (Cashore et al. 2004)—through market adoption. As increasing numbers of marketplace actors "consent to be bound" to a particular set of rules and processes, the easier it is for the governing network to further assert itself (and the harder it is for rival networks to form or govern). In short, RPGNs are most able to compete for influence over the rules of the sustainability game when they are able to secure the resources necessary to sustain a diverse composition of network partners while delivering effective governance at a cost sufficient to gain large-scale market acceptance. With a bar set this high, it is no wonder that these efforts have met mixed performance reviews (Blackman and Rivera 2011).

The sustainable forestry sector is one of the best-established private governance fields in the environmental domain and the competition among systems has been well documented (McDermott et al. 2010). Less attention, however, has been given to understand the common set of rules, norms, processes, and assets that have emerged over the past 20 years that identify RPGNs as such—indicators of the range of possible and acceptable solutions toward sustainable forestry governance. One such institutional characteristic specific to sustainable forestry governance is the drive for third-party certification and the investments made by governance networks to build competitive on-the-ground MRV capabilities at both the landscape (point of origin) and supply-chain (chain of custody [CoC]) levels. The two most prominent forest certification systems are the Forestry Steward Council (FSC) certification and the Programme for the Endorsement of Forest Certification (PEFC). In addition to forest certification, the RPGN family includes other initiatives such as the Forest Footprint Disclosure Project (FFD).

10.2.1 Forest Certification Programs

At the international level, forest certification has been developed as an alternative to regulatory policy in order to promote sustainable forestry (Cashore et al. 2004, 2007), particularly because of the continual degradation of forests due to human activity (Vogel 2008). In general, this instrument is focused on the power of consumers to pull certified, well-managed timber through supply channels and thus create incentives for adoption. With nearly two decades on the market, the area under certification continues to increase—a testament to the forest products supply chains' recognition of social and environmental performance requirements of the global marketplace. Today, approximately one-third of forests managed primarily for the production of wood and non-wood forest products are certified by either FSC or PEFC.* Although certified forest area

---

* In 2012, FSC reported 153 million hectares under certification (www.fsc.org) and PEFC reported 243 million hectares under certification (www.pefc.org). It was estimated in 2010 that nearly 1.2 billion hectares of forest are managed primarily for production (FAO 2010).

worldwide continues to grow, adoption in tropical areas has been disproportionately slow, particularly because implementing certifiable practices are less profitable than conventional (Nasi et al. 2011). Achieving and maintaining forest management certification has associated costs, even under strict regulatory conditions, which is often not the case in tropical areas. Substantial financial investments are needed for preparing species inventories, harvesting plans, training workers in directional felling, low-impact transporting techniques, and tracking and monitoring systems. Despite the lack of large-scale studies on the performance of certification programs in general, it can be argued that they have "introduced a new form of partnership between civil society organizations and leading businesses, shifting the landscape of sustainable production and consumption in important ways" (SCSKASC 2012).

Both FSC and PEFC operate globally and have developed sophisticated networks of independent, third-party, and accredited certification bodies (CBs). These CBs perform regular surveillance and reassessment audits to proactively verify and certify that forest owners or supply-chain companies maintain compliance with the systems' criteria. To ensure independence and impartiality, both FSC and PEFC have developed requirements for CBs to be independent from the standards development process and the entities they certify.

Most of the focus paid to the efficacy of monitoring, reporting, and verification (MRV) of sustainable forestry RPGNs has been on "controversial" cases surfaced by the various watchdog groups and environmental nongovernmental organizations (ENGOs).* Such cases either involve suspected lack of rigor of audit processes of CBs, RPGN's delivery and control over accreditation and reaccreditation services, or the ability of RPGNs to monitor and correct any of these failures. Some comparative studies tend to favor FSC's network over PEFC, with regard to specificity of requirements for CBs and accreditation organizations and control or oversight of the enforcement of requirements (Hirschberger 2005; NEPCon 2012). While this chapter is less interested in the particular comparative performance of one system over the other, it is important to recognize that networks of independent third-party organizations, and their interaction with accreditation and certification (i.e., signaling of conformance to a particular set of rules and practices), are central to the credibility and legitimacy of sustainability RPGNs.

10.2.1.1 Forest Stewardship Council

Established in 1993 by leading environmental groups and their allies, the FSC works to "promote environmentally appropriate, socially beneficial, and economically viable management of the world's forests" (FSC 2013). FSC integrates businesses, communities, and other organizations with well-managed forests through management and CoC standards, accreditation services, and trademark assurance. By November 2012, FSC operated internationally in over 50 countries with 168 million certified hectares (ha) of forests. FSC certification is a one-size-fits-all set of 10 generic principles and criteria for responsible forest management. In addition to the 10 principles, supporting rules and guidelines help to clarify and define the principles' requirements on a regional basis. How well a particular party meets the principles is measured by the FSC standard, which is developed in accordance with the International Social and Environmental Accreditation and Labeling Alliance's (ISEAL) Code of Good Practice for Setting Social and Environmental Standards. ISEAL, created by FSC and other international environmental certification systems, holds

---

* See forestethics.org, fsc-watch.org, globalforestwatch.org, and goodforforests.org, among other less forest-specific NGOs (NRDC, WWF, RFF, etc.) consistently weighing in on forest certification MRV and efficacy.

independent certification systems to a set of criteria defining good social and environmental practices in an industry or product. Independent CBs accredited by FSC issue the certificates to forest organizations. Accreditation Services International (ASI) manages the FSC accreditation program.

The FSC standard outlines three types of certificates that an entity can apply for: (a) forest management certification, (b) CoC certification, and (c) FSC controlled wood certification. The FSC forest management certification is used by forest managers and owners to show that their forest is well managed (i.e., socially beneficial, environmentally sound, and economically viable). The CoC certification is used by those who interact with a forest product throughout its supply chain, from the forest to the hands of the consumer. Manufacturers, traders, processors, and forest managers show consumers and policy enforcers that they are using responsible raw materials by obtaining this certification. The FSC controlled wood certification is used by CoC-certified companies that produce an FSC mixed source. This label conveys that a product may not contain 100% FSC-certified wood, yet the uncertified portion of wood is produced in a socially and environmentally responsible manner.

10.2.1.2 *Programme for the Endorsement of Forest Certification*

PEFC is an international umbrella certification standard that endorses national forest certification systems. It aims to promote and restore trust in SFM and at the beginning of 2013 was the world's largest certification program with over 247 million certified hectares of forests. Originally, the Pan-European Forest Certification Council, PEFC was founded in 1998 by European forestry interest groups. PEFC provides national certification programs the chance to conform to local political, social, economic, and environmental situations while at the same time complying with an internationally recognized standard (PEFC 2013). There are two main benefits to this system. Being certified to a PEFC-endorsed standard means that certification is recognized by other standards. For example, in North America the Sustainable Forestry Initiative (SFI), American Tree Farm System, and the Canadian Standards Association's Sustainable Forestry Management Standard (CAN/CSA-Z809) are all endorsed by PEFC. As a result, each separate entity has access to fiber from 82% of the certified forests in North America (SFIP 2013). Second, the PEFC standard is flexible and adaptable to different circumstances due to the bottom-up approach that requires compliance to both national laws and PEFC requirements. (See Box 10.1 for detailed examples of national forest certification programs endorsed by PEFC.)

National systems wishing to become PEFC certified must undergo third-party assessment. PEFC places high importance on separating its own activities in standards development from the certification and accreditation processes so that the independence of these processes can be maintained and bias is avoided. As a result, PEFC has adopted requirements for certification and accreditation as defined by the International Organization for Standardization (ISO) and the International Accreditation Forum. Independent third parties certify national systems to be PEFC compliant on an annual basis.

Three standard labels are available for identifying PEFC-certified products: (a) PEFC certified, (b) PEFC certified and recycled, and (c) promoting SFM. PEFC certified identifies products that are made with wood from controlled sources and that have a content of at least 70% PEFC-certified forests. PEFC certified and recycled identifies products that are made with wood from controlled sources and/or that have a PEFC-certified fiber content or a postconsumer recycled material content of at least 70%. The third label, promoting SFM, is intended for educational or promotional use about the PEFC.

BOX 10.1   EXAMPLES OF NATIONAL FOREST CERTIFICATION PROGRAMS ENDORSED BY PEFC

AUSTRALIAN FOREST CERTIFICATION SCHEME

The essential elements of the Australian Forest Certification Scheme, which commenced with the drafting of the Australian Forestry Standard (AS 4708) in 2000, were fully developed during 2002–2003 to provide a national forest certification scheme based on Australia's conformity assessment framework. Australian Forestry Standard Limited is a not-for-profit public company, which owns the standard development functions and manages the elements of the Australian Forest Certification Scheme. As of March 2012, over 10 million hectares were certified. Furthermore, additional standards, rules, and procedures have been created to address CoC, logo use, and compatibility assessment with other international certifications programs.

BRAZILIAN FOREST CERTIFICATION PROGRAM (CERFLOR)

CERFLOR was formally launched in 2002 by the Ministry of Development, Industry and Trade as the official Brazilian program for forest certification. The process of discussion and development of principles, criteria, and indicators took place between 1996 and 1999 by the Brazilian Society for Silviculture and the Brazilian Association for Standardization. In 2001, the scheme was expanded to include CoC and was recognized by PEFC in 2005 after an independent audit. As of April 2012, a total of 1.5 million hectares was reported to be certified by CERFLOR, less than half of the acreage under FSC certification in Brazil. Approximately 3.5 million hectares of Brazilian forests are certified by FSC.

CHINA NATIONAL FOREST CERTIFICATION SCHEME (CFCC)

Although FSC and PEFC certification schemes have been in operation in China since 2002 and 2006, respectively, in 2000 the Chinese government started the process of building a national forest certification system to better address its national socioeconomic conditions. For this purpose, it created the Forest Certification Department of the State Forestry Administration and the China Forest Certification Council (CFCC) to administer the scheme's overall development, based largely on FSC principles. The scheme was officially launched in 2007 by CFCC, and in 2009, SFA established the Zhonglin Tianhe Forest Certification Center (ZTFC) as the first CB in the country accredited by the Certification and Accreditation Administration of China. In late 2011, CFCC was recognized by PEFC. It is expected that CFCC will gain wide support with government backing and lower costs of local auditors. ZTFC expected that more than 3 million hectares of forests would be CFCC-certified by the end of 2011; however, PEFC had not reported aggregates as late as May 2012.

SUSTAINABLE FORESTRY INITIATIVE: NORTH AMERICA

SFI Inc. is an independent, nonprofit organization that works with conservation groups, local communities, resource professionals, landowners, and others to promote SFM in the United States and Canada. SFI, formed in 1999 by the American

Forest and Paper Association, is known for providing flexibility and adaptability to local conditions. Although the SFI became independent in 2007, it still retains partnerships with large forest products and paper manufacturers in North America and attains most of its influence from upstream suppliers of forest and paper products. Offering forest certification, CoC, and certified sourcing labels, SFI program requirements are audited by independent, third-party CBs for SFI standard compliance before logo use is granted. Endorsed by PEFC, the SFI has 250 program participants and 74 million hectares certified.

10.2.2 Forest Footprint Disclosure Project

While FSC and PEFC are the largest and most established RPGNs, other initiatives have also emerged. Although established in 2008, the FFD has received important support from the private sector. A project managed by the Global Canopy Foundation and supported by the British government, it is modeled on the successful Carbon Disclosure Project and aims to create transparency and provide quality information to investors. Seventy-five financial institutions, totaling over US $7 trillion in assets, have endorsed the FFD as a risk and transparency indicator. The project defines "forest footprint" as the total amount of deforestation caused directly or indirectly by an organization or product. Participating companies are asked to disclose how their operations and supply chains are impacting forests worldwide, based on exposure to five commodities in their operations and/or their supply chains. Timber, soy, palm oil, cattle products, and biofuels are the commodities evaluated on their potential to be sourced from recently deforested land. As with other market-based mechanisms, the FFD expects companies disclosing their forest footprint to minimize their business risks, receive increasing levels of endorsement by investors, gain experience to address emerging regulation, and capitalize on new market opportunities (Campbell et al. 2011). Because supply chain risk, especially from an environmental perspective, is a relatively new discipline, few standards or protocols currently exist to provide an avenue for monitoring or verification.

Every year, the FFD project sends out information requests to global corporations on behalf of FFD endorsing financial institutions. The FFD directs its requests to companies divided in 10 market sectors making them easier to compare. Furthermore, it enables information sharing on a global supply chain level, perhaps giving seemingly different market sectors a common denominator and encouraging coordination. In June of 2012, FFD announced plans to merge with the Carbon Disclosure Project, a nonprofit organization with which the FFD is operating in parallel. The merger, which will not be complete until 2014, claims to result in the world's largest, most comprehensive natural capital measuring system, touching on carbon, water, and forest resources (CDP 2013).

10.3 Transnational Regulatory Forestry Policies

International environmental agreements (IEAs) have been considered necessary to address harmful environmental consequences that are caused by the transnational trade of environmental resources, mainly focusing on the trade of tropical species to developed countries. The Convention on International Trade in Endangered Species of Wild Fauna and Flora (CITES), the International Tropic Timber Agreement (ITTA), and the more recent Convention on Biological Diversity (CBD) are among the key environmental agreements that to some degree influence the world supply of forest products. Formal IEAs have existed for almost four decades, with questionable success in halting illegal logging or improving forest management. Several environmental agreements influence forest production and trade, with a particular focus on procurement of timber from legal sources. They are now explained.

10.3.1 Convention on International Trade in Endangered Species of Wild Fauna and Flora

One of the oldest agreements is the CITES (CITES 1973), and although not directly focused on illegal logging, it is considered to be one of the most successful efforts safeguarding the trade of endangered species (Aikman 2003). CITES is an international agreement of 175 governments with the objective to ensure that international trade in specimens of wild animals and plants does not threaten their survival. CITES is a cooperative effort to trace the international trade of about 30,000 species of animals and plants helping to ensure that products made from them are from legal and sustainable sources.

Around 200 species of trees are included in the CITES appendices, and trade in their products is regulated to avoid utilization that is incompatible with their survival. CITES includes valuable and commonly traded timber species such as afrormosia (*Pericopsis elata*) from Africa, ramin (*Gonystylus* spp.) from Southeast Asia, and big-leaf mahogany (*Swietenia macrophylla*) from Central and South America. Big-leaf mahogany, for example, was included in 2002 on Appendix II of CITES, requiring exporting countries to verify the legality and scientific certification of each shipment, to assure that the harvest was non-detrimental to the survival of the species. Despite the legal strength of this international agreement, the inadequate implementation and little use of its measures hinder its objective (Aikman 2003). As with forest certification, a must requisite for this policy to work is to separate illegal timber from the supply chain, something difficult to accomplish without a strong and closely monitored CoC tracking system. Another way trade is controlled under this convention is by giving importing countries the power to reject the entry of the timber if CITES documentation does not follow the established procedures, including proper consultation with scientific authorities.

10.3.2 International Tropical Timber Agreement

Another major transnational regulatory policy is the ITTA, first entered into force in 1985 and later renegotiated in 1994 and in 2006. This agreement is overseen by the International Tropical Timber Organization (ITTO), an intergovernmental organization created under the umbrella of the United Nations (UN) to promote conservation and sustainable use and trade of tropical forest resources. With ITTA's latest update, it became the first legally binding international instrument to explicitly address illegal logging (ITTA 2006). Its core objective is the promotion of responsible tropical forest management, yet many critics have suggested that the agreement lacks the appropriate mechanisms to address harvesting of illegal tropical timber as it is primarily focused on timber trade as opposed to SFM (Srivastava 2011). Specifically this agreement includes several definitions, although "SFM" is not mentioned. Furthermore, the only specific international body mentioned as one to coordinate with is the UN Conference on Trade and Development, suggesting its potential bias toward tropical wood production and trade. Another interesting aspect is found in Article 34 of the agreement about nondiscrimination that indicates: "Nothing in this agreement authorizes the use of measures to restrict or ban international trade in, and in particular as they concern imports of, and utilization of, timber and timber products" (ITTO 2013). This nondiscrimination policy seems contradictory with CITES objective of banning the trade of endangered species. Although ITTO promotes the agreement as a conservation and trade agreement, it seems that trade has a prominent role.

10.3.3 United Nations Convention on Biological Diversity

A third IEA is the UN CBD, signed by multiple states in 1992. Article 3 of the convention stipulates that signatory countries maintain full sovereignty to exploit their own resources, based on each country's own set of environmental policies. However, the CBD overall objective seeks to protect ecosystems, including forests, by requiring signatory countries to take steps to limit activities that threaten extinction of species or degradation of ecosystems within their territory. Specifically, Articles 8 and 9 ask to create and enforce laws and regulations to protect threatened species, establish special protection areas, and conduct environmental impact assessments of development projects.

In its most recent Conference of Parties (COP 11th), developed countries agreed to double their financial commitments to help developing countries meet their targets, particularly to deter the rapid loss of biodiversity and the services provided by ecosystems. However, most of the forest industry remains suspicious of the CBD, which is generally seen as a tool of environmentalists. Furthermore, the absence of the United States in the convention is a major limitation to its potential success due to its prominent role in timber production, consumption, and trade (Schloenhardt 2008).

Within this convention, the *Expanded Program of Work on Forests* promotes forest law enforcement and related trade, including the revision of legislation to address illegal activities, capacity building for effective law enforcement, and development and implementation of CoC systems (CBD 2012). However, in general, the CBD has few binding measures to suppress illegal timber trade. Parties have discussed adopting more comprehensive measures to address this problem though no outcomes emerged. While the success of this agreement on SFM or curtailing illegal logging is not clear, it has promoted the development of national and regional schemes.

10.4 Transnational Voluntary Market-Based Policies

International multilateral organizations including the various organs of the UN, financial institutions, and even private financing institutions have adopted voluntary environmental and social standards to assess proposals related to various development projects. This development is in line with the expressed commitment of these various agencies to sustainable development. Although in some cases not directly involved in forestry projects, but because of the extent and scope of their influence, these institutions can significantly influence the adoption of sustainable management practices and procurement protocols by the forest sector and may combat, for example, illegal logging (Ozinga and Mowat 2011).

10.4.1 Equator Principles

The Equator Principles (EPs) is a voluntary set of standards based on social and environmental sustainability performance standards stipulated by the International Finance Corporation (IFC) and on Environmental Health and Safety guidelines of the World Bank Group. The EPs are applicable on any loan at or above US $10 million. Currently, there are 73 participating banks that have adopted the EPs, resulting in 90% of the global project finance industry addressing sustainability in principle (EP 2013), including projects related to forestry and the forest products supply chain. Since 2002, the World Bank's policy has been such that investments in private forest companies meet strict requirements, with guidelines calling for "mandatory independent certification of private company forest harvesting and management operations."*

Because of the influence and credibility of IFC and the World Bank, the EPs have gained widespread acceptance, encouraging continual participation and effective enforcement. Specific to the forest sector, keeping in line with the IFC guidelines, many development banks have committed to the following: (a) not to finance commercial logging operations or the purchase of logging equipment for use in primary tropical moist forest; (b) finance only preservation and light, nonextractive use of forest resources in forest areas of high ecological value; and (c) finance plantations only on non-forested areas (including previously planted areas) or on heavily degraded forestland (EP 2013).

Project financing guidelines also stipulate to avoid financing projects that compromise any relevant IEA or law. There is provision to safeguard and protect the legal and social rights of indigenous and project-affected communities through representation, information sharing, and policy participation.

10.4.2 World Bank and Forestry Financing

In a similar manner, the World Bank (hereafter Bank) does not directly manage forests, local governments, or projects, but it influences the forest sector through its investments. The Bank first became involved with forestry financing in the 1990s when it was a widely held belief that the poor were the main cause of deforestation in moist tropical areas. Initially, the Bank adopted strict policies that protected forests by discouraging any investments within these impoverished communities. However, through the late 1990s and 2000s, an overwhelming body of scientific research emerged indicating that the leading cause of forest loss is not logging, illegal or otherwise, and that neither population nor poverty alone constitutes the sole and major underlying causes of land-cover change worldwide (Lambin et al. 2001; Geist and Lambin 2002; DeFries et al. 2010). In response, the Forest Strategy and Forest Policy were reformulated in 2002 to encourage investments with poor forest communities and to expand beyond moist tropical forests into other forest ecosystems throughout the world. The Bank's *Forest Strategy* and *Forest Policy* now require that all Bank operations take into account the impacts of the investment on forests and by extension the livelihood of those who depend on forests.

In 1998, the Bank committed to a global target of certifying 200 million hectares (ha) of forest by 2005, with help from the World Wildlife Fund (WWF). Between 1998 and 2005, however, only 22 million hectares of forests in Bank client countries could be certified. This alliance also published a systemic framework to evaluate the variety of forest certification systems that are available to forest managers, called the Forest Certification Assessment

---

* World Bank Operational Policy 4.36.

Guide (FCAG). The Bank requires a certification system to incorporate reliable and independent assessment procedures.

10.4.3 International Green Purchasing Network

The International Green Purchasing Network (IGPN) was created in 2005 to create a sustainable society by promoting global green purchasing activities and the development of environmentally preferable goods and services (IGPN 2013). IGPN Council is represented by members from Japan, Malaysia, Korea, India, and also from Sweden and North America but includes other affiliated networks such as those of China, Thailand, Singapore, New Zealand, the Philippines, Indonesia, and Taiwan. Activities of the network include collecting and providing information on examples of global green purchasing, on best business practices involving environmentally preferable products, and the latest green purchasing trends. Among the different projects developed by the International Council for Local Environmental Initiatives (ICLEI), Local Governments for Sustainability (Box 10.2) stands out as the sustainable procurement guidelines developed for the UN Environmental Program specifically for office IT equipment, office furniture, and office stationery (paper, paper consumables, writing implements, and toner cartridges). These guidelines suggest that all wood comes from legal sources and preferably from those managed in a sustainable manner. It specifically indicates that products carrying the Nordic Swan, German Blue Angel, Japan's Eco Mark, FSC, PEFC, or any other sustainable forest standard where the percentage of certified wood is indicated will be accepted as proof of compliance (UNEP 2013).

10.4.4 Transnational Forest Carbon and Conservation Policies

Born under the climate change regime, the reduction of emissions from deforestation and forest degradation, forest conservation, sustainable management of forests, and

---

BOX 10.2   INTERNATIONAL COUNCIL FOR LOCAL ENVIRONMENTAL INITIATIVES

Local governments also have significant procurement leverage. Although multiple other efforts occur at individual governments, through ICLEI–Local Governments for Sustainability, over 1220 local governments from 70 different countries are committed to address the current challenges they face around sustainable development. Founded in 1990 under the name "International Council for Local Environmental Initiatives," ICLEI–Local Governments for Sustainability is based on the idea that global programs require local solutions and governance. It unites local governments that operate through chapters in member countries and includes governments from cities, towns, and counties. This program stands out because it takes into consideration environmental concerns, in addition to social and economic ones. Among socioeconomic aspects are educating consumers on buying only what they really need or educating and supporting fair trade initiatives. Under the "sustainable-procurement.org" platform, ICLEI–Local Governments for Sustainability provides information on sustainable procurement such as best practice cases, policy updates, and practical tools.

enhancement of forest carbon stocks (REDD+) is an international effort to create a financial value for the carbon stored in forests. There is particular interest in the REDD+ incentive system in an attempt to link payments to performance and thus offer developing countries substantial financial benefits for improving the management and protection of their forests in cost-effective ways. It is expected that REDD+ will influence forest management and conservation by helping southern countries develop reference scenarios, adopting a REDD+ strategy, and designing monitoring systems.

REDD+ has mainly been discussed in international climate negotiations, with particular interest to provide tropical developing countries with financial incentives—typically from more developed countries—to reduce deforestation and degradation rates. It is predicted that financial flows for greenhouse gas emission reductions from REDD+ could reach up to US $30 billion a year (UN-REDD 2013). This flow of funds could reward a meaningful reduction of carbon emissions and also support development in southern countries, where deforestation and forest degradation is problematic. Nevertheless, whether this funding takes the form of bilateral aid, market-based carbon trading, or something else has not yet been agreed.

Today there are two major programs in place addressing the development of future systems of financial incentives for REDD+, both under the UN Framework Convention on Climate Change and under the Climate Investment Funds. First is the UN–REDD+ Program, a UN collaborative initiative launched in 2008 to assist developing countries to prepare and implement national REDD+ strategies. Also established in 2008, the Forest Carbon Partnership Facility (FCPF) complements the UN Framework Convention on Climate Change negotiations by demonstrating how REDD+ can be applied at the country level. Both efforts are channeled through several continental development banks and the World Bank. The US $300 million FCPF approved Readiness Preparation Grants with the Democratic Republic of Congo, Ghana, Costa Rica, Mexico, Indonesia, and Nepal in 2011 (FCP 2011).

Another transnational initiative is the UN–REDD Program, created jointly among the UN's Development Program, Environmental Program, and FAO. The UN–REDD Program also works closely with other organizations such as the FCPF, the Forest Investment Program, the UN Forum on Forests, the Global Environment Facility, the UN Framework Convention on Climate Change, and the ITTO. The UN–REDD Program has committed over US $120 million for capacity development, governance, engagement of indigenous peoples, and technical needs, across 42 countries in Africa, Asia-Pacific, and Latin America and the Caribbean (UNDP 2012).

Most financial support of REDD+, however, has come from national governments. Developed nations have pledged more than US $7.3 billion to help developing countries get up to speed on REDD+, $4.3 billion of which is slated to be delivered by the end of 2012 (EMP 2012). While there is legitimate concern around the implementation and accountability of these funds, a number of financial tracking initiatives have recently emerged. According to the Tropical Forest Group, a US-based NGO, over US $335 million of the United State's $1 billion fast start finance REDD+ pledge was deployed as of 2011 (Table 10.1; TFG 2012). If financing proceeds as expected, REDD+ has a high potential to overcome illegal logging, as its roots reside in issues such as opportunity cost of land, cost of SFM and certification, lack of trained staff, inefficient wood processing, weak institutional capacity, and deficient or nonexistent monitoring systems. However, if carbon stocks are valued and traded internationally, standing forests, particularly under full conservation modalities, will remain untouched, influencing the availability of illegal timber from tropical forests.

TABLE 10.1

US REDD+ Funding (2008–2011)

US REDD+ Funding (Status and Region)	Projects	Total
Funding status		
Spent	115	$136,722,527
Authorized for specific program/grant	212	$84,021,588
Deposited to a country	27	$10,584,407
Pledged to a specific country	42	$97,076,000
Unknown	2	$6,791,986
Regional Status		
Africa	46	$57,905,317
Asia	122	$122,490,976
Latin America and the Caribbean	241	$154,800,215
Grand total	409	$335,196,508

Source: Developed from data provided by Tropical Forest Group, http://www.usreddfinance.org/redd, accessed August 2012, 2012.

10.5 State-Based Regulatory Policies

Although potentially occurring in other countries, recently, the United States and the European Union (EU) have adopted domestic regulations to ameliorate the persistent availability and trade of illegal wood in the world market. The US regulation is known as the Lacey Act, and the EU regulation, which enrolls all 27 of the politically united member states, is called the European Union Timber Regulation (EUTR).

10.5.1 Lacey Act

In 2008, the United States became the first country to legally ban the import of illegal wood. The US Congress made an amendment to the 100-year-old Lacey Act originally enacted to prohibit the transportation of illegally captured animals or wildlife products across US state lines and later from other countries too. The new law extends to include timber, paper, and other forest products. Previously, unless a plant was endangered and protected by CITES, there was no system in place to combat their trade, regardless of the legality status of their harvesting. Now, under the amended Lacey Act, plant products that are illegally harvested in a country of origin and that are brought into the United States for further manufacturing or for final sale are considered illegal. Lacey Act also requires importers to declare the country of origin and species name of all harvested plants contained in their products. The law is valid through the US supply chain. It must be noted, however, that "legally sourced" is defined based on the laws of the country of origin.

The Lacey Act is different from other US regulations in that it is fact based rather than document based (EIA 2013). Whereas documentation is a part of Lacey Act enforcement, an equal amount of emphasis is placed on "due care" of the market player or operator to know what is taking place within their supply chain, similar to the later discussed

EU "due diligence" approach. "Due care" is not ultimately about providing extensive documentation and certification, rather it also requires importers to investigate their supply chain and discern where imported plant products originate. Third-party certification is a good way to demonstrate due care, but it is not a way to become exempt from the Lacey Act since the Lacey Act is more concerned with the legality of plant products moving through the supply chain. Even if a market player is not the one to violate a foreign law, if the operator is handling tainted plant products, then the operator may be found guilty by association. The Lacey Act, implemented by the US Department of Agriculture's Animal Plant Health Inspection Service (APHIS), shares responsibility for investigating illegal plant cases with the US Department of the Interior's Fish and Wildlife Service (FWS). Also, the Department of Homeland Security, which controls US customs and monitors the borders through Customs and Border Protection, will assist with enforcing the Lacey Act. Box 10.3 outlines the context for the amendments to Lacey Act that are currently underway.

BOX 10.3   LACEY ACT AND GIBSON GUITAR CORP.

In 2009 and 2011, the US Department of Interiors FWS conducted two separate raids on Gibson's Tennessee facility to investigate evidence of Lacey Act violations. Allegations later revealed that Gibson had illegally imported mahogany and rosewood from Madagascar in 2009 and a certain classification of Indian rosewood and ebony in 2011 (Abbey et al. 2006). The allegations were settled in August of 2012 when Gibson Guitar Corp. entered into a criminal enforcement agreement with the FWS. With Gibson conceding to pay a penalty of $300,000 and an additional community service payment of $50,000 to the National Fish and Wildlife Foundation, the FWS agreed to defer further criminal prosecution against the company (USDOJ 2012). Gibson also agreed to implement an internal compliance program and to withdraw claims to the Madagascar ebony seized in the raids on its Tennessee facility—valued at an estimated $261,800.

The recent enforcement of the Lacey Act has brought to light several policy concerns, triggering interest in amending the legislation. Both policy makers and industry stakeholders alike have expressed concern over what role government agencies can play in enforcing and interpreting other countries' laws. Businesses in particular have argued that the declaration requirements for plants and plant products are too burdensome, especially when dealing with finished and/or composite materials (Sheikh 2012). Musicians and furniture retailers have been pushing to exempt products imported before 2008.

In response to the criticism, two bills have been introduced to Congress to amend the Lacey Act. HR 3210 (2011) proposes to purge the legislation of penalties against those who unknowingly possess illegal wood and/or products imported prior to 2008. It would also lessen declaration and reporting requirements. Executives of the furniture retailer IKEA and the American Forest and Paper Association have said they are in favor of exempting products imported before 2008 but do not support this amendment as a whole (Bewley and Bureau 2012). A second bill, HR 4171, was proposed in early 2012 and would repeal the requirement that importers comply with foreign environmental laws.

10.5.2 European Union Timber Regulation

While the EU has significantly advanced the implementation of policies to sustainably manage forests within its territory, it has also attempted to foster SFM elsewhere. Due to its prominent role both as an exporter and an importer of forest products, the EU is well placed to substantially influence international forest products trade.

Effective for all 27 EU countries on March 3, 2013, the EUTR (EU) 995/2010 is meant to ban products made with illegal wood from entering into the EU (EUR-LEX 2013). The regulation puts the responsibility on operators to ensure that their timber supply is in compliance with EUTR, which "includes provisions to facilitate the traceability of wood products within the EU back to their first placing on the EU market" (EUFLEGT 2013) This policy distinguishes itself from EU Forest Law Enforcement, Governance and Trade (FLEGT) Action Plan, a voluntary policy discussed in the following section, since it legally binds EU members to follow the policy rather than allowing member states to voluntarily adopt the regulations.

Passed due to a demand from EU member states and stakeholder groups, the EUTR aims to ban illegal wood at the point where it is initially placed on the EU market. The legislation applies to most wood products, including pulp and paper, and requires that operators practice what is called "due diligence," a system that is designed to create awareness and minimize the risk of handling products made from illegally logged forest material.

Specifically, the EUTR defines three requirements applicable to operators who place timber products on the EU market for the first time: (1) a prohibition to place illegally harvested timber or products derived from such timber on the EU market (the definition of legal timber is based on the law of the country of harvest); (2) operators who are placing timber on the market for the first time, either imported or domestically harvested, have to exercise "due diligence," so they need to have access to information on the source of the timber and need to take reasonable steps to ensure that their supply is from legal timber; and (3) after placing timber on the market for the first time, the other operators, called "traders" in the regulation, have to keep records about whom they bought from and to whom they sold the timber (EUFLEGT 2013).

Central to the EUTR legislation is the due diligence system, which comprises three main components: (a) information sharing, (b) risk assessment procedures, and (c) risk mitigation procedures. While a great deal of responsibility is placed on the operator to demonstrate due diligence to EU officials and traders, responsibility is also placed on traders to ask for information regarding sourcing, and they must keep the information on file. Operators who initially place a forest product on the market must provide information that demonstrates due diligence. This includes the type of timber, both the trade name and the scientific name of the material; the country of harvest and, where applicable, the subnational region and concession of harvest; the quantity of material; the name of both the supplier and the client; and documentation indicating compliance with applicable legislation.

A second step in the due diligence system is that operators must demonstrate a risk assessment. A risk assessment involves researching and analyzing the risk of illegality. This can be evaluated through checking if the supply is compliant with applicable legislation, often times employing a label of third-party verification scheme. It should be noted, however, that purchasing forest products that are third-party certified does not mean automatic compliance with the EUTR unless it is FLEGT certified. Verifying if the country has a prevalence of illegal harvesting or if there are any EU/UN sanctions against the country is another way to assess risk of timber illegality. Finally, observing the complexity of the supply chain is an additional way for suppliers to conduct a risk assessment. If the supply chain is complex, chances of illegality may increase simply due to difficulty concerning traceability. If the operator determines that the risk to its supply chain is not negligible, the next step is to apply risk mitigation, through, for example, additional information or third-party verification.

Enforcement of the EUTR is left up to each member state's designated competent authority (CA). The CA will verify that the operators are responsible for EUTR compliance through examining its due diligence documentation, records, and inspections. If an operator is found to be out of compliance with the EUTR, the CAs have the authority to seize the timber material or to prohibit marketing and/or trade. Penalties are defined by member states and are in proportion to the environmental damage, value of the timber, and tax losses.

For those operators that are concerned about EUTR compliance, they can make use of a monitoring organization (MO) to assist them with a due diligence system. MOs create a due diligence system, allow importers the right to use it, and are subject to the CA's inspection. While the operator is still individually accountable for their own supply chain, an MO is useful to verify compliance with the EUTR for companies who may not have the time, expertise, or funding to create a due diligence system of their own. MOs will be able to validate the proper use of their system and take appropriate action for information where needed. There are several advantages to allowing MOs to participate in a policy such as the EUTR. Importers will be able to use a common system, increasing the amount of information that is accurate and available. Also, exporters will be able to consult a common system, which means they are not confronted with numerous clients or tracking systems.

10.6 State-Level Voluntary Market-Based Policies

Market-based policies provide countries with instruments to implement, nationally or in cooperation with other countries, mechanisms to reduce negative environmental externalities (e.g., illegal deforestation). In the following section, we discuss the EU FLEGT Action Plan, a policy that provides a framework for partnership between countries that produce and export timber and those countries that import and consume timber. Furthermore, we discuss global developments in Green Public Procurement (GPP) programs.

10.6.1 European Union Forest Law Enforcement, Governance, and Trade Action Plan

In 2005, the European Council adopted the FLEGT Action Plan, a policy that provides a framework for partnership between countries that produce and export timber and those countries that import and consume timber. The agreement is voluntary but once signed is enforced through European Customs. When a FLEGT partnership is formed between two countries, the first step is to create a licensing system to prevent any unlabeled imports from entering the European country. The agreements provide a guarantee of legality since the EU member states will only accept timber shipments accompanied by a FLEGT license. Partnering countries can receive both financial and advisory assistance from FLEGT to set up the licensing system and an enforcement plan, which is independently monitored.

By adopting FLEGT, EU member states are encouraged to examine their own domestic legislation and identify opportunities for further regulations to address illegal imports. Opportunities recommended by FLEGT include governmental procurement policies, voluntary industry initiatives within the supply chain, and the encouragement of financial institutions to analyze flows of finance to the forest industry.

The European Forest Institute recently conducted a progress report on the FLEGT Action Plan for the years 2003–2010 (Hudson and Paul 2011). According to this report, FLEGT policies provided over 600 million euro in distributions to support timber-producing countries and promote responsible trade in timber procurement programs. It is suggested that more attention should be given to conflict timber, use of existing legislative instruments to prevent illegal logging, and attentiveness to responsible financing and investment. The report also concludes that there has been a shift in market values and perceptions about the forestry industry within the EU due to the increased interest in cleaning up its supply chains. Some improvements include (a) the increasing acceptance of voluntary codes of conduct and procurement policies, notably in the timber, paper, and building industries, (b) increasing adoption of forest certification programs, (c) new CoC initiatives by timber trade federations and large-scale importers of tropical timber, and (d) a more general acceptance to check the legal origin of purchased timber products. FLEGT timber policies have continued to evolve. The voluntary program began to evolve in 2008 when EU policy makers passed a regulation making it illegal for EU member countries to import certain types of timber from countries that had entered into a Voluntary Partnership Agreement. In 2010, the European Parliament and European Council advanced regulations even further by prohibiting the sale of timber that has been illegally logged in its country of origin. These regulations would eventually give way to the creation of the EUTR.

Overall, FLEGT has increased transparency of the global forestry market supply. While FLEGT promotes certification, it is uncertain whether it will create its own certification and licensing program or if it will promote another label in the future. Forest certification could help differentiate illegal timber from legal timber and help with the expansion of FLEGT-like regulations to countries other than EU members.

10.6.2 Green Procurement Programs

Many governments have adopted in recent years GPP programs. These programs are often preferred because they can be developed and implemented more rapidly than most other policy options. In the absence of policies able to internalize the external costs and benefits of products, governments could account for them in their own purchasing decisions. In some circumstances, however, given the buying share between governments and the private market, this instrument could lead to increases in the environmental costs (Marron 1997). Governments worldwide have acknowledged their important role at the end of the supply chain and are exercising their power in order to stimulate innovation among suppliers to develop and provide environmentally preferable goods and services (Nader et al. 1992). Featured recommendations of these programs for forest products procurement encompass buying only legal wood, certified material if possible, from local sources, products easily disassembled or recycled, products with high expected durability, and products that through use of life-cycle assessments (LCAs) in design are less energy or carbon intensive.

While the cost of green procurement programs could limit their implementation, the EU commissioned a series of studies in 2006–2007 and found no difference in cost between average products and environmentally preferable products (EC 2013). Also the average administrative cost of implementing and managing the program was considered negligible for procurement authorities at 223 euros per 1000 inhabitants in a city or region. Furthermore, it is expected that companies themselves can find it practical to change their business lines in order to supply the government and their other customers with environmentally preferable products. Initial commitments from companies to change their forest product lines took place in the early 1990s with the WWF's creation of "buyers groups." These groups, consisting of retailers and other wood products purchasers, were formed to stimulate demand for FSC-certified wood products (Meidinger 2011). Buyers groups still exist today both as nonprofit/private partnerships and as public/private partnerships. By making a commitment to purchase only certified wood or to trace the origin of its wood products, members of buyers groups avoid negative publicity and gain a competitive advantage in a growing certified products market (Gadow 2000). Home Depot, for example, given the pressure from environmental groups and customers, has adopted a purchasing policy with a preference for wood from certified, well-managed forests and is also a member of the Sustainable Forest Products Global Alliance, a public/private buyers group (Home Depot 2012). Similar examples can be seen across many of the world's largest wood buyers such as Lowe's and IKEA, influencing thousands of customers, suppliers, and competitors. In Box 10.4, we present some relevant procurement programs functioning worldwide, which may have international, national, or local scope of operation and may be created and administered either by governments or by organized groups.

10.7 Integration of Various Public, Private, and Multilevel SFM-Focused Governance Initiatives

It is important to note that these various governance policies/instruments together achieve more than their sum. Stand-alone effectiveness of each system has its inherent limitation. For example, with the possible exception of CITES, IEAs have not been able to adequately halt the infiltration of illegally logged forest products into global markets. Similarly, privately governed forest certification schemes have made only modest improvements to forestry practice and continue to represent a relatively small fraction of harvested timber. In light of these shortcomings, we have witnessed a resurgence of national and multilevel regulatory efforts targeting the most egregious, and largely illegal, forest management practices.

But, we also observe that interaction among these various systems is increasing. The center place in this interaction is claimed by RPGNs as we observe more frequent interaction between RPGNs and all other systems. To explain this center place, we borrow from Peters and Pierre (1998) and assert that interaction across multilevel and multiactor governance types is facilitated by the need for each to secure "resources that would not be at its disposal were it to remain on its own side of the (presumed) divide." Granted that forest sector RPGNs has not been able to wholly influence market behavior, it has been much more successful in developing capabilities, competencies, and the institutional infrastructure necessary for identifying and segregating forest products that conform to management criteria across supply chains. The institutional legitimacy that certification programs have gained forms the basis for a new interaction across multiactor, multilevel forest governance. Especially, forest sector RPGNs have developed MRV capabilities— resources individually unavailable to governments, transnational organizations, or private governance networks. We argue that these capabilities have been paramount to the initial efforts of RPGNs' ability to operate in lieu of governmental regulation, but now they are proving to be critical to efficient implementation of new national and transnational

BOX 10.4 EXAMPLES OF NATIONAL GREEN PURCHASING PROGRAMS

ECO-BUY AUSTRALIA

ECO-Buy was initially established in 2002 with funding from the Australian Department of Sustainability and Environment and Sustainability Victoria. As of 2007, it operates as the independent, not-for-profit center, supporting both businesses and government agencies to green their purchasing through services and resources such as the Sustainable Procurement Assessment Tool. Born as a local initiative, ECO-Buy is now a national program and is planning to extend to New Zealand.

EUROPEAN COMMUNITY GREEN PURCHASING POLICY

In 2008, the EU created the GPP as a voluntary instrument whereby public authorities seek to procure goods and services with a reduced environmental impact throughout their life cycle. The policy develops clear, verifiable, justifiable, and ambitious environmental criteria for goods and services, based on a life-cycle approach and scientific evidence. In total, 18 sectors have been identified for implementing GPP. Forest products included are paper, furniture, and construction materials. Approximately 26% of the contracts signed by public authorities in the EU-27 (2009–2010) included all GPP criteria. Furthermore, 55% (38% of total value) included at least one core GPP criterion, indicating that some form of green procurement is being conducted at a large scale.

GREEN PROCUREMENT, JAPAN

In 2000, Japan signed the "Law Concerning the Promotion of Procurement of Eco-friendly Goods and Services by the State and other Entities," which requires all governmental and other specified institutions to purchase goods and services on their ability to reduce environmental impact throughout the product life cycle, have a minimal cost, and do not interfere with international trade regulations. In 2006, the term "legal wood" was incorporated in the list of designated procurement items, including measures to preferentially purchase products manufactured using wood with verified legality and sustainability. Later, in 2011, Japan adopted the "Basic Policy on Promoting Green Purchasing," which specifically encompasses wood-related products such as paper, engineered wood products, and other lumber materials.

GREEN PROCUREMENT IN LATIN AMERICA

The Latin American and Caribbean Network for Sustainable Production and Consumption, created by the United Nations Environment Program (UNEP), aims to strengthen the capacities and regional identities toward more sustainable lifestyles. The network works through a Council of Government Experts conformed by governmental professionals and technicians involved on this topic in each country. In addition, ComprasResponsables.org, administered by Fundación Centro de Gestión Tecnológica (CEGESTI), supports sustainable consumption by public agencies in Guatemala, El Salvador, Costa Rica, and Panamá. This program contains product specifications for paper and timber products. FSC-certified wood is mentioned in the documentation, due to its popularity in the region, although this certification does not have exclusivity within the procurement program.

forestry and forest products policies. While rules, criteria, and indicators continue to be established through government regulation, RPGNs are increasingly being looked to for MRV (SCSKASC 2012). In turn, RPGNs gain through integration with legal and regulatory policies access to global financial resources and economies of scale associated with national and transnational policies.

RPGNs also interact with state-level voluntary market policies, such as FLEGT, particularly because auditors of forest RPGNs have created multiple mechanisms toward legality assurance. For example, the Rainforest Alliance's SmartWood program, affiliated with FSC, has launched standards for the Verification of Legal Origin (VLO) and Verification of Legal Compliance (VLC) (Donovan 2010). The growing interaction between RPGNs and state-level voluntary market policies is evinced through a November 2011 FSC release declaring that revisions would be made to its FSC controlled wood, forest management, and CoC standards so that the terminology used in their standards match the terminology in FLEGT. Figure 10.1 diagrammatically outlines these various interactions among RPGNs and other voluntary/regulatory and state-level/transnational initiatives.

There is also a growing convergence between RPGN conformance audits and legal verification among some producer governments, in the sense that they see RPGN conformance as a potential surrogate for legality assurance and are considering treating some private sector certification schemes as equivalent to verification of legality (Brown and Bird 2008). "Legality verification" is becoming increasingly important as a design element of regulatory programs, particularly the Lacey Act in the United States and EUTR in Europe (Gulbrandsen and Humphreys 2006). As previously noted, these regulations

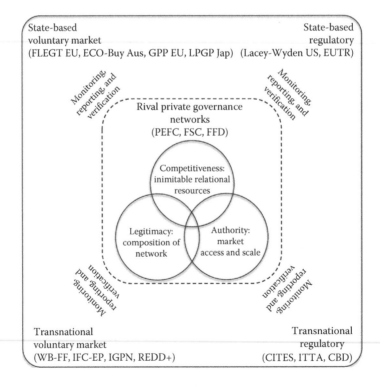

FIGURE 10.1
Integrative role of RPGNs' monitoring, verification, and labeling resources in the implementation of state-based and transnational policies.

require importers to show "due care" and "traceability," respectively, to ensure that forest products are harvested in compliance with laws and regulations in their originating jurisdiction. While legality assurance and conformance to RPGN sustainability criteria are not entirely consistent concepts, numerous arguments have been presented for their convergence. Benefits to convergence include avoiding duplication of bureaucratic requirements and associated costs or reducing conflicts associated with national sovereignty that arises when outsiders interpret foreign governments' laws. Furthermore, it could reward first movers who have invested in certification. In practice, FSC is already seeking to ensure compliance with the EUTR by requiring that European Commission-accredited MOs allow the offering of an FSC due diligence system to their clients. By doing this, FSC is attempting to ensure that their legality requirements align with those of the EUTR (FSC 2011). Similarly, CoC certification, under PEFC guidelines, implements risk analyses, external assessments, and on-site inspections to act as safety checks that ensure the legality of the wood. The PEFC standard offers both the EUTR and the Lacey Act an opportunity for the industry to practice due diligence and seek out certified products that ensure legality (PEFC 2012).

The integration of transnational forestry policies, like REDD+, with nation-based and private governance is also happening through the MRV requirements of "safeguards" in the REDD+ context. Again, FSC has identified integration with REDD+ as a strategic priority, specifically advocating independent third-party MRV. Similar interaction can be seen in transnational voluntary programs, like the World Bank–WRI Global Forestry Alliance (FSC 2012). In fact, the World Bank has been able to influence independent certification schemes to change and conform to Bank standards (Contreras-Hermosilla 2007).

Interaction among policy systems is also occurring. For example, we observe efforts for greater harmonization between FLEGT and REDD+. Several developing countries are engaged in both REDD+ preparation and FLEGT's negotiation or implementation of a Voluntary Partnership Agreement. Countries designing a REDD+ strategy or a system to ensure the legality of forest products face a number of common challenges: unclear legal and regulatory frameworks—particularly regarding land use and access to resources, difficulty in engaging some forest-dependent stakeholders, poorly developed information systems and transparency mechanisms, corruption, and weak law enforcement and judicial systems (Van Orshoven et al. 2012). Therefore, synergies may exist in terms of both governance and verification. The FLEGT negotiation process is proving to be a powerful tool to enhance stakeholder participation and brings significant market opportunity associated with verification and licensing. Through multilateral REDD+ negotiations, political attention has refocused to forests, providing access to global financial resources. While it is unclear the extent to which harmonization with RPGN MRV protocols influences these discussions, the possibility of multiple policies sharing MRV resources provides a conceptual opportunity for increased implementation efficiencies at a minimum.

10.8 Conclusion

SFM has been in the forefront of new multilevel, multiactor governance and continues to evolve. This chapter provides an overview of these various governance organizations and systems. As readers would note, the development of rules, codes, and norms for SFM has spanned across various levels and actors, perhaps more than any other governance field.

As public and private governance organizations and institutions jockey for authority and legitimacy to restrict illegal logging, promote forest resources with lower environmental impact, and protect the environmental, social, and economic values that forests provide, we see an increase in interaction among them. We have noted some of the preliminary observations but understanding the complexity of this interaction deserves significant additional attention to better elucidate the role of competition and collaboration dynamics of forest governance and their influence on policy outcomes. While we begin to highlight a number of areas where governance across levels and actors might find opportunities for harmonization through MRV capacities developed through private governance networks, it remains to be seen what form might emerge from this interaction. The presentation of these findings suggests a "hybrid" interaction among governance entities, through an explicit or implicit division and sharing of functions. However, the dynamics of this integration require both time and investigation to determine whether the authority of governmental policy supersedes the ability of RPGNs and transnational organizations to govern or whether a more symbiotic relationship might emerge.

References

Aikman, P. 2003. Using a 30 year old treaty to combat the illegal timber trade. *International Forestry Review*, 5(3): 307–313.

Blackman, A. and Rivera, J. 2011. Producer-level benefits of sustainability certification. *Conservation Biology*, 25(6): 1176–1185.

Brown, D. and Bird, N. 2008. Convergence between forest certification and verification in the drive to legality assurance, in *Legal Timber: Verification and Governance in the Forest Sector*, pp. 255–262. London, U.K.: Overseas Development Institute.

Buchanan, A. and Keohane, R.O. 2006. The legitimacy of global governance institutions. *Ethics & International Affairs*, 20(4): 405.

Campbell, T., Crosbie, L., McCoy, K., and Mitchell, A. 2001. Forest footprint disclosure project annual review 2010. Global Canopy Foundation. www.forestdisclosure.com (Accessed August 2012).

Cashore, B., Auld, G., Lawson, J., and Newsom, D. 2007. The future of non-state authority on Canadian staples industries: Assessing the emergence of forest certification. *Policy and Society*, 26(1): 71–91.

Cashore, B., Auld, G., and Newsom, D. 2004. *Governing through Markets: Forest Certification and the Emergence of Non-State Authority*. Yale University Press, New Haven, CT.

CBD. 2012. Secretariat of the convention on biological diversity. 2004. Expanded programme of work on forest biological diversity. Element 2, Goal 1, Objective 4 (Accessed August 2012).

Contreras-Hermosilla, A. 2007. *Forest Law Enforcement and Governance Program Review of Implementation*. Washington, DC: International Bank for Reconstruction and Development/The World Bank.

DeFries, R.S., Thomas, R., Maria, U., and Matthew, H. 2010. Deforestation driven by urban population growth and agricultural trade in the twenty-first century. *Nature Geoscience*, 3(3): 178–181.

Donovan, R.Z. 2010. *Private Sector Forest Legality Initiatives as a Complement to Public Action*. New York: Rainforest Alliance.

FAO. 2010. *Global Forest Resources Assessment 2010*. Main Report, Food and Agricultural Organization of the United Nations, Rome, Italy, ISBN: 978-92-5-106654-6.

FSC. 2013. Retrieved from www.fsc.org (Accessed January 15, 2013).

Gadow, K. 2000. Ecological and landscape considerations in forest management: The end to forestry? *Sustainable Forest Management*. pp. 45–50. Dordrecht, the Netherlands: Kluwer Academic.

Geist, H.J. and Lambin, E.F. 2002. Proximate causes and underlying driving forces of tropical deforestation. *BioScience*, 52(2): 143–150.

Gulbrandsen, L.H. and Humphreys, D. 2006. International initiatives to address tropical timber logging and trade: A report for the Norwegian Ministry of the Environment. Lysaker, Norway: Fridtjof Nansen Institute.

Hirschberger, P. 2005. The effects of PEFC-certification: An analysis of audit reports of PEFC Germany. Commissioned by the WWF European Forest Programme. http://assets.panda.org/downloads/theeffectsofpefccertificationfinal.pdf. (Accessed August 2012).

Hooghe, L. and Marks, G. 2001. *Multi-Level Governance and European Integration*. Lanham, MD: Rowman & Littlefield. p. 253.

Hudson, J. and Paul, C. 2011. FLEGT action plan progress report 2003–2010. http://ec.europa.eu/development/icenter/repository/flegt_ap_progress_report_2011-03-11_en.pdf, (Accessed November 2011).

IGPN. 2013. Retrieved from www.igpn.org (Accessed January 15, 2013).

Lambin, E.F., Turner, B.L., Geist, H.J., Agbola, S.B., Angelsen, A., Bruce, J.W., Coomes, O.T. et al. 2001. The causes of land-use and land-cover change: Moving beyond the myths. *Global Environmental Change*, 11(4): 261–269.

Marron, D.B. 1997. Buying green: Government procurement as an instrument of environmental policy. *Public Finance Review*, 25(3): 285–305.

McDermott, C.L., Cashore, B., and Kanowski, P. 2010. *Global Environmental Forest Policies: An International Comparison*. Londaon, U.K.: Earthscan. p. 373.

Meidinger, E.E. n.d. Look who's making the rules: The roles of FSC and ISO in international environmental policy. Errol Meidinger, J.D. http://webz.law.buffalo.edu/faculty/meidinger/scholarship/fsciso.html. Accessed November 2011.

Nader, R., Lewis, E.J., and Weltman, E. 1992. Shopping for innovation: The government as smart consumer. *The American Prospect*, 11: 71–78.

Nasi, R., Putz, F.E., Pacheco, P., Wunder, S., and Anta, S. 2011. Sustainable forest management and carbon in tropical latin America: The case for REDD+. *Forests*, 2(1): 200–217.

NEPCon. 2012. Comparative analysis of the PEFC system with FSCTM Controlled Wood requirements. (Prepared for FSC International Center). www.nepcon.net.

Ozinga, S. and Mowat, H. 2011. Strategies to prevent illegal logging. In: *A Handbook of Globalisation and Environmental Policy: National Government Interventions in a Global Arena. Eds. Wijen, F., Zoeteman, K., and Pieter, J.* Cheltenham, U.K.: E. Elgar.

PEFC. 2012. Retrieved from www.pefc.co.uk/index.php/news-page/news/news-detail/item/270-eutr-issues-high-on-agenda-at-uk-stakeholder-day (Accessed January 15, 2013).

PEFC. 2013. Retrieved from www.pefc.org (Accessed January 15, 2013).

Peters, G. and Pierre, J. 1998. Governance without government? Rethinking public administration. *Journal of Public Administration Research and Theory*, 8(2): 223–243.

Prabhu, R., Colfer, C.J.P., and Dudley, R.G. 1999. CIFOR guidelines for developing, Testing and selecting criteria and indicators for sustainable forest management. The Criteria & Indicators Toolbox Series.

Schloenhardt, A. 2008. The illegal trade in timber and timber products in the Asia-Pacific Region. Australian Institute of Criminology, Research and Public Policy Series No. 89.

SCSKASC. 2012. Steering committee of the state-of-knowledge assessment of standards and certification. *Toward Sustainability: The Roles and Limitations of Certification*. Washington, DC: RESOLVE, Inc.

Smith, T.M. and Fischlein, M. 2010. Rival private governance networks: Competing to define the rules of sustainability performance. *Global Environmental Change*, 20(3): 511–522.

Srivastava, N. 2011. Changing dynamics of forest regulation: Coming full circle? *Review of European Community & International Environmental Law*, 20(2): 113–122.

TFG. 2012. Retrieved from www.usreddfinance.org/redd

UNDP. 2012. Retrieved from http://mptf.undp.org/factsheet/fund/CCF00 (August 2012).

UNEP. 2013. Retrieved from www.unep.fr/scp/sun/facility/reduce/procurement/guidelines.htm (January 15, 2013).

UN-REDD. 2013. Retrieved from www.un-redd.org (January 15, 2013).

USDOJ. 2012. Retrieved from www.justice.gov/opa/pr/2012/August/12-enrd-976.html (August 2012).

Van Orshoven, C., Dawson, T., Leal, I., and Othman, M. 2012. Forest governance and the interrelation between FLEGT and REDD+, FSSP Newsletter, Vietnam Ministry of Agriculture and Rural Development, Vols. 32–33.

Vogel, D. 2008. Private global business regulation. *Annual Review of Political Science,* 11: 261–282.

11

Green Building and the Global Forest Sector

Chris Knowles and Arijit Sinha

CONTENTS

11.1 Introduction

Global economic recovery promises an upward trend in construction sector growth. Some estimates suggest that the global construction sector will grow between 3% and 5% annually for the next decade, primarily driven by a steep increase in construction activities in emerging economies. While construction activities are integral to economic development, they also bear a profound impact on the physical environment. Within the overall construction sector, the building subsector is especially notable since buildings account for 40% of CO_2 emissions (Environmental Information Administration 2008) and 40% of raw material consumption (Roodmen and Lenssen 1995).

Meeting the need for new buildings without jeopardizing the physical environment has necessitated the development of a mechanism that seeks to minimize the adverse impact that buildings and building activities have on the environment. Broadly speaking, the concept of green building captures this mechanism. More specifically, green building is a term used to refer to the environmentally conscious design, construction, operation, and reuse or removal of the built environment. In order to mitigate environmental impacts, the green building movement has primarily focused on five major areas: energy efficiency, indoor environmental quality, water efficiency, resource efficiency, and the construction process. Overall, green buildings can potentially reduce energy use by 30%–50%, CO_2 emissions by 35%, water usage by 40%, and waste output by 70% (Diamond et al. 2006).

Ever changing, the green building movement has evolved from being a niche market to become a significant trend around the globe. Predictions suggest that the green building segment will soon compose more than 20% of the global construction market (IBTimes 2011).

In this chapter, we first provide an overview of the global construction sector. We then introduce some basic tenets of green buildings—basic vocabulary, what green buildings are, how the concept evolved, and some key drivers. We then map the green building landscape including a description of major codes and rating systems in the world. In what follows, we document and critically discuss the role of wood in green building systems. This section is followed by a pinpointed discussion of material type selection in the various rating criteria. In conclusion, we contemplate what may lie ahead as the green building field continues to evolve.

11.2 Global Construction Market: An Overview

The global economic crisis that began in 2008 has had a major impact on construction across the globe. Annual construction growth slowed to less than 1% in 2011 down from an annual rate of approximately 5% in 2006 (Garcia 2011). The most developed countries experienced the sharpest decline. As of 2011, estimates show that the building construction industry accounts for approximately $8 trillion per year, equaling approximately 15% of total global economic output (Garcia 2011). The United States and China are the two largest markets. During 2011, China surpassed the United States as the largest construction market in the world, with China representing approximately 15% of the value of the total global building construction industry and the United States representing approximately 14%. Japan is the third largest construction market representing approximately 7% of the global construction industry. The remaining top markets include Germany, Spain, France, Italy, India, and England. Combined these six markets represent approximately 21% of the industry. The construction market in India is growing rapidly; however, it is only about 1/5 the size of the market in China (Garcia 2011).

While the economic crisis resulted in a slowdown in the construction industry, the outlook for the medium and long term is promising (Garcia 2011). The global construction sector is projected to see growth over the next decade, with estimates ranging from 3% to 5% annually (Betts et al. 2009; Garcia 2011). Some economists predict that the next 5 years will see somewhat stronger growth as the global economy moves into recovery from the current economic downturn (Rider Levett Bucknall 2011; Construction Shows 2012). Brazil, Russia, India, and China (BRIC economies) are projected to experience more rapid growth than developed economies and are expected to more than double the size of their current construction markets (Construction Shows 2012; The World Bank 2012). While China was the largest construction market in 2011, it is expected to fall behind the United States over the next few years as the economic recovery in the United States begins to accelerate. However, China is expected to retake the position as the largest construction market in the world during the next decade representing approximately 20% of the total global construction market by 2020 (Garcia 2011). India is projected to have one of the strongest construction markets in the world over the next decade with a growth rate of 9% (Garcia 2011). This rapid growth will likely move India into the top five global construction markets. Brazil is also forecast to have strong growth in construction through about 2015 in response to hosting the FIFA World Cup in 2014 and the Olympics in 2016. By 2020, the global construction industry is projected to increase to about $13 trillion annually and will represent approximately 15%–18% of the total global economic activity. As noted, this growth will largely be driven by growth in emerging economies where the construction industry is expected to more than double by 2020, compared to a projected 35% growth within developed economies.

Projections show that China will pass the United States to become the largest economy in the world in approximately 2015 and India is projected to surpass Japan to become the world's third largest economy within the next 10–15 years. Additionally, projections show that half of the world's 10 largest economies will be emerging countries, resulting in a significant shift in the structure of the global economy in the coming decade. Also, countries differ in their raw material preference for construction. For example, while wood is a primary residential construction material in the United States and Japan, concrete, masonry, and steel are more important in China and India. A potential replacement of the United States and Japan by China and India as the dominant markets will therefore have ramifications for the global demand for different building materials. It must be noted that the construction boom in China (a market that utilizes large volumes of both steel and concrete) during the mid- to late 2000s resulted in a global concrete and steel shortage. As China, India, Brazil, and other countries that rely primarily on concrete and steel continue to grow, it is likely that shortages in these materials will be more common.

The growth in these markets, because of their primarily utilizing nonrenewable building materials, may likely have profound environmental consequences. It is within this anticipation that green buildings have surfaced as a promising solution for reducing environmental impact of construction activities. While green building mechanisms fall short of providing adequate incentives to encourage the use of renewable materials over nonrenewables, incorporation of life-cycle assessment (LCA) and other relevant tools will likely lead the current green building mechanism to evolve and provide incentives for incorporating increased use of renewable materials. Already, there are efforts underway across Europe, North America, and Japan to build taller buildings using wood, with some proposed projects involving buildings having 20 or more stories. Recent successful projects in Australia and Italy have demonstrated that wood can be used successfully in mid-rise construction.

11.3 Introduction to and Evolutionary Path of Green Buildings

Green buildings (or sustainable designs as they are sometimes called) have been defined and characterized in a variety of ways. In this chapter, we use the definition adopted by the US Office of the Federal Environmental Executive that defines green building as "the practice of (i) increasing the efficiency with which buildings and their sites use energy, water, and materials, and (ii) reducing building impacts on human health and the environment through better siting, design construction, operation, maintenance, and removal—the complete building life cycle" (OFEE 2003). With increasing concerns over the impact of buildings on the environment, concerned building designers came together to devise ways to minimize the environmental impact of buildings and building activities. Multiple initiatives developed at approximately the same time around the globe and finally resulted into what we now refer to as green building rating systems. Rating systems (also referred to as rating standards) are specifications that are created by a consortium referred to as a voluntary standard body; and their adoption is voluntary for a user. In contrast, building codes are created by associations such as the International Code Council. These building codes stipulate the basic minimum requirements for a building and are mandated by law.

The early 1990s saw the development of several green building rating systems around the world. Currently, primarily because of differences in climate and construction techniques among different regions of the world, there are hundreds of green building rating systems around the globe representing local, regional, national, and international markets. While this context specificity enhances their relevance, the lack of a universal system makes cross comparisons among rating systems challenging. As a result, several of the most common systems are increasingly using common metrics (Reed et al. 2009). Major rating systems include the Building Research Establishment's Environmental Assessment Method (BREEAM), Leadership in Energy and Environmental Design (LEED), Green Globes, and Green Star. Some of these systems focus only on individual buildings, whereas others also focus on neighborhoods and communities. Each of the major systems has gained market recognition, aiming to reduce the environmental impacts of a building through construction, operation, use, demolition, and disposal phases of a building or, in other words, its entire life cycle. These rating systems generally approach this goal through reduction of energy use, implementation of techniques to save resources and reduce waste, reduction of water use and management of wastewater, utilization of more environmentally friendly materials, and improvement in indoor environmental quality.

Looking back, we often think of green building as a phenomenon that evolved over the last 30–40 years. However, attempts to increase the energy efficiency and reduce the impact of buildings on the environment—in their modern form—can be traced back to the nineteenth century through utilization of roof ventilators and underground cooling chambers in significant buildings in London and Milan (BDC 2003). However, the availability of inexpensive energy and the technological advancements for heating and cooling indoor environments led to the development of a design paradigm in the early to mid-twentieth century that led designers to reduce their focus on the impacts that buildings have on the environment. Not only did this undermine previous developments, it also delayed the adoption of new technologies or design techniques that could help minimize the impacts of buildings on the environment. This design paradigm was probably most evident in North America. In the 1970s, the OPEC oil embargo of 1973 began to change the way society thought about energy. As an offshoot of this change, the American Institute of Architects (AIA) formed an energy task force that led to the formation of the AIA Committee on Energy. The AIA Committee on Energy focused on the utilization of passive and technological (active) solutions for reducing energy use in buildings. Developments in policy realms were also paralleling these voluntary initiatives. In 1977, for example, the Council of European Communities set community targets for reduction of primary energy consumption. In the United States, the federal government created the Department of Energy (BDC 2003).

The focus on energy efficiency continued through the 1980s. During this period, discussions around energy efficiency began to focus on reducing energy consumption of buildings. Passivhaus, one of the first ratings to focus on reducing energy consumption, was developed in 1988 in Germany. The Passivhaus Institut was formed in 1996 to further develop this standard. The first major global green building rating system, BREEAM, was developed in the United Kingdom in 1990. Others including the French Haute Qualité Environnementale (HQE) rating system and the US LEED rating system were also developed in the early 1990s. These green building rating systems provided guidelines for the design and construction of buildings and covered both existing and new buildings. While primarily focusing on reducing energy consumption, initial green building rating systems also provided guidance for reducing overall environmental impacts of buildings by also focusing on building site and material and water use. This genre of early rating systems formed the basis for many of the green building rating systems that would develop after the early 1990s.

Developments in policy realms continued alongside. The US Environmental Protection Agency and the US Department of Energy jointly created in 1992 the ENERGY STAR program as a voluntary system for identifying and promoting energy-efficient products (ENERGY STAR 2012). These ENERGY STAR-labeled products helped provide designers with easy-to-interpret information about energy consumption of many appliances and components utilized in the construction of buildings. The ENERGY STAR program has evolved to currently cover over 60 home and office product categories including major appliances, heating and cooling equipment, office equipment, lighting, and home electronics.

Throughout the 1990s, interest in green building continued to spread internationally and culminated in 1999 with the formation of the World Green Building Council which in its first meeting had representatives from Australia, Canada, Japan, Spain, Russia, United Arab Emirates, United Kingdom, and the United States (World Green Building Council 2012). In 2002, the World Green Building Council was formally incorporated with the primary role of formalizing international communications focusing on the exchange of information and ideas pertaining to green building, helping leaders in the green building industry gain access to emerging markets, and providing a voice for green building. As of 2011, The World Green Building Council works with green building councils in 65 countries as well as involvement from associated groups in 24 countries (Table 11.1), making it the largest international organization influencing developments in the green building field.

In addition to Green Building Councils, there are other international organizations that are helping shape the green building field globally. In particular, the Initiative for a Sustainable Built Environment (iiSBE) and the Sustainable Building and Climate Initiative (UNEP_SBCI) of the United Nations Environment Program are two key international consortiums that provide worldwide leadership in green/sustainable building practices. The overall aim of these organizations is to provide information to all stakeholders regarding current developments in the field, encourage dialogue and technology transfer among their members, develop tools and strategies to promote green/sustainable building practices, and establish globally recognized thresholds for various criteria such as energy, materials, and water efficiency. These consortiums have also been leading an initiative to develop sustainability assessment programs applicable to the building sector.

Global initiatives for assessing progress toward creation of sustainable design are also developed. These include the Green Building Challenge (GBC) and the United Nations Program-sponsored Sustainable Building and Climate Initiative (UNEP_SBCI). Through research contributions from over 20 countries, GBC was established in 1995 by iiSBE (Todd et al. 2001). The creation of GBC led to the development of a framework for rating tools, and the Green Building Tool (GBTool). GBTool was later adopted by several countries for developing their rating systems—for example, the National Australian Built Environment Rating System (NABERS) is largely based on the GBTool. The GBC ended in 2005, but the iiSBE has retained contact with former GBC member groups and has since developed the Sustainable Buildings tool (SBTool), which considers the wider impacts of buildings using an LCA approach. SBTool remains a third-party framework, which the various regionally applicable rating systems can refer to and be developed from (Larsson 2012).

TABLE 11.1

World Green Building Council Membership by the Type of
Green Building Council

Established	Emerging	Prospective
Argentina	Bulgaria	Austria
Australia	Chile	Bahrain
Brazil	Guatemala	Costa Rica
Canada	Hungary	Croatia
Colombia	Indonesia	Czech Republic
France	Italy	Dominican Republic
Germany	Jordan	Finland
Great Britain	Malaysia	Georgia
India	Russian Federation	Greece
Japan	Turkey	Hong Kong
Mexico		Kenya
Netherlands		Korea
New Zealand		Kuwait
Northern Ireland		Lebanon
Peru		Mauritius
Poland		Montenegro
Romania		Morocco
Singapore		Nigeria
South Africa		Pakistan
Spain		Palestinian Territory
Sweden		Panama
Taiwan		Philippines
United Arab Emirates		Qatar
United States of America		Saudi Arabia
		Serbia
		Slovenia
		Sri Lanka
		Switzerland
		Syrian Arab Republic
		Ukraine
		Uruguay

11.4 Drivers of Green Building

Several factors have contributed to the growth of green building. First of all, the development of new green building standards and the expansion of previously developed standards into new markets around the globe are providing design professionals with specific guidelines for the construction of new green buildings. Secondly, green building rating systems offer the potential of reduced environmental impacts while promising higher returns on investment and lower operating expenses vis-à-vis conventional buildings. According to the US Green Building Council (USGBC), green buildings demand rents as much as 10% higher than conventional buildings that are less energy efficient (Guma et al. 2011). Green buildings have also been shown to have higher levels of occupancy rates (PCA 2008). Thirdly, legislative incentives are driving growth of green building. Incentive programs, including acceleration of permit issuance, tax credits, and rebates, help offset some of the increased up-front costs of designing and constructing a green building. Fourthly, legislative mandates requiring new buildings to be designed and constructed to meet an equivalency of a green building rating, without actually needing to be certified under a rating system, has become common in recent years. Increasing energy prices is also driving green building. Further, green buildings offer wellness benefits to occupants. They can mitigate sick building syndrome (SBS)—a term used to describe the negative influence of buildings on health and productivity of the inhabitants. Finally, a growing concern over the impact that humans have on the environment is another key driver behind the growth of green buildings. Bernstein and Bowerbank (2008) summarize that the top three reasons for green building adoption include doing the right thing, encouraging sustainable business practices, and reducing energy consumption.

11.4.1 Global Landscape of Green Building Field

As green buildings have become popular, rating systems have also proliferated across the globe. While each system has certain unique features, significant overlaps exist among the various systems. Table 11.2 shows the various assessment categories contained in four of the major international green building ratings systems—LEED (United States), Green Star (Australia), Deutsche Gesellschaft für Nachhaltiges Bauen (DGNB) (Germany), and Comprehensive Assessment System for Built Environment Efficiency (CASBEE) (Japan). These systems have significant commonalities in their various assessment categories and that the differences are largely terminological in nature. Essentially, all four systems offer specifications regarding building site, energy use,

TABLE 11.2

Comparison of the Assessment Categories of Four Major International Green Building Rating Systems

Rating System	LEED	Green Star	DGNB	CASBEE
Major Assessment Categories	Sustainable sites	Land use and ecology	Ecology and site	Local environment
	Indoor environment quality	Indoor environment quality	—	Indoor environment
	Energy and atmosphere	Energy	—	Energy efficiency
	—	Transport	Technology	—
	Water efficiency	Water	Processes	—
	Materials and resources	Materials	—	Resource efficiency
	Regional priority	—	Sociocultural and functional	—
	—	Emissions		—
	Innovation in design	Innovation		—
	—	—	Economics	—
	—	Management	—	
Ratings Levels	Certified silver, gold, platinum	4 star, 5 star, 6 star	Bronze, silver, gold	Level 1, level 2, level 3, level 4, level 5

indoor air quality, materials use, and water efficiency. Another commonality is that each system contains different hierarchical levels of rating with each level indicating a distinct level of compliance. For example, the lowest level within the LEED system—the LEED Certified rating—indicates meeting the minimum requirements of the LEED rating system, whereas the LEED Platinum level indicates compliance with the highest level of efficiency and design specifications.

Overlap in measurement and assessment categories notwithstanding, there is significant variation in the way different systems assess a building and aforementioned overlaps must not lead a reader to believe that different systems provide equivalent ratings for a given building (Reed et al. 2009).

While these four dominant systems reasonably capture an overarching picture of the rating system classification worldwide, region-specific developments are important for understanding holistically how the various rating systems are playing out in different regional contexts.

We divide our discussion into the following regions: North America, Europe, Asia Pacific, Middle East/North Africa, Africa, and South America.

11.4.1.1 North America

United States: It is worth a note that there are only two green building codes in the United States, the International Green Construction Code (IGCC) and CALGreen. However voluntary systems are abundant. Currently, there are more than 40 green building initiatives (GBIs), each with its unique rating system. LEED, National Green Building Standard (NGBS), and the Green Globes top the list both by popularity and market share.

LEED is the major rating system for nonresidential construction in the United States. The LEED rating system is administered by the USGBC, which has the responsibility of developing and modifying the LEED standard. The certification of projects is, however, administered through the Green Building Certification Institute (GBCI). Since its inception in 2000, LEED has undergone several revisions and most green building professionals agree that it has improved considerably. While LEED V4.0 was approved in June 2013, it is in draft form and a date of adoption had not been released at time of publication. Therefore, the current LEED standard, LEED 2009, will be discussed in the following text. LEED 2009 contains specific rating systems applicable to the following segments: (1) new construction (NC); (2) existing buildings, operations and maintenance; (3) commercial interiors; (4) core and shell; (5) retail; (6) health care; (7) homes; and (8) neighborhood development. Each of these rating systems is composed of 100 points (or credits), which are divided among these five categories: sustainable sites (26 points), water efficiency (10 points), energy and atmosphere (35 points), materials and resources (14 points), and indoor environmental quality (15 points). Additionally, up to 10 bonus points are possible through innovative design and consideration of regional priorities. For all projects, each of these five categories in LEED 2009 has some mandatory prerequisites that do not count toward points. Points are assigned in an additive way to show incremental conformance with the standard. As noted before, the LEED system rates buildings at four levels—certified, silver, gold, and platinum.

In the residential buildings segment, the NGBS leads the market. The NGBS was developed as a result of several initiatives, led by independent organizations within their jurisdiction, for developing local rating systems in the residential building segment. Some prominent initiatives include the green building movements in Denver, Colorado, and Kitsap County and King County, Washington; the Baltimore suburban builders association program; the Atlanta-based earth craft houses program in Georgia; the Austin Green builder program in Texas; and the Wisconsin green built program. The National Association of Homebuilders (NAHB) took a proactive note of these increasing initiatives and issued guidance to its 800 state and local association to help them create their own local green building programs. Encouraged by its enormous success and feeling a need for standardization, NAHB, in partnership with International Code Council, designed a green building program, in 2008 called the National Green Building Program (NGBP) that, in turn, led to the development of the NGBS (NAHB 2010).

First published in 2008, the NGBS is the very first ANSI standard on sustainable/green building for residential construction. Any municipality can adopt it for incorporating green building standards that go beyond standard building codes. The standard has the following assessment categories: lot and site development; energy, water, and resource efficiency; indoor environmental quality; homeowner education; and global impact. The standard and the certification process, overseen by the NAHB Research Center, has bronze, silver, gold, and emerald as its four threshold levels.

Canada: In Canada, LEED is not the primary rating system and is just beginning to make inroads into the Canadian market. Canada's Green Building Council, which provides rating systems for new construction, core, and shell, commercial interiors, existing buildings, homes, and neighborhoods, implements the LEED rating system in Canada.

The major rating system in Canada is the Green Globes rating system, which was developed in 2000. This system is based on the BREEAM rating system (explained later) of the United Kingdom. This rating system is used in both Canada and the United States. In Canada, the Green Globes system is managed and administered by the Building Owners and Managers Association of Canada, whereas in the United States it is managed and administered by the GBI. Under this standard, there are seven categories: (1) project/environmental management, (2) site, (3) energy, (4) water, (5) resources, (6) emissions, and (7) indoor environment. The Green Globes system has two separate categories for new construction projects and for projects involving continual improvement of existing buildings. This is a thousand-point rating system and certification eligibility includes at least scoring 35% of possible points. Based on their scores, buildings can receive any of the four different certification levels—from one to four globes.

In addition to LEED and Green Globes, two other important initiatives in Canada deserve space. The Commercial Building Incentive Program (CBIP) and the Green Building Challenge are two key programs that are enabling Canadian sustainability initiatives to focus not only on building construction and operations but also on decommissioning and end-of-life issues surrounding the building sector in Canada. These two initiatives are not rating systems, but since they allow for comparison between energy and environmental performance potentials, their role in rating system selection is paramount.

Mexico: Using local materials for building construction and cooling buildings with fountains is a distinguishing feature of the Mexican building sector. Notwithstanding such traditional environment-friendly practices, a green building council was created in 2005. The council helps the government in shaping building sector policies and regulations. Cities in Mexico are beginning to adopt stringent building codes and offer green building programs. The Mexican rating system, which is based on the LEED rating system, is called the Sistema de Calificación de Edificaciones Sustentables (SICES). SICES is a climate-specific rating system and its assessment categories include materials, site management, energy efficiency, indoor air quality, and water conservation. Commercial buildings are utilizing green roofs, renewable energy, and water catchment systems.

11.4.1.2 Europe

The WGBC recognizes 25 countries in Europe that have green building councils. With its strong focus on zero net resource consumption and passive solutions, Europe is widely recognized as a global leader in minimizing the use of resources and energy.

United Kingdom: The BREEAM system, recognized as the first major green building rating system in the world, was developed in 1990 by the Building Research Establishment (BRE), which is a part of the Foundation for the Built Environment. BREEAM uses the following assessment categories: management of resources, energy used, health and well-being of occupants, pollution, transport, land use, ecology, materials, and water. BREEAM measures the following types of buildings: courts, new housing (in accordance with the Code for Sustainable Homes), refurbished housing, healthcare facilities, industrial, multi-residential, prisons, offices, retail, education, communities, and other buildings. BREEAM is a performance evaluation and scoring tool, meaning that long-term monitoring of building performance is conducted to assure adherence to the standard. This is different from most other green building rating systems because they do not include long-term monitoring. The designers, architects, and engineers utilizing the system are encouraged to innovate and demonstrate code-compliant exemplary performance. A unique feature of BREEAM is that scores are weighted against the importance of a provision in the overall environmental impact of the building. Many green building rating systems (e.g., Green Globes in Canada) are based on BREEAM. BREEAM also has international schemes for other European countries and the Middle East.

The Code for Sustainable Homes is another important rating system in the United Kingdom. It must be noted that it is essentially a voluntary rating system, and not a code—as the name may imply at a first glance. This system designed for new homes is used in England, Northern Ireland, and Wales. Its assessment categories include energy use/carbon dioxide emission, water use, materials use, surface water runoff (flooding and flood prevention), waste, pollution, health and well-being, management, and ecology. The Code for Sustainable Homes refers to BREEAM on several criteria.

Germany: The German Sustainable Building Council provides a certification system called the DGNB certificate. It is applicable for new office and administration buildings, new retail buildings, new industrial buildings, new educational buildings, modernized offices and administration buildings, new residential buildings, new hotels, city districts, and existing office and administrative buildings. DGNB is also developing rating systems for several other types of buildings, including new hospitals, new laboratory buildings, and new parking structures. The certification system has three levels (gold, silver, and bronze) based on six rating criteria, namely, ecological quality, economic quality, sociocultural and functional quality, technical quality, process quality, and site quality.

In addition to this rating system, in 2001 the German federal government developed "guidelines for sustainable building." These guidelines ensure that all stages of building life cycle and demolition are included in the design and performance evaluation process. By having a special emphasis on LCA and performance monitoring criteria, the DGNB system is one of the few green building rating systems that monitors building performance.

Also, because of Germany's special focus on energy efficiency, the Passivhaus standard was developed with an objective of achieving energy efficiency in the building sector. The Passivhaus standard is recognized as the world's leading standard for energy-efficient construction. Passivhaus-certified buildings utilize as little as 10% of the energy of a typical central European building. The Passivhaus standard has gained popularity outside of Germany and the standard is now applied across Europe and North America.

France: The rating system used in France is called Haute Qualitié Environnementale (HQE). HQE applies to the following categories of buildings: commercial centers, hotels, schools, houses, residential, offices, healthcare, sports, and occupational. The system consists of 14 operational targets that are arranged in 4 "families": eco-construction, eco-management, comfort, and health. This is one of the two global programs that directly focuses on occupants' health.

11.4.1.3 Asia Pacific

The WBGC recognizes 16 countries in this region that have national green building councils. It must be noted that the development of these green building councils has generally been backed up with significant financial and political commitments by the various national governments.

Japan: Japan utilizes the Comprehensive Assessment System for Building Environmental Efficiency (CASBEE) rating system. Developed in 2001, this system extends to a building's life cycle, including predesign, new construction, existing buildings, and renovation. CASBEE distinguishes the environmental load from a building's environmental performance and thus adds a new dimension to the green building rating system concept. A building's environmental load is assessed using the following criteria: (1) energy, (2) resources and materials used, (3) reuse and reusability, and (4) off-site environment. For evaluating a building's environmental quality and performance, however, CASBEE uses the following criteria: (1) indoor environment, (2) quality of services, and (3) outdoor environment. Each criterion is scored from 1 to 5 (1 = meeting minimum requirements, 3 = meeting average technical and social levels, 5 = high achievement).

By combining environmental load with environmental performance, a Building's Environmental Efficiency (BEE) score is calculated. BEE scores can be graphically plotted. Buildings with low environmental loads and high environmental performance are considered the best.

Australia: Green Star is the primary rating system in Australia, which unlike many rating systems also includes a performance monitoring component. This rating system is also used in the United Kingdom, the Netherlands, and Germany and extends to the following types of buildings: education, healthcare, industrial, multiunit residential, office, office interiors, retail center, office design, and office as built. Green Star also has pilot rating tools for convention center design, public buildings, and custom buildings.

The monitoring and operation of environmental performance is measured using the NABERS. NABERS measures an existing building's environmental performance during operation in water, waste, energy, and indoor environment categories. Measurement and monitoring of building performance allows for an in-depth understanding about how well the tools and techniques utilized in the design and construction of a building reduce its overall environmental impact.

India: The Indian Green Building Council (IGBC) provides several different rating systems. One of the primary systems, LEED India, evaluates sustainable site development, water savings, energy efficiency, materials selection, and indoor environmental quality for both new construction and the core and shell of buildings. Green Homes is another rating system that applies to individual homes, high-rise residential apartments, gated communities, row houses, and retrofit of existing residential buildings. Other systems—though in their pilot stage yet—include the Green Townships rating system (for large developments and townships) and the Green Factory Building rating system (for industries).

China: No discussion of the current state of green building would be complete without a reference to China. China is predicted to develop each year for the foreseeable future a new city the size of Chicago. The new 5-year plan in China includes green buildings as one of the ways to achieve reduced energy consumption and carbon emissions targets. The plan, however, does not provide any specific prescriptive measures for builders and designers. As a result, many Chinese designers have adopted green building rating systems created in other countries such as the LEED and BREEAM. The number of projects certified under these systems has been growing rapidly. In 2006, China's Ministry of Housing and Urban–Rural Development launched the Chinese green building standard, known as the three-star system. This system consists of six assessment categories: land efficiency, energy efficiency, water efficiency, resource efficiency, environment quality, and operational management. Further, each category composes mandatory, regular, and premium items. The system applies to new construction of residential and public buildings through the Green Building Design Label and through the Green Building Label to buildings that have been occupied for at least 1 year. The rating is determined through an accumulation of points in each of the six categories with ratings of one, two, or three stars. The star rating is determined by the minimum score of the categories, not by the total scores. Ratings of one and two stars can be awarded by provincial green building offices; however, the three-star rating can only be awarded by the Chinese Green Building Council. In 2008, China created, through public–private partnership, the China Green Building Council, which is now responsible for management of the system. In 2010, the USGBC and the Chinese Green Building Council signed a memorandum of understanding with an aim to promote green building and carbon reductions in the building sector.

11.4.1.4 Middle East/North Africa

The WGBC recognizes nine countries in this region that have national green building councils. Specifically, the Middle East region has a strong commitment to achieving internationally recognized standards of green building with the UAE and Saudi Arabia being among the top 10 countries in LEED certification. Dubai, UAE, has, since 2008, mandated LEED standards for new construction. A region-specific version of the BREEAM system is also adopted by several countries, including UAE and Qatar. Many speculate that the Arab Spring movement of 2011 will further drive the development of green building in this region.

11.4.1.5 Africa

The WGBC recognizes six countries in this region that have national green building councils. The primary focus area of the green building movement in this region is provision for affordable housing. While as an aggregate this region is one of the least developed regions globally, it is interesting to note that South Africa ranks among the top global markets for LEED certification. It has also adopted a version of the Australian Green Star system The BREEAM rating system is also utilized widely across Africa. Kenya is in the process of forming a green building council, while Nigeria, Zambia, Ghana, Gambia, and Rwanda have shown strong interests in moving toward green buildings.

11.4.1.6 South America

The WGBC recognizes 22 countries in this region with national green building councils, making it one of the most active regions in the world. A major focus of green building in this region is the development of affordable housing. The region is currently experiencing a surge in green building primarily driven by Brazil's hosting the 2014 World Cup.

Brazil: The Brazilian Green Building Council was formed in 2007. It works with the public and private sectors to educate property owners, designers, builders, and the government. The Brazilian Green Building Council recognizes that because of a relatively low environmental awareness and because of green building being a fairly new idea, education and awareness will be key to its success. As a result, the council has produced a Sustainable Construction Manual for public authorities to implement green building practices. The Brazilian Green Building Council adopted a version of the LEED rating system that is specific to the local context and is a member of the LEED International Program.

Argentina: Argentina has been embracing green building ideas during the last few years through several initiatives that include banning incandescent light bulbs, requiring thermal insulation in the capital city of Buenos Aires, and also requiring environmentally friendly air conditioning. Energy efficiency and sustainable concepts are also being taught in schools. Overall, green building development is in its nascent phase but much progress has already been made. The Argentina Green Building Council was created in 2009 with a focus to educate people and bring green building practices into the construction industry. The LEED rating system has been utilized in Argentina and the Argentina Green Building Council is a member of the LEED International Program.

Chile: LEED is gaining popularity in Chile as the leading green building rating system. The Chile Green Building Council is a member of the LEED International Program and has adopted a LEED rating system specific to the local context. The Costanera Center, South America's tallest building, is aiming for LEED Gold certification.

11.4.2 Current Trends and the Future for Codes and Standards

While it is not clear what the future holds for green building rating systems, it is certain that green building rating systems are moving from being conformance based to becoming performance-based systems, which we feel is important for achieving the goal of minimizing the impact of buildings on the environment. Currently, most green building systems focus on a conformance score, while fewer systems focus on performance. LEED, for example, provides a checklist of provisions, each fulfilled earns associated credit, which are added together to achieve a final score. Hence, currently, the LEED system is an additive system implying that the more points a project scores, the better it will perform from an environmental perspective. However, since the LEED system is based on a methodology that does not monitor building performance, there is no measurement to verify the performance of LEED-certified buildings. Several studies (Torcellini 2004; Bowyer 2008; Bribian et al. 2011) have challenged the performance of buildings certified under the LEED system by questioning if the LEED program actually helped in reducing energy consumption and improving energy performance of a building. The lack of focus on performance has caused significant variation in the ways the various systems assess a building. A study by BRE (2011) showed high levels of variability in results when the same buildings were assessed using the BREEAM, CASBEE, LEED, and Green Star systems, with Green Star and LEED giving higher ratings to the same building than the BREEAM system. This discrepancy is attributable, in part, to BREEAM being a performance-based system while others are conformance-based. Many countries around the world have adopted performance-based systems while also having a performance monitoring protocol in place. These countries, for example, the United Kingdom (BREEAM), Germany (DGNB), and Australia (NABERS), are ahead of the curve when implementing sustainable practices. Calls for performance-based systems have been made in the United States through several initiatives, many of which advocate for a time-bound performance goal. The Architecture 2030 Challenge, for example, encourages a 50% baseline impact reduction for buildings by the year 2030.

The USGBC is also beginning to recognize this discrepancy and is deliberating about introducing a weighted scoring system that will allow for provisions with higher weights for indicators with greater environmental impact. The draft of the recently approved LEED V4.0 shows a move toward a more performance-based system but a comprehensive performance-based approach was not part of this version. A movement toward performance-based systems is also evidenced by the creation and publication of the International Green Construction Code (IgCC) by the International Code Council. IgCC, also called the commercial version of NGBS, is striving to become a building code and is approaching many jurisdictions for its adoption by 2015. As is the case with any other building code, however, the final decision to adopt lies with each jurisdiction. IgCC has two parts, of which one is similar to a green building rating system (i.e., LEED or BREEAM), while the other part deals with code and jurisdictional requirements. IgCC is placing significant emphasis on material selection and the use of alternative sustainable materials. Further, the material selection component in IgCC includes considerations such as material reuse, recycled content, recyclable materials, bio-based materials, and indigenous materials. Moreover, IgCC also provides third-party verified sustainable attributes of building products, called Validation of Attributes Reports (VAR). The VAR provides a material selection benchmark for green building standards and codes. IgCC is striving to be the basis of all green building rating systems in the future.

Some evidence also suggests that green building will likely move toward being mandated in the future. In the United States, this process has already seemed to have begun through the adoption of the Green Building Standards Code (CALGreen) by the California Building Standards Commission. CALGreen requires all new buildings in the state of California to be more energy efficient and environmentally responsible. This code, which came into effect on January 1, 2011, aims to achieve major reductions in greenhouse gas emissions, energy consumption, and water use. Historically, California has led the way in the United States for enacting new laws related to the built environment and occupant health. The adoption of CALGreen is considered as one of the major steps for a mandatory adoption of green building systems across the United States. Already, many US states and counties require any new building within their jurisdiction to be equivalent to LEED Silver or above. The state of Oregon, for example, requires all major renovations and new construction of state-funded buildings to be LEED Silver or equivalent. This equivalency provision necessitates that buildings are designed and constructed to meet the LEED Silver requirement without necessarily incurring any costs associated with certification.

In recent years, the green building movement has gained significant inroads into all aspects of the built environment. The process has already started move from to building codes. It is only a matter of time before counties in the United States begin adopting one of the available standards or codes for new construction in both commercial and residential sectors.

Internationally, different systems deal with the issue of environmental impacts differently—some are more performance oriented, while others are conformance oriented. Some systems, such as the Swedish P-mark, Canadian R-2000, and the US ENERGY STAR, focus on evaluating the performance of only selected aspects, such as energy efficiency and quality of materials and workmanship, while having, for example, only pass and fail as the two assessment categories. Despite their limited focus, these systems have the advantage of focusing on conventional practices and are, therefore, more apt for immediate application. Other systems, such as the United Kingdom's BREEAM, Australia's NABERS, and Japan's CASBEE evaluate the environmental impact of a building. Another performance-based standard that deserves mention is the Living Building Challenge (LBC) launched by Cascadia Green Building Council. It has presence in the United States, Canada, Mexico, Australia, and Ireland with projects emerging in other countries as well. The LBC challenges project and design teams to demonstrate a building's performance on sustainability measuring indices. The challenge has extremely stringent performance criteria for certification. As a result, so far only eight buildings in the United States and one in Canada that have been certified under this program.

Their relative focus notwithstanding, all the various systems mentioned earlier are important because they introduce different methods of assessment that lead to more efficient building performance. A building is an extremely complex system, and the performance assessing method needs to reflect that complexity and at the same time should be flexible enough to accommodate technical, economical, and social needs. However, performance-based systems better represent the overall results because the levels of priorities and importance of specific issues are established by introducing weighted scores. Therefore, we emphasize that more impetus be given for employing performance-based systems in conjunction with stringent performance evaluation criteria.

11.4.3 Green Building Standards: Wood as a Building Material

Wood is a common building material used in both structural and nonstructural applications. In some countries, for example, the United States and Japan, wood is a major building material, representing approximately 90% of the single-family residential construction market. As previously noted, material and resource utilization is an integral component of most green building rating systems. Criteria used to evaluate the "greenness" of various materials (steel, concrete, wood, etc.) often include characteristics such as percentage of recycled content and the distance the material is sourced and/or manufactured from the construction site. Among these, wood is the only material that requires external validation or certification (LEED 2009), despite social and environmental impacts associated with other materials (Bowyer 2007, 2008). In the case of wood, many green building rating systems include incentives (or points) to ensure that a project utilizes wood sourced only from well-managed forests. Systems do not, however, agree on what qualifies as wood from a well-managed source. Illustratively, the LEED system only provides a point for wood sourced from forests certified by the Forest Stewardship Council (FSC), whereas the Green Globes system provides points for wood sourced from forests certified by multiple forest certification schemes including FSC, Sustainable Forestry Initiative (SFI), the Canadian Standards Association (CSA), and the Programme for Endorsement of Forest Certification (PEFC). This difference between the ways these two rating systems provide credit for wood has led to accusations by many in the North American wood products industry that the USGBC and the LEED system have a bias against wood products. The certified wood credit (as it has become known) in the LEED system has been reviewed on multiple occasions and each time the USGBC membership has voted against any change. Recent evidence also suggests that fewer LEED projects are now seeking certified wood credits. Watson (2011) reports a significant decline in the use of FSC wood in LEED projects. While this decline is reported across all LEED rating systems, it is most dramatic in the LEED for new construction system, where only 26% of projects specified certified wood in 2011, down from 41% of projects in 2010 and 38% in 2009. This report does not provide a reason for the decrease in certified wood use, but Knowles et al. (2011) explain this trend by concluding that designers often do not specify the use of FSC-certified wood. Despite this declining trend, the long-term outlook for certified wood use in the LEED system is, however, positive with predictions of more than 3 billion board feet specified in 2020 and more than 6.4 billion in 2030, up from 880 million board feet specified in 2011 (Watson 2011). LEED V4.0 has eliminated certified wood as an independent credit, moving it as one of several options under the "Building product disclosure and optimization—sourcing of raw materials" credit.

The newly developed IgCC, under its bio-based materials section, also provides criteria for sourcing wood from a well-managed source. The IgCC is flexible with respect to wood products and allows for sourcing of wood certified under a wider range of certification programs, including SFI, FSC, and any system conforming to PEFC standards.

LEED also assigns extra credits for materials that are "rapidly renewable" (USGBC 2013). The criterion of rapid renewability with respect to wood is a 10-year turnaround period. In other words, these extra credits can be earned by utilizing tress with a rotation time of 10 years or less. Bamboo, which is technically considered a non-wood forest product (NWFP), qualifies for the rapidly renewable materials credit. It is for this reason that under the LEED 2009 rating system bamboo flooring is preferred over, for example, maple flooring. It must also be noted that this preference has been heavily challenged (Bowyer 2007) and the debates continue whether "rapidly renewable" category must be changed to "renewable" (YPFPG 2008).

The current version of the LEED V4.0 system includes allocation of points for materials that meet the criteria of the USDA Biopreferred Program. As currently written, this allocation will replace the current extra credit allocated to rapidly renewable materials and will thus provide wood the same advantages as other renewable materials.

While the certified wood credit receives the most attention in wood-related discussions within green building rating systems, credits for wood utilization can also be obtained through securing wood from a local source. Alike "well-managed forests," the definition of "local" also varies by system. Under the LEED system, in order to qualify for credits associated with local sourcing, a percentage of all materials, including wood, must be sourced and manufactured within 500 miles of the job site. Under the LBC, on the other hand, the definition of local is dependent on the weight of the product with the heaviest materials needing to be sourced within 500 km of the job site and the lightest materials within 2000 km. Exceptions to this criteria are provided for building components that actively contribute to the performance of the building, with a 5,000 km limit and renewable energy with a 15,000 km limit.

In LEED V4.0, the 500 mile limit from LEED 2009 has been replaced by a clause defining local sourcing as products (extracted, manufactured, and purchased) within 100 miles of the project site.

Wood and other forest products face two challenges for acceptance in the green building industry. First, for non-FSC-certified wood and wood products to meet the LEED material and resources, requirements will become more important as LEED enters into the residential construction sector where wood is the predominant structural material of choice. Similar challenge exists in nonresidential commercial buildings also where LEED is already the market leader. Within the green building sector, the forest products industry must think outside the box and try to increase their share in the structural material category by creating a niche for new structural products. Similarly, often overlooked nonstructural products and architectural finish materials offer reasonable opportunities because of their having direct impact on environmental air quality and also on thermal performance of the building. Given the low impetus for wood utilization within the existing green building rating systems, the forest products industry must innovate by going beyond the conventional products and avenues and must venture into finding a niche for its products to be competitive.

11.5 Green Building Systems and Materials

Previous studies have expressed concern over the criteria that green building rating systems use for material selection (e.g., Bowyer 2008; Knowles et al. 2011). Among others, a widely expressed concern is that none of the standards allocate points based on life-cycle performance of the products. This issue fundamentally emanates from the fact that the material selection criteria do not form an important part of rating systems. This is problematic both because materials used are often paramount in quantity and thus have greater environmental implication and also because these materials remain part of a building throughout all stages of its life cycle and therefore need to be recycled or disposed of at the end of a building's life. It is an anomaly that most green building systems do not consider the full life-cycle costs and impacts of materials as a criterion for material selection and consider all materials equally.

While some life-cycle studies have shown that wood has less embodied energy than concrete or steel because it is a biological, renewable material (Puettmann and Wilson 2005), and also because the raw materials to make cement and concrete are products of energy-intensive mining (PCA 2002; vanOss and Padovani 2002; Rajendran and Gambatese 2007), most rating systems consider concrete and wood as equal. In fact, the LEED 2009 system provides more opportunities for steel to obtain points than wood and concrete, because of its recyclability and percentage of recycled content (USGBC 2009). In spite of being recyclable, steel has been shown to have higher environmental impacts than wood because of mining- and furnace-based extraction (IISI 2000). Moreover, steel is seldom recycled into the same product—it is generally downcycled into a lower-value/lower-strength product. Bowyer (2008) argues that such an importance given to steel is a serious error from an environmental standpoint. Also, Knowles et al. (2011) note that the decision about structural materials selection is driven primarily by building code and cost, with environmental impact of materials rarely factoring into the decision.

Overall, we argue that green building rating systems need significant improvement in the way they deal with materials. We emphasize that the life-cycle costs of materials must be considered as a criterion for material selection. In fact, ICC is currently deliberating on including full building LCA as an amendment to the IgCC. LCA's incorporation into IgCC will mean a paradigm shift in processes concerning material selection and utilization since it will require manufacturers of various materials to declare the environmental impacts of their products over their entire life cycle. The LEED V4.0 system now includes LCA as an option for assessing materials. Notwithstanding the improvement in the system that LCA promises, readers must note that LCA is also criticized by many because of its several assumptions and boundary conditions. Until LCA is standardized, comparisons between two competing materials cannot be made with a great degree of confidence.

11.6 Issues, Challenges, and Roadblocks

There are many obstacles to further adoptions of green building. Bernstein and Bowerbank (2008) identify higher initial costs, different budget accounting (i.e., accounting for life-cycle costs), lack of public awareness, and lack of trained/educated green building professionals as prime obstacles to green building. These barriers have differing effects. For example, lack of public awareness was identified as the top barrier in Africa, while lack of trained/educated professionals was the top obstacle in Asia and the Middle East/North Africa. High level of poverty is considered a significant concern in Africa and South America.

Also, the green building movement continues to face challenges in market penetration worldwide. Except for state-funded buildings, green building is a voluntary commitment throughout the world. In the absence of a mandated adoption, the green market will evolve through a slow process primarily dictated by market mechanisms. Green buildings are often associated with higher initial costs vis-à-vis conventional buildings because of their design, construction, and mechanical devices and electrical systems costs. Such higher costs may hamper a rapid market expansion. Some argue that these costs will gradually go down as the newer practices and technologies are developed and adopted by a larger market. We feel that more precise comparisons that currently exist between initial costs and longer-term energy savings will motivate many environmentally conscious end users to invest in green buildings. We also feel that some policy incentives, such as tax breaks, tax credits, and lower interest rates on loans for green building projects, are important to further support the movement. Most countries, especially in Africa, Asia, and South America, do not provide any incentives for stakeholders to invest in green buildings. Government and industry together should drive the movement by generating more information and disseminating new information among various stakeholders such as building designers and the general public. Given the unique local characteristics of each country, a quest for global systems is not feasible. Instead, we call for a global set of benchmark parameters that are responsive to local needs, are performance-based, and seek to reduce the variability that currently exists among the various rating systems.

References

Bernstein, H. and A. Bowerbank. (2008). *Global Green Building Trends: Market Growth and Perspectives from around the World*. McGraw-Hill Construction, pp. 48, New York City, New York.

Betts, M., G. Robinson, C. Burton, A. Cooper, D. Godden, and R. Herbet. (2009). *Global Construction 2020: A Global Forecast for the Construction Industry over the Next Decade to 2020*. Global Construction Perspectives and Oxford Economics, London, UK, http://www.ricsamericas.org/files/editor/file/News/Executive_Summary_FINAL2.pdf. As viewed January 4, 2012.

BRE, 2011. Building Research Establishment. http://www.bre.co.uk/mediacentre.jsp (Accesed January 25, 2011).

Bowyer, J.L. (2007). The green building programs-are they really green? *Forest Prod J*, 57(9): 6–17.

Bowyer, J.L. (2008). The green movement and the forest products industry. *Forest Prod J*, 58(7/8): 6–13.

Bribian, I.Z., A.V. Capilla, and A.A. Uson. (2011). Life cycle assessment of building materials: Comparative analysis of energy and environmental impacts and evaluation of the eco-efficiency improvement potential. *Building Environ*, 46: 1133–1140.

Building Design and Construction (BDC). (2003). White paper on sustainability: A report on the green building movement. 47pp. http://www.ofee.gov/Resources/Guidance_reports/Guidance_reports_archives/fgb_report.pdf. As viewed November 15, 2011.

Construction Shows. (2012). BRIC countries power recovery in global construction market. http://www.constructionshows.com/bric-countries-power-recovery-in-global-construction-market/ (Accessed November 5, 2012).

Diamond, R., M. Opitz, T. Hicks, B. Vonneida, and S. Herrera. (2006). Evaluating the energy per-
formance of the first generation of LEED-certified commercial buildings, in: *ACEEE Summer
Study on Energy Efficiency in Buildings*, American Council for an Energy-Efficient Economy,
Washington, DC, pp. 3-41–3-52.

EIA Annual Energy Outlook, Environmental Information Administration. (2008). Assumptions to
the Annual Energy Outlook, Energy Information Administration, 2008; cited in Green Building
Facts, U.S. Green Building Council, Washington, DC, November 2008.

Energy Star. (2012). http://www.energystar.gov/index.cfm?c=about.ab_index. As viewed July 23,
2012.

Garcia, T. (2011). *The Global Construction Industry: What Can Engineers Expect in the Coming Years?*
Plumbing Systems and Design. Rosemond, IL, December 2011. pp. 22–25.

Guma, A., C. Pyke, and C. Leitner III. (2011). Current trends in green real estate: Summer 2011
update. http://www.costar.com/webimages/webinars/CoStar-Webinar-CurrentTrendsin
Green20110621.pdf (Accessed October 25, 2012).

IBTimes. (2011). Green building industry to hold 20% share by 2013. *International Business Times*.
August 17, 2011. Available at: http://www.ibtimes.co.uk/articles/20110817/green-building-
industry-hold-share-2013.html. As viewed March14, 2012.

International Iron and Steel Institution (IISI). (2000). *Worldwide LCI Database for Steel Industry Products*.
Brussels, Belgium: IISI.

Knowles, C., C. Theodoropoulos, C. Griffin, and J. Allen. (2011). Oregon design professionals views
of structural building products: Implications for wood. *Can J Forest Res,* 41(2): 401–411.

Larsson, N. (2012). User guide to the SBTool assessment framework. iiSBE. www.iiSBE.org/system/
files/SBTool2012 User guide 26oct12.pdf. As viewed July 25, 2013.

LEED. (2009). *LEED for New Construction and Major Renovation*. Washington, DC: USGBC.

NAHB. (2010). *National Green Building System*. National Association of Home Builders. Washington,
DC.

Office of the Federal Environmental Executive (OFEE). (2003). The federal commitment to green
building: Experiences and expectations. September 18, 2003. http://www.ofee.gov/Resources/
Guidance_reports/Guidance_reports_archives/fgb_report.pdf. As viewed December 1, 2011.

Portland Cement Association (PCA). (2002). *Environmental Life Cycle Inventory of Portland Cement
Concrete. Appendix: Life Cycle Inventory of Portland Cement Manufacture*. Skokie, IL: PCA.

Portland Cement Association (PCA). (2008). *Benchmarks 2008. Survey of Operating Costs. Victorian
Shopping Centres*. Sydney, New South Wales, Australia: PCA.

Puettmann, M.E. and Wilson, J.B. (2005). Life-cycle analysis of wood products: Cradle-to-gate LCI of
residential wood building material. *Wood Fiber Sci,* 37: 18–29.

Rajendran, S. and Gambatese, J.A. (2007). Solid waste generation in asphalt and reinforced concrete
roadway life cycles. *J Infrastruct Syst,* 13(2): 88–96.

Reed, R., A. Bilos, S. Wilkinson, and K. Schulte. (2009). International comparison of sustainable rating
tools. *J Sust Real Estate,* 1(1): 1–22.

Rider Levett Bucknall. (2011). Recovery imminent for global construction markets. http://rlb.com/
index.php/australia-and-new-zealand/article/recovery-imminent-for-global-construction-
markets/

Roodman, D.M. and N. Lenssen (1995). *A Building Revolution: How Ecology and Health Concerns are
Transforming Construction*, Worldwatch Paper 124, Worldwatch Institute, Washington, DC p. 5.

Todd, J.A., D. Crawley, S. Geissler, and G. Lindsey. (2001). Comparative assessment of environmental
performance tools and the role of the Green Building Challenge. *Build Res Inf,* 29(5): 324–335.

Torcellini, P.A., M. Deru, B. Griffith, N. Long, S. Pless, and R. Judkoff. (2004). Lessons learned from
the field evaluation of six high-performance buildings, in: *ACEEE Summer Study on Energy
Efficiency of Buildings*, American Council for an Energy-Efficient Economy, Washington, DC,
pp. 3-325–3-337.

USGBC. (2013). LEED 2009 for new construction and major renovations. www.usgbc.org/sites/
default/files/LEED 2009 RS_NC_04.01.13_current.pdf. As viewed July 25, 2013.

van Oss, H.G. and A.C. Padovani. (2002). Cement manufacture and the environment. Part I: Chemistry and technology. *J Ind Ecol*, 6(1): 89–105.

Watson, R. (2011). Green building market and impact report 2011. GreenBiz Group. Available at http://www.greenbiz.com/research/report/2011/11/07/green-building-market-and-impact-report-2011. As viewed March15, 2012.

The World Bank. (2012). Global economic prospects 2012a: Uncertainties and vulnerabilities. The International Bank for Reconstruction and Development/The World Bank. Vol. 4. January 2012. 165pp. Available at http://siteresources.worldbank.org/INTPROSPECTS/Resour ces/334934–1322593305595/8287139–1326374900917/GEP_January_2012a_FullReport_ FINAL.pdf.AsviewedMarch15, 2012.

World Green Building Council. (2012). http://www.worldgbc.org/site2/about/wgbc-history. (Accesed November 5, 2012).

YPFPG. (2008). Assessing USGBC's policy options for forest certification & the use of wood and other bio-based materials. A summary report prepared by the Yale program on forest policy and governance, New Haven, CT, February 25, 2008.

Websites:

BREEAM (2011): http://www.breeam.org/

DGNB (2011): http://www.dgnb.de/_en/

Green Globes (2011): http://www.greenglobes.com/

Guidelines for sustainable building (2011): available at http://pcc2540.pcc.usp.br/Material% 202004/Germany_guideline_SB.pdf

HQE (2011): http://assohqe.org/hqe/

IGBC (2011): http://www.igbc.in/site/igbc/index.jsp

International Initiative for a Sustainable Built Environment (iiSBE) http://www.iisbe.org/

NAHB (2011): http://www.nahbgreen.org/index.aspx

United Nations Programme—Sustainable Building and Climate Initiative (UNEP_SBCI) http:// www.unep.org/sbci/index.asp

USGBC (2011): http://www.usgbc.org/

12

Assessment of Global Wood-Based Bioenergy

Francisco X. Aguilar, Michael A. Blazier, Janaki Alavalapati, and Pankaj Lal

CONTENTS

12.1 Introduction

Climate change, a significant dependence on fossil fuels for energy, the need for sustainable economic development, and the current economic slowdown that affected the housing market have created new opportunities to utilize woody materials as a source of renewable energy. Wood has been used as a source of energy for millennia, but recent developments in technology, improvements in energy conversion, and the potential net benefits of its use compared to nonrenewable fossil fuels have motivated the use of wood as an integral part of comprehensive renewable energy portfolios in many countries. Governments at all levels have taken a proactive approach to promote the use of woody biomass as feedstock for energy and biofuels production. As technology improvements are made and conversion processes become economically viable, the evolution of wood-based bioproducts production will likely continue on an upward trajectory. This chapter provides an assessment of the landscape for wood-based feedstocks, bioprocessing methods and products, developments in global markets, and challenges associated with greater wood energy utilization.

12.2 Sources of Wood-Based Feedstocks

Forests cover 4 billion ha globally and contain an estimated 421 billion Mg of biomass on the forest floor (Parikka 2004; FAO 2011). This abundant biomass can potentially serve as feedstock for bio-based electricity production, biofuels, and other bioproducts. In addition, for forest biomass, wood feedstocks can come from by-products from the production of conventional forest products such as paper, lumber, and furniture, waste wood from urban settings, postharvest woody residuals from logging operations (logging debris, damaged trees), trees cut in thinning operations, chips, and short-rotation purpose-grown trees (Figure 12.1).

FIGURE 12.1
Overstock hardwood stand with saw timber and abundance of small-diameter and dead trees.

By-products from forest product manufacturing facilities are currently the most readily accessible sources of woody biomass as feedstocks, but most of this material in developed countries is used as combined heat and power (CHP) at the facilities where it is produced. Increasing global demand for wood-based biofuels and bioproducts will necessitate increased utilization of other sources of woody material. The following sections discuss these sources of wood-based feedstock supply.

12.2.1 Manufacturing Residuals and Waste Wood

Approximately 30%–70% of log input into sawmills, plywood plants, and paper mills becomes waste (Parikka 2004; Smeets and Faaij 2007). The proportion of residues generated from these forest product manufacturing facilities is dependent on the type of operation, equipment efficiency and maintenance, and wood properties. Mill waste is highly desirable as bioenergy and bioproduct feedstocks because it is clean, concentrated, and low in moisture content relative to trees, slash, and logging debris. However, as much as 97% of these residuals are currently utilized to provide power on site in the manufacture of conventional forest products (Hubbard et al. 2007). This suggests that the price of alternative power sources would have to be competitively low in order to shift this resource to other bio-based uses offsite.

Given this caveat, several sources of manufacturing residuals can potentially be used as feedstocks for bioenergy and bioproducts. Coarse residuals, which include slabs, edgings, offcuts, and veneer clippings, are desirable feedstocks. However, these materials are commonly used as raw material for pulp and composite panels such as particleboard and medium-density fiberboard.

Cores from plywood peeler logs are another potential source of woody material. However, in most cases, in developed countries, spindleless lathes result in small-diameter cores used for landscape timbers. Larger cores can potentially be further processed into lumber products, a higher-value alternative to chipping for bio-feedstock.

Planer shavings, produced in surfacing sawn wood, can serve as feedstock for densified fuels (fuels manufactured from wood compressed into uniform-sized particles such as pellets, fuel logs, and briquettes) production. Sander dust generated during sanding of sawn wood and panel products is generally burned to generate electricity, but it could be used as biofuel and bioproduct feedstock. Sawdust, which is commonly used for particleboard manufacture, can also be used to produce densified fuels (Parikka 2004; Figure 12.2).

In the paper sector, black liquor, a caustic mixture of dissolved lignin and spent pulping chemicals from kraft paper production, is conventionally burned in recovery boilers to produce steam that drives pulp mill processes and to recover spent pulping chemicals. Recently, processes have been developed to produce syngas from black liquor via gasification (LeBlanc 2009). Syngas and gasification will be further discussed later in the chapter.

12.2.2 Urban Wood Waste

Urban wood waste is woody material typically discarded from urban and suburban areas. While forest products generate biofuel and bioproduct feedstocks during their manufacture, they also create potential feedstocks at the end of their life span. These tertiary residues include scrap wood from demolition debris, damaged pallets and crates, discarded wood, discarded furniture, and waste paper (Bungay 2004; Smeets and Faaij 2007). These waste materials are advantageous as biofuel and bioproduct feedstocks because they are dry and have more cellulose per unit weight than trees or postharvest residuals (Bungay 2004).

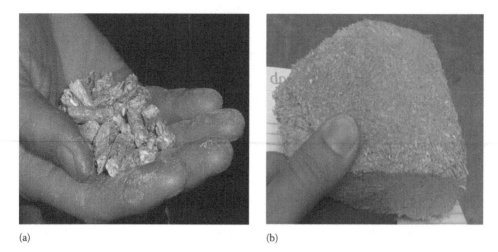

(a)   (b)

FIGURE 12.2
Woody biomass is often manufactured into (a) pellets and (b) briquettes. They have higher energy content because of lower moisture and can be more easily transported and stored.

Other potential sources of urban wood waste include tree biomass cleared during utility and right-of-way maintenance, municipal park and private residence maintenance, urban land clearing, and residues from commercial nurseries. This type of urban wood waste is more heterogeneous in moisture content, cellulose content, and cleanliness than discarded wood and paper. Opportunities for using urban wood waste as biofuel and bioproduct feedstocks are increasing as municipalities seek recycling outlets for this material. However, economies of scale, funding, start-up costs, and costs of collection, equipment, and material separation have limited utilization of urban resources (National Renewable Energy Laboratory 1996).

Increasing utilization of urban wood waste has potential environmental benefits. Landfill space is saved by reducing the quantity of waste disposed. In a national audit of landfills in Australia, it was determined that nearly 50% of all municipal waste loads disposed into landfills contained wood waste (Taylor et al. 2010). In the United States, it has been estimated that nearly 550 ha are consumed by urban wood wastes per year (Solid Waste Association of North America 2002). Taylor et al. (2010) estimated that 1.2–3.2 million Mg of CO_2 equivalent of greenhouse gas (GHG) could be displaced annually if all of Australia's waste wood were co-fired to produce electricity. Furthermore, when wood is used to displace high-sulfur bituminous coal for electricity generation, sulfur emissions to the atmosphere are reduced (Solid Waste Association of North America 2002).

12.2.3 Post-Harvest Residuals

Logging residues in the form of branches, foliage, stumps, and roots that are typically left behind after forest harvest operations can serve as sustainable biofuel and bioproduct feedstocks. The value of harvest residuals as bio-feedstock depends on tree species, harvesting system, and tolerance of conversion processes for moisture content of debris and amount of soil mixed with the debris.

It has been estimated that globally 24%–60% of a harvested tree is left to rot at harvested sites or to make logging easier (Parikka 2004; FAO 2010a). Logging residues are available in substantial quantities in many countries. For example, in the United States, it has been estimated that 36 million dry Mg of logging debris is available annually and that electricity generation from this material could displace 17.6 million Mg of carbon emitted from coal-powered power plants, an amount equivalent to approximately 3% of carbon emissions from the United States electricity sector (Gan and Smith 2006). Langerud et al. (2007) estimated that 27 million dry Mg of logging debris was annually available in Norway.

Use of logging debris as energy feedstocks is particularly substantial in Nordic countries. Forest residues are the primary bioenergy source for the Swedish energy system (Egnell 2011). Wood chip production from forest residues increased 22-fold in Finland between 1995 and 2003 (Hakkila 2006). Harvesting of treetop biomass can be conducted in tandem with conventional roundwood logging or as a stand-alone operation. At the logging deck, treetop biomass is chipped using waste processing machines called hoggers rather than returned to the harvested area as often conventionally conducted to reduce nutrient movement from the site. Hoggers can be track-mounted and self-propelled or mounted to a semitrailer and towed by truck. A common feature of hoggers is a rotor head that grinds the biomass into chips called hog fuel, an unprocessed mix of coarse chips and wood fiber.

Logging debris is relatively nutrient dense because the highest concentrations of nutrients in trees are in foliage and roots. Leaving crown mass, stumps, and roots to decompose on site after tree harvesting has been a recommended practice for minimizing nutrient losses from forests. Harvesting this debris as feedstocks for biofuels and bioproducts has created concerns about nutrient losses and associated losses in forest productivity. Reduced growth rates of trees planted after logging residue removal have been found in Finland, Norway, Sweden, the United Kingdom, and the United States (Scott and Dean 2006; Walmsley et al. 2009; Egnell 2011; Helmisaari et al. 2011). Productivity losses tend to be site specific, with most losses occurring on sites with inherently low nitrogen or phosphorus. Fertilization after logging residue harvest can maintain tree growth at levels comparable to or in excess of those of trees planted after retention of logging debris (Scott and Dean 2006; Helmisaari et al. 2011). Increasing levels of logging debris harvest thus will likely increase fertilization requirements for maintenance of forest productivity. Harvesting logging debris may also increase silvicultural need for competition suppression in succeeding rotations, because residue harvesting has been associated with increased competition from colonizing vegetation (Stevens and Hornung 1990; Walmsley et al. 2009).

Wood chips from forest residues are predominately produced from aboveground biomass, but they can be produced from belowground biomass (stump wood) as well. Stump wood harvesting is most commonly conducted in Finland, which has subsidized the practice due to the perceived benefit of reducing root rot in subsequent rotations (Hakkila 2004; Flynn and Kumar 2005). Stump harvesting has also been practiced in parts of the Pacific coast of the United States and Canada and in eastern England (Greig 1984; Thies and Westlind 2005; Hope 2007). In Finland, stump wood chips for heating and power plants account for approximately 14% of forest chips consumed (Ylitalo 2006). Karjalainen et al. (2004) estimated that up to 9 million m^3 year^{-1} of chips from stump wood can be produced across the European continent.

Stump wood harvesting has the potential to yield as much as 100 m^3 ha^{-1}, which is equivalent to 100–200 MWh ha^{-1} (Hakkila and Aarniala 2004; Flynn and Kumar 2005). Stump and root wood are advantageous as bioenergy feedstocks because they are relatively dry and homogeneous raw materials, and stump wood does not absorb water during storage as readily as logging residues. Chips produced from stumps also have a higher calorific value than those derived from other forest residues (Eriksson and Gustavsson 2008). Stump harvesting also has the benefits of improving the efficacy of machine planting of seedlings in the subsequent rotation and reducing pests and diseases that are typically manifested in rotting stumps (Walmsley and Godbold 2010).

Stump wood must be split into pieces to accelerate drying, increase its bulk density to facilitate transport, and facilitate comminution. Stumps are harvested as soon as possible after tree harvest using an excavator to uproot and split the stumps. After excavation and splitting, stumps and roots are moved and piled on site using a standard forwarder to allow rain to wash away soil that serves as a fuel contaminant (Laitila et al. 2008; Walmsley and Godbold 2010). Efficiency of on-site transporting (forwarding) stump and root wood is lower than that of logging residues from treetops, roundwood, and whole trees (Ranta 2002; Väkevä et al. 2003; Laitila et al. 2007). Stumps are stored on site for up to a year to foster soil washing and drying of the wood; the stumps are then crushed into wood chips using stationary or mobile crushers (Walmsley and Godbold 2010).

12.2.4 Thinning

Thinning trees as biofuel and bioproduct feedstock can be integrated into traditional forest management regimes. Silvicultural thinning can be used to harvest healthy small-diameter trees traditionally either cut and left to decompose on site in pre-commercial thinning operations or harvested as pulpwood in merchantable thinning. Pre-commercial thinning is often employed in forest management to reduce intraspecific competition for site resources (light, nutrients, and water) by reducing stand density. Stands that require pre-commercial thinning are often stands with high incidence of volunteer trees that arise between rows of planted trees from seed of adjacent forest stands. These trees can be harvested as wood fuel as an alternative to leaving them on site. Due to economic trends, in some regions, paper markets have substantially declined or disappeared. Declines in this market can lead to declines in forest productivity because without thinning forests have higher incidence of insects and disease as intraspecific competition weakens trees. Harvesting such trees as bioproduct and biofuel feedstocks would thus enhance forest health in addition to providing a substitute source of mid-rotation forest revenue (Hubbard et al. 2007, FAO 2010a; Figure 12.3).

Salvage thinning to remove defective, sick, dying, and dead trees damaged by insects, disease, fire, wind, or ice can also provide wood-based feedstock (Bungay 2004; Hubbard et al. 2007; FAO 2010a). Such trees are relatively inexpensive and are typically left to rot on site during thinning operations in which trees merchantable as pulpwood or composite paneling are harvested. It would be relatively inexpensive to collect these damaged trees during thinning (Bungay 2004). Thinning such defective trees from stands also reduces fire risks while leaving behind enough trees to support biodiversity (Hubbard et al. 2007; FAO 2010a). In some regions, fire suppression programs have led to overstocked forests at high risk of wildfires, and thinning these stands is of particular importance (Bungay 2004).

As markets for wood-based biofuels and bioproducts develop, planting practices can be altered to facilitate early thinning of small-diameter trees. Planting more trees than conventional planting guidelines suggest can foster relatively early thinning of small-diameter feedstock. A dual-cropping forest management approach developed by Scott and Tiarks (2008) in the Mid-South United States fosters early thinning of trees as biofuel feedstock. In this system, trees to be grown as conventional forest products are planted as seedlings and trees to be harvested as biofuel feedstock are direct-seeded between rows of seedlings. Direct seeding of trees for biofuel reduces the cost of densely planting trees as biofuel feedstocks. Trees from direct seeding are harvested relatively early in the rotation; in the southeast United States, harvest occurs in the fifth year after planting. After this single harvest of biofuel feedstock trees in the rotation, the remaining trees are managed for conventional forest

FIGURE 12.3
Feller buncher conducting an integrated harvest. Removal of small-diameter trees is often done in conjunction with the harvest for higher-value trees to improve cost efficiency.

products. Scott and Tiarks (2008) found that the presence of the direct-seeded trees early in the rotation had no effect on the volume growth of remaining trees by age 22.

12.2.5 Short-Rotation Woody Crops

Short-rotation woody crop (SRWC) systems have been defined as a silvicultural system based upon clear-felling cycles ranging from 1 to 15 years, with productivity achieved through intensive management practices that frequently include planting of genetically elite material, vegetation control, fertilization, mechanical site preparation (subsoiling, bedding), and coppice regeneration (Drew et al. 1987; Dickmann 2006). In SRWC systems, all trees in the plantation are grown as biofuel or bioproduct feedstocks. There are several logistical and economic advantages of SRWC systems relative to other biofuel crops such as soybean and corn. Trees in SRWC can typically be harvested at any time of year, whereas annual agricultural crops can be harvested at the end of the growing season. Trees in SRWC systems can be carried for extra seasons if necessary in order to reach target yields, which can moderate the impacts of drought, insects, disease, and other stressful growing conditions. This flexibility in harvesting reduces inventory holding costs, minimizes storage-related degradation of biomass qualities, and allows better matching of supply and demand (Hinchee et al. 2009).

There are several tree species suitable for SRWC systems. For example, there are many species and hybrids of *Eucalyptus* and *Populus* grown globally as SRWC as pulpwood and biofuel feedstock, with several Eucalyptus species and hybrids being among the fastest growing in tropical climates and *Populus* species and hybrids among the fastest growing in temperate climates (Hinchee et al. 2009). Species of *Paulownia* and *Salix*, as well as American sycamore (*Platanus occidentalis* L.), sweet gum (*Liquidambar styraciflua* L.), and loblolly pine (*Pinus taeda* L.), have all been explored in SRWC systems. A disadvantage of loblolly pine and similar species is that it cannot regenerate via coppicing, which increases its establishment costs over multiple rotations.

At present, many SRWC systems do not reach productivity rates adequate to sustain renewable energy production. English et al. (2006) projected that SRWC growth rates of 18–22 dry Mg ha^{-1} year^{-1} are needed for the long-term feasibility of a renewable energy production facility dependent on biomass. Growth rates of several *Eucalyptus* species and *Eucalyptus* hybrids when grown in tropical and temperate climates on suitable sites meet or exceed this projected threshold, with growth rates of 22–34 dry Mg ha^{-1} year^{-1} (Hinchee et al. 2009). However, *Eucalyptus* species have a relatively high water demand, and on droughty sites growth rates of 7 Mg ha^{-1} year^{-1} have been shown (Harper et al. 2010). Growth rates of *Populus*, *Salix*, sweet gum, American sycamore, and loblolly pine on productive sites ranged 2–18 dry Mg ha^{-1} year^{-1} (Mead 2005; Davis and Trettin 2006; Hinchee et al. 2009).

A primary limitation to growth potential of species in SRWC is the quality of wood material. Many species grown as SRWC currently achieve substantial growth rates only on the most inherently nutrient- and/or water-rich sites. Breeding efforts are underway to drastically increase growth rates through modification of genes that increase growth, stress tolerance, and adaptability. Some breeding efforts involve the development of transgenic trees with genes inserted for increased growth, cold tolerance, nitrogen use efficiency, and/or water use efficiency (Hinchee et al. 2009). Such attributes are necessary to provide greater overall growth and uniformity of growth over a wider array of site conditions. In addition, breeding efforts are focused on altering the chemical composition of wood (particularly lignin content) in order to improve conversion efficiency of diverse biofuel and bioproduct processes (Hinchee et al. 2009).

Weed competition is also a major impediment to optimal growth rates of SRWC. Many SRWC species are highly intolerant of competition, which necessitates virtually continuous competition suppression until canopy closure. In the absence of herbicides labeled for use within SRWC plantations, directed spray with glyphosate is necessary. To facilitate cost-efficient directed spray of glyphosate, rows must be navigable for ground-based spraying equipment. Otherwise, backpack-mounted sprayers are needed for directed spray, which is relatively time-consuming, inefficient, and costly. Studies focused on extending herbicide labeling for SRWC species are ongoing, with particular efforts made for herbicides that can provide broad-spectrum vegetation suppression without damaging SRWC species.

There are also environmental concerns associated with SRWC systems. Intensive suppression of all completion vegetation raises concerns about wildlife habitat, soil, and water quality. The invasiveness potential of species planted outside of their native range as SRWC is a concern due to the economic and ecological impacts of invasive plant species. Similarly, there are often intensive regulatory constraints to planting of transgenic trees due to concerns of the potential for biotech trees to breed with native trees.

12.3 Bioprocessing Methods and Products

Given the considerable potential woody biomass that holds for fulfilling renewable energy needs, it is important to determine how current and likely suitable wood-to-energy conversion technologies can potentially impact the future of wood-based biomass use for bioenergy and biofuels. Wood-based biomass can be converted into bioenergy or biofuels using a number of different processes. Broadly speaking, wood-to-energy or fuel conversion technologies can be grouped into two main categories: thermochemical technologies—including torrefaction, direct combustion using wood pellets and wood chips, gasification, and pyrolysis—and biochemical processes to produce biofuels such as ethanol. Thermochemical conversion involves breakdown of woody biomass into a semi-processed phase to produce ethanol using heat and pressure while applying catalysts. Biochemical processes convert biomass into sugars using either enzymatic or chemical processes to produce transport fuels such as ethanol.

12.3.1 Torrefaction

Under torrefaction, water in wood is removed through a controlled carbonization process that produces low-mass, high-energy material known as "torrefied biomass" or "biocoal." The torrefaction process results in a dry product that can produce an energy-dense fuel with all the logistical benefits of a low-volume, high-energy product. During this process, biopolymers such as cellulose partially decompose, giving off energy that can be used as a heating fuel for the torrefaction process. Interest in torrefaction has been growing in industrial sectors including power utilities, cement, and steel. This technology has gained a lot of traction in European countries like the Netherlands in the last few years with large-scale investments by energy companies. For example, a torrefaction plant is being constructed by Topell Nederland through investments approximating 15 million euros. This plant is expected to start production of a maximum of 54.4 Gg of biocoal year^{-1} (Netherland Agency 2011). In the United States, Integro Earth Fuels LLC has constructed a pilot plant in Gramling, South Carolina, that is one of the first in United States to produce 27–36 Mg of torrefied wood over a period of 3–4 months.

12.3.2 Pellets

Wood pellets are compressed by-products from wood biomass such as sawdust and wood-chips; pellets are used for domestic heating and for CHP plants. As high-density fuel, wood-based pellets allow cost-efficient transportation as well. Pellets have high energy content (about 40% higher than wood chips with 30% moisture content by mass and more than 300% higher energy content by volume), uniform size and shape (which facilitates automated handling), and economical attractiveness (Alavalapati et al. in press). Pellet production facilities can be adjusted for varying scales of demand and wood supply.

Demand for wood pellets has emerged in European countries such as Denmark, Italy, Belgium, and the Netherlands, with 10.8 million metric tons consumed in 2010 (Gibson 2011). By 2020, pellet consumption of the region is projected to grow to 23.8 million metric tons, and global pellet consumption is projected to be 44.9 million metric tons by 2020 (Gibson 2011). The global number of pellet production facilities has significantly increased recently, especially in Europe and North America. Traditional wood pellet exporters such as Canada and Russia are facing competition from US plants. Some of the largest pellet producers in the world are being established in South United States, with 24 mills contributing about 46% of the country's 2 million Mg annual capacity (Spelter and Toth 2009; Pellet Fuels Institute 2010). Rather than using only sawdust from mills for producing pellets, these new plants also use whole trees and chips.

Given the significant price advantage coal and natural gas currently have for producing electricity at a much lower price per megawatt hour, further development of this technology may help make pellets a commercially viable bioenergy option. The International Energy Agency (IEA) (2011) notes that 4 million Mg of pellets were sold globally in 2008.

12.3.3 Gasification and Pyrolysis

Advanced thermal technologies such as gasification and pyrolysis are technically feasible. Gasification is a high-temperature process in which wood biomass is used to generate heat, electricity, methanol, ethanol, and syngas (hydrogen). If the gasification process includes a devolatilization and conversion of biomass in a steam environment, it can produce a medium-calorific gas that can be transformed into fuel for combined cycle power generation (Guo et al. 2007). Otherwise, the syngas is converted to ethanol or hydrocarbon chemicals and fuels. Nexterra has commercial gasification units in Tolko and Kroger, British Columbia, Canada, using wood waste as a fuel source. A similar wood-based gasifier is being built at the University of South Carolina in Columbia, South Carolina, United States, by Nexterra.

Pyrolysis is a type of gasification technique that, at higher temperatures in the absence of oxygen, converts biomass to bio-oil (via fast pyrolysis) and charcoal (by torrefaction). Bio-oil can be used as fuel in heating, for electrical applications, and for production of chemical commodities (Faaij and Domac 2006). Converting wood-based biomass to bio-oil increases energy density, which translates to improved transportability. Its main disadvantages are its low heating value, poor ignition performance, and thermal instability (Jackson et al. 2010). The existing pyrolysis plants worldwide are not yet commercially viable for large-scale production.

12.3.4 Biochemical Processes for Wood-Based Fuels

Processed biodiesel and ethanol are the primary liquid fuels that can be derived from biochemical processes. The biochemical conversion process involves the breakdown of biomass into sugars using enzymatic or chemical processes to later apply fermentation to produce ethanol. In this conversion practice, there are two stages. The first stage involves the use of thermal, acid, alkaline, and biological pretreatments, and the second stage is an acid or enzymatic treatment. Separation of sugar and its further processing can be conducted separately or simultaneously. Lignin from this separation process may be used in biopower production. While separate processing allows for optimal temperature processing, ethanol yields are lower and are more expensive (Dwivedi and Alavalapati 2009; Jackson et al. 2010).

Although several hydrolysis techniques have gained momentum in the last decade, efficiency and cost issues have hindered their commercial viability. An integrated enzymatic process could contribute to cost reductions for hydrolysis techniques, but such processes have not progressed past the laboratory stage. While several such techniques have seen increased attention, there still remains a notable research need to make the process economically viable.

Capital and biomass inputs are cost-reduction issues that need further attention to increase the scale of biofuel and bioproduct plants. If a large plant is set up, then the transportation cost of procuring wood-based biomass might increase per unit cost and/or lead to procuring lower-quality feedstock. The scale of a plant depends not only on cost issues but also on the purpose for which it is being built. For example, Van Loo and Koppejan (2008) suggest that small CHP plant facilities with lower conversion efficiency (10%) can be used where heat is the primary product and power the secondary product, while higher-efficiency facilities (25%) can be used in cases where electricity is the primary product.

The US Department of Energy (DOE) set 2012 commercialization targets for research and development that included reducing the selling price of ethanol by 2012 to $1.07 from the prevalent $1.61 per gallon, increasing ethanol yield per dry ton from 56 gallons in 2005 to 67 gallons in 2012, and reducing 2005 capital and operational costs by 35.5% and 65.3%, respectively. The DOE also set a feedstock cost target for 2012 of $35 per dry ton. Efforts are underway to achieve these targets, but no technological breakthroughs have yet been developed to reach these targets. Uncertainties for potential technologies still exist in a number of areas including production costs, project-specific factors, and the ability to meet required environmental standards (Alavalapati and Lal 2009; Dwivedi and Alavalapati 2009). In addition, harvesting and transportation cost attributes of woody bioenergy feedstocks from harvest to final user location have to be competitive with fossil fuels. This cost-competitiveness need requires reductions in capital costs through technological improvements, reduced feedstock costs, and continuous and long-term flow of woody biomass supply.

Recognizing the need for federal funding for liquid fuels research and development in the United States, the DOE has consistently supported initiatives through discretionary funding. For example, out of $217 million direct appropriations allocated by DOE for liquid fuels in 2009, the cellulosic component comprised $215 million (Bracmort et al. 2010). Other federal agencies such as the US Department of Agriculture (USDA) are contributing toward cellulosic biofuels research as well. For example, USDA contributed $100 million in 2009 for cellulosic biofuels research, 80% of which was targeted at commercialization of thermochemical technologies alone (Bracmort et al. 2010).

The European Union (EU) imposed a target of 10% renewable energy target for transportation sector by 2020. In this target, special emphasis is placed on woody biomass-based fuels whereby their contribution toward meeting renewable energy goals is counted as double their real contribution (European Commission 2009). Countries like Finland are aggressively promoting wood-based energy use by establishing investment subsidy programs where as much as 40% of total construction costs are cofinanced by the government. The wood-based energy and fuel plants in the country have been major beneficiaries as they procured 60% of total federal funding allocated in 2006 (European Review Energy Council 2009).

12.4 Biomass Preparation Processes

The woody biomass supply chain is a complex system that encompasses the identification, collection, and transport of biomass raw material to biofuel and bioproduct manufacturing facilities. In regions with a well-established forest products industry, this woody biomass supply chain is relatively well developed in terms of expert personnel and equipment. This supply chain can be expanded to accommodate wood-based biofuel and bioproduct markets. Conversely, the lack of such a supply chain in regions without a well-established forest products industry is a substantial impediment to the development of wood-based biofuel and bioproduct markets.

Economies of scale may dictate that biorefineries be large and centralized. As a result, several methods exist to collect and comminute (fragment into smaller pieces) biomass into a form that can be transported to a centralized location. These include the common woody biomass preprocessing operations of chipping, grinding, or shredding. Woody biomass may also be compacted into cylindrical bales or bundles that can be handled like roundwood. In-woods "prefining" through mobile pyrolysis units can also condense bulky, low-value forest biomass into higher-value, low-bulk feedstocks for the bioenergy and biochemical industry.

Collecting, transporting, and delivering woody biomass to a centralized biorefinery are challenges for producing wood-based biofuels and bioproducts. Scattered, low-bulk feedstocks and infrequent harvest accessibility of forests result in high handling and transportation costs. Handling includes not only in-woods activities, such as harvesting, skidding or forwarding, chipping, and loading; it also includes operations at the end-use point, such as weighing, grinding, dumping, screening, and storage. In the following section, we describe these processes that take place in producing wood-based biomass.

12.4.1 Conventional Harvesting Equipment

Conventional harvest, handling, and transport infrastructure for woody biomass is ubiquitous in regions with well-developed forest products industries. The preferred harvesting method in these regions is often whole-tree, one-pass timber harvesting in which roundwood and biomass are harvested simultaneously. This system has proven to be the most cost-effective extraction method when terrain does not preclude use of this equipment. An alternative to whole-tree harvesting is the cut-to-length system commonly used in Europe. In cut-to-length systems, all delimbing and bucking is done at the stump, where logs are cut into mill-specified lengths. Cut-to-length harvest systems require less labor and have less potential to disturb soil than whole-tree harvest systems, but they have greater purchase and repair costs (Meek and Plamondon 1996; Hartsough et al. 1997; LeDoux and Huyler 2001). Globally, 65% of mechanical tree harvesting is conducted by whole-tree harvest systems and 35% is conducted by cut-to-length systems (Ponsse 2005).

Conventional timber harvesting equipment consists of equipment for tree-felling (feller buncher in whole-tree harvest systems, harvester in cut-to-length systems), transporting felled trees to loading areas (skidder in whole-tree harvest systems, forwarder in cut-to-length systems), delimbing trees (gate delimber in whole-tree systems, harvester in cut-to-length systems), and a loader. At least in the near term, these conventional harvesting systems can be used for collection and transport of some woody biomass sources for biofuel and biochemical production.

A desirable woody bioenergy system is one in which conventional harvest operations are modified in such a way that residue is captured without reducing production or value of conventional products. A one-pass, whole-tree method of harvesting can be employed to fell trees and transport them to a logging deck. Within this system, the energy wood (thinned trees) is cut and transported to the logging deck at the same time as trees for conventional forest products. The felling equipment would pile energy wood separately from roundwood for the skidding equipment. At the logging deck, whole trees are delimbed, de-topped, and merchandized for roundwood. The remaining limbs, tops, and boles, as well as the energy wood, serve as biofuel and bioproduct feedstock. When combined with roundwood harvest, this type of feedstock collection and transport method may minimize costs. However, collecting smaller material and what would traditionally be cull trees can add an additional 15% to the cost of delivery and sorting at the log deck (Holley et al. 2010; Figures 12.4 and 12.5).

12.4.2 Biomass Harvesters

Harvesting costs of SRWC are high if conventional forest harvesting equipment is used. When harvested, all trees in SRWC are small in diameter. Feller bunchers and skidders

FIGURE 12.4
Skidder hauling tree-length material. During harvest this is one of the most cost-efficient practices to reduce costs associated to removing woody biomass.

FIGURE 12.5
Loader sorting material to be used for timber products from woody biomass.

or harvester and forwarder combinations conventionally used for harvesting larger-diameter trees (trees of pulpwood size or larger) are inefficient at harvesting small-diameter trees (those below pulpwood size). Due the expense of using such equipment to harvest small-diameter trees, the cost of harvesting SRWC can be as high as 70% of the total cost of SRWC management. To substantially reduce this harvesting expense, woody biomass harvesters have been developed. These biomass harvesters are similar to sugarcane harvesting systems, with the harvester driven over multiple rows of trees to chip them internally. Chips are dispensed via a chute into a tractor-driven trailer running parallel to the biomass harvester.

12.4.3 Chipping

In-woods chipping provides a way of processing woody material into an acceptable fuel for some applications and improves bulk density, homogeneity, and handling characteristics of harvest residue to facilitate transportation of woody biomass. Chipping can be conducted independently or in conjunction with a roundwood harvest operation. Small-diameter trees and debris (tops and slash) are converted into chips that can be burned for electricity in boilers. Such chips are typically a mixture of wood and bark colloquially termed "hog fuel." Forest product manufacturing facilities increase their use of hog fuel to generate electricity as fossil fuel prices increase. These chips can also serve as feedstock for conversion into other biofuels and biochemicals. In-woods chipping, when done in combination with a conventional roundwood harvest, requires a chipper, an extra loader and trailer, and additional labor. These extra requirements can increase the cost of logging and transportation of woody biomass by 30% relative to conventional roundwood logging and transportation (Holley et al. 2010).

Once material is chipped, it is essential to use or convert the chips for fuel as soon as possible to prevent excessive energy value loss from decomposition and to reduce the risk of self-ignition associated with storing green chips (Johansson et al. 2006). In studies of in-woods chipping in the Mid-South United States, the maximum storage time for chips was 60 days. Chipping costs for loose residue at or near the harvest site are higher than that of chipping at a biomass storage terminal, power plant, or biorefinery. Chipping costs may be reduced by as much as 65% at a centralized facility with larger, more efficient equipment and utilizing parasitic loads from an electrical producing facility as opposed to diesel-powered grinders (Holley et al. 2010).

12.4.4 Bundling

Logging debris collection can be integrated with forest harvesting using machines that compact debris into cylindrical bales. This process, called bundling, is commonly used in Nordic countries, and its use is increasing in other regions (FAO 2010a). Bundling can be integrated into whole-tree harvest systems by using a slash bundler machine at the logging deck and bundling treetop biomass after delimbing, and bundling can be integrated into cut-to-length harvesting by using a track-mounted mobile slash bundler that follows the harvester to compress slash at the stump into bundles carried back to the logging deck by the forwarder. Bundling creates a compressed and uniform composite residue log (CRL) approximately 0.6 m in diameter and 3 m long from harvest residues and other small-dimension wood. Bundles are efficiently handled and transported with conventional roundwood equipment, but simple modifications to trailers may be required depending upon CRL length and makeup. Slash bundlers can produce 18–26 bundles per hour. A blend of pine and hardwood offers the best economic advantages for bundling because of the amount of available material, relative ease of handling, and energy content. A benefit of bundling is the storage characteristics of CRLs. Bundles can be left on logging sites to season for up to 1 year. During storage, energy value increases as bundles dry. For example, the calorific heat content of green CRLs collected in the Mid-South United States has been measured as 10 MJ kg^{-1}. By comparison, seasoning CRLs at the logging site for 11 months reduced moisture content by 25%–30% and increased calorific heat content to

17 MJ kg^{-1}. The low- or no-cost seasoning of CRLs on site allows transportation of bundles with higher energy value per kg, which reduces transport cost per MJ (Holley et al. 2010).

12.4.5 Grinding

Woody biomass can be reduced to small particles by grinding. Grinding machines reduce woody biomass particle size by repeatedly pounding biomass into progressively smaller pieces through a combination of tensile, shear, and compressive forces. Grinders are derivatives of hammer mills that typically consist of high-speed rotating hammers, and they can accept roundwood as well as short, non-oriented pieces including stumps, tops, brush, and large forked branches. Grinders are often used to produce feedstocks for direct firing in electrical production facilities and for conversion in fast pyrolysis reactors to pyrolysis oil (described later). Grinding can be done in-woods or at a centralized facility. However, in-woods grinders require more energy per kilogram of output than chippers, which makes them more expensive to operate. Additionally, excessive soil contamination can increase internal wear of grinders. As with chipping, ground material must be used within a short time frame to prevent energy loss from decomposition, and costs can be reduced by as much as 65% when performed at a centralized facility (Holley et al. 2010).

12.5 Global Markets for Wood-Based Bioenergy

Woody materials, including solid wood (fuelwood), bark, sawdust, wood chips, wood scrap, pellets, briquettes, and paper mill residues, constitute an important source of renewable energy around the world. Wood energy has been used for thousands of years for cooking and heating. In many of the world's developing countries, it remains the primary source of energy and in much of Africa total consumption of fuelwood is still increasing driven by population growth (FAO 2008). Traditional use of fuelwood is consumed mainly in the household sector (Hillring 2006). Wood energy in the form of heat, electricity, and liquid fuels represents about 10%–15% of the world's primary energy supply (Hillring 2006), of which about 25% is consumed in industrialized nations and 75% in developing countries (Parikka 2004). Luo (1998) has estimated that fuelwood consumption in China accounts for about 40% of energy consumed in rural households in China, the remaining coming from crop stalks and manure. In India, Pandey (2002) reports that most of household energy needs (61.5%) are mainly met by fuelwood, followed by crop residues (14.0%) and manure (10.6%). In the United States, wood provides about 2% of current annual energy consumption, most of it consumed in rural areas. However, it accounts for 46% of the energy generated from all types of biomass and 25% of all renewable energy consumed annually (USDOE Energy Information Administration, EIA 2009). In the EU, Sweden already uses the highest proportion of renewable energy in relation to final energy use of any country in the region at 44.7%. Wood fuels in particular supply about 67% of the total amount of biofuels used for heat production in Sweden. The use of biofuels (mainly logging residues and solid forest products industry by-products) in the Swedish district heating sector has increased by more than fivefold since 1990 (Swedish Energy Agency 2010). Based on data from the United Nations Economic Commission for Europe, wood is the main source of bioenergy in a region that includes the United States, Canada, the EU-27, Israel, various Central Asian nations, and the Russian Federation (UNECE FAO 2011).

National and regional renewable energy targets rely heavily on greater use of wood for energy generation. The EU2020 targets in Europe and the US Energy Independence and Security Act, among other laws, require increasing amounts of energy from renewable sources (including wood) in the power, heating, and motor vehicle sectors. As discussed by Aguilar et al. (2011), industrial consumption of wood energy is closely linked to its output and attainment of renewable energy targets for biomass will be linked to a healthy wood products manufacturing industry. Over time there is a strong correlation between industry output, in particular in the pulp and paper industry, and wood energy consumption. To an extent, future growth in wood energy markets will be highly dependent on a healthy wood products manufacturing sector.

Increasingly, a larger share of wood energy is traded in regional, national, and international markets. However, it is worth mentioning that a large proportion of wood energy is still consumed locally in small-scale or noncommercial uses. As a case in point, in the United States about 20% of wood energy is used in homes for heating and cooking, and most of it is sourced locally. In the UNECE region, the residential section is the primary consumer of wood energy, surpassing uses in industrial and power generation. Globally, biomass is the main source of combustible renewables (IEA 2007). In this section, developments and prospects for wood energy markets consumed by the wood products industry, electric power producers, and commercial businesses are highlighted.

12.5.1 Emerging Markets for Wood-Based Bioenergy

Driven in part by a combination of high fossil fuel prices, public policies, and advancements in technologies, wood energy consumption has been in the rise in recent years (Aguilar et al. 2011). Biomass markets have globalized over the last decade but are still in an emerging stage. They have been driven by imbalances between supply and demand, and trade will play an important role in moderating demand and supply fluctuations among different regions. The EIA (2011) suggests that the development of pelletization, pyrolysis, and torrefaction technologies will speed up the development of markets as they increase energy density. Trade in biomass and biofuels can mobilize currently unused or underutilized woody biomass resources and trigger investments in biomass-rich regions by providing access to international markets.

The most common form of wood energy traded in the global market is wood pellets. The IEA (IEA 2007) identified the United States, Sweden, and Canada as the largest global industrial pellet producers, with capacities exceeding 3.5 million Mg of wood pellets. Other countries with production capacities ranging from 200 to 600 thousand Mg include Austria, Germany, Italy, Estonia, Latvia, the Russian Federation, Poland, and Denmark. Cocchi (2011) estimates that in 2010 global wood pellet production reached 14.3 million Mg, while consumption was close to 13.5 million Mg, corresponding to an increase of over 100% compared to 2006 (Cocchi 2011).

Europe has one of the most active wood pellet manufacturing sectors. The EU 2020 targets for renewable energy and GHG emissions reduction are among the main drivers behind growth in EU wood energy consumption. The EU 2020 targets mandate that at least 20% of energy consumption be met by renewable energy sources (European Commission 2011). Research conducted by Sikkema et al. (2011) suggests that additional demand for woody biomass triggered by EU 2020 targets could reach 305 million tons of wood. Additional supply of woody biomass could come from 45 million Mg from increased harvest levels, 400 million Mg from the recovery of slash from altered forest management, the recovery of waste wood through recycling, and the establishment of woody energy crops.

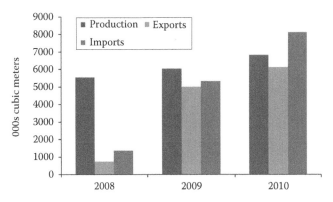

FIGURE 12.6
Total production, exports, and imports of pellets in the EU-27 region (000s cubic meters). (From Eurostat 2011.)

Any short-term shortages in wood pellets to meet renewable energy targets within the EU may be bridged by imports from nearby regions such as Northwest Russia. Long-term wood energy deficits within the EU could be supplied from North America and the Russian Federation. According to Sikkema et al. (2011), approximately 650 pellet plants in Europe produced more than 10 million Mg of pellets in 2009. Figure 12.6 summarizes total wood pellet information for total production, exports, and imports in the EU-27 region. Over the 2008–2012 period, there was a marked increase in regional wood pellet production accompanied by exports. The greatest rate of growth was observed in the volume of wood pellets imported into the EU-27.

Increasing demand for wood pellets has triggered considerable investments in manufacturing infrastructure in countries around the world. To meet EU-27 as well as local demand, there are almost 200 pellet-producing companies in the Russian Federation, of which two produce over 100,000 Mg per year. Corporations and partnerships have been founded, and the trend of increasing production capacities per plant is continuing. Large, capital-intensive companies have replaced small ones (Aguilar et al. 2011). However, profit margins in the Russian pellet market are relatively low, resulting in many production plants closures in 2010–2011. The market potential of pellets nonetheless remains high. Companies that have secured a stable feedstock supply are less prone to bankruptcy. Pellet production is spreading to Russia's inland regions that have underutilized coproducts and low-value forest stands.

US and Canadian pellet plants have also invested in large infrastructure in response to growing, policy-driven demand from European power plants (Sikkema et al. 2011). An analysis conducted by Forisk Consulting (2012) suggests that 452 energy projects using woody biomass in the United States will be in operation by 2022, which could result in increased wood use of 49.7 million moisture-free Mg (99.4 million green Mg). Forisk Consulting identified wood energy projects for electric power, CHP, thermal generation, liquid fuel, and pellet production in the United States. Forisk Consulting applied several screens with limitations based on readiness of technology and status of development. If the project has received/secured/signed two or more of the following, then it passes the status screen: financing, air quality permits, engineering, procurement and construction contracts, power purchase and interconnection agreements for electricity facilities, and supply agreements. The screens are a way to assess the likelihood that projects will complete the development process and actually produce bioenergy. Figure 12.7 summarizes the combined volume of woody biomass, expressed in million moisture-free Mg, forecasted

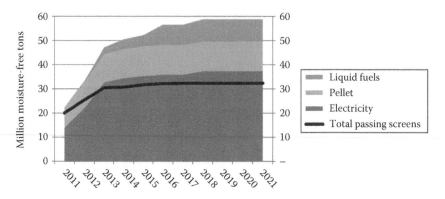

through 2021 (Forisk Consulting 2011). The greatest share in the country is expected to be used for the generation of electricity (primarily locally), followed by the manufacturing of pellets (for domestic and European markets).

The industrial wood pellet sector is expected to grow given the expectations for expanding global demand, primarily driven by the EU. Estimates from different sources suggest that by 2016 total industrial wood pellet demand from Europe might range from 30 to over 80 million dry Mg. As a result, APX-ENDEX and the Port of Rotterdam in the Netherlands have created a wood energy commodity contract exchange market (APX-ENDEX 2010). Figure 12.8 shows recent market trends for industrial wood pellet prices as reported by APX-ENDEX based on delivery CIF Rotterdam and net caloric value of 17 MJ kg^{-1}, with less than 10% water content. M+1 represents price traded per ton for the upcoming month, Q+1 is next quarter price, and Y+1 captures prices for the upcoming

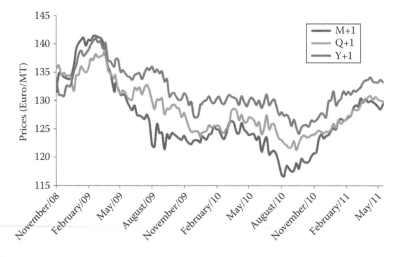

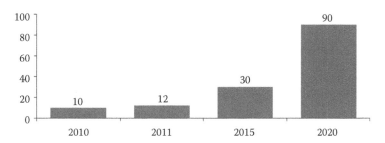

FIGURE 12.9
Global market of industrial wood pellets (million metric tons). (Adapted from APX-ENDEX, *Industrial Wood Pellet Prices*, APX-ENDEX, Amsterdam, the Netherlands, 2011.)

year (e.g., for 2011 it represents trading prices for 2012). Recent market trends show that the market price reached its highest level in January–March 2009, after which it declined to its lowest level recorded over the third quarter of 2010 and then began to increase again well into 2011 (Aguilar et al. 2011).

The APX-ENDEX estimates significant growth in global trading of wood pellets. By 2020 this number is expected to exceed 80 million Mg, with the largest share traded in the EU. This is an eightfold increase in wood pellet trading relative to 2010 (Figure 12.9).

Most wood pellet production is currently located within North America, Europe, and the Russian Federation. Prospects for region-level potentials from a growing wood products sector in South America (primarily Brazil, Chile, and Argentina) and East Asia (mainly China) suggest that in the coming years new players will enter the market. Analysis conducted by Cocchi (2011) suggests that some of the countries with the greatest annual surplus in forest growth include the United States, China, Japan, Ukraine, France, Russia, Italy, Poland, Sweden, and Chile. Much of this surplus in growth could be manufactured into wood pellets and traded in global markets. Africa may be able to enter the global market in the future. Africa has 16% of the global forest area and about 35% of aboveground woody biomass. While the entire continent faces tremendous challenges associated with illegal deforestation, an underdeveloped wood products sector suggests the potential growth in wood energy production capacity in sub-Saharan countries including South Africa, Mozambique, and Tanzania (FAO 2010b). South Africa has the most developed wood products industry in the continent and large areas of forest plantations. Good manufacturing and transportation infrastructure have already facilitated the creation of companies exporting woodchips and pellets.

12.5.2 Standardization and Certification

Given the prominence of wood pellets in wood energy trading, it is likely that schemes addressing product standardization and sustainability will first be adopted by this industry and then expanded to other products. Hence, we concentrate on advances made by the wood pellet industry in this section. The fast growth in markets for wood pellets has demanded the use of standardized product specifications to facilitate trade. The International Organization for Standardization (ISO) is working on a program, named ISO/TC 238, for the "standardization of terminology, specifications and classes, quality assurance, sampling and sample preparation and test methods in the field of raw and processed materials originating from arboriculture, agriculture, aquaculture, horticulture and forestry to be used as a source for solid biofuels" (ISO 2012). Currently, there are 22 countries participating in the development of this standard.

In the EU, the European Pellet Council (EPC) has developed the ENplus system of certification. The goal of the ENplus system for wood pellets is to secure the supply of wood pellets with clearly defined and consistent quality for heating purposes (European Pellet Council 2011). The certification addresses consistency in the quality of processes involved in the production and logistics of manufactured and delivered wood pellets. The ENplus certification system contains requirements for (a) wood pellet production and quality assurance; (b) wood pelletized product (EN 14961–2); (c) labeling, logistics, and intermediate storage; and (d) delivery to end customers. The EPC organized within the European Biomass Association (AEBIOM) has received the license rights to the ENplus system by a contract with the developer of the system, the German Pellet Institute (DEPI). It will pass on this right to national associations that will organize the introduction of ENplus into their respective countries or geographic areas (European Pellet Council 2011).

To address questions about the sustainability of woody materials going into the production of industrial pellets, governments and private organizations are working on guidelines for the measurement of their impacts. For example, the United Kingdom's Renewable Heat Incentive for wood-fuelled heating is scheduled to develop sustainability criteria as part of a consultation process to measure the impact of biomass energy utilization both on the forest resource and in terms of GHG emissions. Sustainability criteria is expected become mandatory from 2013 onward (UK Department of Energy and Climate Change 2011).The APX-ENDEX has also developed draft sustainability guidelines for the market characteristics of industrial wood pellets (APX-ENDEX 2011). APX-ENDEX will require all deliveries to have proof that the product originates from feedstock obtained in a sustainable manner. Cargo delivered under this specification must be accompanied by a proof of sustainability. APX-ENDEX has currently listed several certification schemes (Green Gold Label, Laborelec-SGS, Drax Biomass) but will likely change this guideline once a common standard accepted by all parties involved is adopted. ENplus-certified producers have to document the origin of the raw material and inform the inspection body at the yearly audit regarding the share of raw material coming from certified sources (FSC, PEFC, or equivalent systems). The inspection body integrates this information in an audit report. In view of the importance of GHG mitigation, industrial pellet producers must be able to state the amount of GHGs emitted as a consequence of pellet production (European Pellet Council 2011).

12.6 Issues and Challenges

12.6.1 Public Policy and Existing Markets

An assortment of public policies at various levels (local, state/provincial, national, and regional) has been adopted to address concerns associated with heavy dependence on nonrenewable fossil fuels and to enhance national energy independence (Aguilar and Saunders 2010). Alongside these objectives, public policy supporting renewable energy aims to internalize into energy systems the external costs and benefits of energy production (i.e., external costs of GHG emissions and the advantage of renewable fuels generating lower net GHG emissions). The failure to incorporate true costs of energy generation is an important reason why most bioenergy approaches are not yet economically competitive with conventional fossil fuels (IEA 2009). IEA (2009) lists other complementary objectives to renewable energy policy including addressing climate change effects, encouraging environmental protection, triggering economic rural development, fostering technological progress, and improving cost effectiveness.

A complex system of public mandates and monetary incentives and disincentives has been adopted to support greater wood energy utilization. A general categorization of public policies may distinguish between those that (a) create markets for woody feedstock and wood energy generation and/or (b) incentives relying on existing markets to improve cost-competitiveness of wood energy generation, distribution, and consumption.

12.6.1.1 Creation of Markets for Woody Feedstock and Wood Energy Generation

Governments may adopt rules and regulations requiring the use of renewable energies. Policies under this group include the adoption renewable energy purchasing goals and Renewable Portfolio Standards (RPS) that mandate a certain percentage of energy purchased or produced that has been generated from renewable feedstocks, including woody biomass. Another tool that relies on the creation of markets for GHG emissions and the allowance of emission permit trading is what is commonly referred to as "cap-and-trade" systems. Cap-and-trade system is founded on the principle of reducing GHG emissions in the most cost-efficient manner (i.e., emissions reduction is done at the least-cost possible to society by the most efficient companies).

The United States' Federal Energy Policy Act of 2005 (U.S. Public Law No. 109–58) created the Federal Green Power Purchasing Goal that required that "to the extent it is economically feasible and technically practicable," the total amount of renewable electric energy consumed by the federal government should be at least 5% from 2010 to 2012 and 7.5% after 2013. At the state level, by January 2012, 30 states in the United States and the District of Columbia have adopted enforceable RPS or other mandated renewable capacity policies, and other seven states have voluntary goals for renewable generation (EIA 2012). In the EU, EU 2020 targets mandate that at least 20% of energy consumption should be met by renewable energy sources (European Commission 2011).

The EU Emission Trading System (ETS) launched in 2005 is a central part of the EU policy to reduce GHG emissions. It is the first and biggest international scheme for the trading of GHG allowances, covering some 11,000 power stations and industrial plants in 30 countries. The ETS currently operates in 30 countries (the 27 EU member states plus Iceland, Liechtenstein, and Norway). At the end of each year, each a company must surrender enough allowances to cover all its emissions; otherwise, fines are imposed. If a company reduces its emissions, it can keep the extra allowances to cover its future needs or sell them to another company in need of allowances. The number of allowances is reduced over time so that total GHG emissions decline. In 2020, EU emissions will be 21% lower than in 2005 (European Commission 2011). To date, biomass energy is deemed to be carbon neutral, which has promoted its use in the EU to reduce GHG emissions as part of national renewable energy action plans.

12.6.1.2 Financial Incentives Based on Existing Energy Markets

Financial incentives may be classified by those (a) promoting the demand/supply of renewable energy feedstocks, (b) reducing start-up costs through cost-share programs, (c) lowering the cost of capital necessary to generate renewable energy, and (d) providing financial incentives based on energy output (Aguilar and Saunders 2010). Instruments used to promote sustained demand and/or supply of renewable feedstocks include tax credits for the use of woody biomass, exemption from sales taxes of wood for heating, monetary incentives for purchasing renewable energy, and ability to sell or receive credit from renewable energy (net metering). Another type of financial incentive is provided through start-up cost-share programs to carry out feasibility studies and cofinance initial construction of facilities to recruit energy industries. Instruments to lower costs of capital investments include access to public and private grants; loans with zero, below market, or low fixed-interest rates; government-sponsored energy bonds; subsidies to interests on loans; and property tax credits for the acquisition of equipment to generate renewable energies. Financial incentives based on energy output include tax deductions or credits, feed-in tariffs, or rebates based on energy generation (e.g., payments made per renewable kWh produced from renewable feedstocks (Aguilar and Saunders 2010).

Programs in the United States and the EU illustrate the use of different tools to incentivize wood energy generation through monetary incentives at different scales. The Biomass Crop Assistance Program in the United States provided an incentive in the form of a government payment per unit of biomass supplied for the production of bioenergy (US Department of Agriculture 2011). Also in the United States, Aguilar et al. (2011) mentioned that the use of energy conservation bonds created by the Energy Improvement and Extension Act of 2008 has been an effective policy instrument for financing energy projects powered by renewable energy, including biomass. Under this program, borrowers pay back only the principal of the bonds, and the bondholders receive federal tax credits in lieu of the traditional bond interest. In the case of renewable fuel standards in the United States, the Energy Independence and Security Act of 2007 requires 60.5 billion L of cellulosic fuel to be used in vehicle fuels by the year 2022. Public financial support has been instrumental for large pellet operations in the EU to cover high start-up costs of capital, operation, and maintenance of renewable energy equipment and pellet fuel feedstock procurement (Sikkema et al. 2010). The United Kingdom's feed-in tariffs scheme encourages deployment of small-scale (<5MW) low-carbon electricity capacity by guaranteeing a fixed payment per kWh of electricity generated (Office of the Gas and Electricity Markets 2011).

It is noticeable that most public incentives have targeted electricity production from woody feedstock (Aguilar et al. 2011). Other sectors that also consume a large share of wood energy, such as the residential sector, have not received as much government support and should not be neglected. For instance, research by Song et al. (2012) stresses the importance of the residential sector in the United States and the potential of greater wood energy use in particular in rural areas.

Concerns: Impacts of energy policy on the wood products industry and carbon neutrality of wood energy.

Public policy implementation has been instrumental in the recent growth of production of wood-based feedstock as well as the generation of energy from woody materials. However, there are also direct and indirect impacts that are important to point out. Policies promoting use of woody biomass that previously had little to no commercial value will certainly have effects on current wood product markets. Additional removals of materials from forestlands can also result in undesired impacts on the forest resource (Aguilar and Garrett 2009). More recently, several reports have brought into question the carbon neutrality of wood-based energy.

There are growing concerns about the impacts of EU policies and initiatives on the traditional wood products industry's (e.g., woodworking, furniture, pulp, and paper) competitiveness and profitability. In a preliminary draft opinion by the European Economic and Social Committee's (EESC) Consultative Commission on Industrial Change, Zbořil and Pesci (2011) stress that the woodworking and furniture sector faces growing competition for wood from the renewable energy sector, due to subsidies and other measures promoting the use of biomass, of which wood makes up a major share. Zbořil and Pesci (2011) mention that EU policies make the use of wood for energy a more profitable activity than using it for solid wood products, thus deviating materials from the wood products to the energy industry. Public financial support from the EU to the use of wood for energy has affected market availability and price of forest-based raw materials for nonenergy manufacturing businesses. Hence, the EESC has called for a detailed study to evaluate the impact of public support on the supply of woody raw materials.

Although the electricity sector is a major beneficiary of public policy support, it has faced recent increased scrutiny because of GHG emissions. Whether the use of woody feedstock is deemed a carbon neutral option to power generation is under debate. The US Environmental Protection Agency (EPA) has been working on guidelines to restrict emissions from certain stationary sources—such as electric power plants. As part of that development, the EPA has suggested the possibility that emissions from biomass should be treated the same as fossil emissions. Moreover, the EPA has also recognized uncertainty about the carbon offset benefits of wood and other biomass sources (EPA 2010). EPA is proposing studies that will include a review of available technical information on biogenic emissions and will seek to develop accounting options for CO_2 from stationary sources that satisfy principles of predictability, practicality, and scientific soundness. Revisions to the carbon neutrality of wood energy can have major consequences on present and future energy policies.

An important aspect regarding the use of wood for energy and its net level of GHG emissions has been whether woody materials come from existing forests or land-use changes that displace forests or agricultural lands (e.g., energy plantations that replace natural forests or farming land). Direct or indirect land-use change can have a significant impact on GHG emissions savings associated with bioenergy and potentially reduce its contribution to climate change goals (European Commission 2010). Melillo et al. (2009) stress that a global GHG policy that protects forests and encourages best management practices can dramatically reduce emissions associated with bioenergy production.

In summary, great strides have been made to implement public policy to encourage the use of woody materials for energy. Most public programs have targeted the electricity generation sector through different mandatory targets and market-based tools. Renewable energy policy to be most effective must be part of a long-term vision for bioenergy and its role in accomplishing environmental, social, and economic objectives. An effective public policy for wood energy should identify specific national or regional strengths and specific features of the industrial sector. Successful deployment of bioenergy requires a sound national policy strategy, but energy planning and incentives at local administrative levels are also necessary (IEA 2009). Public policy should be flexible and be able to incorporate new scientific knowledge as it is generated. Nonetheless, long-term continuity and predictability of policy support is instrumental for the sustainability of wood energy efforts (IEA 2009; Aguilar et al. 2011).

12.6.2 Life-Cycle Analyses of Wood-Based Bioenergy

In addition to issues of energy security and rural development potential, additional support for the emerging bioenergy sector comes from the potential environmental benefit of reducing GHG emissions. It is increasingly being recognized that environmental effects of wood-based bioenergy production are linked to net GHG reductions. Life-cycle analyses (LCAs) of different biofuels (including woody biomass) suggest overall GHG reductions

(Birdsey et al. 2006; Blottnitz and Curran 2007; Eriksson et al. 2007; Gustavsson et al. 2007). However, Searchinger et al. (2008) argue that life-cycle studies have failed to factor in indirect land-use change effects and suggest that using US croplands or forestlands for biofuels results in adverse land-use effects elsewhere, thus harming the environment rather than helping it.

The LCA includes all stages of product development and utilization, from planting of trees to the use of wood-based energy by the consumer and the disposal of all wastes. Achieving a competitive GHG reduction performance by biofuels will require due attention be given to material and energy use in the production and harvesting phases of the biofuel life cycle. The US Energy Independence and Security Act of 2007 specifies that the use of advanced biofuels (such as cellulosic biofuels) should lead to a minimum 50% reduction in GHG emissions relative to 2005 levels. Countries such as New Zealand are trying to set a much lower target stipulating that biofuels must lower GHG emissions by at least 35% over their life cycles than fossil fuels (New Zealand Sustainable Biofuels Bill 2009). In reality, however, accounting for all the GHG footprint of wood-based bioenergy or biofuels is a complex process. Irrespective of the conversion technology adopted, the processes for producing wood-based energy have multiple phases along the wood-to-energy supply chain that might influence the climate change impact or net energy balance of biofuel/bioenergy produced from wood-based biomass. Land-use issues such as retention levels of biomass on site, use of fertilizers and irrigation, species selection, and rotation length can affect the GHG footprint of wood-based biofuels (Biomass Research and Development Initiative [BRDI] 2008). In addition, given interlinkages between the global market, the development of bioenergy facility in one region might lead to land-use changes elsewhere, thus further complicating the GHG accounting process.

The GHG changes related to land-use change are dependent on several factors, which are site- and time-specific and hence differ across geographic regions. Gnansounou and Dauriat (2005) through a worldwide review of LCA studies showed that net energy gain (energy output by input ratio) from lignocellulosic biofuels ranged from 95% to 460%. A study undertaken in Brazil reported 330% net energy gain (Macedo 2004), while another study undertaken in France estimated net energy gain to be just 80% (Ecobilan 2002). This specificity and variability introduces uncertainty to data used in LCAs, making selection of a representative figure over geographically large studies questionable. There is uncertainty in GHG emissions due to land management activities, particularly for N2O (Crutzen et al. 2008). Measurement protocol for GHG emission and sequestration is prone to uncertainty. For instance, geographic misspecification may result in failure to address leakage associated with regions outside of the study area (Thompson et al. 2011).

The identification of the starting point in LCA indicating where GHG impact accounting begins is important because it determines what is considered and what is not (BRDI 2008). Indirect land-use change effects, on the other hand, are much more difficult to assess, and today there is no generally accepted methodology for determining such effects. A satisfactory methodology might be incorporated into the LCA of wood-based energy at a later date to account for such indirect impacts.

12.6.3 Sustainability Guidelines for Wood-Based Bioenergy

Forests not only produce wood for traditional uses, they also provide ecosystem services such as clean water, habitat for flora and fauna, biodiversity, hunting, fishing, and other recreational opportunities. In light of many policy initiatives favoring renewable technologies worldwide, it is likely that the demand for larger harvests and higher removal intensities might increase. Many sustainability concerns are being raised about wood biomass utilization for energy production.

Reduction of soil nutrients and soil compaction from increased woody biomass harvest could likely decrease forest productivity (Burger 2002; Minnesota Forest Resources Council [MFRC] 2007). Intensive biomass removal might affect water quality by increasing erosion, runoff, and waterway sedimentation (Neary 2002). Intensive forest management may also degrade forest habitat conditions, negatively affecting flora and fauna and adversely impacting biodiversity (Amacher et al. 2008). Land-use changes from natural forest to managed plantations might adversely affect imperiled species in certain locations. Intensive woody biomass removal might also have some negative implications for community relationships, aesthetics, and public perceptions about forestland as an integral component of forested ecosystems. The degradation of *Miombo* forests in South Africa is an example where intensive use of wood for energy through charcoal production has led to conversion of forestland to shrubland (Kutsch et al. 2011). Reduced biomass use in forest systems can result in recovery of degraded forests as well. This has been documented through a long-term Austrian forest study undertaken by Gingrich et al. (2007). However, high grading of stands generally observed during some timber harvesting might be eliminated with biomass harvesting.

European countries such as the Netherlands and the United Kingdom have recognized that overharvest of woody biomass for bioenergy/biofuel production could threaten the environment and economy. The Dutch government has even set up a commission to develop sustainability criteria for bioenergy production and to incorporate them into national renewable energy policies (Cramer Commission 2007). The New Zealand government in their Sustainable Energy Strategy Document for 2050 mandates that sustainable biofuels must not reduce indigenous biodiversity or adversely impact high-value conservation land (Ministry of Economic Development 2007). In the United States, many states are in the process of creating new guidelines or improving existing ones for forest biomass harvesting. Most of the existing guidelines are geared toward regulating timber harvesting for wood products rather than for wood-based bioenergy production (Saleh 2007). On the other hand, in many Asian countries such as China, India, the Philippines, Thailand, and Malaysia, the policy focus has been geared toward increasing production of biofuels, largely ignoring sustainability issues (O'Connell et al. 2009).

Sustainability concerns go beyond establishing woody biomass harvesting guidelines; concerns range from production processes to consumption processes—feedstock production, harvesting, transport, conversion, distribution, consumption, and waste disposal—and include issues of job creation and societal benefit distribution. Failure to identify sustainability indicators for wood-based biomass use for energy may result in negative impacts for local economies, ecological integrity, and unintended ancillary impacts on implicated people (Lal et al. 2011). Existing certification systems, such as the Forest Stewardship Council, American Tree Farm System, Bra Miljoval (2009), Sustainable Forestry Initiative, and GreenPower (2010), must revisit their criteria and indicators to safeguard site productivity, water quality, biodiversity, and reduced GHG emission for wood-based biomass use for energy production.

Potential impacts on forest ecosystems at local and regional levels will most likely challenge the research community to use research findings to update existing certification systems with guidelines on how, when, and where forest biomass removals should be conducted and how efficiently biomass can be converted to energy. Multi-stakeholder efforts such as the Roundtable on Sustainable Biofuels and Global Bioenergy are developing a set of global, science-based sustainability criteria for wood-based bioenergy. Lal et al. (2011) report nine criteria that are necessary to the pursuit of sustainable woody biomass removal: (1) forest productivity, (2) land-use change, (3) biodiversity conservation, (4) soil quality, (5) hydrologic processes, (6) profitability, (7) community benefits, (8) stakeholder participation, and (9) community and human rights.

12.6.4 Public Perception of Wood Energy

Determining the potential of the biofuel/bioenergy sector to generate renewable energy requires insight into public perception. Determination of public perception may be enabled by understanding how various stakeholders respond to the introduction and expansion of a wood-based biofuel/bioenergy sector. Stakeholders in the woody bioenergy/biofuels supply chain may include (1) local residents; (2) land developers; (3) agriculture, energy, commerce, and environmental professionals; (4) investors; (5) researchers; (6) farmers; (7) loggers; (8) pulp and paper mill representatives; and (9) nongovernmental organizations. Attributes that might aid in positive stakeholder perception include (1) enhancement of energy security, (2) reduction of GHG emissions, (3) sustainable feedstock supply, (4) carbon market emergence, (5) government commitment, (6) improved forest health, and (7) rural economic improvement. Factors that may result in unfavorable perceptions include (1) negative impacts on the traditional forest products industry, (2) reduced forest health, and (3) a lack of economic competitiveness fossil fuels and bioenergy from other nations.

Stakeholder perceptions and rankings may be inventoried and analyzed for priorities to effectively integrate such information into relevant policies (e.g., Alavalapati et al. 2009). Alternatively, the strengths, weaknesses, opportunities, and threats (SWOT) framework (e.g., Dwivedi et al. 2009), in which different stakeholders assign varying weight to the different attributes of bioenergy according to their preferences, can be used to incorporate public perceptions into policies governing wood-based biofuels.

Aguilar and Garrett (2009) conducted a survey of US public and private forest sector stakeholders to explore perceptions on items associated to the use of woody materials for energy. Based on an analysis of the views from State Foresters, State Energy Biomass Contacts, and National Council of Forestry Association Executives, Aguilar and Garrett (2009) suggest that these stakeholders deem that woody biomass from public and privately owned lands is instrumental to the development of a wood-for-energy industry and discouraged the exclusion of woody biomass from public lands from energy uses. A definition of woody biomass should not be restricted to plantations managed for biomass only but biomass treatments should be integrated as part of professionally coordinated practices that can improve forest health and stand quality. The adoption of woody biomass as a feedstock for renewable energy generation can result in more locally generated energy and generate new work opportunities among forestry professionals. Costs of harvesting and transportation can hinder the emergence of energy efforts based on woody biomass; hence, research investments in new techniques and harvesting equipment are necessary. An industry structure with scales and distribution that allows for minimum transportation distances will be another fundamental factor to this new industry. Insufficient additional revenues from sales of woody biomass can limit the willingness of landowners to open lands for biomass treatments.

Aguilar and Saunders (2011) investigated the perception of various stakeholders toward the adoption of different policy instruments promoting wood energy use in the United States and compared these views between respondents in South United States and outside the region. The study detected differences between respondents in the South United States and non-South United States in regard to the capacity of subsidies and grants and rules and regulations to be the most cost-efficient forest policy tool. At the average, respondents in South United States have less favorable views toward subsidies and grants and rules and regulations than respondents from the rest of the country. Overall, analysis of general policy instrument categories suggests that education and consultation was the best positioned to meet ecological, economic, social, and political objectives. Subsidies and grants were deemed to be the least adequate policy tool to meet energy and resource management objectives.

In the EU, Kainulainen (2012) suggests that European forest landowners have a strong favorable attitude toward the use of wood for energy. Many landowners see the creation of new markets as an opportunity for additional revenues that can justify better and more intense forest management. European forest landowners also see the benefit of coupling renewable energy objectives with reduction in GHG emissions that wood energy can effectively meet. In the United States, Gruchy et al. (2012) surveyed forest landowners in the state of Mississippi and suggest that most woody biomass feedstocks from pine plantations would be available for the production of bioenergy. Joshi and Mehmood (2011) based on data gathered from landowners in Arkansas, Florida, and Virginia concluded that their willingness to harvest wood for energy was influenced by their ownership objectives, size of the forest, and structure and composition of tree species, among others factors. Although a majority of the landowners in the study area regarded nation's dependence on imported fuel as critical for both national security and healthy economy, there is also a reported need for better education about the use of wood for energy.

The European Panel Federation (Döry 2012), nonetheless, has expressed concerns over additional market pressures for fiber demand and the deviation of wood from material (e.g., panels) to energy uses (pellets). This industrial organization has called for a revision of current subsidies for renewable energy in the EU and that only energy facilities with highest efficiencies and lowest emissions engage in power generation.

Support or lack of it for wood energy facilities is also linked to the size of facilities. Sierra Club in the United States in a guidance statement indicates that it believes that biomass projects can be sustainable, but that many are not. "We are not confident that massive new biomass energy resources are available without risking soil and forest health, given the lack of commitment by governments and industry to preservation, restoration, and conservation of natural resources" (Sierra Club, n.d.). Mainville (2011) has raised concerns over the impact of large wood pellet manufacturing facilities in Canada. Demand for large amounts of wood fiber can result in standing trees being targeted for raw materials.

The use of wood for energy can also pose ecological and human risks. Ellul (2012) reported that increase in wood removals from European forests and more use from less conventional sources will be expected to meet EU2020 targets. Increase in harvest residues and stump extraction may pose a high ecological threat to European forests. Potential options to ameliorate such risk include imports of sustainably certified wood and management of protected areas. While wood energy provides an important source of renewable energy, the potential impacts on human health must also be considered. Residential use of woody biomass, and combustion in particular, can be a significant source of pollutants. FAO (2010) estimates carbon emissions from wood-fired stoves at 1 g kg^{-1}, from crop-residue-fired stoves at 0.75 g kg^{-1}, and from dung-fired stoves at 0.25 g kg^{-1}. Particular matter in residential use such as PM2.5 can be of particular concern. Olendrzyñski (2012) has reported that PM2.5 has been linked to a premature death of about 350,000 people in 25 EU member states in the year 2000. However, advances in the design, performance, and dissemination of biomass cookstoves have been significant in recent years (REN 2012). Some advanced wood-burning cookstoves emit less particulate matter and carbon monoxide than liquefied petroleum gas.

12.7 Final Remarks

A combination of higher fossil fuel prices and public policy has fostered the growth in the use of wood to generate energy at an industrial scale. The development of technology to densify energy content of woody feedstocks has been a key driver for the greater use of wood for energy. While the use of wood has been presented as a preferable alternative to the combustion of fossil fuels, life-cycle assessment reports the benefit of the use of woody residues as part of traditional forest management compared to dedicated plantations that may displace land currently used for farming or forestry. Important global markets have developed in recent years, mainly trading of wood pellets. Global demand for industrial wood pellets has been mainly driven by demand from the EU under EU2020. Stakeholder groups deem the use of woody biomass to provide an important opportunity to encourage more active forest management, revitalize local rural economics, and provide local sources of energy. New technological developments will be necessary to reduce the cost per unit of energy generated from wood and make it more competitive. A consistent and long-term public policy will be instrumental in providing adequate market signals to promote investments in new technology and energy conversion facilities.

Wood energy is an important component of a resilient energy system and it will continue to be in the future. Energy has become an integral part of the array of products offered by the wood products industry. Growth in wood energy consumption will depend on a combination of improvements in the economics of biomass growth, harvesting, transportation, and conversion that are directly affected by public policy and also competing energy sources. The potential of any bioenergy strategy will also be highly influenced by local context including location relative to supply and demand; infrastructure, climate, and soil; land and labor availability; and social and governance structures (FAO 2008).

References

Aguilar, F.X. and H.E. Garrett. 2009. Perspectives of woody biomass for energy: Survey of state foresters, state energy biomass contacts, and national council of forestry association executives. *Journal of Forestry.* 107(6):297–306.

Aguilar, F.X., Gaston, C., Hartkamp, R., Mabee, W., and K. Skog. 2011. Chapter 9: Wood energy markets. In: *United Nations Forest Products Annual Market Review 2010–2011.* pp. 85–97. Online at http://live.unece.org/fileadmin/DAM/publications/timber/FPAMR_HQ_2010–2011.pdf

Aguilar, F.X. and A. Saunders. 2010. Policy instruments promoting wood for energy uses: Evidence from the continental U.S. *Journal of Forestry.* 108(3):132–140.

Aguilar, F.X. and A. Saunders. 2011. Attitudes toward policy instruments promoting wood-for-energy uses in the United States. *Southern Journal of Applied Forestry.* 35(2):73–79.

Aguilar, F.X., Song, N., and S. Shifley. 2011. Review of consumption trends and public policies promoting woody biomass as an energy feedstock in the U.S. *Biomass and Bioenergy.* 35:3708–3718.

Alavalapati, J. and P. Lal. 2009. Woody biomass for energy: An overview of key emerging issues. *Virginia Forests Fall* 2009:4–8.

Alavalapati, J., Lal, P., Susaeta, A., Abt, R.C., and D. Wear. In press. Forest biomass-based energy. In D.Wear, and J. Greis (Eds.) Southern Forest for Futures Project. General Technical Report. Southern Research Station, NC: U.S. Department of Agriculture, Forest Service (In Press).

Alavalapati, J.R.R., Hodges, A.W., Lal, P., Dwivedi, P., Rahmani, M., Kaufer, I., Matta, J.R., Susaeta, A., Kukrety, S., and T.J. Stevens. 2009. Bioenergy roadmap for Southern United States. Southeast Agriculture & Forestry Energy Resources Alliance (SAFER), Southern Growth Policies Board, NC.

Amacher, A.J., Barrett, R.H., Moghaddas, J.J., and S.L Stephens. 2008. Preliminary effects of fire and mechanical fuel treatments on the abundance of small mammals in the mixed-conifer forest of the Sierra Nevada. *Forest Ecology and Management*. 255:3193–3202.

APX-ENDEX. 2010. Press Release: APX-ENDEX to launch wood pellet trading in 2011. November 24, 2010. http://bepex.org/uploads/media/Press_Release_Wood_Pellets.pdf (Accessed December 10, 2011).

APX-ENDEX. 2011. *Industrial Wood Pellet Prices*. Amsterdam, the Netherlands: APX-ENDEX. http://www.apxendex.com (Accessed December 12, 2011).

Biomass Research and Development Initiative [BRDI]. 2008. Increasing feedstock productions for biofuel: Economic drivers, environmental implications, and the role of research. http://www.usbiomassboard.gov/pdfs/increasing_feedstock_revised.pdf (Accessed September 5, 2011).

Birdsey, R.A., Pregitzer, K., and A. Lucier. 2006. Forest carbon management in the United States: 1600–2100. *Journal of Environmental Quality*. 35:1461–1469.

Blottnitz, V.H. and M. A. Curran. 2007. A review of assessments conducted on bio-ethanol as a transportation fuel from a net energy, greenhouse gas, and environmental life-cycle perspectives. *Journal of Cleaner Production*. 15(7):607–619.

Bra, M. 2009. Criteria for good environmental choice-labelling: Electricity. http://www.naturskyddsforeningen.se/upload/bmv/english/bmv-electricity-crit.pdf (Accessed September 5, 2011).

Bracmort, K., Schnepf, R., Stubbs, M., and B.D. Yacobucci. 2010. Cellulosic biofuels: Analysis of policy issues for congress. CRS Report for Congress, Washington, DC, RL34738.

Bungay, H.R. 2004. Confessions of a bioenergy advocate. *Trends in Biotechnology*. 22(2):67–71.

Burger, J.A. 2002. Soil and long-term site productivity values. In J. Richardson, R. Bjorheden, P. Hakkila, A.T. Lowe, and C.T. Smith (Eds.), *Bioenergy from Sustainable Forestry: Guiding Principles and Practice*, pp. 165–189. Dordrecht, the Netherlands: Kluwer Academic Publishers.

Cocchi, M. 2011. *Global Wood Pellet Industry Market and Trade Study*. IEA Bioenergy. 190pp.

Cramer Commission. 2007. Testing framework for sustainable biomass. Final report from the project group sustainable production of biomass. http://www.lowcvp.org.uk/assets/reports/070427-Cramer-FinalReport_EN.pdf (Accessed September 4, 2011).

Crutzen, P.J., Mosier, A.R., Smith, K.A., and W. Winiwarter. 2008. N_2O release from agro-biofuel production negates global warming reduction by replacing fossil fuels. *Atmospheric Chemistry and Physics Discussions*. 8:389–395.

Davis, A.A. and C.C. Trettin. 2006. Sycamore and sweetgum plantation productivity on former agricultural land in South Carolina. *Biomass and Bioenergy*. 30:769–777.

Dickmann, D.I. 2006. Silviculture and biology of short-rotation woody crops in temperate regions: Then and now. *Biomass and Bioenergy*. 30:696–705.

Döry, L. 2012. European panel federation and wood energy. UNECE/FAO policy debate on wood energy: Is it good or bad? 8 May 2012, Geneva, Switzerland. Available at: www.unece.org/energy-debate-2012.html

Drew, A.P., Zsuffa, L., and C.P. Mitchell. 1987. Terminology relating to woody plant biomass and its production. *Biomass*. 12:79–82.

Dwivedi, P. and J.R.R. Alavalapati. 2009. Stakeholders' perceptions on forest biomass-based bioenergy development in the southern US. *Energy Policy*. 37:1999–2007.

Dwivedi, P., Alavalapati, J.R.R., and P. Lal. 2009. Cellulosic ethanol production in the United States: Conversion technologies, current production status, economics, and emerging developments. *Energy for Sustainable Development*. 13:174–182.

Ecobilan. 2002. Energy balances and GHG of motor biofuels production pathways in France. Prepared for ADEME and DIREM, by Ecobilan PWC, France.

Egnell, G. 2011. Is the productivity decline in Norway spruce following whole-tree harvesting in the final felling in boreal Sweden permanent or temporary? *Forest Ecology and Management*. 261:148–153.

Ellul, D. 2012. European Forest Sector Outlook Study II-the Wood Energy Scenario. UNECE/FAO policy debate on wood energy: Is it good or bad? 8 May 2012, Geneva, Switzerland. Available at: www.unece.org/energy-debate-2012.html (Accessed January 30, 2013).

Energy Information Administration. 2012. Most states have renewable portfolio standards. Online at http://205.254.135.7/todayinenergy/detail.cfm?id = 4850

English, B.C., De La Torre Ugarte, D.G., Jensen, K., Hellwinckel, C., Menard, J., Wilson, B., Roberts, R., and M. Walsh. 2006. 25% renewable energy for the United States by 2025: Agricultural and economic impacts. University of Tennessee Agricultural Economics. http://www.agpolicy.org/ppap/REPORT%2025x25.pdf. (Accessed February 1, 2012).

EPA. 2010. Prevention of significant deterioration and title V greenhouse gas tailoring rule. See www.epa.gov/nsr/documents/20100413final.pdf. (Accessed July 28, 2013).

Eriksson, E., Gillespie, A., Gustavsson, L., Langvall, O., Olsson, M., Sathre, R., and K. Stendahl. 2007. Integrated carbon analysis of forest management practices and wood substitution. *Canadian Journal of Forest Research*. 37(3):671–681.

Eriksson, L.N. and L. Gustavsson. 2008. Biofuels from stumps and small roundwood—Costs and CO_2 benefits. *Biomass and Bioenergy*. 32:897–902.

European Commission. 2009. Directive 2009/28/EC of 23 April 2009 on the promotion of the use of energy from renewable sources and subsequently amending and repealing directives 2001/77/EC and 2003/30/EC.

European Commission. 2010. Report from the Commission on indirect land-use change related to biofuels and bioliquids. 14pp.

European Commission. 2011. Energy 2020. A strategy for competitive, sustainable and secure energy. 24pp. Online at http://ec.europa.eu/energy/publications/doc/2011_energy2020_en.pdf (Accessed July 28, 2013).

European Pellet Council. 2011. Handbook for the certification of wood pellets for heating purposes. 33pp. Online at http://www.enplus-pellets.eu/downloads/enplus-handbook/ (Accessed July 28, 2013).

European Review Energy Council. 2009. RES 2020 monitoring and evaluating the RES directives implementation in EU-27 and policy recommendations report: Renewable Energy Policy Review Finland. Available at http://www.cres.gr/res2020 (Accessed July 28, 2013).

Eurostat. 2011. External trade detailed data: EU27 trade since 1998 by CN8. Available via: http://epp.eurostat.ec.europa.eu/portal/page/portal/eurostat/home (Accessed July 28, 2013).

Faaij, A. and J. Domac. 2006. Emerging international bioenergy markets and opportunities for socio-economic development. *Energy for Sustainable Development*. 1:7–19.

FAO. 2008. Forests and energy—Key issues. Forestry Paper 154. 56pp.

FAO. 2010a. Criteria and indicators for sustainable woodfuels. Food and Agriculture Organization of the United Nations. FAO Forestry Paper 160.

FAO. 2010b. National, regional and global markets for biofuels. In: What woodfuels can do to mitigate climate change. FAO Forestry Paper 162, pp. 17–24. Food and Agriculture Organization of the United Nations.

FAO. 2011. The state of forest resources: A regional analysis. In State of the World's Forests. Food and Agriculture Organization of the United Nations. http://www.fao.org/docrep/013/i2000e/i2000e00.htm (Accessed January 30, 2012).

Flynn, P. and A. Kumar. 2005. Trip Report: Site Visit to Alholmens 240 MW Power Plant, Pietarsaari, Finland. August 29 to September 2, 2005, funded by BIOCAP Canada Foundation and the Province of British Columbia. 13pp. Online at www.biocap.ca/files/reports/MPB_Study_Finland_Trip_Report.pdf

Forisk Consulting. 2011. *Wood Bioenergy*. 3(4). http://www.forisk.com/UserFiles/File/WBUS_Free_201105.pdf (Accessed July 28, 2013).

Forisk Consulting. 2012. *Wood Bioenergy US*. 4(1). Online at http://www.forisk.com/Forisk-Bioenergy-Research-v-42.html (Accessed July 28, 2013).

Gan, J. and C.T. Smith. 2006. Availability of logging residues and potential for electricity production and carbon displacement in the USA. *Biomass and Bioenergy*. 30:1011–1020.

Gibson, L. 2011. Conference addresses burning questions. *Pellet Mill Magazine*. 1(2):25–28.

Gingrich, S., Erb, K-H., Krausmann, F., Gaube, V., Haberl, H. 2007. Long-term dynamics of terrestrial carbon stocks in Austria: A comprehensive assessment of the time period from 1830 to 2000. *Regional Environmental Change*. 7:37–47.

Gnansounou, E. and Dauriat, A. 2005. Energy balance of bioethanol: A synthesis, Swiss Federal Institute of Technology & ENERS Energy Concept. Available at http://www.eners.ch/downloads/eners_0510_ebce_paper.pdf (Accessed July 28, 2013).

Green Power. 2010. National green power accreditation program: Program rules. http://www.greenpower.gov.au/~/media/Business%20Centre/Become%20 Accredited/National%20GreenPower%20Program%20Rules%20Version%206%20Jan%202010.pdf (Accessed September 4, 2011).

Greig, B.J.W. 1984. Management of East England pine plantations affected by *Heterobasidion annosum* root rot. *European Journal of Forest Pathology*. 14:392–397.

Gruchy, S., Grebner, D., Munn, I.A., Joshi, O., and A. Hussain. 2012. An assessment of nonindustrial private forest landowner willingness to harvest woody biomass in support of bioenergy production in Mississippi: A contingent rating approach. *Forest Policy and Economics*. 15:140–145.

Guo, Z., Sun, C., and Q. Grebner. 2007. Utilization of forest derived biomass for energy production in the U.S.A.: Status, challenges, and public policies. *International Forestry Review*. 9(3):748–758.

Gustavsson, L., Holmberg, J., Dornburg, V., Sathre, R., Eggers, T., Mahapatra, K., and G. Marland. 2007. Using biomass for climate change mitigation and oil reduction. *Energy Policy*. 35:5671–5691.

Hakkila, P. 2004. *Developing Technology for Large-Scale Production of Forest Chips. Wood Energy Technology Programme 1999–2003*. Helsinki, Finland: National Technology Agency.

Hakkila, P. 2006. Sustainable production systems for bioenergy: Impacts on forest resources and utilization of wood for energy. *Biomass and Bioenergy*. 30:281–288.

Hakkila, P. and M. Aarniala. 2004. Stumps—an unutilised reserve. Online at www.tekes.fi/eng/publications/kannotengl1.pdf (Accessed July 29, 2013).

Harper, R.J., Sochacki, S.J., Smettem, K.R.J., and N. Robinson. 2010. Bioenergy feedstock potential from short-rotation woody crops in a dryland environment. *Energy Fuels*. 24:225–231.

Hartsough, B.R., E.S. Drews, J.f. McNell, T.A. Durston, and B.J. Stokes. 1997. Comparison of mechanized systems for thinning ponderosa pine and mixed conifer stands. *Forest Products Journal*. 47(11/12):59–68.

Helmisaari H., Hanssen, K.H., Jacobson, S., Kukkola, M., Luiro, J., Saarsalmi, A., Tamminen, P., and B. Tveite. 2011. Logging residue removal after thinning in Nordic boreal forests: Long-term impact on tree growth. *Forest Ecology and Management*. 261:1919–1927.

Hillring, B. 2006. World trade in forest products and wood fuel. *Biomass and Bioenergy*. 30:815–825.

Hinchee, M., Rottmann, W., Mullinax, L., Zhang, C., Chang, S., Cunningham, M., Pearson, L., and N. Nehra. 2009. Short-rotation woody crops for bioenergy and biofuels applications. In D. Tomes, D. Songstad, and P. Lakshmanan (Eds.), *Biofuels: Global Impact on Renewable Energy, Production Agriculture, and Technological Advancements*, pp.139–157. New York: Springer, LLC.

Holley, A.G., Taylor, E.L., and M.A. Blazier. 2010. A comparison of woody biomass comminution systems in the Western Gulf region. *The Forestry Source*. 15(11):12–13.

Hope, G.D. 2007. Changes in soil properties, tree growth, and nutrition over a period of 10 years after stump removal and scarification on moderately coarse soils in interior British Columbia. *Forest Ecology and Management*. 242(2/3):625–635.

Hubbard, W., Biles, L., Mayfield, C., and S. Ashton (eds.). 2007. Sustainable forestry for bioenergy and bio-based products: Trainers curriculum notebook. Southern Forest Research Partnership, Inc. www.forestbioenergy.net (Accessed January 31, 2012).

International Energy Agency (IEA). 2007. *Global Wood Pellets Markets and Industry: Policy Drivers, Market Status and Raw Material Potential*. Amsterdam, the Netherlands: International Energy Agency. 120p.

International Energy Agency (IEA). 2009. Bioenergy—A sustainable and reliable energy source. IEA Bioenergy: ExCo: 2009:06. 107pp. http://www.ieabioenergy.com/libitem.aspx?id=6479 (Accessed July 27, 2013).

International Energy Agency (IEA). 2011. Technology roadmap: Biofuels for transport. Organization of economic cooperation and development international energy agency. http://www.iea.org/papers/2011/biofuels_roadmap.pdf (Accesed July 29, 2013).

International Organization for Standardization. 2012. TC 238–Solid biofuels. Online at http://www.iso.org/iso/iso_catalogue/catalogue_tc/catalogue_tc_browse.htm?commid=554401& published = on & development = on

Jackson, S., Rials, T., Taylor, A.M., Bozell, J.G., and K.M. Norris. 2010. Wood to energy: A state of the science and technology report. University of Tennessee and U.S. Endowment for Forestry and Communities Report. Knoxville, TN: The University of Tennessee.

Johansson J.L.J., Gullberg, T., and R. Bjorheden. 2006. Transport and handling of forest energy bundles—Advantages and problems. *Biomass and Bioenergy*. 30:334–341.

Joshi, O. and S. Mehmood. 2011. Factors affecting nonindustrial private forest landowners' willingness to supply woody biomass for bioenergy. *Biomass and Bioenergy*. 35(1):186–192.

Kainulainen, A. 2012. Forest owners position on wood energy – Central Union of agricultural producers and forest owners. UNECE/FAO policy debate on wood energy: Is it good or bad? 8 May 2012, Geneva, Switzerland. Available at: www.unece.org/energy-debate-2012.html

Karjalainen, T., Asikainen, A., Ilavsky, J., Zamboni, R., Hotari, K.-E., and Röser, D. 2004. Estimation of energy wood potential in Europe, Working Paper of the Finnish Forest Research Institute, Vantaa, Finland. Online at www.metla.fi/julkaisut/workingpapers/2004/mwp006.htm (Accessed July 29, 2013).

Kutsch, W.L., Merbold, L., Ziegler, W., Mukelabai, M.M., Muchinda, M., Kolle, O., and R.J. Scholes. 2011. The charcoal trap: Miombo forests and the energy needs of people. *Carbon Balance and Management*. 6:5. doi:10.1186/1750–0680–6–5.

Lal, P., Alavalapati, J., Marinescu, M., Matta, J.R., Dwivedi, P., and A. Susaeta. 2011. Developing sustainability indicators for woody biomass harvesting in the United States. *Journal of Sustainable Forestry*. 30(8):736–755.

Laitila, J., A. Asikainen, and Y. Nuutinen. 2007. Forwarding of whole trees after manual and mechanized felling bunching in pre-commercial thinnings. *International Journal of Forest Engineering*. 18(2):29–39.

Laitila, J., Ranta, T., and A. Asikainen. 2008. Productivity of stump harvesting for fuel. *International Journal of Forest Engineering*. 19(2):37–47.

Langerud, B., Stordal, S., Wiig, H., and M. Orbeck. 2007. Bioenergy in Norway: Potentials, markets and policy instruments. (in Norwegian). OF-rappert nr.17/2007, 98pp.

LeDoux, C.B. and N.K. Huyler. 2001. Comparison of two cut-to-length harvesting systems operating in eastern hardwoods. *International Journal of Forest Engineering*. 12(1):53–59.

Luo, Z. 1998. Biomass energy consumption in China. *Wood Energy News*. 12(3):3–19.

Mainville, N. 2011. *Fuelling a BioMess: Why Burning Trees for Energy Will Harm People, the Climate and Forests*. Montreal, Quebec, Canada: Greenpeace. 38pp.

Macedo, I. 2004. *Assessment of Greenhouse Gas Emissions in the Production and Use of Fuel Ethanol in Brazil*. São Paulo, Brazil: Government of the State of São Paulo.

Mead, D.J. 2005. Forests for energy and the role of planted trees. *Critical Reviews in Plant Sciences*. 24:407–421.

Meek, P. and J.A. Plamondon. 1996. Effectiveness of cut-to-length harvesting at protecting advance regeneration. Technical Report. TN-242. FERIC, Pointe Claire, Montreal, Quebec, Canada. 12pp.

Melillo, J., Reilly, J.M., Kicklighter, D.W., Gurgel, A.C., Cronin, T.W., Paltsev, S., Felzer, B.S., Wang, S., Sokolov, A.P., and C.A. Schlosser. 2009. Indirect emissions from biofuels: How important? *Science*. 326(5958):1397–1399.

Minnesota Forest Resources Council (MFRC). 2007. Biomass harvesting guidelines for forestlands, brushlands and open lands. Retrieved from http://cemendocino.ucdavis.edu/files/17408.pdf (Accessed July 3, 2011).

National Renewable Energy Laboratory (NREL). 1996. An assessment of management practices of wood and wood-related wastes in the urban environment. NREL/TP-430–20696, 78pp. http://www.nrel.gov/docs/legosti/fy96/20696.pdf (Accessed July 27, 2013).

Neary, D.G. 2002. Hydrologic values. In J. Richardson, R. Bjorheden, P. Hakkila, A.T. Lowe, and C.T. Smith (Eds.), *Bioenergy from Sustainable Forestry: Guiding Principles and Practice*, pp. 190–215. Dordrecht, the Netherlands: Kluwer Academic Publishers.

Netherland Agency. 2011. Bioenergy status document 2010. Published by Ministry of Economic Affairs, Agriculture, and Innovation Publication no. 2DENB1107. http://www.mvo.nl/Portals/0/duurzaamheid/biobrandstoffen/nieuws/2011/04/BioEnergy%20Status%20Document%202010-UK.pdf

New Zealand Parliament. 2009. Sustainable biofuel bill. Available at http://www.parliament.nz/resource/0000082892 (Accessed June 23, 2011).

O'Connell, D., Braid, A., Raison, J., Handberg, K., Cowie, A., Rodriguez, L., and B. George. 2009. Sustainable production of bioenergy: A review of global bioenergy sustainability frameworks and assessment systems. Australian Government. RIRDC Publication No. 09/167. Rural Industries Research and Development Corporation, Canberra, Australia.

Office of the Gas and Electricity Markets. 2011. FIT payment rate table with year 1 & 2 retail price index adjustments. Online at http://www.ofgem.gov.uk. (Accessed July 29, 2013).

Olendrzyñski, K. 2012. Wood burning leads to emissions of air pollutants and is harmful to human health and the environment. UNECE/FAO policy debate on wood energy: Is it good or bad? 8 May 2012, Geneva, Switzerland. Online at http://www.unece.org/energy-debate-2012.html

Pandey, D. 2002. *Fuelwood Studies in India: Myth and Reality*. Jakarta, Indonesia: Center for International Forestry Research. 93pp.

Parikka, M. 2004. Global biomass fuel resources. *Biomass and Bioenergy*. 27:613–620.

Pellet Fuels Institute. 2010. Pellet fuels institute homepage. http://www.pelletheat.org/3/residential/fuelAvailability.cfm.(Accessed July 6, 2011).

Ponsse, O. 2005. The cut-to-length harvesting system. Online at www.ponsse.com.

Ranta, T. 2002. Logging residues from regeneration fellings for biofuel production—A GIS based availability and cost analysis. *Acta Universitatis Lappeenrantaensis*. 128:182.

REN21. 2012. Renewables 2012. Global Status Report. Paris, France, 172pp.

Saleh, D.E.S.A.M. (2007). Harvesting forest biomass for energy in Minnesota: An assessment of guidelines, costs and logistics (Doctoral dissertation, University of Minnesota). http://www.ftp://ftp.dnr.state.mn.us/pub/Biomass%20Harvest/Biomass%20related%20research/Biomass%20Literature/dalia_complete%20dissertation.pdf (Accessed February 4, 2012).

Scott, D.A. and T.J. Dean. 2006. Energy trade-offs between intensive biomass utilization, site productivity loss, and ameliorative treatments in loblolly pine plantations. *Biomass and Bioenergy*. 30:1001–1010.

Scott, D.A. and A. Tiarks. 2008. Dual-cropping loblolly pine for biomass energy and conventional wood products. *Southern Journal of Applied Forestry*. 32(1):33–37.

Searchinger, T., Heimlich, R., Houghton, R.A., Dong, F., Elobeid, A., Fabiosa, J., Togkoz, S., and T.H. Yu. 2008. Use of U.S. croplands for biofuels increases greenhouse gases through emissions from land use change. *Science*. 319:1238–1240.

Sierra Club. n.d. Sierra club conservation policies: Biomass guidance. Online at http://www.sierraclub.org/policy/conservation/biomass.aspx

Sikkema, R., Junginger, H.M., Pichler, W., Hayes, S., and A.P.C. Faaij. 2010. The international logistics of wood pellets for heating and power production in Europe. *Biofuel Bioproducts and Biorefining*. 4:132–153.

Sikkema, R., Steiner, M., Junginger, M., Hiegl, W., Hansen M.T., and A. Faaij. 2011. The European wood pellet markets: current status and prospects for 2020. *Biofuels, Bioproducts and Biorefining*. 3:250–278.

Smeets, E.M.W. and A.P.C. Faaij. 2007. Bioenergy potentials from forestry in 2050. *Climatic Change*. 81:353–390.

Solid Waste Association of North America. 2002. Successful approaches to recycling urban wood waste. General Technical Report FPL-GTR-133. Madison, WI: U.S. Department of Agriculture, Forest Service, Forest Products Laboratory. 20pp.

Song, N., Aguilar, F.X., Shifley, S., and M. Goerndt. 2012. Factors affecting wood energy consumption by U.S. households. *Energy Economics*. 34:389–397.

Spelter, H. and D. Toth. 2009. The North American wood pellet sector. Research Paper RP-FPL-656. Madison, WI: USDA Forest Service, Forest Products Laboratory. 21pp. www.fpl.fs.fed.us/products/publications/specific_pub.php?posting_id = 17545

Stevens, P.A. and M. Hornung. 1990. Effect of harvest intensity and ground flora establishment in inorganic-N leaching from a Sitka spruce plantation in north Wales, UK. *Biogeochemistry*. 10:53–65.

Swedish Energy Agency. 2010. Energy in Sweden. 141pp. Online at http://www.energimyndigheten.se/en/Facts-and-figures1/Publications/ (Accessed July 28, 2013).

Taylor, J.A., Herr, A., and D.R. Farine. 2010. Wood waste disposed of in landfills in Australia: An evaluation of the amount and potential use for bioenergy. *Proceedings of the Third International Symposium on Energy from Biomass and Waste*. Venice, Italy, November 8–11, 2010.

Thies, W.G. and Westlind, D.J. 2005. Stump removal and fertilization of five *Phellinus weirii*-infested stands in Washington and Oregon affect mortality and growth of planted Douglas-fir 25 years after treatment. *Forest Ecology and Management*. 219:242–258.

Thompson, W., Whistance, J., and S. Meyer. 2011. Effects of US biofuel policies on US and world petroleum product markets with consequences for greenhouse gas emissions. *Energy Policy*. 39:5509–5518.

U.K. Department of Energy and Climate Change. 2011. Renewable Heat Incentive. https://www.gov.uk/renewable-heat-premium-payment (Accessed July 28, 2013).

UNECE/FAO. 2011. Joint wood energy enquiry (JWEE) 2009. Geneva, United Nations. Available at www.unece.org/forests/jwee.html

US Department of Agriculture. 2011. The biomass crop assistance program (BCAP)—Final rule provisions. http://www.fsa.usda.gov/FSA/newsReleases?area = newsroom&subject = landing& topic = pfs&newstype = prfactsheet&type = detail&item = pf_20101021_consv_en_bcap.html

USDOE Energy Information Administration. 2011. Annual energy outlook. 2011. Report DOE/EIA-0383(2011). http://www.eia.gov/forecasts/aeo/ (Accessed June 13, 2011).

U.S. Public Law 109–58. 2005. Energy Policy Act of 2005. August 8, 2005. Available online at www.epa.gov/oust/fedlaws/publ_109–058.pdf (Accessed March 10, 2008).

Väkevä, J., A. Kariniemi, J. Lindroos, A. Poikela, J. Rajamäki, and K. Uusi-Pantti. 2003. Puutavaran metsäkuljetuksen ajanmenekki (Time consumption of roundwood forwarding). Metsätehon raportti 123 (in Finnish).

Van Loo, S. and J. Koppejan (Eds.). 2008. *Handbook of Biomass Combustion and Co-Firing*. London, U.K.: Earthscan.

Walmsley, J.D. and D.L. Godbold. 2010. Stump harvesting for bioenergy: A review of the environmental impacts. *Forestry*. 83(1):17–38.

Walmsley, J.D., Jones, D.L., Reynolds, B., Price, M.H., and J.R. Healey. 2009. Whole tree harvesting can reduce second rotation forest productivity. *Forest Ecology and Management*. 257:1104–1111.

Ylitalo, E. 2006. Puupolttoaineiden käyttö energian tuotannossa 2005. (Use of wood fuels in Finland 2005). Metsätilastotiedote 820. 8pp. (in Finnish).

Zbořil and Pesci. 2011. Preliminary draft opinion of the Consultative Commission on Industrial Change (CCMI) on the opportunities and challenges for a more competitive European woodworking and furniture sector. CCMI/088 European woodworking and furniture sector. Online at www.toad.eesc.europa.eu (Accessed July 29, 2013).

Section IV

Capability Development and Strategic Imperatives for the Forest Sector

13

Current and Future Role of Information Technology in the Global Forest Sector

Taraneh Sowlati

CONTENTS

13.1 Introduction

Information technology (IT) has advanced rapidly during the past two decades and has impacted all aspects of our lives. IT, or its extended synonym information and communication technology (ICT),* refers to hardware, software, and telecommunication networks used for collecting, storing, processing, and transmitting information (World Bank 2012). New developments in IT, especially the advent of the Internet, have created significant opportunities for business competitiveness. IT has been recognized as a major enabler of business growth and innovation (Koellinger 2008; OECD 2011).

IT investments are important at the firm, industry, and national levels and can improve productivity at all these levels (Harder 2001; OECD 2011). A number of studies

---

* Throughout this chapter, we use the term IT (not ICT) for simplicity.

in Organization for Economic Cooperation and Development* (OECD) countries concluded that IT investments were the main source of economic productivity growth in
these countries (OECD 2011). For example, during the past decade, more than half of
the labor productivity in the total economy in Denmark, Switzerland, and Belgium was
associated with IT investments (OECD 2011). A recent study (Dimelis and Papaioannou
2010) analyzed the potential effects of foreign direct investments and IT investments on
productivity growth of 42 developing and developed countries using panel data during the period of 1993–2001. Their results showed a positive and significant impact of
IT investments on productivity growth in both types of countries; however, the impact
was more pronounced among developing countries with the highest effect observed in
China and India. Several additional studies have found a positive relationship between IT
investments and productivity growth at the firm level (Brynjolfsson and Hitt 1996; Li and
Ye 1999) and industry level (OECD 2011; Hetemäki and Nilsson 2005). Empirical evidence
has also shown a positive correlation between IT capital investments and labor productivity in 114 countries (Harder 2001).

Although IT investments have increased in both developed and developing countries,
the diffusion of IT differs among countries and regions. While IT investments have accelerated among OECD countries during the past two decades, the share of IT investments
was particularly high in the United States, Sweden, Denmark, Great Britain, and New
Zealand (OECD Factbook 2011–2012). IT investments accounted for about one-third of
capital investments in the United States in 2009 (OECD Factbook 2011–2012). In Canada, IT
investments have shown an average annual growth of 6.9% from 1981 to 2010 (CSLS 2011).

The utilization and role of IT differ among industry sectors. Although all sectors employ
some form of IT in their operations and activities, service industries tend to be more concentrated on IT use, with higher portions of their total investments being IT investments
compared to goods-producing industries. Figure 13.1 presents the share of IT investments
in total investments by industry in Canada and the United States in 2009 (CSLS 2011). IT
investments in "Information and Cultural Industries," "Management of Companies and
Enterprises," and "Professional, Scientific, and Technical Services" account for more than
50% of their total investments. "Wholesale Trade" and "Finance and Insurance" also have
a high share of IT investments relative to total investments. Goods-producing industries
are much less IT intensive. For example, IT investments in "Agriculture, Forestry, Fishing
and Hunting" and "Mining and Oil and Gas Extraction" account for less than 5% of these
sectors' total investments.[†]

Previous research has shown low levels of IT investment and adoption specific to the
forest sector (Kozak 2002; Vlosky et al. 2002; Vlosky and Smith 2003; Arano 2008) and
relative to the manufacturing sector on average (Karuranga et al. 2006; Hewitt et al.
2011). However, increased efficiency and productivity have resulted for those forest
sector companies that have adopted IT into their business practices (Hetemäki and
Nilsson 2005). For example, a primary reason for improvements in labor productivity of

---

* "The mission of the OECD is to promote policies that will improve the economic and social well-being of
  people around the world" (www.oecd.org). It has 34 member countries in North and South America, Europe,
  and the Asia-Pacific region.
† Note that based on the North American Industry Classification System (NAICS), data on forestry and logging
  (NAICS 113) are reported within "Agriculture, Forestry, Fishing and Hunting," while data on wood products
  manufacturing (NAICS 321), paper manufacturing (NAICS 322), and furniture and related product manufacturing (NAICS 337), which include wood furniture as well, are reported within the "Manufacturing" data.
  Therefore, the IT investment share reported for forestry in Figure 13.1 does not represent that of the forest
  sector, which is aggregated in the manufacturing sector.

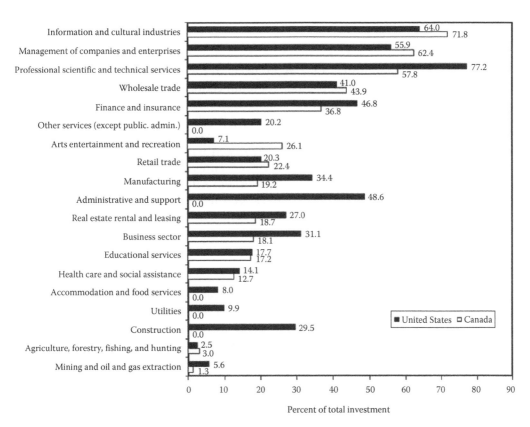

FIGURE 13.1
Total IT investment as a share of total investment by industry in Canada and the United States, 2009. (From CSLS (Centre for the Study of Living Standards), Database of information and communication technology (ICT) investment and capital stock trends: Canada vs. United States, accessed on January 3, 2012, from http://www. csls.ca/data/ict.asp, 2011.)

paper mills in OECD countries was the increased use of IT in those firms (Hetemäki and Nilsson 2005, Chapter 2).

Rapid advances in IT have brought many changes to the forest industry from forest management to raw material supply, logistics, production, marketing, and customer service of forest products. IT has been used in the forest sector to automate and improve operations and processes within firms and between exchange partners (Hetemäki and Nilsson 2005; Stennes et al. 2006) and aid managers in their decision making (Janssen et al. 2004). IT applications have been used in a wide range of areas including forest planning, product design and development, price and market forecasting, production planning and scheduling, and supply chain management (SCM) to name a few (Hetemäki and Nilsson 2005). Utilization of new and advanced IT technologies and applications, such as global positioning systems (GPSs), satellite photography, enterprise resource planning (ERP), customer relationship management (CRM), and e-business applications, has impacted and will continue to transform the global forest sector.

The rest of the chapter is organized as follows. First, IT applications in forestry, including forest management and forest products, are explained. Next, e-business adoption in the forest sector is discussed. And finally, cloud computing and its benefits for small- and medium-sized forest products firms are described.

13.2 IT in the Forest Sector

IT applications have been utilized for many years in forest management and services, forest products, and in exchange partner communications in supply chains (Hetemäki and Nilsson 2005). In forest management and services, IT applications have been used for planning, assessment, and optimization to provide support for decision making (Hetemäki and Nilsson 2005; Weintraub et al. 2007; D'Amours et al. 2008, Chapters 2 and 5). In forest product, IT applications have been used to automate processes and improve efficiency (Hetemäki and Nilsson 2005; Stennes et al. 2006) and aid managers in decision making (Janssen et al. 2004). IT applications in forest products cover applications including accounting, human resources, product design and development, sales and marketing, production planning and scheduling, logistics, and SCM (Hetemäki and Nilsson 2005).

13.2.1 IT in Forest Management and Forest Services

Strategic forest planning has a long time horizon with unique decision influences to consider including broad categories of values (economic, social, and environmental), different potential forest uses, and different viewpoints from a wide range of public and private stakeholders. The use of IT-based multicriteria decision-making approaches (Weintraub et al. 2007), decision support systems (DSSs), expert systems, artificial intelligent approaches, and optimization models (Weintraub and Epstein 2002; Ronnqvist 2003; D'Amours et al. 2008) has all significantly impacted the quality and accuracy of decisions in forest management.

For example, numerous models have been developed to support forest managers and public forest organizations in decision making on harvesting and road construction and maintenance (Kent et al. 1991; Andalaft et al. 2003). In another example, FORPLAN in the United States and FOLPI in New Zealand are two optimization softwares that have been used for strategic forest planning (Ronnqvist 2003). FORPLAN was designed and developed by the USDA Forest Service to help forest planners evaluate and analyze the economic and production trade-offs of different forest management plans (Kent et al. 1991). FOLPI (Garcia 1984; Johnson 1986) was developed by the New Zealand Forest Service to determine optimal harvesting strategies. Other examples are the use of a multicriteria decision-making method for land-use allocation in a forest reserve in the Philippines (Bantayan and Bishop 1998), a multicriteria decision-making approach used to assist in managing a public forest in Finland (Laukkanen et al. 2004), and a DSS developed by Lexer et al. (2005) to facilitate consultation given by forest authorities to small forest landowners in southern Austria. This DSS supports forest planning, taking into account landowner expectations and preferences in evaluating forest management alternatives. In a more recent application, Spectrum, developed by the USDA Forest Service (2012), is a computer tool used to assist decision makers in forest planning by displaying the outcomes of management options and strategies in terms of their environmental and socioeconomic implications.

Mathematical models are also used in tactical harvest and transportation planning to help decide which areas to harvest, when to harvest them, and optimal road locations. At the operational harvesting level, which has a time horizon less than a year, other salient decisions include which bucking pattern to use, which logs to send to each destination to optimally satisfy demand, and which machinery to use for harvesting (Epstein et al. 1999). Daily scheduling of trucks for transporting logs from the forest roadside to different destinations is also done using mathematical simulation modeling (Weintraub et al. 1996).

Many Chilean and Brazilian companies have been using OPTICORT and PLANEX, which are linear programming optimization systems, to support tactical harvest planning decisions (Weintraub and Epstein 2002). OPTICORT (Epstein et al. 1999) supports decisions on which areas to harvest, which bucking patterns to use, which products to send to each destination, and which harvesting machinery and trucks to use. Decisions on the allocation of harvesting machinery to the scheduled harvest areas and construction of secondary roads are supported in PLANEX (Epstein et al. 1999).

In forest management, information about forest area, volume of trees, yield and growth, soil type, topography, etc., are all needed for overall planning and administration. Advances in IT have led to less expensive and faster computer systems, reliable mass storage devices, advanced real-time graphical representations of decisions, and high-quality scanning devices to collect, store, and analyze vast amounts of data and information.

Geographic information systems (GISs) and GPSs have radically changed forest management. GIS is a system of hardware and software used for storage, retrieval, mapping, and analysis of geographic data. GPS is a radio navigation system that allows land, sea, and airborne users to determine their exact location, velocity, and time 24 h a day, in all weather conditions, anywhere in the world. The capabilities of today's system render other well-known navigation and positioning "technologies"—namely, the magnetic compass, the sextant, the chronometer, and radio-based devices—impractical and obsolete. GPS is used to support a broad range of military, commercial, and consumer applications. GIS, GPS, and remote sensing have played important roles in forest management by capturing and mapping geographic locational data. Advanced applications are used to determine forest acreage, stand volume, crown cover, and health. These technologies improve the ability to collect high-quality data and develop improved forest plans (Hetemäki and Nilsson 2005, Chapters 2 and 5).

Today, many forestry organizations and governmental agencies rely on GIS, GPS, and remote sensing technologies to support forest planning, valuation, fire risk assessment, and management. Various independent companies provide image data retrieval and image processing services to the forest sector. One example of these companies is Satellite Imaging Corporation (Satellite Imaging Corporation 2012a), which has performed data and mapping projection projects in Africa, Europe, North America, South America, the Middle East, and Southeast Asia (Satellite Imaging Corporation 2012b). The data and information provided by remote sensing and GIS can be used for a variety of purposes in applications such as fire and emergency mapping, forest management, environmental analysis, land cover and detection, and forest protection (Satellite Imaging Corporation 2012c).

Other technologies, such as bar coding and radio frequency identification (RFID), have been employed for identifying individual standing trees, tracking logs from harvest to domestic or export mills, and tracking rough, semifinished, and finished forest products through the chain of custody from the mill to the consumer. Log identification and tracking can help reduce the illegal log trade.

In addition to applying technology in log trade, many forest products companies have received chain-of-custody certification for their products. In order to have an effective chain-of-custody system, logs and processed wood products must be tracked and identified through supply chains. There are a variety of marking or labeling options for logs and processed wood products ranging from painted markings to paper or plastic bar-coded tags to RFID tags and smart cards (Dykstra et al. 2002). Durable, weatherproof plastic bar-coded labels are currently the most utilitarian and cost-effective methods available to use in chain-of-custody systems, since they can be scanned electronically and information can be transmitted through the supply chain to buyers and sellers. Bar codes are stapled to logs or processed wood products and only a small percentage of labels gets detached during transportation. RFID tags are too expensive to be used in chain-of-custody systems for high piece count wood products (Dykstra et al. 2002). However, in the past few years, specialized RFID tags have been developed that can be embedded into trees or logs, manually or by machine. Some of these tags are made of biodegradable materials, so they can be safely destroyed in the primary breakdown process or be ground to make chips for pulp and paper production (Wasserman 2011).

In one example, in 2009, the Forestry Department of Peninsular Malaysia (FDPM) completed a trial of timber tracking and forest management using RFID. The impetus for this trial was the European Union (EU), an importer of Malaysian forest products that requires transparent control and monitoring processes to be instituted by topical wood product export trade partners to combat illegal logging. One requirement of the agreement with Malaysia is a national timber-tracking system to improve traceability in the supply chain. In the trial, all trees tracked through the chain of custody were tagged using RFID tags either stapled or nailed to a tree or log. None of the tags were damaged during the process, despite the fact that they lacked special protective coverings. Four RFID-enabled hand-held computers, running data-capture software, were used in the field to read a unique ID reference number encoded to each tag. Finally, RFID-enabled devices were utilized to confirm the ID number at the various checkpoints along the supply chain as trees were felled and as logs were processed (Friedlos 2009).

13.2.2 IT in Forest Products

The Internet has dramatically increased the implementation of e-business and e-commerce in the forest products industry. The majority of forest products firms have used IT applications in a wide range of activities including front office and back office business activities. Front office applications refer to applications that enable the company to communicate and connect with suppliers, partners, customers, and clients such as marketing, sales, and customer service applications. Back office applications are applications related to internal processes in an organization such as product design and development, production, accounting, and human resource management.

Table 13.1 shows the classification of business software applications based on International Data Corporation's (IDC) software taxonomy (Heiman and Byron 2005). Consumer applications, used for education, entertainment, personal finance and taxes, identified in ICD's software taxonomy are not shown since they are for personal, not business, use.

IT applications can also be classified based on inter- and intraorganizational business processes (Kalakota and Robinson 2003). Interorganizational (between businesses) IT applications include those applications that help companies manage their suppliers and orders and sales and distribution of products. Collaborative applications and supply chain applications are examples of interorganizational IT applications. Intraorganizational (within a business) IT applications contain applications that enable managers to communicate with employees and help them better manage the internal operations. Examples include computer-aided design (CAD), computer-aided manufacturing (CAM), and material requirements planning (MRP).

All the IT business software applications shown in Table 13.1 including collaborative, content, enterprise resource management (ERM), SCM, manufacturing, engineering, and CRM applications are used in forest products. Therefore, IT applications in forest products cover both inter- and intraorganizational applications.

TABLE 13.1

Classification of Business Software Applications Based on IDC's Software Taxonomy

Business Software Applications	Description
Collaborative applications	Applications for electronic collaboration of groups of users within one or more organizations. These applications enable users to share information, share and manage files, and assign and coordinate tasks. Examples include e-mail, shared folders/databases, group calendar and schedule, and conferencing applications.
Content applications	Software applications to build, edit, manage, and store digital works in any format, such as office suite and web content management software.
Enterprise Resource Management (ERM)	Applications to automate business processes including people, finances, capital, materials, and facilities, such as payroll, procurement, and order management applications.
Supply Chain Management (SCM)	Applications that automate and optimize supply chain processes of the business from raw material suppliers to manufacturers, distributors, transportation companies, and customers. Examples include logistics, inventory management, and production planning software applications.
Operations and manufacturing	Applications that automate and optimize business processes related to manufacturing and services activities. Examples include applications on automating quality control and Materials resource planning (MRP).
Engineering	Applications that automate business processes related to planning, design, and execution of manufacturing and construction activities, such as Computer aided manufacturing (CAM) and Computer aided design (CAD) systems.
Customer Relationship Management (CRM)	Applications that automate sales, marketing, and customer support.

Source: Heiman, R.V. and Byron, D., IDC's Software Taxonomy, 2005.

13.2.2.1 Primary Wood Products

In primary wood manufacturing,* the main focus of IT applications has been on automating business processes; improving quality control, production planning, and logistics; and enhancing productivity. For example, in sawmills, the manual sorting and inspection of logs have been replaced by automated systems. Automated systems, resulting from IT developments and enhancements in these mills, increase accuracy, flexibility, speed, and recovery and decreased labor costs (Hetemäki and Nilsson 2005). Improvements in scanning and computer technologies have enabled sawmills and plywood mills to obtain and store log geometry information and use this information in calculating mill throughput and log quality by species. Computer-controlled optimization and scanner technology are used in these mills to determine the optimum bucking, primary breakdown, edging, trimming patterns, peeling solutions, and veneer clipping (Wood Products Online Expo 2012). Data from scanners are transmitted to the optimizers where the information is used to determine the optimum cutting/trimming solutions. While sawmill optimization is often used to obtain the highest volume output, a planer mill uses an optimizer for advanced defect detection based on board value, which is typically derived from commodity market price

---

* The wood industry is normally classified into the primary and secondary wood manufacturing. Primary wood producers, e.g., sawmills, and engineered wood manufacturers, convert forest raw products such as logs into lumber and other finished or semifinished products. Secondary wood producers manufacture lumber and other semifinished wood products into more value-added products, such as cabinets, windows and doors, and furniture.

tables. Many companies supply scanning, optimization, and control packages to fully automate the bucking, sawing, and trimming processes. Simulation tools have also been used in sawmills, veneer, plywood, and oriented strand board (OSB) mills. These tools do not all provide optimal solutions, but rather they are often implemented to support decision making. A variety of commercial simulation and optimization software products, such as WOODSIM™ for simulating log supply operations, SAWSIM® for simulating the sawing of logs in a sawmill, and PANELSIM™ for simulating and optimizing veneer, plywood, LVL, I-joist, and OSB operations (Halco Software Systems 2012), are available in the market.

The use of IT in primary wood product mills has also provided a wide range of opportunities for improving the logistics and transportation of products. Weintraub et al. (1996) developed a computerized truck scheduling system, called ASICAM, that has been used by several Chilean, Brazilian, and South African forest companies. The use of this system by Chilean forest companies resulted in 15% savings in total costs (Weintraub and Epstein 2002).

Automation of business processes in manufacturing, including forest products manufacturing, has brought enormous value to companies and their customers. IT applications and software products provide powerful tools to automate complex business processes. For example, plant operations and SCM are two key areas that Microsoft® delivers tools for automation. These tools and solutions increase productivity, improve decision making, and enhance operational excellence in companies. Universal Forest Products (UFPs), a leading manufacturer of lumber products in the United States, implemented Microsoft solutions to automate its supply chain and forecasting processes. Microsoft® business process automation solutions enabled UFP to integrate its data collection and report generation and reduce its inventory levels (Microsoft 2012).

In other examples, IT solution provider SAP® integrates sawmill production, supply, and demand (SAP 2012a), and FOREST AX, developed by Sierra Systems specifically for the forest products industry, is a business information management solution that manages the entire supply chain from harvesting of trees to delivery of products (Sierra Systems 2012). It provides real-time inventory management, accurate costing, and production optimization capabilities for sawmills and pulp and paper mills.

ERP systems, which integrate all aspects of operations including production planning, inventory control, material purchasing, accounting, financing, marketing, and distribution, have also been used in forest products companies. The Accenture Advanced Enterprise Solution for Forest Products is a comprehensive ERP solution for forest companies to increase their performance (Accenture 2012). In 2007, Roseburg Forest Products with 15 manufacturing facilities implemented the JD Edwards Enterprise One ERP web-based system in six of its plants with the help of Strategic Solutions NW, a technology and management consulting company (Strategic Solutions NW 2012).

13.2.2.2 IT in Pulp and Paper

The future consumption of paper products depends on factors such as economics, consumer preferences, IT developments, and environmental issues (Hetemäki and Nilsson 2005, Chapter 6). Overall, the increasing use of information and communication technologies in personal and business activities has boosted the consumption of office papers (Hetemäki and Nilsson 2005, Chapter 1). The trend in OECD countries is toward an increase in use of electronic communication and a concurrent decline in print communication; however, this trend is not expected for the near future in non-OECD countries due to their lower economic conditions and lower IT utilization rate compared to OECD countries

TABLE 13.2

Production of Printing and Writing Paper (Tons)

	1980	2010
World	41,213,200	112,288,662
Africa	337,500	902,132
America	17,214,200	26,829,587
Asia	7,227,500	48,882,765
Europe	16,191,000	35,178,178
Oceania	243,000	496,000

Source: FAOSTAT, Food and Agriculture Organization of the United Nations, accessed on January 29, 2012, from http://faostat.fao.org/Desktop Default.aspx?PageID=626&lang=en#ancor, 2012.

(Hetemäki and Nilsson 2005). Global production of printing and writing papers increased from about 41 million tons in 1980 to more than 112 million tons in 2010 (FAOSTAT 2012). Table 13.2 shows the increase in production of printing and writing papers in different continents during the past three decades.

The global consumption of office papers increased in all types of organizations and home offices due to the use of personal computers, fax machines, copy machines, and printers, which became more affordable over time. Despite the popularity of devices that provide books and periodicals on demand electronically, such as tablets, and e-readers such as Amazon's Kindle™ and Barnes & Noble's Nook™, many people still read documents on paper rather than on screen. In a survey of 3000 people by the Pew Research Center, 88% of e-book-reading respondents indicated that they still read printed as well as e-books (Indvik 2012). Although some suggest that the idea of "paperless office" is far from reality (Hetemäki and Nilsson 2005; Sciadas 2006, Chapter 6), "There's a secular shift to paperless. It's an overarching mind-set," according to Matt Arnold, an analyst with Edward Jones & Co. (Moore 2012).

The increasing use of the Internet has affected the newspaper industry negatively. In the United States, newspaper circulation, readership, and concurrent newsprint paper consumption have declined since the 1990s (Hetemäki and Nilsson 2005, Chapter 6). The same trend was observed in most OECD countries, while newsprint demand has grown in non-OECD countries because of their low IT utilization rate (Hetemäki and Nilsson 2005, Chapter 6).

In the paperboard and packaging sector, IT has provided a unique opportunity for stimulating product innovation and market development of "active and intelligent packaging" products (Hetemäki and Nilsson 2005, Chapter 7). The rapid advances in IT, combined with the need for the forest industry in developed countries to produce new value-added products to sustain economic growth, have led to increased research and development on fiber-/forest-based products for consumer goods packaging (Wood and Fiber Product Seminar 2009).

Today, consumers expect better performance and less waste not only from products but also from the packaging of products. This expectation is due to the global concerns regarding the environmental impacts of human activities and increased attention to sustainability. Moreover, the requirement to have new embedded functionalities in packages, such as displaying information on quality, has opened up a market for active and intelligent packaging. Active packaging, also referred to as packaging systems, is the packaging that enhances product life and its safety by controlling a product's conditions.

Electronic sensors or chemicals are used in intelligent packaging for storing and transmitting product information. Intelligent packaging improves logistics and SCM and thereby generates added value for customers. RFID applications have also been developed and are being integrated into packaging materials to store and transmit data on product and inventory information. Some large retailers, such as Walmart, have implemented RFID for shipments of products from their suppliers to their distribution centers and stores (Hetemäki and Nilsson 2005, Chapter 7).

There is an opportunity to use new forest-based materials in active and intelligent packaging (Wood and Fiber Product Seminar 2009) and an opportunity for the packaging industry to provide new packaging solutions (Hetemäki and Nilsson 2005, Chapter 7). The paperboard and packaging sector in developed countries, which have advanced IT capabilities, is more likely to be able to adopt new electronic packaging systems than those in developing countries. Stora Enso, which is a global paper, wood products, and packaging company, has active R&D collaboration efforts to adapt new techniques in packaging applications, such as IT-based techniques to produce packages with new functionalities (Stora Enso 2012).

IT developments in this sector are similar to other forest products sectors. In pulp and paper mills, IT has been used for quality control, production planning, product scheduling, order management, distribution, CRM, SCM, ERP, and human resources management (Janssen et al. 2004). SAP* products for pulp and paper (SAP 2012b) automate and support sales, production planning, maintenance, and transportation planning processes as well as financial and human resources management. Optimization models have also supported the operational planning decisions at pulp and paper facilities. Optimization of paper roll cutting at paper mills and process control optimization at pulp mills are two examples (Ronnqvist 2003).

A survey of pulp and paper mills in North America (Janssen et al. 2004) focused on information management systems (IMS). Pulp and paper mills used a variety of business software packages, such as SAP, JD Edwards, System Software Associates (SSA) Infinium, Oracle Financials, and Epicor eBackOffice, to store data and information more efficiently and used them in problem solving and decision making. Examples of IMS applications used by pulp and paper mills in the study included product tracking, maintenance, online cost tracking, and lost production tracking systems. Operations-related applications such as those for quality control, production planning and scheduling, and order management were used in most mills. However, the use of IMS in SCM, CRM, and ERP applications was low (Janssen et al. 2004). Integration of IMS creates a database to support different applications and provides information for a variety of decision makers. Most surveyed mills (80%) had integrated their IMS and indicated that respondents had a thorough understanding of the business process[†] required for IMS integration. Improved process data flow, data transparency, improved decision making, increased shipment reliability, improved product quality, and reduced shipment time were among the listed benefits of IMS integration by respondents. Lack of funding, time constraints, low acceptance of system integration by personnel, lack of training, and lack of top management support were listed by respondents as barriers to IMS integration.

---

[*] SAP, established in 1972, is one of the largest software suppliers in the world that provides software, solutions, and services for improving business processes. Today, many corporations, including IBM and Microsoft, use SAP products to run their businesses.

[†] Business processes can be classified into (1) management processes, such as strategic management; (2) operational processes, such as purchasing, manufacturing, advertising, marketing, and sales; and (3) supporting processes, such as accounting, call center, and technical support (www.wikipedia.com).

13.2.2.3 IT in Secondary Wood Products

In secondary wood products manufacturing, various IT applications are used for product design and development, production planning and scheduling, material resource planning, forecasting, and SCM. Similar to manufacturing of other products, manufacturing of wood products benefits from the utilization of CAD and CAM systems. CAD systems are used in the design stage for documentation, organization, storage of information, and generation and analysis of drawings. CAM systems assist in planning, controlling, and managing operations during the manufacturing stage. CAD and CAM systems are used by kitchen cabinet, window and door, construction, furniture, and packaging industries to facilitate the design, development, and production of products (Hetemäki and Nilsson 2005, Chapter 8; Assadi and Sowlati 2009; Assadi et al. 2009).

There are many companies that provide manufacturing and design software solutions specifically for the secondary wood products industry. For example, Planit and Microvellum are global leaders in developing CAD/CAM software specifically for the wood industry. CADCode Systems focuses on software solutions from order entry to design and manufacturing for wood products manufacturers. 20-20 Technologies provides software solutions for furniture (including kitchen cabinet) design and manufacturing. Wellborn Forest Products, a cabinet manufacturer, uses 20-20 Technologies ERP solution to automate its business processes (Reuters, July 15, 2008). Jefferson Industrial Software offers management and operations software, called J-MOS and J-MOS lite, for the woodworking industry to reduce their operation costs. These tools automate job tracking and job costing for custom wood millworking (Jefferson Industrial Software 2012).

USC Forest Group, an international distributor of wood products, implemented Microsoft® CRM to automate its sales activities in 2003. The benefits realized by USC Forest Group include increased sales, improved customer service, reduced cost, and improved access to customer and vendor information (Microsoft Business Solutions 2003).

An analysis of software packages available for the kitchen cabinet industry (Hewitt 2011) revealed that two-thirds of the functionalities in these software products were related to engineering and manufacturing, while functionalities related to content, collaborative, and CRM applications were not that common. In a survey of software companies conducted by Frayret and Rousseau (2004), only nine respondents had supply chain and logistics management solutions specifically for the forest products industry. Scoopsoft® (Canada), IFS (Sweden), and Savcor Group (Finland) were the three companies that provided ERP systems for the forest products industry. It was mentioned by the authors that major software providers such as Microsoft® or Oracle had not developed any solutions specific to the forest products industry. However, there have been some attempts by large companies to address this gap. Today, for example, SAP® for mill products offers software solutions for furniture, pulp, paper and packaging, and timberlands and solid woods (SAP 2012a–c). SAP® solutions for furniture manufacturing provide software solutions to automate operations and optimize the supply chain (SAP 2012c).

In order to remain competitive, many forest products companies have implemented optimization and simulation models to integrate and optimize the processes within their firms and along the supply chain. Advances in both hardware and software made it possible to develop and run large-scale models for analysis and optimization of business processes. One successful example of optimization models is that of Aydinel et al. (2008). They developed mathematical models to optimize the allocation of customer orders and their transportation for a leading forest products company in North America. This forest company implemented the models for their production and transportation planning because of the significant cost savings shown from the optimized models compared to their existing practices. A user interface was also developed for the sales staff to use the models in allocating the customer orders to different production mills and decide on their transportation mode, size, and carrier simultaneously (on a weekly basis) in order to minimize the total cost of production and transportation.

13.2.3 Discussion

In the forestry sector, the use of new information technologies has advanced the collection of high-quality forest data, improved forest administration, and planning and has provided a means to reduce illegal logging. This trend will continue in the future as more sophisticated technologies, in terms of storage, graphics, etc., with more affordable prices will enter the market. IT applications in both primary and secondary wood manufacturing have helped companies automate processes and increase operational efficiency.

Wood manufacturing companies can further benefit by investing in and implementing software applications. Integration of inter- and intracompany software systems in the organizational supply chain is necessary to have an efficient and actionable flow of information and materials through production systems and value chains (Janssen et al. 2004).

Access to accurate online data is also necessary to support tracking, analysis, and decision making in the companies. Compatibility of new IT systems with ones currently in use, often called legacy systems, is also necessary to actualize IT advantages (Janssen et al. 2004; Assadi and Sowlati 2009; Assadi et al. 2009).

Lack of highly skilled personnel and lack of finances can impede the ability of many forest companies, especially small- and medium-sized firms, to invest in IT and benefit from IT in their operations and planning (Hewitt et al. 2011). Increased funding is also necessary for IT research in relevant forest sector areas at universities and colleges to train students and to create collaboration between industry and research groups. Moreover, training of the current workforce is also needed as advances in IT and the potential benefits they offer continue to grow rapidly.

13.3 E-Business

The use of IT and e-business applications can provide benefits to inter- and intraorganizational business processes. E-business is defined as the use of electronic networks such as the Internet in conducting business transactions and exchanging business information (Glover et al. 2002). Businesses are increasingly using electronic technologies to streamline their processes, increase sales, reduce costs, and provide improved customer support. On average, in OECD countries, 94% of businesses with more than 10 employees are connected to the Internet (OECD 2011). While this figure is lower in non-OECD countries, in some non-OECD countries such as Brazil, Malta, Latvia, and Serbia, more than 85% of businesses with more than 10 employees are connected to the Internet (OECD 2011).

Even though e-business and e-commerce are often used interchangeably, e-business has a broader scope than e-commerce. E-commerce is the use of electronic networks to buy and sell products or services and is a subset of e-business. It can be classified based on the type of participants (providers and buyers) in the transactions into four main categories of business to business (B2B), business to consumer (B2C), consumer to business (C2B), and consumer to consumer (C2C) e-commerce. B2B implies that both the provider and consumer involved in the transactions are businesses. B2C involves individual consumers as the buyers (Heizer and Render 2006).

There are a growing number of companies buying and selling products and services over the Internet. Figure 13.2 shows the percentage of businesses in different OECD countries engaged in e-commerce (OECD 2011). E-commerce has grown immensely during the past decade as shown in Figure 13.3 in Australia, e-commerce growth in 2008 was eight times higher than its level in 2001.

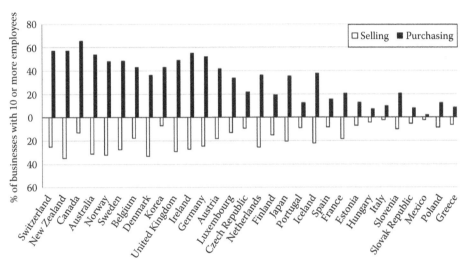

FIGURE 13.2
Businesses selling/purchasing over the Internet by OECD country in 2007. (From OECD, The future of the Internet economy, accessed on January 10, 2012, from http://www.oecd.org/dataoecd/24/5/48255770.pdf, 2011.)

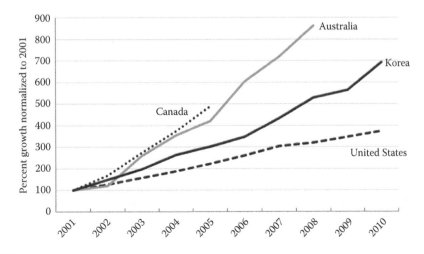

FIGURE 13.3
Growth of e-commerce in Australia, Canada, Korea, and the United States, 2001–2010. (From OECD, The future of the Internet economy, accessed on January 10, 2012, from http://www.oecd.org/dataoecd/24/5/48255770.pdf, 2011.)

E-business, in addition to e-commerce, includes the use of IT in business processes such as marketing, customer relations, and logistics (Ouellet 2010). E-business applications can be divided into three groups: (1) internal business systems, such as ERM and CRM; (2) enterprise communication and collaboration, such as e-mail and content management system; and (3) electronic commerce, such as online marketing and SCM.

The rapid growth of e-business is due to its potential benefits for businesses and customers. The benefits for businesses include expanded sales opportunities, improved communication and customer service, lower costs, and improved supply chain and human resources management (Glover et al. 2002). At the same time, customers benefit from more competition between businesses to get lower prices, improved service, more selection of products and services, and more customized information.

In order to achieve these benefits, businesses require reengineering their current processes and integrating e-business processes with them, which is not an easy task. They also need to integrate e-business strategies into their overall operations strategy. Organizations conducting e-business face some unique risks in addition to those common to all businesses. These unique risks are related to security and can result from organization's IT infrastructure vulnerability to external attack. Some security concerns for e-businesses include maintaining robust firewalls, keeping the business and customer information private, making sure the data received is identical to the data sent (data integrity), and assuring the authenticity of connected entities (Glover et al. 2002, Wikipedia).

13.3.1 E-Business in the Forest Sector

E-business has continued to grow in all industry sectors although adoption in the forest sector has lagged other sectors. A review of literature highlights the low IT adoption rate of the forest products industry (Karuranga et al. 2006; Hewitt et al. 2011). The forest sector has been conservative in applying Internet-based technologies (Shook et al. 2002). Lower-order e-business applications, such as e-mail and static website, were mainly used by forest products firms (Karuranga et al. 2006; Stennes et al. 2006; Arano 2008) relative to higher-order e-business applications, such as those for SCM processes. IT adoption rates by forest products firms have been found to be positively correlated with firm size (Kozak 2002; Shook et al. 2002; Vlosky et al. 2002; Stennes et al. 2006; Arano 2008) and the extent of participation in exporting products (Stennes et al. 2006; Arano 2008).

Information security (Vlosky and Smith 2003; Stennes et al. 2006; Arano 2008) and availability of technically skilled employees (Vlosky and Smith 2003) were cited as main concerns in e-business implementation. Developments of a long-term and comprehensive plan in using the Internet to conduct business and formal training programs for employees were recommendations by surveyed forest products firms that had implemented e-business (Vlosky and Smith 2003).

Inventory management is an essential part of every supply chain. Vendor-managed inventory (VMI) is a supply chain strategy that helps reduce the inventory in the supply chain (Gronalt and Rauch 2008). VMI requires collaboration of the suppliers and their customers (retailers, distributors, manufacturers, or end customers). In this approach, instead of purchase orders, the customer sends its demand information to the supplier. The supplier manages the customer's inventory and handles the replenishment process based on demand information. In order to share information between two parties, safe, affordable, and fast data collecting and transmitting technologies are needed. Therefore, successful implementation of VMI systems requires advanced information technologies and applications.

Electronic data interchange (EDI) and bar-coding systems have been used to facilitate VMI. EDI is the electronic transmission of business information and documents in standard formats between exchange partner organizations (Glover et al. 2002). It allows companies to exchange business information such as those related to purchasing, shipment, inventory, and payment. Forest products companies started to adopt this technology in 1990s mainly due to the request of their customers (homecenter retailers) (Dupuy and Vlosky 2000). Increasing data accuracy and reducing costs were among other reasons the forest sector implemented EDI. As an example, Alliance Forest Products (www.allianceforestproducts.com), which produces wood pallets and wood packaging products, offers VMI programs to its customers.

In order to facilitate collaboration between forest companies using the Internet and improve supply chain efficiency, papiNet (www.papinet.org) was originated, initially as a paper sector information exchange. papiNet is an international paper and forest products initiative that provides papiNet standards. The papiNet standard, a set of XML business documents, is a free and open standard for communicating business transactions in forest products supply chains. Today, many large paper and forest products firms in Europe and North America, such as Södrä, Stora Enso, Tembec, and Weyerhaeuser, are papiNet members.

Buying and selling of logs and lumber through the Internet were among the first e-commerce activities in forestry (Johnson 2001). E-marketplaces provide a platform on the Internet for companies to exchange business information or to find new suppliers or buyers for their products. TALPX Inc.'s Internet exchange was among the first forest products e-marketplaces that provided close to 27 billion board feet of lumber and panel products for trading online in 2001 (Johnson 2001). The dot.com crash, slow IT adoption rate in the forest sector, lack of human interaction in e-marketplaces, and lack of focus on the business process were mentioned as the failure factors of forest products e-marketplaces (Shook et al. 2004).

Presently, Fordaq (www.fordaq.com) is an international e-marketplace for forest products. Many large wood companies in Europe are members of Fordaq. Member companies can check the inquiries and offers of other member companies, communicate their own offers and enquiries, find new suppliers or clients, access the online directory, receive industry news, and compare prices. Another successful application of e-business in forest products was that of the Weyerhaeuser. Weyerhaeuser used an Internet-based platform software, DoorBuilder, for order taking and product tracking in its door manufacturing facility in Wisconsin to turn this facility around and make it profitable again (Glover et al. 2002). The use of DoorBuilder significantly reduced errors in ordering, production, and delivery and increased on-time delivery from 40% to 97%.

13.3.2 Discussion

Previous studies on e-business adoption of forest products firms revealed that firms in this sector have been slower in investing and implementing IT in their business processes compared to other industry sectors. In addition, IT investments in e-business have been lower for smaller companies than larger firms. It was concluded in most of the conducted studies that information and communication applications were more common in forest companies than the use of Internet for purchasing, order tracking, and SCM. Based on the ICDT model (Angehrn 1997), most forest products firms are in the virtual information and communication spaces, while few have been benefited from the new channels provided in distribution and transaction spaces (Hewitt et al. 2011).

These observations are not surprising since most wood products firms are small- and medium-sized enterprises (SMEs) that have been shown to be slower than larger businesses in adopting new technologies including information and communication technologies. In a 2003 survey (OECD 2004), SMEs across multiple industry sectors with fewer than 250 employees in 19 European countries were asked about the reasons for not selling products using the Internet. The main reasons were that e-commerce was not suitable for their type of products or type of business, lack of IT-skilled employees, concerns about unbalanced cost and benefits of e-commerce, and technology issues such as online security. The same issues were highlighted in previous studies as the reasons for low e-business adoption in the forest sector. Security of sensitive information, lack of employee technical skills, and cost were mentioned as the main concerns of the US lumber industry for using the Internet to conduct business (Vlosky and Smith 2003).

Moreover, wood products are not the same as other products or services (e-books, banking services, etc.) that are suitable for distributing easily using an Internet platform. However, this type of web-based trading platform can potentially be developed and used by the wood products industry, particularly in higher-value segments such as hardwood lumber, furniture, and other secondary, value-added products. Retail sales already take place through online ordering but industrial business to business commodity sales have yet to be successful. Therefore, similar to the finding of the OECD 2003 survey (OECD 2004), the type of product is one of the reasons hindering forest products firms from benefiting from Internet-based distribution channels. Another factor for low e-business adoption in the forest sector, especially in purchasing and selling online and SCM applications, is the culture in the sector in that managers prefer human interactions rather than virtual interactions through the Internet (Vlosky and Wesbrook 2002). This was highlighted as one of the main factors of forest sector dot.com failures in an analysis of forest e-marketplaces (Shook et al. 2004).

13.4 Future of IT Applications: The Cloud

Cloud computing refers to delivery of IT services over a network, typically the Internet. It is an emerging and newly commercialized delivery model that allows consumers to access IT services using any device, such as desktops, laptops, smart phones, and tablets, with Internet access (Conway 2011). In this delivery model, which resembles the delivery of water and electricity, the IT services, such as software applications, data, and storage services, are delivered to the consumers without the need for the end user to understand the infrastructure and components required for providing the service (Wikipedia, Convey 2011). The term cloud computing represents both the delivered service over the network and the software and hardware at the datacenters that provide those services. The hardware and software at datacenters are also referred to as the cloud (Armbrust et al. 2009). The distinguishing features of cloud computing are (1) the availability of computing resources on demand, (2) the omission of upfront commitments by the users as the service is provided and managed by the cloud provider, and (3) the flexibility to use computing resources as much as needed (Conway 2011; Prince 2011). An example of cloud computing is Google Mail. A user requires a device and Internet access to connect to the e-mail service. Google is the cloud provider and the server and e-mail management software are on the cloud (Internet) (http://gmail.com). Google Docs is another example of cloud computing. It is a free web-based office suite (word processor, spreadsheet, and presentation application) and storage service. Users can develop, edit, and store documents online using this service. It also provides the ability to collaborate with others on the documents in real time (http://www.google.com/google-d-s/documents/).

Based on a recent report by Gartner (2009), global revenue from cloud computing would reach $148.8 billion in 2014. The US share of the worldwide revenue in cloud computing was estimated to be 60% in 2009, but it could be reduced to 50% in 2014 as other countries increase their presence in this market space. Western Europe is expected to account for 29% and Japan for 12% of this market by 2014 (Gartner 2009).

Cloud computing can be deployed using four different models: public cloud, private cloud, community cloud, and hybrid cloud (Conway 2011; Mell and Grance 2011). Cloud computing infrastructure can be owned by an organization that provides services to the public or group of industries, referred to as the public cloud model. This type of cloud reduces user's capital expenditures on computing resources but poses the risk of data/information security since the infrastructure is shared by other companies. Microsoft Azure is an example of the public cloud. On the other hand, in the private cloud model, the infrastructure is owned or leased by an organization and the service is only used by itself. This type of the cloud does not require a lower IT capital expenditure; however, it offers a higher level of data security than the public cloud. Microsoft has its own private cloud. In a community cloud model such as Google Gov, the infrastructure is shared between several organizations and provides services to a specific community. The capital cost of this model is higher and the data/information security it provides is more secure than the public cloud model. A hybrid cloud model is a combination of the aforementioned cloud models. Using a hybrid model, an organization can choose the best combination of cloud models for deployment. For instance, an organization can use a private cloud for sensitive data and critical applications and use a lower cost public cloud model for less sensitive data.

Cloud computing services are typically grouped into three categories: (1) infrastructure as a service (IaaS), (2) platform as a service (PaaS), and (3) software as a service (SaaS) (Arasaratnam 2011; Conway 2011). IaaS offers network equipment, servers, and datacenter space to the end users. In PaaS, the provider includes hardware, operating systems, storage, and network capacity into the service. SaaS provides software applications and the hardware infrastructure to the end users on demand. Sales Force CRM is an example of SaaS.

The growth in cloud computing resulted from cost-cutting benefits for participating companies because they pay for the computing resources on a per-use basis (Conway 2011). On a pay-per-use basis, companies do not incur start-up costs for computing resources and do not require purchasing expensive software or hardware. They also have the flexibility to increase or decrease their use and deployment of cloud services quickly, which is not possible in a traditional computing model. In a cloud model, software updates are automatic through the web. Businesses save time and money since cloud computing deployment is faster than traditional computing, and the cloud provider is responsible for immediate support during any emergency situation such as network outages. Group collaboration is facilitated in cloud computing since documents and applications are on the cloud not on a single machine. Remote access to the documents and information is possible from any device anywhere that increases employee's mobility and accessibility. It also means that end users can access their data and documents at anytime from anywhere and can store them safely (Armbrust et al. 2009). In addition, IT staffs in companies do not require performing routine maintenance of computing services, tasks borne by cloud service providers (Conway 2011; Prince 2011).

Although cloud computing provides various benefits, businesses should be aware of related issues. The business and its IT department should have a clear strategy and vision in order to be successful and benefit from cloud computing. This, to some extent, relates to the maturity of the business (Conway 2011). Another issue is the security of data/information since they are stored in the cloud. Many companies are cautious about deploying cloud computing because residing sensitive data outside the internal infrastructure increases the inherent risk level of data breaches. Data ownership is another reason that hinders some companies from cloud computing deployment (Arasaratnam 2011; Conway 2011). There are other risks to consider as well. In cloud computing, services are supplied through the Internet. Therefore, there could be difficulties for companies in areas with poor or unreliable connectivity. Most of the current web-based applications do not include all the functionalities and features provided in desktop-based applications, for example, Microsoft PowerPoint (non-cloud) offers more functionality than Google Presentations (cloud based). Most cloud computing services target SMEs and may not provide large enterprises the same support and reliability that they can get from their own internal IT infrastructure and services. Moreover, the cost-benefits of cloud computing compared to traditional computing may not exist for large enterprises (Conway 2011). In addition, it is not easy to have multiple cloud service providers since they each have their own method of interacting between the cloud and the users. To date, there is no standard cloud architecture.

13.4.1 Cloud Computing in Forestry

To date, the literature does not contain examples of cloud computing by forest sector companies. However, an example of a cloud computing service relevant for the forest sector is provided by Safertek Software (www.saferteck.com), which offers its UFactory software to its customers as SaaS for a small monthly fee. UFactory is a business management software tool for mill products manufacturing companies that automates business processes.

In forestry, one of the organizations that has reported the use of cloud computing is the Ministry of Forests, Land and Natural Resources in British Columbia, Canada (Berg 2011). Each year, the ministry determines allowable annual cut (AAC), which is the volume of trees that can be harvested in a sustainable manner in the province. Historically, the ministry used six computer servers and ran the analysis model for several weeks to calculate the AAC. In order to speed up processing time, the ministry began using the cloud computing service provided by Windows Azure instead of purchasing additional servers. This has helped the ministry to dramatically decrease processing time (to less than 1 h) with a concurrent decrease in IT costs.

Cloud computing was also the subject of two recent studies in forestry. Jiang et al. (2009) explained an idea of a special cloud computing to provide professional service for forest pests' control. They proposed a cloud computing structure that was neither a private cloud nor a public cloud. It would be for a particular field provided by a "professional Cloud Computing provider and special administrations in the field." The suggested benefits were for forestry professionals to collect and cloud-share data on forest pests to better craft pest control systems over a wide geographic range. In another forest-based example, a forest fire management system was designed using cloud computing services (Tang et al. 2011). Cloud computing was used to collect and analyze ecological data and provide information for forest fire management.

13.4.2 Discussion

The implementation of cloud computing presents unique opportunities and benefits for small- and medium-sized firms (SME) (and even large companies) in terms of cost savings in IT infrastructure and applications and a reduction in required IT personnel. It appears to be a viable solution for small- and medium-sized forest products companies since they have relatively limited resources (money and IT-skilled personnel) to participate in Internet trade exchange. However, most cloud computing applications do not provide the same level of features and options that are available for desktop applications and most are tailored toward SMEs rather than larger firms. The former shortcoming will diminish over time with future advances in this area; this may be the case for the latter shortcoming as well. Another major challenge for forest companies to implement cloud computing services is the security of data/information, something repeatedly mentioned in previous research as a major factor in not implementing e-business in forest sector firms.

Moreover, knowledgeable managers and technical personnel who have expertise on emerging IT applications and are capable of aligning company cloud computing strategies with its overall business strategy and vision are necessary for success in the evolving IT arena. This again emphasizes the importance of training highly qualified graduate students in universities and colleges who wish to enter the forest industry.

13.5 Conclusions

The forest sector has been slower in IT adoption in comparison with other sectors. It was among the last industries to move toward e-business (Johnson 2001). Forest products companies have a lower level of e-business sophistication than those in other sectors. They are involved in buying and selling online, but they are less active in SCM applications. Larger forest products firms are better equipped to invest in new technologies including IT than smaller firms. However, with the decrease in prices of hardware and software over time and new technologies such as the cloud computing, smaller firms will have the opportunity to reap IT benefits as well. Prices of computers, monitors, and printers have been declining over time making it more affordable for small- and medium-sized firms to invest in IT.

As has been the case in all industry sectors globally, IT has had a defining impact on the forest sector. It has been used in the forest sector to automate processes, improve efficiency and productivity, and support decision making. Remote sensing, GIS, and GPS applications have improved the quality and utility of data used in forest management and planning. The use of computers and wireless data communication systems in harvesting has improved the bucking and sorting processes by considering the daily log demand from different mills. Truck transportation of logs has become more efficient by matching data collected on log inventories at the roadside and sawmill locations and demand using satellite navigation systems. IT applications have also been essential during the design, development, and manufacture of products in primary and secondary wood facilities. The distribution of products and customer support have benefited from IT developments as well. Nevertheless, there are significant opportunities for further IT adoption and additional improvements in the forest sector.

New mathematical and simulation models should be developed to better optimize and improve processes at every step in the forest sector supply chain from the forest to the sort yard, to production facilities such as sawmills, pulp and paper mills, and secondary wood manufacturing facilities. Although IT adoption in SCM and DSSs has historically been limited, the industry will gain by implementing IT applications that integrate business activities and support collaborative business relationships along the supply chain. This will allow the industry to reduce supply chain costs and simultaneously increase customer service. Future advances in IT will bring further opportunities for the forest sector, regardless of country of origin, to compete in a global economy. Companies that are forward-looking and that invest in this future will be well equipped to change IT-driven potential into reality.

References

Accenture. 2012. Accenture advanced enterprise solution for forest products—A proven approach for accelerating high performance. Accessed on April 13, 2012 from http://www.accenture.com/SiteCollectionDocuments/PDF/AdvancedEnterpriseSolutionforFP_FINAL.pdf.

Andalaft, N., Andalaft, P., Guignard, M., Magendzo, A., Wainer, A., and Weintraub, A. 2003. A problem of forest harvesting and road building solved through model strengthening and Lagrangean relaxation. *Operations Research*, 51(4):613–628.

Angehrn, A. 1997. Designing mature Internet business strategies: The ICDT model. *European Management Journal*, 15(4):361–369.

Arano, K.G. 2008. Electronic commerce adoption in west virginia's primary and secondary hardwood industries: Preliminary results, *Proceedings of the Southern Forest Economics Workers Annual Meeting*, March 9–11, 2008, Savannah, GA, pp. 72–81.

Arasaratnam, O. 2011. *Introduction to Cloud Computing*. Wiley Online Library, Hoboken, NJ.

Armbrust, M., Fox, A., Griffith, R., Joseph, A.D., Katz, R., Konwinski, A., Lee, G. et al. 2009. *Above the Clouds: A Berkeley View of Cloud Computing*. UC Berkeley Reliable Adaptive Distributed Systems Laboratory, Berkeley, CA. February 10, 2009.

Assadi, P. and Sowlati, T. 2009. Design and manufacturing software selection in the wood industry using analytic hierarchy process. *International Journal of Business Innovation and Research*, 3(2):182–198.

Assadi, P., Sowlati, T., and Paradi, J.C. 2009. Multi-criteria evaluation of design and manufacturing software packages considering the interdependencies between criteria: The ANP approach. *International Journal of Information and Decision Sciences*, 1(4):397–410.

Aydinel, M., Sowlati, T., Cerda, X., Cope, E., and Gerschman, M. 2008. Optimization of production allocation and transportation of customer orders for a leading forest products company. *Mathematical and Computer Modelling*, 48(7/8):1158–1169.

Bantayan, N.C. and Bishop, I.D. 1998. Linking objective and subjective modelling for land use decision-making. *Landscape and Urban Planning*, 43:35–48.

Berg, G. 2011. BC forest minister speeds up modelling times from 6 weeks to under an hour with Windows Azure platform. July 21, 2011. http://bepublic.ca

Brynjolfsson, E. and Hitt, L. 1996. Paradox lost? Firm level evidence on the returns to information systems spending. *Management Science*, 42(4):541–558.

Conway, G. 2011. *Introduction to Cloud Computing*. Innovation Value Institute, National University of Ireland, Ireland. White Paper, January 2011. Accessed on February 2, 20112 from http://ivi.nuim.ie/publications/IVI-Exec-Briefing-Intro-to-Cloud-Computing.pdf

CSLS (Centre for the Study of Living Standards). 2011. Database of information and communication technology (ICT) investment and capital stock trends: Canada vs United States. Accessed on January 3, 2012 from http://www.csls.ca/data/ict.asp

D'Amours, S., Ronnqvist, M., and Weintraub, A. 2008. Using operational research supply chain planning in the forest products industry. *INFOR*, 46(4):265–281.

Dimelis, S.P. and Papaioannou, S.K. 2010. FDI and ICT effects on productivity growth: A comparative analysis of developing and developed countries. *European Journal of Development Research*, 22(1):79–96.

Dupuy, C.A. and Vlosky, R.P. 2000. Status of electronic data interchange in the forest products industry. *Forest Products Journal*, 50(6):32–38.

Dykstra, D.P., Kuru, G., Taylor, R., Nussbaum, R., Magrath, W., and Story, J. 2002. *Technologies for Wood Tracking: Verifying and Monitoring the Chain of Custody and Legal Compliance in the Timber Industry*. World Bank/WWF Alliance for Forest Conservation and Sustainable Use, Washington, DC.

Epstein, R., Morales, R., Seron, J., and Weintraub, A. 1999. Use of OR systems in the Chilean forestry industry. *Interfaces*, 29(1):7–29.

FAOSTAT. 2012. Food and Agriculture Organization of the United Nations. Accessed on January 29, 2012 from http://faostat.fao.org/DesktopDefault.aspx?PageID=626&lang=en#ancor

Frayret, J.-M. and Rousseau, A. 2004. Supply chain management software solutions in the forest products industry and the potential of agent-based technology. FORAC Research Consortium.

Friedlos, D. 2009. Malaysian forestry department studies RFID. *RFID Journal*, November 2. Accessed July 1, 2012 from http://www.rfidjournal.com/article/view/5350/1

Garcia, O. 1984. FOLPI: A forestry-oriented linear programming interpreter. *Proceedings of IUFRO Symposium on Forest Management Planning and Managerial Economics*, University of Tokyo, Tokyo, Japan, pp. 293–305.

Gartner. 2009. Gartner says worldwide cloud services market to surpass $68 billion in 2010. Accessed on February 7, 2012 from http://www.gartner.com/it/page.jsp?id=1389313

Glover, S.M., Liddle, S.W., and Prawitt, D.F. 2002. *e-Business—Principles and Strategies for Accountants*, 2nd edn. Pearson Education Inc., Upper Saddle River, NJ.

Gronalt, M. and Rauch, P. 2008. Vendor managed inventory in wood processing industries—A case study. *Silva Fennica*, 42(1):101–114.

Halco Software Systems. 2012. Principle software products. Accessed on April 21, 2012 from http://www.halcosoftware.com/software/index.html

Harder, P. 2001. Canada's competitiveness in the global economy. Presentation to the *Public Policy Forum*. Industry Canada. January 17, 2001, Ottawa, Ontario, Canada.

Heiman, R.V. and Byron, D. 2005. IDC analyze the Future. February 2005, IDC#32884, volume 1. www.idc.com

Heizer, J. and Render, B. 2006. *Operations Management*, 8th edn. Pearson Education Inc., Upper Saddle River, NJ.

Hetemäki, L. and Nilsson, S. 2005. *Information Technology and the Forest Sector*. IUFRO World Series, Vol. 18, Vienna, Austria.

Hewitt, R. 2011. Evaluation of strategic software investments in the Canadian cabinet industry. Master's thesis. University of British Columbia, Vancouver, British Columbia, Canada.

Hewitt, R., Sowlati, T., and Paradi, J.C. 2011. Information technology adoption in United States and Canadian forest industries. *Food Products Journal*, 61(2):161–169.

Indvik, L. 2012. Americans reading more ebooks on computers than ereaders, phones. Accessed on July 25, 2012 from http://mashable.com/2012/04/06/americans-ebooks-computers/

Janssen, M., Laflamme-Mayer, M., Zeinou, M.H., and Stuart, P.R. 2004. Survey indicates mills' need to exploit it systems with new business model. *Pulp and Paper*, June 2004, 78(6):46–51.

Jefferson Industrial Software. 2012. What is J-MOS system. Accessed on July 25, 2012 from http://www.j-mos.com/WhatIsIt.html

Jiang, S., Fang, L., and Huang, X. 2009. An idea of special cloud computing in forest pests' control. *Lecture Notes in Computer Science*, 5931:615–620, doi: 10.1007/978–3–642–10665–1_61.

Johnson, H. 2001. E-commerce continues to make gains in the forest industry, building a base on the buying and selling of logs and lumber. *Logging and Sawmilling Journal*, December/January 2001. http://forestnet.com/archives/Dec-Jan-01/index.html (Accessed August 1, 2013).

Johnson, S.E. 1986. Forest, regional and sector planning models in New Zealand. *The Forestry Chronicle*, December, 62(6):537–541.

Kalakota, R. and Robinson, M. 2003. *Encyclopedia of Computer Science*, 4th edn. John Wiley & Sons Ltd., Chichester, U.K.

Karuranga, É., Frayret, J.M., and D'Amours, S. 2006. E-business in the Quebec forest products industry: Perceptions, current uses and intentions to adopt. *Journal of Forest Products Business Research* (2):4.

Kent, B., Bare, B., Field, R.C., and Bradley, G.A. 1991. Natural resource land management planning using large-scale linear programs: The USDA Forest Service experience with FORPLAN. *Operations Research*, 39(1):13–27.

Koellinger, P. 2008. The relationship between technology, innovation, and firm performance–empirical evidence from e-business in Europe. *Research Policy*, 37:1317–1328.

Kozak, R.A. 2002. Internet readiness and e-business adoption of Canadian value-added wood producers. *Forestry Chronicle*, 78(2):296–305.

Laukkanen, S., Palander, T., and Kangas, J. 2004. Applying voting theory in participatory decision support for sustainable timber harvesting. *Canadian Journal of Forest Research*, 34:1511–1524.

Lexer, M.J., Vacik, H., Palmetzhofer, D., and Oitzinger, G. 2005. A decision support tool to improve forestry extension services for small private landowners in southern Austria. *Computer and Electronics in Agriculture*, 40 (1):81–102.

Li, M. and Ye, L.R. 1999. Information technology and firm performance: Linking with environmental, strategic and managerial contexts. *Information and Management*, 35:43–51.

Mell, P. and Grance, T. 2011. *The NIST definition of Cloud Computing*. National Institute of Standards and Technology. Special Publication 800-145. January 2011, U.S. Department of Commerce, Gaithersburg, MD.

Microsoft. 2012. Business process automation in manufacturing. Accessed on July 27, 2012 from http://search.microsoft.com/en-U.S./results.aspx?q=BPA+Manufacturing&x=17&y=14

Microsoft Business Solutions. 2003. Microsoft business solutions crm helps ucs forest group open doors to increased sales success. Accessed on April 13, 2012 from http://www.accountingmicro.com/docs/UCSForestGroupcasestudy.pdf

Moore, M. 2012. Economic downturn, "paperless" society take toll on Staples sales. *Boston Business Journal*, June 4. http://www.masslive.com/business-news/index.ssf/2012/06/economic-downturn-paperless-society-take.html (Accessed August 1, 2013).

OECD. 2004. ICT, E-business and SMEs. Accessed on January 15, 2012 from http://www.oecd.org/dataoecd/6/9/31919255.pdf

OECD. 2011. The future of the internet economy. Accessed on January 10, 2012 from http://www.oecd.org/dataoecd/24/5/48255770.pdf

OECD Factbook. 2011–2012. Economic, environmental and social statistics. Science and technology. Investments in ICT. Accessed on January 10, 2012 from http://www.oecd-ilibrary.org/docserver/download/fulltext/3011041ec073.pdf?expires=1327432126&id=id&accname=freeContent&checksum=DD711BEB14886FBF1AE19398EF8AB6A0

Ouellet, S. 2010. *The Deployment of Electronic Business Processes in Canada*. Business Special Surveys and Technology Statistics Division Working Papers. Statistics Canada, Ottawa, Ontario, Canada.

Prince, J.D. 2011. Introduction to cloud computing. *Journal of Electronic Resources in Medical Libraries*, 8(4):449–458.

Ronnqvist, M. 2003. Optimization in forestry. *Mathematical Programming*, 97(1/2):267–284.

SAP. 2012a. SAP for mill products. Timberlands and solid wood. Accessed on April 10, 2012 from www.sap.com/industries/millproducts/timberlands_solid_wood.epx

SAP. 2012b. SAP for mill products. Pulp and paper. Accessed on April 10, 2012 from www.sap.com/industries/millproducts/pulp_and_paper.epx

SAP. 2012c. SAP for mill products. Furniture. Accessed on April 10, 2012 from www.sap.com/industries/millproducts/furniture.epx

Satellite Imaging Corporation. 2012a. Forestry management products. Accessed on April 13, 2012 from www.satimagingcorp.com/svc/forestry_management.html

Satellite Imaging Corporation. 2012b. About satellite imaging corporation. Accessed on April 13, 2012 from http://www.satimagingcorp.com/about.html

Satellite Imaging Corporation. 2012c. Forestry remote sensing. Accessed on April 13, 2012 from www.satimagingcorp.com/svc/forestry.html

Sciadas, G. 2006. Our lives in digital times. Statistics Canada. Accessed on January 28, 2012 from http://www.statcan.gc.ca/pub/56f0004m/56f0004m2006014-eng.pdf

Sierra Systems. 2012. Creating competitive advantage from stump to stick. A ceo's guide to forest AX: A sierra systems industry solution. Accessed on April 13, 2012 from http://www.sierrasystems. com/Documents/Library/MicrosoftDynamicsServices/White%20Paper%20-%20 Creating%20Competitive%20Advantage%20From%20Stump%20to%20Stick.pdf

Shook, S.R., Vlosky, R.P., and Kallioranta, S.M. 2004, Why did forest industry dot. coms fail? *Forest Products Journal*, 54(10):35–40.

Shook, S.R., Zhang, Y., Braden, R., and Baldridge, J. 2002. Pacific Northwest secondary forest products industry. *Forest Products Journal*, 52(1):59.

Stennes, B., Stonestreet, C., Wilson, W.R., and Wang, S. 2006. e-Technology adoption by value added wood processors in British Columbia. *Forest Products Journal*, 56(5):24–28.

Stora Enso. 2012. Where is the new research? Accessed on July 30, 2012 from http://www.storaenso. com/research/Documents/where-is-the-new-research.pdf

Strategic Solutions NW. 2012. Roseburg forest products. Accessed on July 26, 2012 from http:// www.ssnwllc.com/Clients/Roseburg.aspx

Tang, J., Gong, T., and Zhou, J. 2011. Research on anti-forest-fire ecological management system with cloud computing. *IEEE Seventh International Conference on Natural Computation*, Shanghai, China. IEEE, Piscataway, NJ. ISBN: 978-1-4244-9950-2

USDA Forest Service. 2012. Spectrum—Long range planning model. Accessed on April 7, 2012 from https://fsplaces.fs.fed.us/fsfiles/unit/wo/emc/imi/IMI_OPEN_TeamRoom.nsf/1ba02e5ee0 ba987c8525663a006cb7ec/9d6ec3fa57ddc68f8725711d007210b7?OpenDocument&ExpandSecti on=2%2C4#_Section2

Vlosky, R.P. and Smith, T.M. 2003. e-Business in the U.S. hardwood lumber industry. *Forest Products Journal*, 53(5):21–29.

Vlosky, R.P. and Westbrook, T. 2002. e-Business exchange between homecenter buyers and wood products suppliers. *Forest Products Journal*, 52(1):38–43.

Vlosky, R.P., Westbrook, T., and Poku, K. 2002. An exploratory study of Internet adoption by primary wood products manufacturers in the western United States. *Forest Products Journal*, 52(6):35–42.

Wasserman, E. 2011. RFID in the forest. *RFID Journal*. Accessed on July 1, 2012 from http://www. rfidjournal.com/article/purchase/8177

Weintraub, A. and Epstein, R. 2002. The supply chain in the forest industry: Models and linkages. Chapter 13 of *Supply Chain Management: Models, Applications and Research Directions*. Kluwer Academic Publishers, Dordrecht, the Netherlands, pp. 343–362.

Weintraub, A., Epstein, R., Morales, R., Seron, J., and Traverso, P. 1996. A truck scheduling system improves efficiency in the forestry industry. *Interfaces*, 26(4):1–12.

Weintraub, A., Romero, C., Bjorndal, T., and Epstein, R. 2007. *Handbook of Operations Research in Natural Resources*. Springer, New York.

Wood and Fiber Product Seminar. 2009. VTT and USDA joint activity. September 22–23, 2009. Accessed on April 24, 2012 from http://www.vtt.fi/inf/pdf/symposiums/2010/S263.pdf#page=29

Wood Products Online Expo. 2012. Equipment and technology. Optimization. Accessed April 21, 2012 from http://www.woodproductsonlineexpo.com/content_menu.php/124/ optimization_optimizers.html

World Bank. 2012. ICT glossary. Accessed on January 5, 2012 from http://web.worldbank.org/ WBSITE/EXTERNAL/TOPICS/EXTINFORMATIONANDCOMMUNICATIONANDTECHN OLOGIES/0,,contentMDK:21035032~menuPK:2888320~pagePK:210058~piPK:210062~theSit ePK:282823,00.html

14

Cross-Cultural Sales, Marketing, and Management Issues in the Global Forest Sector

Ernesto Wagner

CONTENTS

14.1 Introduction

Global trade in forest products has grown significantly in recent years. According to one recent report, global forest products trade doubled in just 6 years from US$ 300 billion in 2002 to 600 billion by 2008 (Pepke 2011). This increase in trade is also characterized by an altered trade-flow pattern that now includes countries that were hitherto absent on the global forest products trade map. The emergence of China as a major supplier and buyer of forest products, forestry land acquisition by several large forest products companies in South America, and opening up of new markets such as in India, the Middle East, and Asia have necessitated, more than ever before, that forest products companies attend and address the challenges of cross-cultural marketing and management issues arising due to a globalized interface.

Two subjects warrant space at the outset. First, there is some ambiguity about the word "cross-cultural." Hofstede (1980, 25) defined culture as the "collective programming of the mind which distinguishes the members of one human group from another... the interactive aggregate of common characteristics that influence a human group's response to its environment." Cross-cultural, then, refers to juxtaposition of any two culturally distinct entities that may be found even within the same country. Second, despite the volumes of literature and plethora of studies conducted to help international managers better understand the transnational context of their work, there is very little "how-to" help available. It is so primarily because the available academic literature has its limitations in differentiating between findings that could be generalized across countries and the ones that are largely country specific. Readers of cross-cultural studies must recognize this limitation especially when scales or measures developed in one context are applied elsewhere.

While some scales—such as the individualistic versus collectivist orientation of a culture—are considered universally applicable, a cross-cultural application of such scales is often flawed because all societies do not as neatly polarize on a phenomenon (Metcalf et al. 2007). For example, the assessment of the preference of a culture to do explicit or implicit types of agreements poses measurement challenge in Mexican context since Mexican managers seem to prefer both implicit and explicit agreements.

In the academic realm, Adler (1983) divides cross-cultural management research into six broad categories—parochial research (domestic management studies), ethnocentric research (replication of domestic management studies in foreign cultures), polycentric research (individual studies of organizations in specific foreign cultures), comparative research (studies comparing organizations in many foreign cultures), geocentric research (how multinational organizations work), and synergistic research (how to manage the intercultural interaction within a domestic or international organization). These different research streams within international business studies have produced knowledge benefitting international managers in making sound decisions in cross-cultural settings. In this chapter, I refrain from taking a theoretically rooted approach to examine cross-cultural issues. While I sometimes allude to relevant theoretical propositions, my main objective is to offer readers a practical and lively portrait of some of the commonly arising issues forest products managers deal within cross-cultural marketing and sales contexts.

The chapter moves as follows: First, I provide a brief account of the globalizing nature of the forest industry. Then I focus on consideration and skills necessary for international managers while also comparing some of their roles and responsibilities in the context of smaller and larger companies. Further, I discuss the various facets of cross-cultural negotiations and provide some examples to distinguish them from within-a-culture negotiations. I wrap up the chapter by discussing characteristics and skills of a successful cross-cultural/international manager.

14.2 Globalized Forest Industry

The pace of globalization of the world economy has increased significantly since 1989, with the collapse of the Soviet bloc; creation of a single Europe, its eventual expansion to 25 member states, and adoption of a single European currency; implementation of the North American Free Trade Agreement (NAFTA); establishment of the World Trade Organization (WTO); and the emergence of several Asian countries as manufacturing hubs (Johnson et al. 2006). With respect to the forest sector, forest products have been transported internationally for centuries. For example, lumber made from *Fitzroya cupressoides* (a Patagonian species similar to the US redwoods) was shipped for centuries from the island of Chiloe in southern Chile to Lima, Peru. The current wave of globalization, however, has dramatically changed the character of international trade in the forest sector. Many forest products companies have now created permanent sales offices in foreign countries. More recently, several companies have even started to own forest land overseas. There are several reasons for this offshore land acquisition and manufacturing. Firstly, small size of domestic markets has led many companies to explore international opportunities (like Stora Enso, a European company that has invested heavily in Latin America, outside their home turf). Secondly, low land prices and outstanding mean annual increments (MAIs) of plantation crops (such as pine or eucalyptus) have fostered overseas investment.

For example, a large proportion of the 1 million hectares of planted forests in Uruguay are owned by international companies.

Similar developments have happened on the manufacturing side. For example, not only companies such as Stora Enso and Arauco (a large Chilean forest products company) own land in Uruguay, they are also investing there over 2 billion US dollars to build a 1.3 million ton cellulose mill. US companies like Louisiana Pacific (LP) now own manufacturing facilities in Chile and Brazil and are successful in selling all that production within South America at advantageous pricing. China currently leads world plywood production, which had reached over 35.4 million m^3 by 2009 (Research Report on Chinese Plywood Industry). Further, a global push for bioenergy production is also impacting globalization of the forest sector as several companies are forming alliances with energy sector companies globally (e.g., the biofuel joint venture between Weyerhaeuser and Chevron).

In addition to economic factors, natural changes have also impacted global forest products trade. According to the State of World Forests 2011 report (FAO 2011), today's competitive environment for forest products is constantly changing, often in unpredictable ways—for example, the mountain pine beetle infestation in West Canada is creating a huge amount of cheap logs that need to be commercialized before they rot. Experts forecast that this overabundance of logs will be followed by a scarcity of pine, which will likely result in skyrocketing prices of lumber and wood in North America (and elsewhere) (International Wood Markets Group–Press Release–April 25, 2012).

In addition to the supply/company side, globalization is also happening on the consumption/customer side. While some regional differences continue to exist in consumption patterns, consumers across the globe are seeking western products. Major global retail chains, such as IKEA, leverage this homogenization of consumption and operate in far-flung locations. More recently, IKEA is setting up to enter India, a market that has traditionally had a fragmented retail sector but the one that promises huge potential for international brands.

In fact, many international companies strive to find and develop new markets outside the traditional markets since the traditional markets are either saturated or the opportunities outside are more promising for long-term growth. Today, demand for wood products from China, India, Mexico, South America in general (e.g., Chile, Colombia, Peru), the Middle East (e.g., United Arab Emirates), and Southeast Asia (e.g., Philippines, South Korea, Vietnam) is driving global markets and providing buffer to many international companies against the fall of the US market in several wood products segments. Mexico and the Middle East import today about 1 million m^3 each of plywood (1 million m^3 equates to 1.13 bsf on a 3/8 in. basis, where bsf is billion square feet). This demand did not exist 15 years ago. To keep this figure in perspective, it is equivalent to over 20% of the entire North American production during 2010 (10.9 bsf 3/8 in. basis) (Crow's Weekly Market Report July 22 2011). It is also noteworthy that the current surge (Second quarter 2013) in wood products prices in the United States owes less to a reviving US housing market and more to the increasing demand outside the United States. This increasing demand has caused a manufacturing capacity shock and has led to price increases.

14.3 Considerations for International Managers

Cultural differences are one of the most important considerations for international managers. Failing to appreciate cultural differences may lead to cultural clashes, which can ruin a business relationship. Culture is known to have a profound impact on people's interaction goals, which in turn influence their styles of interaction and their interpretation of others' behaviors (Hartel et al. 2003). It is therefore foremost that international managers understand that their counterparts may be coming from an entirely different frame of reference in their communication since culture provides structure for the communication process (Birdwhistell 1970). There are numerous skills and competencies that can facilitate cross-cultural communication (Lloyd and Hartel 2003) and are broadly categorized into three factors—attitude, skills, and knowledge. Others have proposed similar typologies. Hofstede (2001), for example, suggests that intercultural communication competence involves awareness, knowledge, and skills. LaFromboise et al. (1993) suggest that in order to be culturally competent, an individual must

1. Possess a strong personal identity
2. Have knowledge of and facility with the beliefs and values of the culture
3. Display sensitivity to the affective processes of the culture
4. Communicate clearly in the language of the given cultural group
5. Maintain active social relations within the cultural group

It must be noted that knowledge of others' culture is only a small part of what is involved in cross-cultural competence. In addition, an individual's ability to step over his/her cultural boundaries to make the strange familiar is a key prerequisite for cross-cultural competence (Byram 1997).

In addition to cultural issues, managing relationships and maintaining coordination with other functional areas are critically important for international marketing managers. Differences of opinions between manufacturing/production and marketing/sales areas are well known and can be particularly visible in large companies. Manufacturing folks generally tend to possess a "you sell what I make" attitude, whereas the sales/marketing folks hold a "you need to manufacture what I sell" attitude. Although tools (e.g., sales and operations planning [S&OP]) have been developed to effectively integrate these two functional areas, differences, sometimes irreconcilable, are commonplace occurrences. Box 14.1 has a real-life example that illustrates how disagreement on small issues may arise between the two functional areas and how an open-minded attitude may be helpful to resolve a conflict.

Cultures differ in assigning importance to different traits. Punctuality is, for example, more valued in North America and Europe than many other cultures. An international manager must be patient with others' pace and sense of time. In the initial phases of business relationship development, an awareness and appreciation of such differences may be helpful for building trust. Small gestures can make big differences. Once, a sales manager in Uruguay stopped his car three times in a 250-mile trip to let his two Japanese customers have a smoke. It was deeply appreciated by the Japanese customers and was a good example of sensitivity to others by the Uruguayan manager, who himself was a nonsmoker. People in Argentina and Southern Brazil drink an herbal tea called "mate" in which they drink sharing a straw both in social and professional settings. Outsiders are also expected to join in.

Cultural setting and the importance of market knowledge and careful negotiation: During a sales meeting, the sales director of a multinational firm was intending to agree to sales volumes and prices for the second half of the coming year. The commercial manager of a national distributor in Argentina started by giving an extended presentation of how sour the political situation was becoming in his country. He argued that only an important price reduction across the product portfolio would allow market share maintenance.

BOX 14.1   MANAGING THE RELATIONSHIP BETWEEN SALES AND PRODUCTION FUNCTIONAL AREAS

An example from a South American operation illustrates the type of issues that can come between production and sales. A discussion between the mill manager and the country sales manager took place regarding inventory management. Mills typically have a "frontier" or dispatch area where the sales and logistics people take "ownership" of the product created and packaged by the manufacturing people. In this particular situation, the mill warehouse was a complete mess. The sales, logistics, and mill people collectively decided to divide the physical warehousing space in lots and each lot had a letter associated with it—a, b, etc. Each letter represented a destination (e.g., Europe could be "a," and bundles of material destined for Europe would have an "a" printed on their labels and would be allocated to the "a" lot, where they would wait for a truck that would take them to the port later in the week).

When the country sales manager visited the mill, he realized that the material was separated in lots, but no letters were to be found anywhere in the warehouse. When asked, the mill manager countered that any fixed lot would subtract flexibility as the volume to each destination varied wildly month to month (which was true). Accordingly, he argued that letters printed on the mill floor would be not good, and indications hanging from the ceiling were not good either. The sales manager countered that it would be impossible for the forklift operators to know where to place a bundle with an "a" on it. As might be expected, the discussion got harsh, as supposedly everybody knew where each lot was, but for the sake of flexibility, it was just a collective "knowledge" with no signs whatsoever. In the end, the sales manager suggested the lots could be indicated with red cones on the floor (movable, no flexibility compromised). And the letter identifying the lot would be nicely printed on tripods standing in front of the piles. If the pile needed to be moved, so did the tripod.

The sales director was in trouble—the longstanding relationship did not allow for even doubting the grim outlook and, after all, a local person supposedly knows the domestic situation better than an outsider. Still, the sales director's intuition was that half of the requested discount would be enough to keep the business—and he needed the market badly. So, he basically commented that such a reduction would force him to sell the usual volume in an alternative market, namely, Germany. Only seconds after, the reduced discount was accepted and the meeting ended. Here, the sales director could not have competed with the domestic knowledge of his counterpart, but he was aware of the cultural setting he was in, so he guessed the needed discount was quite lower. By suggesting switching to an alternative market, he proved he was right.

14.4 Typical Roles and Responsibilities of an International Manager

Trust building is key to successful international marketing, which necessitates in-person contact, at least during initial phases of a trade relationship. Unlike in domestic marketing where trade relationships are often developed and cultivated over telephone conversations, it is very uncommon for international marketing managers to not have personally met their suppliers or customers. Metcalf et al. (2007) argue that in the Unites States, this difference exists primarily because of a heavy reliance on the legal system that would make trust building with trade partners (and, therefore, in-person contact) somewhat less important for domestic marketing.

International managers perform a myriad of functions that vary based on the size of the companies they work for and also on the type of role they are assigned. I have identified and described in the following texts three categories of situations that will provide readers a general and broad picture of the roles and responsibilities of international managers in the forest products sector. First, I outline the roles and responsibilities of an international manager working for a small trading company. The second category includes roles and responsibilities of an international manager working for a large company. Third, I outline the roles and responsibilities of an international manager working for a subsidiary company in a host country location. Readers will note that these three categories also appear in an ascending order of complexity in roles and responsibilities.

A typical day of an international manager of a small trading company constitutes making phone calls to different suppliers across the world. Typically, these managers first identify buyers in the domestic market and then look for international suppliers (located outside the home country) who could potentially meet buyers' demands. Generally, international suppliers can be found using the Internet and initial contact can be set up via emailing. To build trust and to better understand a potential supplier, however, a personal visit is often made. Prices, product specifications, manufacturing capacity, logistics, etc., are discussed and if the trader is satisfied with the various aspects, she or he would enact a first trial container for her or his customer in the home country. Upon a series of successful trials with a supplier spread over several months, a trader often decides to do steady business with this supplier and increases the volume of business with this supplier. Overall, there are fewer continued relationships than the ones that last a trial or two. Adherence to quality and timely delivery are two founding aspects of a longer-term trade relationship. International traders generally remain open to explore businesses with new suppliers and therefore their travel frequency remains high, both for keeping ongoing trade and developing new relationships.

Although an international manager working for a large forest products company also shares some of the roles of a small company trader, their roles have vast differences. While they both need to create successful trading programs, travel significantly, and deal with numerous cross-cultural issues arising between their customers and suppliers, differences in the complexities of their organizations differentiate their roles. Traders in a small company often need to focus on developing trading programs for increasing the volume of their companies' business. In doing so, they always look for entrepreneurial trade partners who may be trusted for product quality and timely delivery. In a large company, however, international managers' first focus remains to abide by their companies' strategy, which may sometimes even trump a potentially profitable deal with those firms that don't fit in their companies' strategy. In a large company, several peers often do similar things and thus coordination with peers is very important. In other words, there are several internal policies issues that international managers of large companies must comply with. Due to such bureaucratic requirements, large companies are also slow in responding to market changes. While the slow decision-making process of large companies might frustrate some international managers, large companies can sometimes make managers' tasks easy. First, these managers do not need to worry about not having enough money to pursue a trading program. Large companies' established reputation and brand name help managers easily develop business relationships much faster. Also, because large companies typically have experts in all fields, international managers can easily find someone to consult with regarding any product or process.

Country sales and marketing managers of large forest products companies working in one of parent companies' subsidiaries abroad have all the responsibilities that small company traders or international managers of a large company located in headquarter do. In addition, they are often required to represent their companies at various forums such as the chambers of commerce, trade fairs, and trade meetings. These latter responsibilities are especially relevant when the subsidiary is located in a small country (e.g., Uruguay). Involvement of country sales and marketing managers in various government and public affairs is relatively lower in larger countries (e.g., Australia, Brazil, Chile, or China). This difference exists because numerous international companies operate in larger countries where opinions or inputs of one international company are not considered as important. It must also be noted that country sales and marketing managers must also be particularly careful with antitrust and politically sensitive issues. These positions are also highly visible to NGOs and local government officials, and therefore these managers are well advised for consulting with their headquarters on social and political matters.

14.5 Negotiations in Cross-Cultural Settings

Product quality and pricing are two most important issues that an international manager should be diligent about and become apt at articulating and negotiating. Countries differ in terms of their quality expectations. Japan is generally considered the most quality-demanding market in the wood products segment. In many countries, price considerations are foremost and quality comes later. US markets typically consider a material acceptable if it meets grading rules. In contrast, customers in China, Chile, Mexico, or Uruguay place less emphasis on the grading rule and rather place a heavier emphasis on the looks of a product relative to a competing one. This is true even in Western European countries that have their own grading rule system but where looks relative to a competing product are of utmost importance. To illustrate this with a hypothetical example, a product from Brazil may pass the grading rule, but if, say, a Chilean product looks better, a Mexican customer might prefer it, so much so that a better-looking product may command a 10%–15% price premium. On the flip side, there are sometimes negative perceptions associated with some countries and even some product species. Right or wrong, such perceptions are hard to change and transactions involving negatively perceived country of origin or species generally suffer a price reduction. An example in this regard is the perception of eucalyptus wood products in Mexico. In Mexico, eucalyptus trees grow crooked and are rather small, so there is a stigma over anything made of eucalyptus—a big challenge for suppliers of veneer from Tasmania, Eucalyptus lumber from Brazil, or Eucalyptus plywood from Uruguay.

Similarly, price negotiations—their extent and the ways they are conducted—considerably vary among countries and cultures. In the United States, pricing guidelines are well established and there are reference resources such as Random Lengths or Crow's that lists weekly prices for a large number of wood products. Random Lengths uses market surveys to arrive at these prices. The popularity and market acceptance of Random Lengths is so high that prices in the United States are typically negotiated "at Random" or, for example,

"$10 per MBF below Random Lengths." In other words, Random Lengths has become a quasi-unit for wood product pricing in the United States. An immediate conclusion is that these price guidelines take challenge out of price negotiations between buyers and sellers as they basically ride the market up or down. Interestingly, several forest products companies in the United States have officially declared that they do not share their own trading pricing with these publications—sometimes those companies have a very important share of the production of a certain wood product, which for sure poses a question over the validity of the price guidelines (Shook and Sisodiya 2011).

On the other hand, publications that offer prices for wood products are available in Europe and some places in South America as well, but they are not typically used for negotiations. In most countries outside the United States, prices are generally arrived at where the seller and the buyer agree, which is an important difference between the United States and the rest of the world. In these international settings a seller can always try to convince a buyer that market conditions are getting tighter and that prices are poised to go up (or not). This all means that in international business quick and accurate information of any events that may shock demand or production capacity are of utmost importance. In this regard, a great example was the fire that destroyed a very large plywood mill in Chile at the beginning of 2012. Chile produces about 1.3 million m^3 of plywood per year, which may look small, but a large proportion of that production is high-grade pine plywood, and actually most of the high-grade pine plywood that is imported in the United States comes from Chile. The fire event in Chile caused a huge surge in pricing, as traders and customers typically overreact to events like this.

China is another very interesting example regarding price negotiations. China is a furniture-manufacturing powerhouse and producing companies inside China are fierce competitors. Access to cheap raw material is the key. In this regard, Chinese companies typically refrain from riding a market up when a specific raw material gets too expensive.

Instead, Chinese companies look for a substitute—they typically all move to that substitute up to when it gets again too expensive and scarce and then they move again to a new raw material. All this happened with Russian logs and lumber. When they got too expensive, Chinese companies moved to Radiata pine logs and lumber from New Zealand and Radiata pine lumber from Chile.

14.6 Characteristics and Skills of a Successful Cross-Cultural/International Manager

International sales management is a versatile task that is significantly different from domestic sales management. Long ago, Black and Porter (1991) very eloquently commented that a successful manager in Los Angeles may not succeed in Hong Kong, since the managerial skills and characteristics leading to a good performance in one location may be ineffective in another.

First and foremost, it is important for international managers to possess a high degree of cultural intelligence, defined as "a system of interacting knowledge and skills, linked by cultural metacognition, which allows people to adapt to select and shape the cultural aspects of their environment" (Thomas et al. 2008, 126). Basically, the notion of cultural intelligence suggests that a person needs to possess cultural knowledge (e.g., to know that Chinese are a collectivist society) and cultural skills (e.g., the ability to behave appropriately in a new cultural setting). In addition to these two qualities, culturally "intelligent" behavior entails a person possessing cultural metacognition, that is, knowledge of and control over one's thinking and learning. A seductively simple interpretation of these characteristics may lead one to think that cultural intelligence is all about cultural mimicry. While this interpretation holds good to a certain extent, Thomas et al. (2008) found that high levels of mimicry is a double-edge sword since it can be perceived as an insincere and even devious act.

Additionally, Maisonrouge (1983) suggested that international managers should speak one or two foreign languages while also having a reasonable understanding of the sociopolitical environment in which they will be working. Knowing the history, geography, economy, sociology, government, and laws of the world's main countries is always helpful for breaking social boundaries. In Maisonrouge's view, an international manager is close to being a Renaissance man—searching for knowledge for the sake of knowledge.

Overall, a successful international forest products manager must have the following characteristics:

- *Mastering more than just one language is a must.* English is an international language, but certain nationalities, notably Spain and Brazil, are very lingual-centric societies, and therefore it is always advantageous to speak Spanish or Portuguese. Knowledge of Russian, Mandarin, and Portuguese has become important with the advent of Russia, China, and Brazil on the global trade map.

- International managers typically deal with either high-level executives or owners of companies abroad. Therefore, mannerisms and the ability to have social conversation on a wide range of issues/topics are necessary for making a good personal impression. Therefore, in addition to technical knowledge of their products and industry, international managers must try to develop a reasonable understanding of general sociopolitical affairs of their own countries as well as of those where they do business.

- Having a cross-cultural training before departure to the work location of the international manager can play an important role (Black and Porter 1991).

- *Having a sound knowledge of trade rules and detailed technical aspects of the products (or services) is extremely important.* Many successful international wood products managers have technical degrees in wood science and engineering or civil engineering, which certainly helps, particularly in instances involving huge claims around product quality aspects. A sound technical understanding of wood is important. An international wood products manager, for example, must know that while a blue stain in pine is not aesthetically appealing, it does not affect its structural integrity.

- Logistics make integral, important components of international trade. It must be emphasized here that it is one of the most crucial technical skills required for an international manager and she or he must have a thorough understanding of and familiarity with transportation issues including inland and international routes. It is imperative to understand implications of decisions such as using trucks versus railways and private versus public port. Besides all this, it is helpful to develop good working relationships with key international shipping companies.

- A good international manager is often an outstanding negotiator, who is able to deal with a compounded effect of different personalities coming from different cultural contexts.

- A successful international manager is continually gathering market intelligence from around the world. In the absence of a global quality–price matrix, customers in the international arena are often hard negotiators and an uninformed seller may easily get tricked. Keeping a close eye on market developments and price fluctuations is important.

- Wood products companies selling internationally need abundant cash flow to pay salaries, logs, steamship lines, etc. A person may say that a company selling domestically faces similar issues. The difference between both such companies is that international sales typically involve letters of credit or other such instruments, so effective payment of a load of lumber may take 3 months before the cash really enters the company's checking account. A good international sales manager needs to be aware of these nuances, so he can balance the different sales terms and keep cash flowing.

- International managers typically develop within a team. The diversity of a multicultural team allows for an eclectic set of perspectives, which can be most helpful in developing a holistic, internationally robust perspective. A successful international manager must essentially be a team player.

14.7 Conclusion

International business is a lot of fun as it continually varies. Timely information is key and good international managers need to be continually reading news, talking to good contacts, and basically reading market signs so as to make smart moves and be among the first to ride a market up and also be among the first to reduce exposure to a market when it is showing signs of weakness. The latter sentence "reduce exposure" to a market is used as customer loyalty is key in international business. Thus, leaving a market altogether because prices are lower at a particular point in time may mean losing a good customer. He will have no alternative but buying from the competition. So, maintaining adequate volumes in key markets, even if prices at a certain point in time are disappointing, may prove very smart in the long term, as a willing to buy customer is, at the end, the most important thing.

International managers are in high demand today. The world is getting smaller and smaller and people with the right set of hard skills (technical knowledge, languages, etc.) and soft skills (be able to relate effectively with people of other nationalities) are scarce. The hard skills can be learned, the soft skills are much trickier, and I hope this chapter shed light on some of the nuances necessary to become a good, effective, international manager.

References

Adler, N. N. J. (1983) A typology of management studies involving culture, *Journal of International Business Studies*, 14(2): 29–47.
Birwhisttell, R. L. (1970). *Kinesics and Context: Essays on Body Motion Communication*, Philadelphia, PA: University of Pennsylvania Press.

Black, J. S. and Porter, L. W. (1991) Managerial behaviors and job performance: A successful manager in Los Angeles may not succeed in Hong Kong, *Journal of International Business Studies*, 22: 99–113.

Byram, M. (1997). *Teaching and Assessing Intercultural Communicative Competence*. Clevedon, U.K.: Multilingual Matters.

Crow's Weekly Market Report Viewpoint (2011) North American softwood plywood market profile by Wade Camp, Bedford, MA, July 22, 2011.

FAO. 2011. *State of the World's Forests 2011*, Rome, Italy, www.fao.org/docrep/013/i2000e/i2000e00.htm (January 2, 2012).

Härtel, C. E. J., Panipucci, P., and Fujimoto, Y. (2003). Fostering diverse workgroups who excel in decision-making, *Australian Journal of Psychology*, 55(Supp.): 127–128.

Hofstede, G. (1980) *Culture's Consequences: International Differences in Work-Related Values*, Beverly Hills, CA: Sage Publications.

Hofstede, G. (2001) *Culture's Consequences: Comparing Values, Behaviors, Institutions, and Organizations Across Nations*, 2nd edn, Thousand Oaks, CA: Sage Publications.

International Wood Markets Group-Press Release—BC. Interior's annual allowable timber harvest to fall from 60 million cubic metres to 40 million cubic metres and will support a smaller forest products industry, Vancouver, British Columbia, Canada, April 25, 2012.

Johnson, J. P., Lenartowicz, T., and Apud, S. (2006). Cross-cultural competence in international business: Toward a definition and a model, *Journal of International Business Studies*, 37(4): 525–543.

Maisonrouge, J. G. (1983) The education of a modern international manager, *Journal of International Business Studies*, 14: 141–146.

Metcalf, L. E., Bird, A., and Peterson, M. F. (2007) Cultural influences in negotiations: A four country comparative analysis, *International Journal of Cross Cultural Management*, 7: 147–168.

LaFromboise, T., Coleman, H. L., and Gerton, J. (1993). Psychological impact of biculturalism: Evidence and theory, *Psychological Bulletin*, 114(3): 395.

Lloyd, S. L. and Hartel, C. E. J. (2003). The intercultural competencies required for inclusive and effective culturally diverse work teams. In: P. James, (ed.) *Cultural Diversity in a Globalising World: Proceedings of the Diversity Conference 2003*. Honolulu, HI. February 13–16, 2003.

Pepke, E. Forests, *Markets, Policy & Practice*, Shanghai, China, 22 June 2011.

Research Report on Chinese Plywood Industry, 2009–2010 China Research and Intelligence, www.shcri.com

Shook, S. R. and Sisodiya, S. R. (2011) Confronting the information vacuum: Price discovery and determination in the forest products industry, by Steven R. Shook Professor of Marketing and Sanjay R. Sisodiya, Assistant Professor of Marketing, College of Business and Economics, University of Idaho, presented at the IUFRO 5.10.00—*Forest Products Marketing and Business Management meeting UNECE/FAO Team of Specialists on Forest Products Markets and Marketing*, June 16, 2011, Corvallis, OR.

State of World's Forests 2011 FAO

Thomas, D. C., Stahl, G., Ravlin, E. C., Poelmans, S., Pekerti, A., Maznevski, M., Lazarova, M. B., Elron, E., Ekelund, B. Z., Cerdin, J.-L., Brislin, R., Aycan, Z., and Au, K. (2008) Cultural intelligence domain and assessment, *International Journal of Cross Cultural Management*, 8: 123–143.

Wood Markets Global Wood Products Conference Press Release A 'Super-Cycle' of soaring lumber prices… May 16, 2011—Vancouver, British Columbia, Canada.

15

Corporate Social Responsibility in the Global Forest Sector*

David Cohen, Anne-Hélène Mathey, Jeffrey Biggs, and Mark Boyland

CONTENTS

---

* David Cohen, University of British Columbia, Anne-Hélène Mathey, Jeffrey Biggs, and Mark Boylan, National Resources Canada, Canadian Forest Services © Her Majesty the Queen in Right of Canada and Taylor & Francis.

15.1 Introduction

Over the past decades, the evolution of the "old economy" into a "new global economy" has brought about numerous changes in how businesses go about conducting their operations. The accompanying explosion of information technology has changed the speed and reach of business operations, not only bringing about new structures at the market, industry, and firm levels but also changing the role and power of consumers as *customers* (Zuboff and Maxmin 2004). In this new business environment, the customer has evolved from "solution demander" to "value demander" and from "client" to "partner" (Prahalad and Ramaswamy 2004; Porter and Kramer 2011). Moreover, there are increased expectations from businesses to be responsive to a wide variety of environmental and social issues. The period characterizing such expectations is often called the corporate social responsibility (CSR) era, and CSR has emerged as one of the most profound topics for both business practitioners and academicians. In recent years, there has been a proliferation of environmental and social initiatives led by the private sector that traditionally were within the purview of government. Overall, businesses are developing new models to respond to emerging expectations.

   Through much of the 1970s and 1980s, the main objective of senior corporate management was to maximize shareholder value to the exclusion of other stakeholder interests (Gordon 2007). However, since the early 1990s, this approach has been challenged. The impetus to this change was associated with the presentation of *Our Common Future* to the United Nation's Environment Program's 14th Governing Council Session in 1987 (Our Common Future 1987). That report provided the foundation for contemporary definitions of sustainability through the quote in the conclusion of Chapter 2 "sustainable development meets the needs of the present without compromising the ability of future generations to meet their own needs." It also contributed to the development of the UN Conference on Environment and Development in Rio de Janeiro in 1992.

   This conference reflected a broad shift in societal expectations for governments and business. It initiated some of the first global declarations to incorporate social and environmental concerns in economic development. Governments signed agreements on sustainable development (Agenda 21), the environment and economic growth (Declaration on Environment and Development), climate change (UN Framework Convention on Climate Change), biodiversity (UN Convention on Biological Diversity), and forests (The Statement of Forest Principles).* It was also one of the first conferences with strong participation of nongovernment organizations (NGOs), with over 2,400 representatives and attendance of over 17,000 people at a parallel forum run by NGOs. The conference instigated changes not only to economic development but also to how business stakeholders were defined and fundamentally how business operated in the global economy.

---

* Downloaded on March 10, 2010, from page 2 of http://www.un.org/geninfo/bp/enviro.html.

These changes have had substantial impact on the forest sector. Indeed, forests and forestry was the only industrial sector singled out in the agreements signed at the conclusion of the Rio meeting. Only one year after the Rio Conference, protests in Clayoquot Sound on the west coast of Vancouver Island in Canada garnered international attention as NGOs ensured that environmental and social values, key concerns of stakeholder groups other than corporate shareholders, were considered (at least) as equally legitimate as returns to capital in forest management decision making. The evolving role of business in sustainability is exemplified by the development and growth of numerous forest certification schemes that have set the groundwork for growth in certification schemes for a variety of products and product groups. The forest sector has also provided a model for a shift from confrontation to increased cooperation among businesses, environmental nongovernment organizations (ENGOs), and indigenous peoples (for a detailed account of ENGO strategies, see Chapter 9 in this book). This focus on the environmental impacts of forestry is illustrated by CSR trends identified by Li and Toppinen (2011):

1. The need for firms to acknowledge and communicate their social and environmental impact to a wider base of stakeholders.
2. The adoption of sustainable forest management policies.
3. Forest companies shape their sustainability strategies to fit specific geographical parameters.
4. Planning environmental communication and adopting risk management strategies.

What is CSR? How has it been evolving? How have CSR and associated concepts been applied, and evolved, in the forest products sector? The purpose of this chapter is to explore these fundamental questions. This section introduces the relevant academic literature that forms the background to the evolution of CSR theory in general, particularly in terms of how it has moved from theory to practice—in the process becoming an integral part of business strategy and operations. Section 2 clarifies the impacts of this evolution in terms of business emphasis and terminology, in order to facilitate the understanding of its integration into actual business practices. Section 3 examines the various strategic postures firms adopt relative to CSR/ESG implementation. Section 4 identifies global and firm level drivers for environmental and social action by firms, including forest products firms. Section 5 projects current actions and trends into a more sustainable future and outlines various approaches used to measure and assess CSR/ESG. Section 6 addresses the implications of CSR for policy making.

15.1.1 Evolution of Theory

CSR is also referred to as corporate responsibility, corporate citizenship, business ethics, stakeholder management, triple bottom line accounting, and, more frequently in the twenty-first century, sustainable business (Van Marrewijk 2003; Sharma and Henriques 2005; Carroll and Shabana 2010). The terminology is loosely defined and meanings shift based on regional, industrial, authorial, or other contexts (Vidal and Kozak 2008b).

CSR evolved from the early twentieth century era of paternalistic philanthropy by wealthy industrialists (Van Marrewijk 2003), which formed the foundation of nonprofit organizations, such as the Rockefeller Foundation, to a much broader concept of sustainable business operations and development. However, the contemporary era of CSR started with the 1953 publication of Howard Bowen's *Social Responsibility of Businessmen*: According to Carroll and Shabana (2010), for the decades following Bowen's (1953) publication, CSR was predominantly an academic exercise with only "slowly emerging realities of business practice." Vogel (2005) considers that "old style" CSR of the 1960s was motivated by social (e.g., labor welfare and social justice) considerations. This era was characterized by growing academic musings about business social responsibility but little uptake by business firms. Eberstadt (1977) argues that the modern CSR movement is trying to restore the historic connection of community and business. However, during the 1970s firms began to respond to a changing business environment, with what Frederick (1978) labeled corporate social responsiveness. Over the decades of 1980s and 1990s, environmental considerations began to be incorporated alongside social elements, due to growing concern over the environmental impact of business growth in a globalizing context (Sharma and Henriques, 2005). During these last two decades, CSR has been linked with financial performance (Orlitzky et al. 2003) as it has evolved from a relatively theoretical field to one closely linked with business, best practice.

A key shift has recently occurred in the CSR realm. For a while it remained a topic of academic discussion generally in terms of socially oriented actions of businesses beyond profit considerations (Vidal and Kozak 2008b), whereas businesses now seek to adopt socially responsible actions to *improve business performance* generally known as the "business case" for CSR. According to Carroll and Shabana (2010), key business benefits of CSR included cost and risk reduction, gaining a competitive advantage, reputational improvement, and value creation through overlapping win–win outcomes. Business acceptance of CSR was further accelerated as global business scandals (e.g., Enron) affected the global business environment for multinational enterprises. As it stands today, CSR has evolved "into a core business function which is central to the firm's overall strategy and vital to its success" (Carroll and Shabana 2010, 93).

In 2003, Van Marrewijk summarized the scientific literature on CSR as a sequence of three approaches:

1. Shareholder approach—The classic view that the social responsibility of a firm is to increase profits for its shareholders (Friedman 1970).
2. Stakeholder approach—Expands beyond shareholder to include other agents that can affect or are impacted by a firm's actions. In this approach, defining stakeholders becomes a critical activity (Freeman 1984).
3. Societal approach—Firms need a social license to operate and to obtain this requires actions beyond profit to meet broad societal goals (Donaldson and Dunfee 1994).

We add a fourth approach that has emerged since Van Marrewijk's 2003 publication, described in recent articles by Porter and Kramer (2006, 2011). They contend that for many forward-looking businesses, future growth and profits will rely on contributing to the solution of key global social and environmental problems. Development of hybrid vehicles and environmentally friendly engineered wood products exemplify Porter and Kramer's contention. A recent report prepared jointly by the Massachusetts Institute of Technology and Boston Consulting Group (2011) suggests that many leading firms are responding to growing local and global social and environmental risks and finding unprecedented business opportunities by doing so. We argue that this emerging integration changes Van Marrewijk's sequence into a interlinked circle as shown in Figure 15.1, as pressures

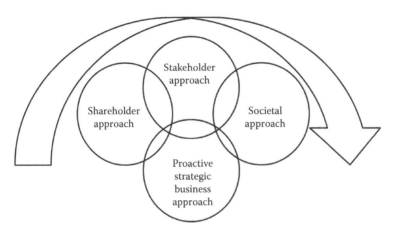

FIGURE 15.1
Evolutionary path of CSR theory.

from stakeholders and the need for social license create opportunities for profit that a shareholder-oriented firm must pursue. Following the Porter and Kramer proposal, CSR is no longer beyond profit, it is now a key component of otherwise traditional business models. We call it proactive-strategic business approach.

15.1.2 Evolution of Forestry Practice

For more than a century, forest products companies have been publicly criticized for the ways they manage forest resources. This criticism intensified after the 1970s and started to cover issues related to both environmental impact and community relations. These debates were further fueled by an increasing appreciation of the role of forests in global environmental issues such as climate change mitigation. The need for the forest sector to respond to various social and environmental concerns coincided with its financial struggles lasting over a decade. Return on capital employed (ROCE) has been low and has fallen below a modest target of 12% annually (IBM Consulting Services 2003). Especially in countries with high economic dependence on the forestry sector, its economic health had direct impact on the regional economy and social well-being (Panwar et al. 2006).

Despite the financial challenges, the forest industry has been generally responsive to these evolving concerns and has shifted its emphasis as a given concern passed through the identification, rationalization, resistance, and eventually recognition as requiring a response. Hansen and Juslin (2011) note the key social and environmental issues over the past 50 years for the forest products industry: 1970s (water and air emissions), mid-1980s (recycling), late 1980s (chlorine bleaching), early 1990s (forest management), late 1990s (forest certification), and twenty-first century (climate change, poverty alleviation, and indigenous peoples). Readers will note that unlike most other industry sectors where social concerns have typically preceded environmental ones, CSR in the forest sector has followed a different evolutionary path.

In addition to temporal differences, regional differences also exist in terms of the forest sector's response to CSR. Panwar et al. (2006) noted that European companies had been formally reporting their CSR activities for a longer time than North American or Asian companies. Recent research by Panwar and Hansen (2009) has identified the social and environmental issues that the US forest products industry must address in fulfilling its social responsibilities. Social issues include encouraging public scrutiny of forest practices, investing in communities, promoting responsible use of wood products, and improving the industry's public image. Environmental issues include promoting sustainable forest practices, adopting green purchasing policies, improving eco-efficiencies, and increasing the use of renewable materials. The same authors (2007) found that overarching CSR issues in India and the United States included global warming, declining biodiversity, deforestation, and climate change. Sharma and Henriques (2005) found that Canadian forest products firms had moved beyond the early stages of sustainability practices (e.g., reducing pollution and increasing eco-efficiencies) to the intermediate stage (e.g., material reuse, sustainable harvests, and redesign of processes). However, they had not moved to the advanced stage (e.g., redefining business and industrial ecosystems that may require relocation to utilize waste generated by other industries). Vidal and Kozak (2008b) found that different regions emphasized different issues in CSR implementation. Africa and Latin America had a greater focus on social issues, while Asia, Europe, and Oceania emphasized environmental issues. Within these broad categories, they found further differences in specific social or environmental activities that were emphasized within each region. In contrast to focusing on land-management practices that are typically emphasized in forest product exporting countries, Mikkila and Toppinen (2008) found that the import-reliant Japanese pulp and paper industry focused more on philanthropy, carbon sequestration, employee loyalty, and compliance to legal environmental requirements.

Lately, the forest sector is being increasingly viewed as a partner in providing solutions to many social and environmental concerns. Emergence of a carbon market, increasing use of wood as a renewable energy source, and an increasing acceptance of the value of the host of ecosystem services that forests provide have made a strong *proactive-business strategy* case for the forest sector wherein socio-environmental concerns, offer new economic opportunities for the forest sector. Scandinavian forest products firms, for example, have been particularly active in using corporate responses to climate change concerns as drivers of new (and often lucrative) profit centers, through green certificate programs.

15.2 Changing Theory and Practice: A Kaleidoscope of Terms

A rapidly changing theoretical landscape of CSR over the past few decades has created a proliferation of divergent terminologies. The initial focus on social concerns broadened through the environmental movement to make CSR increasingly mainstream, to which the proliferation of CSR programs and CSR reports attest (KPMG 2011). As CSR reporting and communication became popular among companies, the corporate world began to wrestle control of the CSR terminology away from academics, creating a semantic confusion over the basic meaning of core concepts. Lately, however, a terminological shift has happened that seeks to shift CSR away from an academic and theoretical field of study to an integral part of business strategy and operations. This terminological shift has at least two notable characteristics. (1) More companies are now reporting their social and environmental performance as "sustainability reports" as opposed to a previous practice of publishing "CSR" reports (Vidal and Kozak 2008a). (2) Companies are preferring the label environmental, social, and governance (ESG) rather than CSR, which is understandable because of an intertwined nature of corporate governance issues with social and environmental performance. More importantly, this terminological shift also replaces CSR responsibility and philanthropic motivation with a matter-of-fact approach that considers ESG issues insofar as they affect business. The new focus does not go *beyond* profit but remains on *generating* profit through ESG practices. Essentially, while CSR language generally focused on motivation and intentions, ESG seems to focus more on actions and results.

15.2.1 What Do "CSR" and "ESG" Mean?

CSR has been defined in numerous ways. From an economic perspective, CSR suggests that firms should internalize social and environmental externalities resulting from their operations (Auld et al. 2008). It focuses on a firm's ethical obligation to society based on sustainability criteria (Wilson 2003). In his metastudy of the current literature, Dahlsrud (2008) found that the definition for CSR by the Commission of the European Communities was used most frequently. According to this definition, CSR is "[a] concept whereby companies integrate social and environmental concerns in their business operations and in their interaction with their stakeholders on a voluntary basis" (Dahlsrud 2008, 7).

Given the way CSR literature evolved in the academic realm, it is a daunting task to ascertain whether CSR refers to a company's intentions, actions, or outcome of its actions. With a notable exception of Wood (1991), who made clear distinction between CSR principles, processes, and outcomes, CSR discourse has largely remained vague about its own boundaries. Putting together ontologically distinct elements of human experience as motive, action, and consequence has led to incoherent terminologies and their inconsistent use. The simultaneous use of motive and action domains to describe CSR is especially problematic since the same action could be considered socially responsible or not depending on the motive. The prevailing ambiguity fails to answer whether a firm that inadvertently created environmental good while seeking profits should be called socially responsible.

In order to reconcile the motive and action domains as they relate to CSR, we offer a framework (Figure 15.2) that visually represents the separation and integration of CSR motives (fundamental drivers of firms' CSR-related decision making) and actions (steps taken by firms). This clearly illustrates how the same action may have different motives depending on the stakeholder's perspective.

One of our observations is that while the different elements of the CSR action domain may often overlap (i.e., an action can be both socially and environmentally, or economically and socially desirable), elements of CSR motive domain *do not*. One motive ultimately trumps another, and a firm comes to a point where it must decide whether its ultimate bottom line is profit generation or environmental soundness. As a result, while one ENGO may engage in a joint project with a firm to achieve a particular action, another may find incompatibilities in motives to pursue the very same actions. This latter ENGO might hold the view, "it is not CSR because the motive is profit-seeking," whereas the firm might be espousing, "it is CSR because the actions are sustainable." Readers will note that while actions are visible, motives remain hidden and often unexamined even by those who pursue them.

In part to avoid the quagmire of determining CSR motives, the ESG concept was proposed and further developed as a finance initiative by the Asset Management Working Group of the United Nations Environment Group (UNEP 2006). ESG typically refers to firm level environmental and social impact management strategies to increase and protect firm value as a component of a traditional business model. The ESG components (i.e., environment, social, and governance) are typically nonfinancial in-and-of-themselves but may affect the financial performance of businesses, in the mid to long term (WBCSD and UNEP 2010). The ESG concept is *not* altruistic—its principal purpose is the financial interest of the firm,

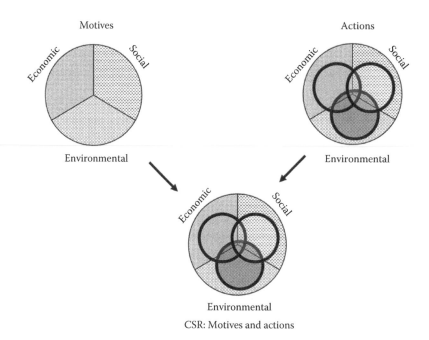

Motives

Actions

Environmental

Environmental

Economic

Social

Environmental

CSR: Motives and actions

FIGURE 15.2
CSR as defined by motive and action components.

more generally the interest of shareholders/owners. A number of studies indicate that improving ESG performance enhances long-term competitiveness (Margolis et al. 2007; Nidumolu et al. 2009; Robinson et al. 2011). ESG is also used frequently in the context of socially responsible investing (SRI), but these terms are not interchangeable: though both have social and environmental objectives, ESG is motivated by improvement to the firm financials, while SRI's motivations are broader, encompassing an idealist focus on societal benefit. Recent research highlights the use of ESG as one of the many investment philosophies driving SRI (Nikolakis et al. 2012).

As risks and opportunities associated with environmental and social impacts have grown, it has become necessary for firms to limit risk and improve competitiveness in these areas as well (Nidumolu et al. 2009). The concept of ESG is useful because it describes the new business environment that firms are facing: a broader group of stakeholders—including the general public, customers, investors, business partners, firm employees, and large customers within supply chains—are paying increased attention to the environmental and social performance of firms. Collectively, these stakeholders have a large influence on critical business functions such as sales, financing, supply chains, and social license. It is becoming clear to managers that failing to adequately respond to this new environment can damage profits and share price and, conversely, that effective responses can be well rewarded, providing a competitive advantage (Berns et al. 2009; Jantzi-Sustainalytics Inc. 2010).

The terms CSR and ESG are often used interchangeably in the literature, which has exacerbated semantic confusion. Since they both share many issues, drivers, and measures, a clear distinction between the two is important, which we propose as follows: CSR refers to a value-laden motivation ("firms with a conscience"), whereas ESG considers environmental and social outcomes insofar as they improve financial outcomes. The emphasis of

ESG is on generating value, while the emphasis of CSR is on operations based on values. That is, CSR has an idealist approach to social and environmental matters, whereas ESG is purely instrumental.

15.3 CSR and ESG: The Emergence of Strategies in a Changing Business Environment

The first two sections charted the theoretical evolution of CSR while clarifying terms in the light of emerging practice. This section investigates in greater detail the alternative strategies to implement CSR/ESG. Firms' CSR strategies can broadly be grouped into three distinct categories: reactive, responsive, and proactive. Companies can move from having adopted one type of strategy to embrace other strategies over time and such a movement will represent a strategic shift for an individual firm (Figure 15.3).

15.3.1 Reactive Strategies

In response to external pressures, many firms develop *reactive strategies*, which are typically designed to control risks while minimizing disruption of operations. Reactive strategies also include regulatory compliance without trying to get ahead of existing regulations. This approach protects companies' social license to operate when and where it is threatened but is based on companies assuming the role of a follower. In response to a catastrophe, a company can make significant short-term gains by showing that they take a given issue seriously. A timely response that is publicly perceived as responsive can pay large dividends (literally and figuratively). However, those gains often plateau once short-term social expectations have been met. Although reactive strategies were most common in the forest products sector during 1980s and 1990s, a recent example in June 2011 involved the toy-maker Mattel and its paper packaging sources when they were targeted by a global Greenpeace ad campaign against rainforest destruction in Indonesia. This campaign prompted Mattel to announce changes in its packaging supplier and its packaging supply chain management just a few days after the campaign was launched.

This strategy is based on the traditional trade-off analysis whereby CSR-related actions are considered as inherently increasing costs and reducing profitability. In other words,

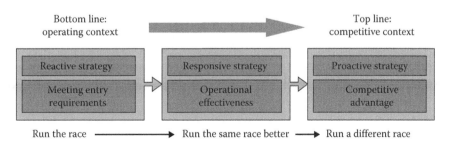

FIGURE 15.3
CSR/ESG strategy continuum. (Adapted from Porter, M.E. and M.R. Kramer., *Harvard Bus. Rev.*, 84(12), 78, 2006.)

firms trade some profits for social license to operate (Jaffe et al. 1995). Firms adopting this strategy would strive for minimizing their CSR-related costs, and therefore, they would initially resist calls for action but then react strongly before costs significantly increase.

15.3.2 Responsive Strategies

Increased awareness of environmental and social issues in the public sphere has led some firms to adopt *responsive strategies*. Responsive strategies are still fundamentally defensive and primarily oriented around risk management but consider a longer-term perspective. For instance, in case of a retail firm, a responsive strategy would involve the modification of its supply chain to minimize the potential of an ENGO action against environmental degradation associated with products this firm carries. The firm would then develop strategies to mitigate that risk and then incorporate the "values" that motivated that shift into their regular public communication. Firms such as Home Depot and Kingfisher set minimum certification standards for the suppliers of wood products they carry as a measure to minimize threats of potential ENGO actions. Another example of responsive strategies would be to actively decrease greenhouse gas (GHG) emissions over time by reducing energy use and/or switching to renewable energy sources in anticipation of future social and/or regulatory requirements. Responsive strategies can have CSR motivations, where senior management wants to "do the right thing" or ESG motivations, where firms seek to protect or enhance market share. Despite the different motivations, the outcome strategy remains the same. For instance, early adopters of forest certification appear to have had CSR motivations, while later adopters appear to be practicing ESG. This illustrates that differing motivations can represent the same responsive strategies and actions as illustrated in Figure 15.2. That is, early adopters shifted the playing field and made it necessary for other firms to engage in the same responsive strategies to protect their operations and markets.

Responsive strategies still base themselves on a trade-off analysis but over a much longer time frame. The trade-off recognizes that mid- to long-term risk-adjusted profits can be maximized by strategically implementing more sustainable activities (Porter and van der Linde 1995).

15.3.3 Proactive Strategies

Recently, the broad integration of ESG issues in the development of corporate strategies (MIT and BCG 2011) has led to the emergence of *proactive* strategies (see Figure 15.4). Rather than react or respond, firms adopting proactive strategies seek to turn environmental and social issues into competitive advantages that can be leveraged to improve financial performance and gain market share. For some firms, this is making explicit the previously unexaminable distinction between a purely profit-oriented motive (ESG) and a more idealistic and value-laden motive (CSR).

One of the best known examples of proactive strategies in CSR realm is of GE Ecomagination, which was formed in 2005 with a sole mission of developing solutions to address challenges related to cleaner and more efficient energy supplies, improved sources of clean water, and reducing air and water emissions. They have focused on developing new solutions to global environmental problems through the development of numerous products and services. "Ecomagination is GE's business strategy to create new value for customers, investors and society by solving energy, efficiency and water challenges— today" (p. 6). Over its first 5 years of operation (2005–2010), it has produced the following

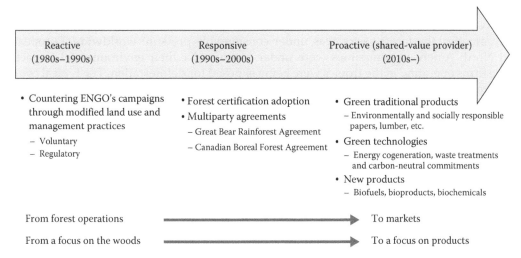

FIGURE 15.4
The evolution of ESG strategies of leaders in the forestry industry.

results: $85 billion worth of new revenue, 22 new product or solution launches, $1.8 billion in R&D research, company energy intensity efficiency increased by 33% since 2004 base year, and $130 million in energy savings (GE 2010).

In the forest products sector, current examples of proactive strategies include the launch of Domtar's Paper Trail resource calculator in June 2011, which allows potential clients to evaluate the environmental impact of the production of various Domtar paper grades and compare these results to those of Domtar's competitors. By taking this proactive step, Domtar not only limits its exposure to negative environmental campaigns by making a large volume of environmental impact data publicly available, it is also developing its brand value based on transparency and superior environmental performance. ESG has become one of their competitive performance criteria. Recent articles by Porter and Kramer (2011) have called this approach "shared value," which "involves creating economic value in a way that *also* creates value for society by addressing its needs and challenges (Porter and Kramer 2011, 64)". It suggests that a firm can best profit over the long term by helping solve society's challenges. Collins Products of Portland, Oregon, is a company that has embraced responsibility concepts for decades. As a mid-sized, private company, it was one of the first in the United States to become Forest Stewardship Council (FSC) certified and now maintains certification on all of its lands. The company uses the Natural Step principles to help mitigate its environmental impacts company-wide and is one of the first mid-size forest products companies to conduct a company-wide carbon footprint audit to gain a better understanding of its total carbon impacts.

Proactive strategies recognize that responsible actions can create profit for a firm by solving social and environmental problems through technological and social innovation. This approach is based on the recognition that with increasing standards of living in many developing countries with large populations (e.g., China, Brazil, and Indonesia) and a growing global population, there is pressure at a global scale on social and environmental systems that will require innovative technological and societal solutions to maintain sustainable systems and development. The larger global firms with big brands "leverage corporate sustainability for competitive advantage, business growth, and increased sales" (Dauvergne and Lister 2011).

The forest industry has been at the forefront of both reactive and responsive strategies and is currently evolving into more proactive strategies as illustrated in Figure 15.4. In the early nineties, the forest sector was under considerable pressure worldwide. European and North American companies were under pressure for their environmental impacts while firms operating in the tropics were being exposed for deforestation. The typical company reaction was either to ignore the pressure until customers forced them to respond or to use well-established public relations firms to counter much of the negative publicity. This proved to be ineffective and firms shifted to a more responsive set of actions. These included agreeing to increased areas of forest and protected from industrial activity in North America and more rigorous laws and restrictions on areas of forest harvest in both North America and Europe. Forest certification was developed and embraced predominantly by firms operating in the developed world, while deforestation in the tropics, the stated target of the first certification schemes, continued unabated. With the growth in public environmental and social concerns and increased opportunities in sustainable business activities, the proactive approach is now being adopted by leading edge firms, those referred to as the "embracers" by MIT Sloan (2011). For forest companies, this means the development of new processes and products that facilitate the substitution of nonrenewable material use with renewable material using wood supplied from well-managed and sustainable forests. It also includes the development of markets for and the provision of forest-based environmental services, increase longevity of wood products to enhance carbon storage in construction, bio-based chemicals and materials, and much more. The research is underway and we anticipate an explosion of new processes and products over the next decade.

15.4 Drivers of Environmental and Social Corporate Activities

Stakeholders of a firm include its customers, employees, surrounding communities, concerned citizens, investment firms, unions, NGOs, certification organizations, and regulators. These groups increasingly put pressure on firms to improve their environmental and social performance. On the policy side, this stakeholder pressure has resulted in increasingly strict ESG-related regulations. Governance-related regulations are most pronounced in Europe, where they have been created to regulate banking, labor, environmental pollution, product composition, and the make-up and use of fuel. While China also has exemplary environmental regulation, these are only effective if implemented and monitored to assure adherence. The Yale Environmental Performance Index indicates that European countries lead the way in performance, but China falls into the weak performer category (Yale EPI Rankings 2012).

Whereas previous stakeholder actions focused on applying external pressure that would force firms to become more socially responsible (e.g., successfully pressuring for more rigorous forest management regulations), the latest wave has demanded that firms voluntarily choose sustainability rather than have it chosen for them (e.g., demands from pension plans for socially responsible operations and the growth of SRI funds). Although external pressure through reputation, distribution, or integration within a larger value chain remains an important driver of CSR/ESG, operational and growth-oriented benefits in the areas of cutting costs through eco-efficiency and pursuing opportunities in new markets and products have more recently become leading reasons for adopting CSR/ESG strategies (Bonini and Gornen 2011).

15.4.1 Global Drivers

Traditional global drivers toward increasing acceptance of ESG include the growth of global value chains, the growing power of large multinational retailers, and the growing public awareness of sustainability issues, particularly environmental issues in developed countries and social issues in lesser developed and developing countries.

15.4.1.1 Global Value Chains

Sustainability requirements (a non-tangible product attribute) of a significant minority of consumers in developed countries (e.g., Europe) impacts the standards of operations and production throughout the supply chain, all the way back to resource extraction, which often takes place in developing countries with relatively lax regulations and/or enforcement. For example, farmers in Africa are rejecting the use of genetically modified seed to ensure continued access to European markets. Wood products from tropical countries cannot be imported into the United States without meeting new environmental and social criteria under the Lacey Act that was amended in 2008 to cover imported wood. The globalization of value and supply chains, as well as markets, puts further pressure on good management: Firms run the risk of market exclusion if the chain and/or distributor they are involved with require them to have good environmental and social performance.

15.4.1.2 Growing Power of Large Retailers

Globally, the power of big box stores all along the value chain has been increasing. These retailers are shifting away from reacting to single-protest initiatives (e.g., the use of giant blowup chainsaw balloons in Home Depot parking lots) to proactively moving toward becoming part of sustainable value chains, which not only improves their sustainability performance but also provides protection from future protest actions. These powerful retailers are driving measurable sustainability indicators back along their value chains. For example, Walmart is requiring additional energy efficiency from its suppliers in China with the expectation that savings will lead to reduced costs of production. Home Depot and Rona in North America and B&Q in Europe have set minimum wood certification standards for purchases of solid wood products.

According to Dauvergne and Lister (2011), there has been a race since 2005 among many of the largest branded multinational corporations to become recognized as "global sustainability champions." These include firms as diverse as Walmart, HP, Nike, and Coca-Cola. Due to a global lack of international standards, these firms are defining, and developing metrics to measure, sustainability "effectively behaving as global environmental regulators."

15.4.1.3 Growing Public Awareness of Sustainability Issues

There is increasing consumer awareness of the potential environmental and social harm associated with their purchases, which is driving firms to focus on sustainability issues (MIT and BCG 2011). The reality is that human wants are expanding, but there is only a limited supply of natural resources. Businesses want to ensure both a continuing demand for their products and a capability to be a continuing supplier, which requires them to produce goods that do not have serious social and environmental negative externalities. The continuous increases in the proportion of waste paper is recycled and the expectation that it will continue in the future are good examples of the efforts to stretch the supply and use of even renewable materials (UNECE/FAO 2012, 34). More recently, forest products companies have been reducing their environmental impact by converting waste into new products such as particleboard, pellets, bark mulch, and nanocellulose crystalline. Current research is looking at producing a perfume fixative from Balsam Fir that would replace ambergris, which the endangered sperm whale excretes and is collected from shorelines.

15.4.2 Firm-Based Drivers

Some of the drivers apply at the firm level but may not be pertinent to all companies. Still, they apply to many firms, particularly large multinationals.

15.4.2.1 Risk Control

Due to the proliferation of market-based campaigns by NGOs, risk control is a growing driver for global companies to become increasingly accountable and be proactive in managing adverse environmental and social impacts. A 3-year campaign in the early 2000s to stop Boise Cascade from selling products produced from endangered forests is an apt example of how NGO campaigns could pose serious threat to a firm. This campaign resulted in the abrupt cancellation of millions of dollars' worth of business for the company, and a more gradual loss of customers, including institutional buyers such as Washington Mutual, LL Bean, Levi-Strauss, Lowes, and Kinko's. During this period, profits dropped and Boise fell from being the largest harvesting company of public lands in the United States to being number eleven. Single catastrophic events such as this campaign, natural and industrial catastrophes, or sudden regulatory changes can destroy value and often directly threaten the continuity of operations. Such events exert acute pressure on a firm to adopt ESG over a short period of time. Shareholders of a firm now expect that firms are vigilant in reducing exposure to such risks, which increasing pressure on managers to embrace ESG.

15.4.2.2 Brand Enhancement/Protection

Environmental and social catastrophes can inflict serious brand damage, while improved outcomes can enhance brand value. In 1993, a massive public outcry over plans to clear-cut forests in Clayoquot Sound in British Columbia, Canada, was supported by ENGOs who convinced several European firms to cancel pulp contracts with MacMillan Bloedel, the company holding the logging rights in this area. In 1996, MB lost 5% of its sales when two of its customers in the United Kingdom, Scott Paper and Kimberly-Clark, canceled existing contracts (Svendsen 2000). These market-based initiatives by ENGOs are not just historic: In 2011, Greenpeace targeted CottonSoft in New Zealand for selling toilet tissue that included endangered Indonesian hardwoods (Greenpeace 2012), a claim CottonSoft denied as they struggled to defend their brand. This campaign may result in the closure of a $20 million dollar plant and the loss of 130 jobs in an economically distressed region, after several large supermarkets stopped carrying these products to avoid the negative publicity associated with this campaign (Adams 2011). Proactive social and environmental actions may provide some protection to a brand from this type of green mail—particularly when announced in conjunction with an ENGO.

15.4.2.3 Enhanced Firm Performance

A meta-analysis of numerous studies exploring links between corporate social performance and firm profitability found "a positive or neutral effect on financial performance" (Arnold 2008). A more recent study found superior share performance for companies in the Dow Jones Sustainability Index, which may be a surrogate for better quality management in general (Robinson et al. 2011). Promoting eco-efficiencies such as reducing energy use and packaging can decrease costs of input resources and waste management (Nidumolu et al. 2009). In Finland, paper mills have used combined heat and power systems since the 1960s and become major electric utilities, supplying about one-third of all electricity to the national grid. UPM, one of the world's leading forest products firms, has stated that by becoming more energy efficient, it decreased its dependency on fossil fuels, decreased its carbon footprint, and increased its revenues from selling both carbon allowances and excess power.

15.4.2.4 Access to Resources and Market

A record of poor environmental/social performance can limit a firm's access to key resources or trigger a consumer boycott of its products. Poor environmental/social performance may also make a firm less attractive to current and prospective employees, particularly knowledge workers who are in short supply. One of the results of the Clayoquot protests was a much-reduced annual allowable cut and the creation of a land-use management planning system, which made accessing timber more difficult, time consuming, and costly for MacMillan Bloedel. The degraded reputation of Asia Pulp and Paper (APP) in Indonesia continues to hurt the credibility of their sustainability claims even when the company pursues third party forest certification.

15.4.2.5 Innovation and New Initiatives

Growing awareness of the impact on the environment and society of an expanding global population and the impressive economic growth of emerging countries is creating new opportunities for businesses to help society address major social problems. Growing a business through responding to social problems can help a company or even a sector reinvent itself as part of the solution rather than part of the problem. Many firms are using social and environmental responsibility as guides toward innovation in new products and processes. From GE Ecomagination, to Unilever's Sustainable Living Plan, to Weyerhaeuser's commitment to superior sustainable solutions, many businesses are using innovation for solving sustainability challenges. Currently, UPM-Kymmene and Domtar are basing their current research and development and future growth on bioproducts and bioenergy sectors.

15.5 ESG Today and Tomorrow: Shaping the Future of Environmental and Social Issues Management?

Linking business with social and environmental solutions is a relatively new concept. It is unrealistic to expect a rapid maturation of strategies and actions. The world Business Council for Sustainble Development (WBCSD) was formed in 1995 and the Forestry Working Group in 1997 has become the Forest Solutions Group today. Its members consist of companies all along the forest products supply chain focusing on sustainable forest management. Its Vision 2050 document identifies the decade from 2010 to 2020 as the "turbulent teens": a dynamic time during which new measures are developed, businesses achieve a deeper understanding of environmental realities, and corporate leadership helps address the needs of all segments of society (WBCSD 2010). WBCSD considers this decade the time period within which the legal and regulatory system evolves to support sustainability, and business models change to help deliver the infrastructure for a sustainable future. During the turbulent teens, the components of ESG have also been shifting based on changing societal values and available metrics. In the following text, we discuss how firms move from simpler to a more evolved approach and set of strategies with respect to CSR/ESG implementation, from low-hanging fruit to proactive long-term strategies.

Implementing CSR/ESG strategies requires significant organizational commitment including leadership by senior level management, buy-in by other managers and employees, knowledge development that crosses functional areas, and resource allocation. Additionally, environmental and social issues must become an integral part of the strategic planning process and not just tactical ad hoc adjustments.

15.5.1 First Steps

CSR/ESG strategy implementation typically focuses on improving resource efficiency and waste management (MIT and BCG 2011). This can be approached by pushing lean operations both down and up the supply chain, focusing on minimizing material and resource use, often known as "eco-efficiency," and encompasses production material, energy, water, and packaging, etc. Many firms have found that implementing eco-efficiencies all along the value chain leads to increased knowledge of the supply chain and improves the management of the chain itself, leading to better overall management of operations. Once a firm starts pushing lean concepts along the supply chain, the immediate improvement to the bottom line ensures continued vigilance in areas where environment and economy intersect (see Figure 15.2). In the forest products industry, this resource efficiency is exemplified by improved lumber recovery factors, computer-controlled pulping to reduce chemical and energy use, producing energy from black liquor, continuous waste reduction, and closing in on energy self-sufficiency. Many forest products firms, particularly in Europe and North America, have moved beyond this first stage and have already reaped the benefits of eco-efficiencies.

During these initial steps, CSR/ESG reporting is typically a mere public relations exercise based on anecdotes and glossy publications, with little quantitative content. Reports provide little information that covers a time frame of more than a single year and few midterm targets. They often follow a perfunctory adherence to some of the simpler concepts given in reporting standards but exhibit a lack of rigor in terms of reliability, balance, and measures. There is also no external assurance of the content.

15.5.2 Transitional Steps

As company commitment and knowledge increase and the benefits of following a sustainability path become apparent, firms transition to a more proactive strategy (see Figure 15.3) by becoming more engaged in pertinent issues. They adopt some of the existing measures and look for having better measures to demonstrate their progress. In regions where environmental aspects are paramount, firms start to explore other aspects of sustainability (e.g., social aspects such as indigenous rights). Similarly, where social aspects are paramount (e.g., many parts of the developing world), firms start focusing on environmental aspects. Firms adopt measures to track changes and start to set short-term targets for areas such as reduction of energy use, water use, waste flows, and improvement in worker health and safety, community relations, employee satisfaction, etc. Firms in this stage often see sustainability as a means for risk management and efficiency gains (MIT and BCG 2011).

During this stage, reporting becomes more quantitative with the adoption of targets for measurable results. This can include reduction in energy and reporting of targets using GHG protocols. Commitments are made in terms of reduction in waste, water, and energy use. Reports are more balanced and include both positive and negative sustainability performance of the firm. There is an increase in timeliness (i.e., regularity of reporting), clarity, balance, and comparability. There is more rigor in following reporting guidelines.

15.5.3 Leading the Pack, the Embracers

The leading edge firms in terms of sustainability—"the embracers"—have a far greater commitment to sustainability, integrating it into their core business model (MIT and BCG 2011). According to Bonini and Gorner (2011), "larger shares of executives say sustainability programs make a positive contribution to their companies' short- and long-term value." These leading firms engage in a proactive strategy and make sustainability a foundation of innovation to create long-term gains. As an example, a joint project between Siemens and the Government of Singapore to develop a lower-cost means to desalinize water, using less energy than the current processes exemplifies sustainability emanated innovation (Economist 2011). Initiatives undertaken by forest companies in Europe and North America to promote building codes that allow high-performance, high-rise wood buildings to reduce GHG footprints is another example.

Embracers report comprehensively using both quantitative and qualitative measures, create defined targets, and conduct regular assessment of progress or lack thereof. As more data get collected and standardized metrics for sustainability variables are developed, there is a gradual move toward integrating sustainability reporting with financial reporting (IIRC 2011). However, as discussed in the next section, developing stronger sustainability metrics/measures is a key step moving forward.

15.6 Approaches and Initiatives for Measuring CSR/ESG

Measurement of CSR/ESG is important because it provides a basis for monitoring, control, and comparison. Leaders today "spend time and effort quantifying the impact on their businesses of things such as brand reputation, employee productivity and the ability to attract and retain top talent" (MIT and BCG 2011). Measurement tools are not only essential to management teams in assessing their firms' performance, they are also helpful to various stakeholders, including investors and policy makers, to make decisions of sorts.

Auld et al. (2008) identify the following commonly used measurement approaches in the CSR/ESG realm: information approaches, market-driven approaches, public–private partnerships, environmental management systems (EMS), and industry association codes of conduct.

15.6.1 Information-Based Approaches

Information-based approaches focus on developing new measures to gather data over time and provide sustainability performance in a consistent reporting standard to facilitate both firm-based changes over time and comparison between various firms. One of the prominent reporting standards, Global Reporting Initiative (GRI), is a network-based nonprofit organization that provides a sustainability reporting framework, which, according to KPMG (2011), is used by over 80% of the largest 250 global companies and 69% of the largest 100 companies in 34 developed and developing countries. GRI has become the de facto standard for sustainability reporting. Information tools also include stock sustainability indices such as the Dow Jones Sustainability Index (DJSI) and Brazil's Business Sustainability Index (ISE). A third category within information tools consists of the third-party lists such as the Greenpeace Guide to Greener Electronics, Corporate Knights Global 100, and Newsweek Green Rankings.

15.6.2 Market-Driven Approaches

These approaches rest on the premise that consumers will reward companies that show environmental/social stewardship. For consumers to compare performance standards, companies get product-/process-related certifications. The forest sector was a leader in using this approach with the emergence of various forest certification programs. In the social arena, there are several fair trade and product quality (such as organic food certification) and workplace conditions programs (such as Social Accountability International's SA 8000:2001 certification). Market-based approaches have their inherent strengths and limitations. For example, increased influence by any single group of stakeholders may have some detrimental effect on overall stakeholder management processes. In case of certification programs, there is often resentment from many stakeholder groups since these programs are often perceived as being dominated by industry associations. The breakdown in meaningful standards that inhibit the effectiveness of the Kimberley Process Certification Scheme to prevent "blood diamonds" from entering mainstream markets is a current example of some of the problems associated with well-intentioned non-state market-driven approaches (Legget 2011). The most effective market approaches include diverse stakeholders in a process to ensure that one group of stakeholders does not subvert the overall intent of any program.

15.6.3 Public–Private Partnerships

Public–private partnerships are collaborations between private firms and state organizations. Key examples include the various business sector partnerships with the Environmental Protection Agency (EPA) of the United States; the UN Global Compact, REDD+*; and a recent agreement between forest products firms, first nations (indigenous peoples of the area), and the government with financial support from international foundations in the Great Bear Rainforest in British Columbia, Canada. Many of these public–private partnerships are developing new models of collaborative management among diverse stakeholders and have had an admirable success record although they are still works in progress. As such,

---

* REDD is the acronym for the UN program on reducing emissions from deforestation, which offers incentives for developing countries to reduce emissions for forested land by reducing deforestation and increasing conservation (see Chapter 10 for details).

continuing efforts and communication are necessary to keep them on track given the diverse perspectives involved in developing and implementing plans.

15.6.4 Environmental Management Systems

Environmental management systems (EMSs) are "structured framework[s] of procedures to implement and achieve a company's environmental performance targets and goals" (Auld et al. 2008). EMSs often are externally developed criteria based on International Organization for Standardization (ISO) standards, such as Europe's Eco-Management and Audit Scheme known as EMAS and ISO 14001. Numerous private firms now offer EMS systems to help companies regularly measure their environmental impacts based on internationally recognized standards. In the forest products sector, many large forest products companies in industrialized countries are certified for ISO 14000, a well-recognized environmental management standard.

15.6.5 Industry Codes of Conduct

Often industry associations lay down written rules that detail the social and environmental expectations of businesses belonging to a specific industry sector. Industry codes of conducts are growing in popularity. Notable industry codes include the chemical industry's voluntary Responsible Care Program and more recently the Sustainable Apparel Coalition (which is also a public–private partnership since the EPA is a member). Industry codes of conduct are in vogue in the forest sector as well. In Canada, for example, no firm can belong to the Forest Products Association of Canada (the primary forest industry association) unless its forest operations are certified by a credible forest certification scheme.

15.6.6 Environmental Product Declarations

Environmental product declarations (EPDs) are considered the next wave of eco-labels within the market-driven approaches. EPDs consist of a standardized report of environmental impacts of products. They are similar to nutritional labels on food packages in that they provide information on specific environmental criteria. EPDs do not rate or rank the products; such comparison is left to consumers who can then compare labels. EPDs use life cycle analysis to evaluate the environmental impacts over time, providing a rigorous (and in principle, replicable) quantifiable standard for comparison (Bergman and Taylor 2011). EPDs are increasingly used in Europe and Asia and are now being introduced to North America (Bowyer et al. 2011; O'Connor et al. 2011).* LEED, the largest green building certification program in North America, is exploring using EPDs to evaluate materials used in construction (Gonchar 2011).

15.6.7 Other Popular Approaches/Initiatives

As public and scientific concern regarding climate change has grown, many firms have implemented strategies to reduce both their energy use and GHG emissions. Since 2001,

---

* The head start of European firms in business and sustainability initiatives results in the dominance of European countries in KPMG's (2011) Leading the Pack category. It is of note that the forestry and pulp and paper sectors also fall in the Leading the Pack category.

an ongoing partnership between the WBCSD and the World Resources Institute has developed an accounting framework "to understand, quantify and manage greenhouse gas (GHG) emissions."* It has become a de facto standard to measure GHG emission in a consistent and transparent manner that allows accuracy, comparability, timeliness, and reliability and incorporates key principles of the GRI reporting principles. Many international organizations such as ISO and the Climate Registry have accepted this standard.

Similarly, the Carbon Disclosure Project (CDP) is a nonprofit organization developed as a collaboration among corporations, government, and NGOs. It currently represents over 550 investors with $71 trillion of assets. A recent survey of the S&P 500 reported that almost two-thirds of 336 respondents reported that climate change issues are integrated into their overall business strategy and disclose GHG emissions targets and progress using the GHG protocols (CDP 2011).

Another recent approach gaining traction is ISO 26,000, a guidance standard that sets principles for organizations to follow for sustainable operations. It is a generic, global, and holistic framework for organizations to become more sustainable. ISO 26,000 addresses seven principles of responsibility: accountability, transparency, ethical behavior, respect for stakeholders' interests, respect for the rule of law, respect for international norms of behavior, and respect for human rights. Each organization must address seven core subjects: organizational governance, human rights, labor practices, the environment, fair operating practices, consumer issues, and community involvement and development. ISO 26000 is a comprehensive guideline that details what needs to be addressed but does not specify how to address or measure the specifics.

15.7 Policy Implications

Involvement in addressing social and environmental issues—firm actions, standard setting, etc.—introduces firms into roles that have traditionally been filled by the government. In this sense, CSR/ESG is changing the context within which governments' business-related policies evolve. The multiplication of the number of providers of environmental and social goods is likely to increase the complexity of the policy making and implementation process (e.g., Cashore 2002). More actors mean more opportunities for diverging goals, more means for reaching them, and the higher potential for conflict among them. Therefore, policy innovation, in particular effective multi-stakeholder processes are needed for coordinated and harmonized efforts. Jurisdiction is another area where CSR/ESG practices have increased complexity. The boundaries between federal, state of province, indigenous communities, and private jurisdictional authority are increasingly muddled by the growth of social and environmental good provisions of private firms. Policies will necessarily have to be crafted carefully to avoid violating emerging jurisdictional norms, while appropriately promoting and defending traditional roles.

Furthermore, while it is possible that the private sector will expand its role in social and environmental good provision through CSR/ESG, these strategies remain in the realm of voluntary firm decisions and are therefore subject to change without any assurance of continuity for the environmental and social goods they deliver. A private sector actor

---

* From http://www.ghgprotocol.org/ downloaded on December 10, 2011.

expanding its role in social and environmental good provision through CSR/ESG is doing so as a corporate decision, not necessarily as an act to ensure long-term general well-being. Therefore, government policy must be continuously cognizant of and vigilant towards the risks private provisioning of social and environmental goods may inflict upon the society (Vogel 2005). In particular, many proactive CSR/ESG activities depend on developing and tapping into *fashionable* environmental and social causes. Changing fashions may conflict with some key social issues. State policies can play a key role in ensuring continuity in such situations.

Even when changing social/environmental fashions are not in play, as firms provide environmental and social services, they implicitly enter into new forms of social contracts whose rules, responsibilities, and arbitrators are yet ill-defined. This new contract may raise stakeholder expectations and a withdrawal may raise discontent. For instance, Walmart, which had for decades promoted the idea of having a family relationship with its employees, rather than pursuing management–union wage/benefit negotiations, suddenly reduced benefits and implemented large-scale layoffs to cut costs. Many employees felt that a social contract had been broken. This led to significant damage to Walmart's brand due to a public perception of unacceptable labor practices. In this case, an ESG strategy designed to enhance the value of the firm was eventually abandoned, resulting in significant social and economic costs. Going further, a firm that halts environmental and social programs or services that have come to be expected by society may present some new problems for governments: To whom does the responsibility for providing these services devolve? This is not to say that CSR/ESG activities are ill-advised—it is to emphasize that they are relatively new phenomena that must be considered carefully by firms before implementation and by governments before promotion.

15.8 Summary and Conclusions

In recent decades, the nature of the rights and responsibilities that characterize the relationship between business and broader society has been changing, particularly in terms of the role that privately owned firms have in responding to social and environmental concerns. Initial attempts to describe these changes were descriptive and highly theoretical and gave birth to a growing literature on CSR. As the practice of CSR has become more common, the business community itself has started to develop competing, often overlapping, terminologies, which have served to introduce confusion into the literature. This chapter has been an attempt to review the relevant literature so as to chart the evolution of this space, while providing a means to examine the evolution of these theories in practice. As a result, the evolution of CSR/ESG strategies (reactive, responsive, and proactive), drivers of adoption (at international and intrafirm levels), and the approaches and tools for institutionalizing CSR/ESG are covered. Throughout, the forest industry has been used to provide relevant examples and highlight particular issues. As the new relationships are emerging between private and public policy, we gave particular attention to the implications of CSR/ESG for policy makers. We anticipate that this complexity will remain the defining characteristic of CSR/ESG for many years to come and that the ability of all stakeholders to navigate this complex space will largely determine whether these new relationships are characterized by new solutions or new conflicts. History suggests that they will likely be remembered for both.

References

Adams, C. 2011. Cottonsoft: Greenpeace claims pulp fiction. *The New Zealand Herald*, November 22, 2011. Downloaded from http://www.nzherald.co.nz/environment/news/article.cfm?c_id=39&objectid=10767797

Arnold, M. 2008. Non-financial performance metrics for Corporate Responsibility Reporting Revisited. Doughty Centre for Corporate Responsibility, Cranfield University School of Management. Bedford, UK. Available at http://www.som.cranfield.ac.uk/som/dinamic-content/research/doughty/wp1.pdf (Accessed September 10, 2010).

Auld, G., S. Bernstein, and B. Cashore. 2008. The new corporate responsibility. *Annual Review of Environmental Resources*, 33:413–435.

Bergman, R. and A. Taylor. 2011. EPD–environmental product declarations for wood products-an application of life cycle information about forest products. *Forest Products Journal*, 61(3): 192–201, 10.

Berns, M., A. Townend, Z. Khayat, B. Balagopal, M. Reeves, M. S. Hopkins, and N. Kruschwitz. 2009. Sustainability and competitive advantage. *MIT Sloan Management Review*, 51(1).

Bonini, S. and S. Gorner. 2011. McKinsey global survey results: The business of sustainability. Downloaded November 10, 2011 from https://www.mckinseyquarterly.com/Energy_Resources_Materials/Environment/The_business_of_sustainability_McKinsey_Global_Survey_results_2867, 16 pages.

Bowen, H. R. 1953. *Social Responsibilities of the Businessman*. New York: Harper & Row.

Bowyer, J., J. Howe, K. Fernholz, S. Bratkovich, and S. Stai. 2011. *Environmental Product Declarations (EPDs) Are Coming: Is Your Business Ready?* Minneapolis, MN: Dovetail Partners, 11 pp.

Carroll, A. B. and K. M. Shabana. 2010. The business case for corporate social responsibility: A review os concepts, research and practice. *International Journal of Management Review*, x: 85–105. DOI: 10.1111/j.1468–2370.2009.00275.

Cashore, B. 2002. Legitimacy and the privatization of environmental governance: How non state market-driven (NSMD) governance systems gain rule making authority. *Governance*, 15(4): 503–529.

CDP. 2011. CDP S&P 500 Report 2011: Strategic advantage through climate change action climate disclosure project https://www.cdproject.net/CDPResults/CDP-2011-SP500.pdf

Dahlsrud, A. 2008. How corporate social responsibility is defined: An analysis of 37 definitions. *Corporate Social Responsibility and Environmental Management*, 15(1): 1–13.

Dauvergne, P. and Lister. J. 2011. Big brand sustainability: Governance prospects and environmental limits. *Global Environmental Change*, doi:10.1016/j.gloenvcha.2011.10.007, in press.

Donaldson, T. and T. W. Dunfee. 1994. Toward a unified conception of business ethics: Integrative social contracts theory. *The Academy of Management Review*, 19(2): 252–284.

Eberstadt, N. N. 1977. What history tells us about corporate responsibilities. Business and Society Review (Autumn). In: *Managing Corporate Responsibility*. A. B. Carroll, ed. Toronto, Ontorio, Canada: Little, Brown and Company, 351 pp.

The Economist. 2011. Desalination: Drops to drink. July 19, 2011 available from http://www.economist.com/blogs/babbage/2011/07/desalination

Frederick, W. C. 1978. From CSR1 to CSR2: The maturing of business and society thought. *Business and Society*, 33(2): 150–164.

Freeman, R. E. 1984. *Strategic Management: A Stakeholder Approach*. Boston, MA: Pitman.

Friedman, M. September 13, 1970. The social responsibility of business is to increase its profits. *New York Times Magazine*, pp. 32–33.

GE. 2010. Ecomagination 2010. Annual Report downloaded on November 21, 2011, from http://files.gecompany.com/ecomagination/progress/GE_ecomagination_2010AnnualReport.pdf

Gonchar, J. 2011. Industrial evolution. From architectural record downloaded on December 10, 2011 from http://continuingeducation.construction.com/article.php?L=5&C=854

Gordon, J. 2007. The rise of independent directors in the United States, 1950–2005: Of shareholder value and stock market prices. *Stanford Law Review,* 59(6): 1465–1568.

Greenpeace. 2012. Get rainforests out of NZ supermarkets. Downloaded from http://www.green-peace.org/new-zealand/en/take-action/Take-action-online/Cotton-Soft/

Hansen, E. and H. Juslin. 2011. *Strategic Marketing in the Global Forest Industries.* Corvallis, OR: Authors Academic Press.

IBM Consulting Services. 2003. Looking good on paper. Downloaded on November 2011 from http://www-935.ibm.com/services/uk/igs/pdf/esr-looking-good-on-paper.pdf (Accessed November 3, 2011).

IIRC. 2011. Towards integrated reporting: Communicating value in the 21st century by the international integrated reporting committee. Downloaded from http://www.theiirc.org/the-integrated-reporting-discussion-paper/ (Accessed November 3, 2011).

Jaffe, A., S. Peterson, P. Portney, and R. Stavings. 1995. Environmental regulation and the competitiveness of U.S. manufacturing: What does the evidence tell us? *Journal of Economic Literature,* 33: 132–163.

Jantzi-Sustainalytics Inc. 2011. *Executive Summary of Sustainability and Materiality in the Natural Resources Sector.*

KPMG. 2011. KPMG international survey of corporate responsibility reporting 2011. Available from http://www.kpmg.com/Global/en/IssuesAndInsights/ArticlesPublications/corporate-responsibility/Pages/2011-survey.aspx (Accessed September 7, 2011).

Legget, T. 2011. BBC News–Global Witness leaves Kimberley Process diamond scheme. Available at http://www.bbc.co.uk/news/mobile/business-16027011?SThisFB (Accessed December 9, 2011).

Li, N. and A. Toppinen. 2011. Corporate responsibility and sustainable competitive advantage in forest-based industry: Complementary or conflicting goals? *Forest Policy and Economics,* 13(2011): 113–123.

Margolis, J. D., H. A. Elfenbein, and J. P. Walsh. 2007. Does it pay to be good? A meta-analysis and redirection of research on corporate social and financial performance. Harvard University Working Paper, Boston, MA.

Mikkilä, M. and A. Toppinen. 2008. Corporate responsibility reporting by large pulp and paper companies. *Forest Policy and Economics,* 10(2008): 500–506.

MIT and BCG (Sloan Management Review and the Boston Consulting Group). 2011. Sustainability: The embracers seize the advantage. 28 pages. Available from http://sloanreview.mit.edu/feature/sustainability-advantage/ (Accessed October 11, 2011).

Nidumolu, R., C. K. Prahalad, and M. R. Rangaswami. 2009. Why sustainability is now the key driver of innovation. *Harvard Business Review,* 87(9): 56–64.

Nikolakis, W., D. H. Cohen, and H. W. Nelson. 2012. What matters for socially responsible investment (SRI) in the natural resources sectors? SRI mutual funds and forestry in North America. *Journal of Sustainable Finance & Investment,* 2(2): 16.

O'Connor, J., P. Lavoie, and L. Mahalle. 2011. Environmental Product Declarations (EPDs): Market assessment for Canadian products. FP Innovations, Vancouver, British Columbia, Canada, 115 pp.

Orlitzky, M., F. L. Schmidt and S. L. Rynes. 2003. Corporate social and financial performance: A meta-analysis. *Organization Studies,* 24(3): 403–441.

Our Common Future. 1987. *Report of the World Commission on Environment and Development Towards Sustainable Development.* Oxford University Press, Oxford 398 pp.

Panwar, R. and E. Hansen. 2009. A process for identifying social and environmental issues: A case of the US forest products manufacturing industry. *Journal of Public Affairs,* 9: 323–336.

Panwar, R., T. Rinne, E. Hansen, and H. Juslin. 2006. Corporate responsibility. *Forest Products Journal,* 56(2):4–12.

Porter, M. E. and M. R. Kramer. 2006. Strategy and society the link between competitive advantage and corporate social responsibility. *Harvard Business Review,* 84(12): 78–92.

Porter, M. and M. Kramer. 2011. Creating shared value. *Harvard Business Review*. January–February: 62–79.

Porter, M. E. and C. van der Linde. 1995. Toward a new conception of the environment-competitiveness relationship. *Journal of Economic Perspectives*, 9(4): 97–118.

Prahalad, C. K. and V. Ramaswamy. 2004. *The Future of Competition: Co-Creating Unique Value with Customers*. Boston, MA: Harvard Business School Press.

Robinson, M., A. Kleffner, and S. Bertels. 2011. Signaling sustainability leadership: Empirical evidence of the value of DJSI membership. *Journal of Business Ethics*, 101: 493–505.

Sharma, S. and I. Henriques. 2005. Shareholder influences on sustainability practices in the Canadian forest products industry. *Strategic Management Journal*, 26(2): 159–180.

Svendsen, A. 2000. Stakeholder engagement: A Canadian perspective in accountability quarterly. Downloaded from http://www.cim.sfu.ca/pages/resources_stakeholder.htm (Accessed Septemper 10, 2010).

UNECE/FAO. 2012. The North American forest sector outlook study 2006–2030. Geneva forest and timber study paper 26. Downloaded from http://www.unece.org/fileadmin/DAM/timber/publications/SP-29_NAFSOS.pdf, 68 pp.

UNEP Finance Initiative Asset Management Working Group. 2006. Show me the money: Linking environmental, social and governance issues to company value. 50 pages. Downloaded November 2006 from http://www.unepfi.org/fileadmin/documents/show_me_the_money.pdf (Accessed November 9, 2006).

Van Marrewijk, M. 2003. Concepts and definitions of CSR and corporate sustainability: Between agency and communion. *Journal of Business Ethics*, 44(2/3): 95–105. Corporate Sustainability Conference 2002: The Impact of CSR on Management Disciplines (May, 2003).

Vidal, N. and R. Kozak. 2008b. Corporate responsibility practices in the forestry sector: Definitions and the role of context. *The Journal of Corporate Citizenship*, 31: 59–75.

Vidal, N. G. and R. A. Kozak. 2008a. The recent evolution of corporate responsibility practices in the forestry sector. *International Forestry Review,* 10(1): 1–13.

Vogel, D. J. 2005. Is there a market for virtue? The business case for corporate social responsibility. *California Management Review*, 47: 19–45.

WBCSD. 2010. Vision 2050. Available for download from http://www.wbcsd.org/vision2050.aspx 80 p. (Accessed November 10, 2010).

WBCSD and UNEP. 2010. Translating ESG into sustainable business values (p. 20) 2010, by the world business council on sustainable development and the United Nations environment program finance initiative. Downloaded from http://www.scribd.com/doc/51913119/Translating-ESG-into-sustainable-business-value-UNEPFI-2010 (Accessed November 10, 2010).

Wilson, M. 2003. Corporate sustainability: What is it and where does it come from? *Ivey Business Journal,* March/April 2003: 1–5.

Wood, D. 1991. Corporate social performance revisited. *Academy of Management Review*. 16(4): 691–718.

Yale EPI Ranking. 2012. Yale University environmental performance rankings 2012. Downloaded September 2012 from http://epi.yale.edu/epi2012/rankings (Accessed December 9, 2011).

Zuboff, S. and J. Maxmin. 2004. *The Support Economy: Why Corporations are Failing Individuals and the Next Episode of Capitalism*. New York: Penguin.

16

Innovation in the Global Forest Sector

Scott Leavengood and Lyndall Bull

CONTENTS

16.1 Introduction

In this chapter, we explore innovation in the global forest sector by first making the case for the "innovation imperative"—what is innovation, how does it differ from innovativeness, and how may firms, as well as the sector and the global economy, benefit from studying and pursuing innovation? We examine innovation in the global forest sector by examining the drivers of innovation and provide examples of the industry's response to these drivers.

Innovation systems (ISs) are defined and analyzed from the perspective of how they are serving to foster innovation as well as areas for improvement. Innovation management in the sector is explored from the standpoint of historical areas of focus with respect to innovation, industry culture and strategic orientation, as well as new product development (NPD) practices and trends in research and development (R&D). We conclude with a discussion of what the future may hold for innovation in the global forest sector.

16.2 "Innovation Imperative"

16.2.1 What Is Innovation?

Recognition of the importance of innovation dates back at least to the mid-nineteenth century when Marx suggested that innovations could be associated with waves of economic growth (Trott 2011). Studying innovation as a concept in isolation is difficult because innovation can be considered to be part of several other areas of study including industrial organization, regional economics, the theory of the firm, management science, sociology, and history. Further, as stated by Wolfe (1994), prior research on innovation has often failed to be explicit regarding what aspects of innovation are being addressed—innovation diffusion, organizational innovativeness, or process theory (i.e., the processes organizations go through in implementing innovations). Perhaps as a result of its complex and multifaceted nature, much of the expansive literature incorporating the concept of innovation is disjointed and focuses on isolated aspects of the topic. Definitions of the concept of innovation have, as a result, often been designed according to different scholars' particular interests.

The *Shorter Oxford Dictionary* (1972) provides a useful though broad definition of innovation that highlights the importance of newness and change: "the act of innovating to change into something new, to renew, the introduction of novelties; the alteration of what is established; something newly introduced; a novel practice, method, etc." Other definitions presented, while acknowledging the importance of newness, also tend to associate the concept with particular areas of scholarly interest. Table 16.1 summarizes a range of these definitions.

While innovation is commonly thought of simply as new products, Betz (1998) suggests that the definition of innovation includes process and service innovations in addition to product. Taking the typology a step further, Utterback (1994) adds administrative and technological innovations. Hovgaard and Hansen (2004) defined three basic types of innovations—product, process, and business systems. In their view, business systems innovations include any innovation that does not fall under product or process innovation. Examples include innovative management and marketing techniques.

In addition to these various definitions of innovation, typologies have also been developed to describe the impact of the innovation on an industry. As described by Betz (1998), discontinuous innovations result in creating or altering industrial structures; continuous innovations reinforce the existing structure. The essence is the degree of "radicalness" of the innovation. Thus, discontinuous innovations are also referred to as radical, next-generation, revolutionary, breakthrough, or disruptive. Conversely, continuous innovations are also referred to as sustaining, evolutionary, or incremental. Betz gives examples of both radical and incremental innovations from the railroad industry: The steam engine was a radical innovation that enabled the development of the railroad (and many other) industry.

TABLE 16.1

Definitions of Innovation

Definition	Author
Whereas an invention is the process of converting intellectual thoughts into a tangible new artifact (product or process), an innovation is the commercial and practical application of these inventions.	Trott (2011)
An idea, practice, or objective that is perceived as new by an individual or other unit of adoption. It matters little, so far as human behavior is concerned, whether or not an idea is objectively new as measured by the lapse of time since its first discovery. It is the perceived or subjective newness of the idea for the individual that determines his reaction to it. If the idea seems new to the individual, it is an innovation.	Rogers (2003)
A discontinuously occurring implementation of new combinations of means of production	Schumpeter (1939)
Innovation is not a single action but a total process of interrelated subprocesses. It is not just the conception of a new idea nor the invention of a new device, nor the development of a new market. The process is all these acting in an integrated fashion.	Myers and Marquis (1969)
Product innovation is defined as the conceptualization, development, operationalization, manufacture, launch, and ongoing management of a new product or service. New means new to the organization and can involve new customers, new uses, new manufacturing, new distribution and/or logistics, new product technology, and any combination of these. Product innovation is inherently interfunctional.	Dougherty (1996)
Industrial innovation includes the technical, design, manufacturing, management, and commercial activities involved in the marketing of a new (or improved) product or the first commercial use of a new (or improved) process or equipment.	Freeman (1982)
The adoption of an idea or behavior new to the adopting organization	Damanpour (1996)

One continuous innovation to the steam engine included development of a two-stroke steam-cycle engine that doubled horsepower. And then in the early twentieth century, diesel engines were the next-generation technology that resulted in making the steam engine obsolete. Johannessen et al. (2001) viewed this "incremental vs. radical" dimension of innovation from the viewpoint of newness but specifically asked the question, "new to whom?" In their view, incremental innovations are new to the adopting firm, whereas radical innovations are new to the industry. Examples might include a forest products firm adding barcodes to their products. While this technology might seem "disruptive" or "radical" to the adopting firm, the prevalence of this technology in industry suggests the technology is more incremental than radical. By contrast, adoption of radio frequency identification (RFID) technology to track products would be seen as a radical innovation given that RFID is rarely used in the forest industry.

16.2.2 Innovativeness

Closely related to the concept of innovation is that of innovativeness. As with innovation, firm innovativeness has been defined in a variety of ways. It has been considered by some simply as firms that adopt innovations (e.g., Attewell 1992; Utterback 1994), while others are more specific in their definitions. Rogers, for example, emphasizes how proactive firms are in defining innovativeness as "the degree to which an individual or other unit of adoption is relatively *earlier* (emphasis added) in adopting new ideas than any other member of the system" (Rogers 2003). Gebert et al. (2003) emphasize organizational capacity for innovation, and incorporated creativity as well, in their definition—"the capacity of an organization to improve existing products and/or processes, and the capacity to utilize the creativity resources of the organizations to the fullest." While Hurley and Hult (1998) emphasize organizational culture in their definition—"the notion of openness to new ideas as an aspect of a firm's culture."

The boundaries between innovation and innovativeness can be fuzzy and some authors use them interchangeably (Damanpour 1991). Hansen et al. (2007) state, however, that separating the concepts is important: "whereas an innovation can be a new process, product or business system, innovativeness is an organizational trait or characteristic. Thus innovativeness is what enables an organization to create or adopt innovations." Similarly, Knowles et al. (2008) define innovativeness as "…the propensity of firms to create and/or adopt new products, processes, and business systems." The key terms in these latter two studies are *trait, characteristic*, and *propensity* as these terms reflect that innovativeness is an attribute of an organization and hence a reflection of organizational culture.

Of course, the "newness" of all innovations eventually fades. That is, innovations have a history reflecting birth, development, growth, and eventually decline. The innovation life cycle represents that history and is outlined in the following section.

16.2.3 Innovation Life Cycle

Understanding the birth, development, and decline of innovation—be it product, process, or indeed that of an entire industry—has been the basis for considerable study and reflection. This section seeks to explain this evolutionary pattern of development, growth, and decline of innovation. Such an understanding is considered critical to the long-term success or failure of firms and industries (Utterback and Suarez 1993).

Utilizing the automobile industry as the demonstrating case, Abernathy and Utterback (Utterback and Abernathy 1975; Abernathy and Utterback 1978) described how industries evolve from birth through to maturity. Building on this and other work incorporating industrial evolution and the link between market structure and R&D, Klepper (1996) developed and tested a model to explain both the evolutionary pattern of industry development as well as the role of firm innovative capabilities and size in conditioning firm R&D spending, innovation, and market structure. Here, we focus only on his outcomes relating to industry evolution. Testing of Klepper's model confirms the following patterns:

1. At the beginning of an industry, two patterns may emerge:
   a. The number of entrants may rise over time or it may attain a peak at the start of the industry and then decline over time, but in both cases, the number of entrants eventually becomes small.
   b. The number of entrants grows initially and then reaches a peak, after which it declines steadily despite continued growth in industry output.
2. Eventually, the rate of change of the market shares of the largest firms declines and the leadership of the industry stabilizes.
3. The diversity of competing versions of the product and the number of major product innovations tend to reach a peak during the growth in the number of producers and then fall over time.

4. Over time, producers devote increasing effort to process relative to product innovation.

5. During the period of growth in the number of producers, the most recent entrants account for a disproportionate share of product innovations.

16.3 Innovation and the Global Forest Sector

The global forest products industry is defined as those businesses whose primary activities depend on the growing, managing, or processing of trees for wood and paper production (Schirmer 2010). Given the wide diversity in the products produced and processing technology used by the different players in the forest sector, it is not surprising that there would be a wide variety in the forces driving innovation and, of course, emphases on innovation in the sector as well. We begin by discussing the benefits innovation can provide individual firms and the sector as a whole and then explore the various drivers for innovation in the forest sector. The section concludes with a discussion of how innovation has been approached in the sector, including examples of product, process, and business systems innovations.

16.3.1 Importance of Innovation/Innovativeness to Firms in the Forest Sector

It is commonly cited that innovation is an effective means for a firm to increase both profits and its competitive advantage (Brown and Eisenhardt 1995; Abernathy and Clark 1998; Damanpour and Gopalakrishnan 2001). However, innovation impacts not only individual firms, but there is a ripple effect through a sector and perhaps the economy as a whole; the more radical the innovation, the more widespread the effect may be. For example, a product innovation like oriented strand board (OSB) spurred development of other product innovations such as new adhesives and processing innovation technologies required to efficiently produce OSB such as equipment to produce the strands, adhesive blenders, mat forming, and hot presses. Further, enabling the utilization of wood species previously deemed unsuitable for structural panel products (e.g., aspen and other *Populus* species) fostered the development of new industrial clusters in North America, Europe, and Latin America.

As with innovation, a firm's level of innovativeness has been linked to its financial performance (Han et al. 1998; Calantone et al. 2002; Hult et al. 2004). Johnson et al. (1997) asserted that "organizational innovativeness is crucial to the very survival of modern organizations," while Utterback and Suarez (1993) state that understanding the relationships between product technology and their enabling processes and corporate strategy and competitiveness are central to understanding the success or failure of firms and even entire industries.

A key benefit of innovation for the forest sector is the potential for additional profits through more efficient utilization of wood fiber (Bull and Ferguson 2006). Additional profits may be obtained this way when firms are able to increase the yield of finished products, that is, convert a greater percentage of their raw materials (trees, logs, lumber, veneer, etc.) into finished goods. The sector has been under increasing levels of competition from both wood and non-wood-based products, creating a situation where companies must innovate or they simply will not survive (Crespell and Hansen 2009). Even disregarding these urgent imperatives, incorporating innovation as part of a company's strategy and fostering

a "climate" for innovation have been shown to impact upon both the company's innovativeness and overarching performance (Han et al. 1998; Calantone et al. 2002; Hult et al.
2004). Innovation is an important driver of firm growth, and being innovative through the
development of new or improved products, processes, or business systems can help a company better identify and satisfy customer needs, stay ahead of competition, and explore
new markets. Competitive markets consider innovation as being necessary for survival,
not just growth (Johnson et al. 1997).

In addition to the need to optimize efficiency and compete with emerging products, the
approach and direction of both product and process innovations that the forest industry
has invested in have also been impacted by a range of environmental, social, and economic
factors. Examples of these factors are described below.

16.3.1.1 Changes in the Types of Supply and Demand of Forest Products

The supply scenario of the wood products industry has changed over the past 30 years. The
size, availability, and cost of accessing timber from native forests have generally decreased
while plantation establishment has increased (Australia 2008; ABARE 2010; FAO 2011).

Unlike its native grown counterpart, much of the rapidly grown plantation wood does
not exhibit the wood qualities deemed necessary for furniture and building construction (Skog et al. 1995). The industry's response to this change in supply from native to
plantation-grown timber has mirrored the change in resource base the industry experienced a decade or two earlier in the shift from larger, old-growth trees to smaller, second-
growth trees. These resource shifts have induced the development of a number of wood
product innovations (Balatinecz and Woodhams 1993; Skog et al. 1995; Byron and Sayer
1996; Sayer et al. 1997; Schuler et al. 2001). Many of these innovations have been developed
as replacements for structural solid wood. Examples include OSB and other strand composites like oriented strand lumber and parallel strand lumber, laminated veneer lumber
(LVL), and wood fiber composite products (Balatinecz and Woodhams 1993; Schuler et al.
2001; Adair 2004).

There are also many other innovations currently in development or in the early stages
of being introduced to the market that are intended to overcome limitations in wood quality. While these innovations may initially appear to be examples of product innovations,
they all are in fact a combination of a process innovation that enables the development
of a product innovation. For example, scientists are developing methods (processes) to
harden plantation softwoods (products) to make them suitable for flooring. Examples of
such innovations include the following:

- Densified wood, for example, viscoelastic thermal compressed (VTC) wood (see
  Kamke and Sizemore 2008)—The VTC method uses a three-step process involving steam, heat, and pressure; the result is collapse of the wood cell walls and an
  increase in stiffness proportional to the increase in density.
- Polymer-impregnated wood—Pressure impregnating wood with a chemical
  such as urea and starch, a polysaccharide (sugars), and furfuryl alcohol; the
  chemical polymerizes and "hardens" inside the wood resulting in increased
  hardness; examples include Everdex by Alowood (EPA 2007) and the Indurite™
  (Franich 2007) process. When furfuryl alcohol is used the process is known as
  furfurylation; the alcohol polymerizes in the wood resulting in increased durability, hardness, and stability (see, e.g., Kebony®, Lande et al. 2004).

- Thermally modified wood—This process uses prolonged high temperatures (e.g., 2–3 h at 185°C–215°C) to alter the material properties of wood such that it is more durable and dimensionally stable (e.g., ThermoWood®, Anonymous 2003).

- Acetylated wood—Reacting wood with acetic anhydride to alter the natural chemistry of the wood and thereby increase its dimensional stability and durability (e.g., Accoya® wood, Rowell 2006).

16.3.1.2 Push for Sustainability

Coupled with this change in the nature of the supply is the perceived pressure to produce so-called sustainable products. Perhaps the most explicit demonstration of this has been the uptake of forest certification. By December 2011, nearly 148 million hectares of forests were Forest Stewardship Council (FSC) certified (www.fsc.org), and 242 million hectares were certified via the Programme for Endorsement of Forest Certification (PEFC, www.pefc.org).

A number of other so-called "green" requirements are adding further pressure on the forest industry to demonstrate its sustainability credentials. These include green building ratings, improved indoor air quality, the desire for building materials to be locally sourced, and a preference for fast growing/"rapidly renewable" plantation species such as bamboo (e.g., *moso bamboo, Phyllostachys pubescens*) and several species of eucalyptus (*Eucalyptus* spp.).

The wood products industry has recognized that products produced from plantation, forest, and mill residues are, at least in part, a means of addressing this push for sustainable products (Ryan 1993; Maloney 1996; Schuler et al. 2001; Kubeczko and Rametsteiner 2002). Several of the new product and process innovations listed at the end of the previous section are in fact intended to serve a dual purpose, that is, modification of plantation-grown wood properties to enable suitability for use as flooring, exterior cladding, etc., and at the same time, to limit pressure on native forests that produce species currently used for these materials. In many cases, a key benefit listed by those marketing the new products is the reduced pressure on tropical forests. As one example, the firm producing Accoya® (acetylated wood) states the material "…takes sustainably-sourced, fast growing softwood and, in a non toxic process that 'enables nature,' creates a new durable, stable and beautiful product—that has the very best environmental credentials… [the wood] matches or exceeds the durability, stability and beauty of the very best tropical hardwoods" (Accoya 2011).

16.3.1.3 Cost of Production

Emerging economies such as China and Vietnam have witnessed spectacular growth in the past decade. As one example in the forest products industry, China's growth in forest product processing is mirrored by the United States' decline (Lihra et al. 2008). Statistics from the US International Trade Administration (ITA 2009) show that during the years of 1999–2006, imports of furniture from China into the United States increased 4.3 times or an annual rate of nearly 24%.

Products in developing countries generally have a low-cost advantage that includes, but is not limited to, the cost of labor. One example of a business systems innovation that has allowed the wooden cabinet industry to grow despite the pressure of low-cost imports is that of mass customization (Lihra et al. 2008). This concept, defined as a trend toward the production and distribution of individually customized goods and services for mass markets (Davis 1987), gives producers who maintain close relationships with their customers a sustainable competitive advantage (Lihra et al. 2008).

At the same time, outsourcing production to low-cost producers may in fact lead not only to cost advantages but potentially to innovation as well. For example, one group of researchers discovered that the "insourcer/vendor" may offer not only cost advantages but also quality improvement and innovation (Maskell et al. 2006). As the authors state, "the quality improvements that offshore outsourcing may bring about evoke a realization in the corporation that even innovative processes can be outsourced."

Lean manufacturing is another business systems innovation that is being embraced by firms in a wide variety of industries, including the forest sector, as a means to maintain or improve competitiveness. The primary emphasis of lean manufacturing is to reduce waste in areas such as spending excessive time setting up a machine, producing excessive inventory, wasted energy, and wasted movement (Womack et al. 1990; Womack and Jones 1996); such an emphasis is particularly attractive in the forest sector given the emphasis on efficient utilization of wood fiber.

16.3.2 Forest Sector's Approach to Innovation

The global forest sector has historically been characterized as being mature and production oriented (Hansen et al. 2007) with little customer focus (Crespell et al. 2006). While it is well documented that innovation and innovativeness are vital to a firm's competitiveness and financial performance (e.g., Calantone et al. 2002; Deshpande et al. 1993) because this academic field of endeavor is relatively new to the forest sector, there have been few published studies on the topic (Hansen 2010; Stone et al. 2011). The research has shown that there has generally been a focus on process innovation and that, unfortunately, forest products managers do not see their companies as particularly innovative (Schaan and Anderson 2002; Välimäki et al. 2004; Crespell et al. 2006; Hansen et al. 2006; Knowles et al. 2008).

The research also shows that the forest industry does not tend to have a systematic approach to NPD (Hansen 2006a), and new products developed by the industry are often the result of attempts to utilize readily available raw materials ("resource push products") rather than from the specific demands in the marketplace (Bull and Ferguson 2006). Bringing together innovation and innovativeness research that has been carried out in the forest sector over the past 5 years, Hansen (2010) provides the following summary:

- Forest products industry managers do not see their companies as particularly innovative, but companies have been found to consistently focus on innovation.
- Overall, large companies are more innovative, which suggests that resources and improved opportunities for networking increase the ability to be innovative.
- Large companies tend to focus on process innovation, while smaller companies use a more balanced portfolio.
- Few forest product companies have structured NPD processes with larger companies more likely than independent mills.
- The industry is weakest at employing marketing-related NPD tools.
- A clearly defined product concept, strong leader, firm size, and education of managers are related to successful NPD.

We explore the innovation culture in the industry, as well as NPD practices, in greater detail in Section 16.4.

16.3.3 Examples of Innovation in the Forest Sector

The forest sector has emphasized development and adoption of processing innovations. The primary drivers for these innovations are to reduce costs and improve operating efficiency and product quality (Peters et al. 2006; Hansen et al. 2007). Some key examples of such successful processing innovations include

- Cut-to-length harvesters—Mechanized harvesting equipment that fells trees and then delimbs and cuts the stems to precise lengths
- 3D log scanners—Determine log shape (e.g., using several laser proximity sensors) as well as the presence and location of internal defects (e.g., using x-ray) for use in optimizing primary log breakdown
- Edger optimizers—Determine shape of rough boards and then adjust the edger (machine that cuts lumber to width) to maximize yield
- Lumber grading—Automatically grade lumber using multiple scanning technologies (e.g., color video, laser, and dielectric)
- Automated stacking and stickering—Systems to stack lumber and place thin boards ("stickers" to allow airflow) between layers of lumber prior to drying
- Chop saw scanning/optimizers—Used in molding and millwork plants to detect the location of knots and other defects and automatically crosscut lumber to remove the defects
- Veneer grading and sorting systems—Scanning systems to detect and remove defects in veneer for plywood and LVL
- Density scanners—Systems that use ultrasound to determine the density and strength of veneer for LVL
- Computer numerically controlled (CNC) machining—CNC routers to cut and shape parts for furniture and cabinets that also include optimization routines to ensure maximum yield of parts from any given board or panel stock
- Online moisture content sensors—Often used in composite panel plants to detect when steam "blows" in the hot press may occur due to high moisture content

Of course, there have been numerous product innovations as well—over the centuries, the industry has progressed from producing a very simple product (i.e., sawn lumber) to a wide variety of composite/engineered products. More recent developments in composite materials include cross-laminated timber (CLT), wood/non wood composites like wood/plastic composites (as discussed in some detail in Chapter 6 of this volume), and wood/cement products (e.g., siding). Veneer-based products include not only plywood, but products like LVL and profile-wrapped products such as wood moldings as well. And of course, there are combinations of innovative materials like LVL and OSB to produce engineered products like wood I-joists.

Many researchers and industry experts have issued the "innovation imperative" for the forest industry to improve its innovation performance and the importance of developing a more balanced innovation portfolio, that is, an emphasis on product, process, and business systems/marketing innovations (Damanpour et al. 1989; Hansen 2010). However, prior to a discussion of recommendations for how the industry can improve its innovation performance, we must first examine specifically how innovation occurs in the sector—what drives innovation, who are the important actors, what are their roles, and how do they interact? For this we shift our attention to innovation systems (IS).

16.4 Forest Sector Innovation System

For many decades, research on innovation approached the subject through a linear view of the process, that is, a sequence of activities that occur in an individual firm. For example, Rogers (2003) proposed a five-step model of innovation that begins with "knowledge," and then proceeds through "persuasion," "decision" (i.e., adopt or reject), "implementation," and "confirmation." However, as research in the field of innovation has progressed, it has become increasingly apparent that innovation occurs following an inherently nonlinear, iterative process (Rametsteiner and Weiss 2006).

Further, it has been said that "innovation is a team sport" (Dougherty and Takacs 2004), and there is ample evidence to support that fact. The legend of the lone genius working in the laboratory and developing a breakthrough innovation is just that, more legend than reality. It often takes years for an invention to progress to the point of being commercialized. Numerous individuals and organizations typically play a significant role in the process. These entities may include industry, academia, various levels of government, and nonprofit organizations that work together to form what is known as an IS. The overall purpose of IS is to produce, diffuse, and use innovations (Edquist 2001).

Edquist (1997) defines a system of innovation as "all important economic, social, political, organizational, and other factors that influence the development, diffusion, and use of innovations." In research specific to the forest sector, Rametsteiner and Weiss (2006) define an IS as "...the set of distinct actors and institutions which contribute to the development and diffusion of innovations in forestry." A key reason for taking a systems view of innovation is recognition that the set of actors are interconnected, and as such, the performance of the system as a whole is determined by the performance of the individuals as well as their interactions.

Kubeczko et al. (2006) state that ISs serve functions related to innovation inputs, the innovation process, as well as use of the innovation. More specifically, these functions include

- Provision of resources (innovation inputs)—Provision of financial (e.g., tax incentives and subsidies), human (e.g., educating a skilled workforce), and knowledge resources (e.g., patents or licenses)
- Management of complexity (innovation process)—Providing information to reduce uncertainties and lessen risks, management of cooperation and conflicts between entities in the IS, and providing technical assistance
- Promotion of use—Marketing of the innovation; procurement policy changes; changes in regulations, standards, and codes

Figure 16.1 presents a graphical view of an IS demonstrating these various functions and how they may interact. Culture and policy are shown in the figure as having an overarching impact on IS. Culture influences IS on multiple levels—national, regional, and organizational. National and regional cultures can also influence policy development and implementation. Policy can have a broad impact on IS such as in how resources are allocated; policy may serve to promote use of innovations and may encourage and enable interfirm cooperation. Firm/interfirm cooperation is shown as a key element of the innovation process; management of cooperation is a particularly important role when the innovation requires cooperation with other firms or nonprofit organizations.

Researchers have identified several distinct categories of ISs (Edquist 1997, 2001; Lundvall et al. 2002; Kubeczko et al. 2006; Rametsteiner and Weiss 2006). Relevant categories for the forest sector include regional innovation systems (RISs) and sectoral innovation systems

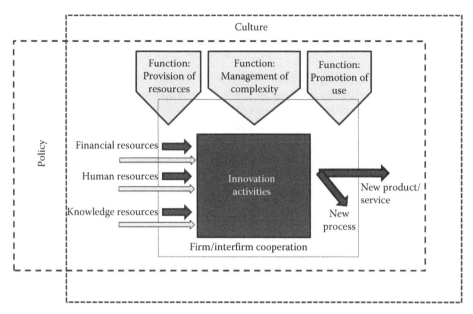

FIGURE 16.1
Functions of actors and institutions in IS. (Adapted from Kubeczko, K. et al., *Forest Policy Econ.*, 8(7), 704, 2006.)

(SISs). As implied by the name, RISs are defined by regional boundaries and function based on fostering innovation in firms located in a specific geographic area. Hansen et al. (2006) state that RISs are related to industry cluster* concepts, although as these authors also state, industry clusters do not explicitly emphasize innovation performance. SISs focus on innovation development in specific industry sectors such as forestry, high tech, and agriculture. By contrast, SISs are not limited by the relative proximity of the firms but rather by the knowledge, technology, and interactions common to the respective sector. We examine the various roles that different entities may play in an IS by exploring one recent innovation (see Box 16.1).

The example in the sidebar underscores how researchers have defined and described IS, that is, the fact that there are numerous functions, actors, and interactions that must come together for an innovation to become a commercial success. And in this greatly simplified version of the innovation process, it may appear that things flow smoothly in a linear fashion from the driver (inspiration), to the innovation process, through to the eventual use of the innovation. However, a more in-depth examination of the case would include all the iterations and hurdles that occurred along the way. Of course, this is the case with all systems—there are always complexities and opportunities for improvement.

Hence, to increase the forest sector's innovation performance, we should ask how, in general, are forest sector ISs performing? What is working and what is not? And what can be done to improve their performance?

We will first begin with what researchers have found is working and then follow with suggestions for specific areas for improvement.

---

* The "cluster approach" emphasizes "…geographically concentrated and interconnected economic activities, linkages to customers and suppliers and limited conformance to standard economic classifications… this approach focuses on the distinct competitive advantage offered by a region and its networked economic activities" (Hovee 2005).

BOX   16.1

Dr. Kaichang Li at Oregon State University (OSU) developed a soy-based, formaldehyde-free adhesive for use in composite wood products (Sherman 2007). Inspiration for the invention came from observing how mussels cling to rocks in a marine environment. Further inspiration was the advent of new regulations from the California Environmental Protection Agency's Air Resources Board (CARB) limiting formaldehyde emissions from composite wood products (CARB 2011). The US Department of Agriculture (USDA) provided funding to conduct research to synthesize an adhesive from soybean protein. When Dr. Li presented this research at a conference, a representative from Columbia Forest Products (CFP, North America's largest manufacturer of hardwood plywood and veneer) approached him expressing interest in funding further development of the adhesive. Hercules Inc., an adhesive manufacturer, was then brought in as a partner on the project. OSU's technology transfer office contracted with a legal firm to patent the adhesive and eventually to license the technology to Hercules Inc., who gave exclusive rights for the adhesive's use to CFP. The adhesive is now in use in all of their hardwood plywood operations in North America.

In summary, this specific innovation (formaldehyde-free adhesive) involved the following IS actors and functions:

1. Government:
   a. Federal—USDA provided funding for the invention and the US Patent and Trademark Office registered the patent.
   b. State of Oregon–Oregon's state system of higher education funds university faculty to conduct research as well as personnel in OSU's Technology Transfer Office.
   c. State of California—Via the California Environmental Protection Agency, California's government in effect served as a key driver in the innovation; the new regulation on formaldehyde emissions was a primary driver for industry demand for "green adhesives."
2. Research organizations—In 2004, the International Agency for Research on Cancer (part of the World Health Organization [WHO]) classified formaldehyde as carcinogenic to humans. CARB cites this classification as a driver for regulating formaldehyde emissions from composite wood products (CARB 2011).
3. Private industry—Two private industry firms (CFP and Hercules Inc.) provided funding for development, technical feedback, and materials; a legal firm patented the invention; in addition, firms in Oregon's forest industry provide funds for research at the university through a tax on harvested timber.
4. Nonprofit organizations—The Forest Products Society (a professional association) organized and hosted the conference where Dr. Li first presented his results; hence, they provided a forum for presentation of the invention as well as networking opportunities.

16.4.1 Forest Sector Innovation Systems: What's Working?

Kubeczko et al. (2006) studied RIS and SIS for the forest sector (with a particular focus on landowners and bioenergy) in seven central European nations. Their research showed that ISs were primarily effective in supporting incremental process innovation in the sector. These findings agree with those of Ukrainski and Kajanus (2011) for the IS in Finland and Estonia. These latter authors further state that this focus on incremental process innovation is not surprising given that forest sector firms are primarily small- to medium-sized enterprises (Weiss 2011); such emphasis is supported in the general business literature as well (e.g., Kaufmann and Tödtling 2002).

Kubeczko et al. state further that the actors in the IS played a key role in the diffusion (as contrasted with development) of innovations. In a case study of the development of biomass district heating in Austria, the authors report the importance of institutional support for the diffusion of this type of innovation; the innovation spread very quickly, particularly in areas where local chambers of agriculture became engaged in the projects by, for example, founding biomass associations. Such diffusion activity involved implementation of information systems and financial policies supportive of biomass district heating.

In the United States, the extension service (often termed "cooperative extension" or "agricultural extension") has played a key role in the IS for nearly a century. Each US state has a land-grant university* that was established to educate citizens in professions such as agriculture, home economics, and mechanical arts (use of tools and machinery). The extension service was formally established in 1914 by forming a partnership between the land-grant universities' agricultural colleges and the USDA to disseminate and demonstrate research results to practitioners. The initial topic of focus was agriculture; however, the focus has broadened significantly to include home economics, forestry, youth development (4-H), small business development, and others. Extension professionals include faculty located on campus that serve as subject matter specialists as well as faculty located in county extension offices; these county-based faculty are typically called "agents" in that they maintain close working relationships with stakeholders and serve as a liaison between the university and the community(ies) they serve.

In his book *Diffusion of Innovations* (2003), Rogers (citing Eveland 1986) states, "It is impossible for anyone to speak ten words about diffusion [of innovations] without two of them being 'agricultural extension'." Extension professionals function primarily in the innovation "process" and "use" functions described earlier—they serve in the innovation process by managing cooperation and conflicts and by providing technical assistance; they fulfill perhaps the most significant role in use of the innovation by disseminating information about the innovation to potential users.

Lastly, with regard to "what is working" in the forestry IS, there are examples of how policy and regulatory changes are serving to drive innovations in the forest sector. For example, the formaldehyde-free adhesive discussed earlier was driven in part by the California Environmental Protection Agency's regulation on formaldehyde emissions from composite wood products. And there are numerous examples of how policy changes related to the use of renewable fuels are driving innovation in energy generation and distribution. Nybakk et al. (2011) present case examples for 14 bioenergy applications in Europe. As one example, the French government's policy to produce 23% of the nation's energy from renewable energy sources by 2020, combined with guaranteed rates paid to producers

---

* The term "land-grant" is used as the federal government granted federally controlled land to the states such that the states could develop or sell the land to raise funds to establish and endow "land-grant" colleges.

for 15–20 years, helped drive innovation in biomass energy production at pulp mills in the country. Similar results have been seen elsewhere such as in the United States where state-level "renewable fuels standards" (RFS) have served as drivers for development of biomass district heating systems for schools and other community buildings (BERC 2011). The national-level RFS in the United States has sparked significant innovations such as a major petroleum company developing a biofuels refinery; as stated by BP's biofuels chief, the RFS "...was a game-changer assuring absolute global primacy in advanced biofuels for the United States. It made clear that there would [be] a market for this advanced technology and that there would be a price mechanism... we are on the path to completing our first cellulosic biofuels plant" (Anonymous 2012).

In summary, research suggests that ISs are having varying degrees of success in promoting innovation in the forest sector around the world. The primary areas of success include fostering the diffusion of incremental process innovations as well as assisting industry with process and product innovations in the IS "innovation process" and "use" functions. In addition, recent policy and regulatory changes have been effective in driving innovation, particularly with respect to woody biomass-based energy. To improve the innovation performance in the forest sector, it is also critical that we explore "what isn't working" in the IS, that is, where there are opportunities for improvement.

16.4.2 Forest Sector Innovation Systems: Opportunities for Improvement

Numerous researchers have suggested specific actions to improve the performance of forest sector IS. For example, Kubeczko et al. (2006) as well as Rametsteiner and Weiss (2006) recommend enhanced focus on radical innovation is needed as well as greater emphasis on innovation development (vs. diffusion). These authors' research suggests that more cross-sector interaction is needed as well as more support for innovators in the early stages of innovation. Such cross-sector interaction includes not just industry to industry interactions but interactions between all the actors in the IS.

For example, researchers around the globe have pointed to a "communication gap" between research/education institutes and industry (Ukrainski and Kajanus 2011; Kaufmann and Tödtling 2002; Lundvall et al. 2002). For example, Chinese furniture industry managers reported seeing little value in educational institutes with respect to innovation; the authors speculate this is likely due to a lack of communication between industry and academia (Cao and Hansen 2006). Similarly, research has shown that universities and public research institutes were not important knowledge sources for forest sector firms in Estonia or Finland (Ukrainski and Kajanus 2011). There is implicit support for this sentiment in Canada as well in that the premise for the study by Van Horne et al. (2006) was to develop a tool to help improve knowledge transfer between Canadian "Centres of Expertise" and industry. Likewise in Europe, of seven central European nations examined by Rametsteiner and Weiss (2006), "Only Slovakian universities and research institutions are regarded as central actors in innovation." And in a paper focusing primarily on differences between the IS in Finland and the United States, Hansen (2010) sums up the situation by stating, "...the academy must do a better job of technology transfer...."

While the extension service in the United States may be regarded as a valuable component of the IS for diffusing innovations, there are opportunities for improvement with respect to developing innovations. For example, from the author's (Leavengood) experience as director of the Oregon Wood Innovation Center, industry assistance is primarily conducted in response to needs expressed by stakeholders, that is, in response to needs as opposed to proactively pursuing development of innovations to address future opportunities.

Research conducted by Van Horne (2006) suggests this is largely the case for Canada's forest sector Centers of Expertise as well. In descriptions of the functions of these centers, the authors state, "...the innovation process often begins with industrial companies that arrive with a problem to fix or an idea to develop." Therefore, one potential area of improvement would be changing the institutions' strategy (and accordingly, their capacity) such that a proactive focus on developing innovations was a core part of the mission in addition to providing problem-solving technical assistance and research.

With regard to opportunities for policy makers, several authors have pointed to an opportunity for bolstering the RIS' use of clusters to foster innovation. For example, Cao and Hansen (2006) believe that industry clustering has been a key driver for the Chinese furniture industry. Further, they state that clusters, with staffing from open-minded young professionals, will become the industry's innovation leaders and the main drivers of the industry's growth. Similarly, Hansen (2010) discusses how the focus on the forest cluster in Finland has served to strengthen the network of researchers and practitioners.

Of course, there are also many opportunities for forest industry leaders to improve the performance of the IS as well. Given that the research has shown that many of these opportunities are related to industry culture, we defer that discussion to the following section.

In summary, opportunities for improvement in the forest sector IS include increasing the emphasis on proactive development of radical innovations and more cross-sector interactions; in particular, enhanced communication between the research/education community and the industry. Use of clusters in RIS appears particularly promising as a focal area for policy makers.

In addition to understanding how the IS is working and impacting on the development of innovations in the forest sector, it is equally important to understand how innovations are managed within those organizations that are (or perhaps are not) doing the innovation.

16.5 Innovation Management in the Forest Sector

Numerous researchers and industry observers have stated that innovation is essential to the long-term success of the global forest industry. However, while it is a simple matter to encourage industry leaders to "be more innovative," it is quite another matter to give specific guidance regarding how to go about doing this. Stendahl (2009) states it well in saying there is a lack of practical advice on how to manage for innovation and to effectively develop new products.

Like most aspects related to management, innovation management includes activities related to developing strategy, establishing the "right" culture, creating metrics, and evaluating progress toward those metrics. Additionally, innovation management includes activities related to R&D and NPD.

16.5.1 Innovation Focus by the Forest Industry

As discussed previously, forest industry researchers have recognized three broad categories of innovation—product, process, and business systems (Hovgaard and Hansen 2004). Research has shown that the forest industry has focused primarily on process innovations with the goal of reducing operating costs and improving product quality; product and business systems innovations have received lesser emphasis (Schaan and Anderson 2002; Crespell et al. 2006; Hansen 2006b).

The industry's emphasis on process innovation has largely been a response to external forces, supply of raw materials in particular. Adequate resource supply in terms of both quantity and quality has been stated as one of the dominant concerns for the industry, especially for the pulp and paper and primary industry (e.g., sawmills, veneer, and plywood mills) (Stendahl et al. 2007). As stated by Välimäki et al. (2004), "…the need for basic resources could normally be avoided or diminished by innovations aimed at a more effective and economic use of the scarce resources." In short, process innovations have enabled firms to maintain competitiveness by reducing operating costs, maximizing product yields, increasing operating efficiency, and improving product quality.

At the same time, the emphasis on reducing costs and improving efficiency are also increasingly recognized as "necessary but insufficient for long-term competitiveness" (Hansen 2006b). While low-cost strategies may maintain competitive position, development of new products and services tends to create new income and employment and thereby foster industry growth (Kubeczko et al. 2006).

Further, as competition increases with firms with lower operating costs, the low-cost strategy ceases to be effective. This has been the case for furniture manufacturers in developed nations competing with firms in developing nations (Lihra et al. 2008). Furniture industry experts cite mass customization—one example of a business systems innovation—as a key to the future success of furniture industries in developed economies such as in North America and Europe (Cao and Hansen 2006; Buehlmann and Schuler 2009; Lihra et al. 2008). Mass customization is just one example of a larger shift in the forest sector from a production to a market orientation (MO).

A firm's orientation (production vs. market) is often discussed as a manifestation of organizational culture (Narver and Slater 1990; Stendahl et al. 2007). In that regard, we will now examine what the research has shown about the innovation culture in the forest industry.

16.5.2 Innovation Culture in the Forest Industry

How a firm approaches innovation, or even perhaps if a firm chooses to pursue innovation, depends in large degree on the culture of the organization (Deshpande et al. 1993). The culture of the forest industry is often referred to via adjectives such as risk-averse, traditional/conservative, commodity focused, and production oriented. On this latter point, the predominant strategic orientations that guide businesses' interactions with customers and competitors may be defined as follows:

- Production orientation (PO)—Pursuing production and other operating efficiencies in order to produce "widely available and relatively inexpensive products and services that will attract consumers" (Noble et al. 2002). Hansen and Juslin (2011) describe the manifestations of this orientation as emphasizing producing large volumes of commodity products at low costs and relying on sales personnel to sell the product; given that products are typically sold through intermediaries, market information within the firm is minimal and there is little knowledge of customers and little if any contact with end customers.

- MO—"the degree to which the business unit obtains and uses information from customers, develops a strategy which will meet customer needs, and implements that strategy by being responsive to customers' needs and wants" (Ruekert 1992). In their definition, Narver and Slater (1990) also include an emphasis on competitors (e.g., via benchmarking) and interfunctional coordination to obtain and disseminate information from customers and competitors throughout the organization.

Many authors have asserted a positive correlation between MO and innovation performance or innovativeness, although questions remain on the nature of the cause-and-effect relationships. For example, Kohli and Jaworski (1990) stated that, because MO involves doing something new as a response to market conditions, it can be viewed as a form of innovative behavior. Agarwal et al. (2003) simply report that "market orientation spurs innovation." Other researchers have reported more complex, but generally positive, relationships. For example, Lukas and Ferrell (2000) state that customer orientation (an aspect of MO) increases the introduction of radical product innovations (what the authors describe as "new-to-the-world" products), competitor orientation leads to a reduction in the launch of line extensions and radical product innovations, and interfunctional coordination leads to an increase in the launch of line extensions and a reduction in introduction of incremental innovations (described as "me-too" products).

Several researchers have commented on the fact that the dominant culture in the forest industry is that of a PO (Crespell et al. 2006; Hansen 2006; Stendahl et al. 2007). Hence, to improve innovation performance, experts suggest the industry shift to a market-oriented culture. As implied in the description of MO, a key component of such a shift rests in how firms interact with their customers and competitors. Acquisition, dissemination, and use of information about the market, customers, and competitors are key areas of focus for market-oriented firms.

BOX 16.2   FOREST INDUSTRY MANAGERS' VIEWS ON INNOVATION CULTURE

Perhaps the best way to describe forest industry managers' views on innovation is to report the views in their own words. Researchers that have interviewed forest industry managers have reported statements such as

> This is a redneck-in-the-woods kind of industry. (Hansen et al. 2006)

> I think our actual technology is probably behind our competitors. You know from the standpoint of degree of automation it probably is… So our kind of belief is why let yourself fall into a trap like that? When it's not a proven technology—in other words, technology is wonderful when it's proven. (Leavengood 2011)

> We manufacturers want to adopt automation…. (Cao and Hansen 2006)

> We really don't [develop new products]. Outside of, like I say if there's stuff other companies are providing that our customers are asking about… There's nothing that we have developed along those lines. It's the industry standard type stuff. (Leavengood 2011)

> [in pursuing new product development] we would see how it **fits** within our operating model. In other words, just to make a new product to make a new product for us doesn't make any sense… It would follow the established channels of distribution that we have, we'd utilize our plant, we're familiar with utilizing it. All that would **fit** and would work. (Leavengood 2011)

MO may drive innovation through obtaining ideas or suggestions for improvements to existing products or services, requests for new product or services, suggestions for new delivery options, etc.; customers and competitors serve as key sources for such ideas and suggestions.

16.5.3 Differences among Industrial Sectors within the Forest Sector

Much of the discussion of the innovation focus and culture of the forest industry to this point has been in broad generalizations of the industry as a whole. However, given the fact that the forest industry includes several sub-sectors—the growing and management of forests, timber harvesting and haulage, sawmilling and processing, furniture and cabinets, etc.—it is reasonable to expect there are significant differences in innovation management among subsectors.

For instance, we have already noted the fact that some firms in the furniture industry have adopted mass customization as a business systems innovation. Schaan and Anderson (2002) examined product and process innovation in the Canadian forest sector and discovered differences between sub-sectors; process innovation was emphasized over product innovation for all sub-sectors there also. However, the difference in emphasis (i.e., ratio of process to product innovations) was higher for logging, sawmilling, wood preservation, veneer, plywood, and engineered wood product manufacturing than for pulp and paper and "other" wood product manufacturing such as millwork, windows, flooring, pallets, and containers (but excluding furniture or cabinets).

There are apparent differences as well between relatively closely related sectors such as furniture manufacturing and kitchen cabinet manufacturing. Lihra et al. (2008) report on differences in performance as well as customization levels within various sub-sectors of furniture and cabinet manufacturing. For wood household furniture manufacturing, the US market grew from $4.2 billion in 1997 to $19.6 billion in 2002. However, by 2002, imports accounted for 63% of the shipments. By contrast, the value of shipments in the United States of kitchen cabinets increased from $9 billion in 1997 to $14 billion in 2002, and imports represent a mere 4% of shipments. The authors cite at least part of the reason for the dramatic difference as due to the level of customization offered to customers. Industry respondents rated cabinet manufacturers the highest with respect to level of customization offered to customers; office furniture, upholstered furniture, and household furniture were all rated significantly lower.

A common approach to explore differences in innovativeness between firms is to compare R&D expenditures (Schaan and Anderson 2002; Diaz-Balteiro et al. 2006; Hansen et al. 2007). Hence, R&D activities and expenses often serve as a proxy for the emphasis a sector (or a firm) places on innovation.

16.5.4 R&D Patterns

"I think in our company we have recently started to realize the R&D function is the only function that can take us there where we want to go" (industry manager quote from Hansen et al., 2007).

To examine the innovation focus of various sub-sectors of the forest industry, we may first begin by comparing what the various sub-sectors invest in R&D. Of course, this is easier said than done. Several researchers have found that industry personnel are not always able to provide data on their firm's investment in R&D (Hansen 2006b; Stendahl and Roos 2008). Comparisons between firms or nations may not always be meaningful, depending on what data are included (e.g., whether or not supplier industries' data are included).

TABLE 16.2

R&D Activities of Innovative Firms in Selected Forest Sector Industries

	Percentage of Innovative Firms Undertaking Given Activity				
	Logging	Sawmills and Wood Preservation	Veneer, Plywood, and Engineered Wood Products	Other Wood Products	Paper
Undertake R&D (%)	26	53	64	55	69
% for whom R&D is conducted in separate/distinct R&D dept	38	31	41	25	53
Contracted to other firms	60	45	46	27	37

Source: Data from Schaan, S. and Anderson, F., *Forest. Chron.*, 78(1), 60, 2002.

Limitations aside, researchers continue to examine R&D funding as an indication of trends—are firms/sectors increasing or decreasing investment in R&D and how do different sectors compare? With respect to increasing vs. decreasing investments, Hansen et al. (2007) report that at the time of their study, several European firms had recently increased investment in R&D, whereas indications were that firms in North America were eliminating the R&D function to cut costs. The authors pose the question as to whether North American firms are accomplishing R&D by outsourcing.

Schaan and Anderson (2002) addressed this question for several forest sector industries in Canada. Table 16.2 indicates that the paper and veneer, plywood, and engineered wood products sectors are more active in R&D in general than are other forest sectors. The data also suggest differences in approach between sectors with respect to conducting R&D in-house vs. outsourcing. For example, few logging firms undertake R&D and even of these, 60% outsource the R&D. The paper industry by contrast is far more active in R&D and over half the firms active in R&D do the work in-house in a separate department.

In addition, Hull (1986) discovered that the Canadian paper industry's investments in formalized R&D processes promote more rapid adoption of process innovations; firms without formalized R&D processes were found to be late adopters of process innovations.

There is additional support for the fact that the paper industry (and perhaps panel industry as well) uses a more formalized approach to R&D. In a study of the Swedish and Finnish sawmilling industry, Stendahl et al. (2007) report that many of the fiber conglomerates (i.e., sawmills that were part of firms also involved in paper or panels) in their study had "development engineers" in centrally organized R&D departments as opposed to "independents" (firms strictly focused on sawmilling).

Few sources provide data on forest sector investments in R&D. In a study examining the North American lumber and panel (structural and non-structural) industries, Hansen (2006) reports that the average investment in R&D across the sectors examined was 1.4% of sales. He further states that this estimate may be high given that Stora Enso, a leading global company, dedicates approximately 0.7% of sales to R&D, large US furniture firms around 1%, and the US steel industry around 0.5% (citing Fruehan et al. 1999; Quesada and Gazo 2003; Anonymous 2004).

In summary, details on R&D in the forest sector are limited. However, the available data suggest that investments in forest sector R&D are increasing in Europe (Finland in particular) while decreasing in North America. Researchers have suggested that North American firms appear to be eliminating R&D departments in favor of outsourcing. This finding reinforces the importance of increasing our understanding of the function of ISs, as discussed previously. In general, the pulp and paper industry uses a more formalized approach to R&D, and research suggests these firms reap the benefits in the form of more rapid adoption of process innovations. There is also evidence indicating that panel producers and engineered wood products firms are more involved in R&D than are sawmills and "other" wood products firms (e.g., millwork, flooring, pallets, and containers) as well as firms involved in logging. Of course, there are close linkages between R&D and NPD.

16.5.5 New Product Development Practices

There is a wealth of information on NPD in the general business literature and a growing body of literature on NPD in the forest industry. As an introduction to NPD in general, we begin by examining various NPD models and tools described by experts. We will then discuss what researchers have discovered regarding the use of these models and tools in various sectors of the forest industry and conclude with recommendations for improving NPD and thereby product innovation performance in the forest industry.

Many NPD models involve "activity stages" (where the tasks are carried out) and "decision stages" (where information is acquired and decisions are made) (Hansen and Juslin 2011). Perhaps the most widely recognized such model is Cooper's Stage-Gate (Cooper 1990), which includes five gates and five stages where the gates essentially represent go/no-go decisions and the stages are the active steps in NPD.

Bumgardner et al. (2001) developed what is likely the most detailed NPD model in the literature for a specific segment of the forest sector; these authors developed a 14-step model for the furniture industry. Of note is the first step—identification of opportunity/ need. As we will discuss later, one key variable in the success of NPD lies in specifically how this first step is conducted.

One of the first studies related to NPD practices in the forest industry suggested that the furniture industry uses a relatively structured approach to NPD (Bumgardner et al. 2001). However, for other sectors of the forest industry, the consensus in the research community appears to be that the forest industry in general does not use a structured approach to NPD. For example, Hovgaard and Hansen (2004) examined small forest products firms in Alaska and Oregon and concluded that firms did not undertake consistent, structured processes for product development. In a study of the North American lumber and panel industry, Hansen (2006) reported that respondents rarely implement a structured NPD process. Similarly, in a study of product development in the Swedish and Finnish sawmilling industry, Stendahl et al. (2007) report that the NPD process is "informal and flexible." And Leavengood (2011) interviewed innovative firms in the forest sector (broadly defined) in the US states of California, Oregon, and Washington; as stated by one interviewee, "…on the product [development] side, it's all kind of loosey goosey–let's try it and see what happens."

Researchers have found a correlation between using a structured approach to NPD and innovativeness. For example, Crespell et al. (2006) found that North American sawmills that were ranked high and medium with respect to innovativeness consistently had more structured NPD processes than low-innovative mills. This was particularly the case for mills following a specialty/custom-made product strategy. However, in a study of small- to medium-sized wood products firms in Australia and New Zealand, Bull and Ferguson (2006) concluded that a project management approach (what the authors termed "governance") similar to the public sector where there are high levels of public and political scrutiny is deemed unfavorable to NPD; flexibility in the management of the product is best. And Stendahl et al. (2007) reported that managers in the Swedish and Finnish sawmilling industry felt that key factors for NPD success included rapid and informal, yet complete and well-defined, development projects led by a strong leader. Hence, at the least, we may conclude that further research is warranted with respect to the structure of NPD processes and perhaps potential trade-offs with degree of structure given the capabilities of the project leaders.

With respect to the use of specific NPD tools, Crespell et al. (2006) assessed the North American softwood sawmilling industry's extent of use of brainstorming, feasibility analysis, focus groups, conjoint analysis, concept testing, prototype testing, computer simulation, in-house product testing, cost estimation/forecasting, financial analysis, test marketing, field testing, and limited rollout. Hansen (2006) evaluated the North American lumber and panel industry's use of the same set of 13 tools. Both groups of researchers discovered that mills used brainstorming, cost estimation/forecasting, and financial analysis with the highest frequency followed by feasibility analysis and in-house testing. However, the firms were not actively using tools that are more market-related, for example, focus groups and test marketing. Further, the business and market analysis tools were found to have a significant effect on business systems innovativeness. In short, the researchers conclude "the marketing, customer led aspects of NPD appear to need special attention."

It is on this particular point that the research on NPD "closes the loop" in a sense with respect to industry culture as discussed previously. Again, researchers have recommended industry leaders embrace a more market-oriented approach to business and the research related to NPD practices reinforces this point. For example, Bull and Ferguson (2006) recommend a market pull vs. technology/resource push; innovations are more successful when "pulled" from the market (i.e., originating from a need expressed by the market) rather than "pushed" to the market based on availability of technology and/or resources. Likewise, Hansen (2006b) stated that successful NPD depends on maintaining close connections with customers during development; firms rely heavily on customers for innovative ideas, that is, customers are significant drivers of innovation. However, the authors go on to say that current industry focus appears to be primarily on existing customers versus new customers; as a result, innovations tend to be incremental as opposed to radical.

On this latter point, one senior manager in the Chinese furniture industry summed up this point well in defining his firm's approach to NPD: "Two styles of new product development—one being market-driven, which means products are developed to meet existing demands… the other being market-leading–products are developed as symbols of new lifestyles and to create demands" (Cao and Hansen 2006). In short, customer focus includes not merely reacting to existing customer demands but proactively seeking to meet needs of new (and existing) customers as well (Leavengood and Anderson 2010).

In summary, the dominant focus of innovation management in the forest sector has been described as reflecting that of a production orientation; to a large extent, firms in the sector have focused on process innovation to reduce costs and improve product quality. Therefore, many experts have recommended that industry and academia strive to shift the industry focus to be more market-oriented and to place greater emphasis on product and business systems innovations.

Specific steps industry leaders can take to change the culture of their organizations include incorporating innovation as an explicit component of company strategy and ensuring that progress is measured. Hiring upper managers from outside the industry as well as younger professionals trained in university programs will also help to bring in new perspectives. Other steps that can be taken include providing employees opportunities to network with other professionals both within and outside the industry. On the latter topic, specific actions might include supporting attendance at professional association meetings and conferences and continuing education programs. Industry personnel can also monitor and support university research in an effort to keep abreast of technology advancements as well as to provide input to innovations in the development stage.

With respect to product innovation, there is evidence to suggest that innovation performance in the forest sector will benefit from efforts to add structure to the NPD process; this may simply be a shift away from "zero emphasis" on the topic. A first step in this direction may be to provide training in the use of NPD tools, particularly those that are market/customer focused.

16.6 Future of Innovation in the Sector

To speculate what the future may hold for innovation in the forest sector, we must ask what will be the drivers for innovation. That is, what forces will the industry be responding to and/or what opportunities will the industry be seeking to capitalize upon? While there are myriad forces/opportunities, we can begin by reexamining the current industry and societal trends in light of how they will and, in fact, currently are driving product, process, and business systems innovations.

16.6.1 Response to Current Trends

The key trends that will continue to impact the forest industry of the future include changes in supply and demand of forest products and the push for sustainability. Establishment of plantations as a source of "rapidly renewable" fiber is a key trend affecting the supply of raw materials to the forest industry, both in terms of quantity and quality of the fiber. This trend may be seen as a response to a combination of a growing population's demand for wood and fiber products as well as a link to a societal push for sustainability. Innovations in products and processes will continue in order to enhance the suitability (e.g., hardness or exterior durability) of plantation-grown wood for a variety of end uses such as flooring, decking, and structural materials.

However, there is another aspect related to "changes in supply and demand" that will drive innovation in the forest sector. Efforts to address climate change are leading to increased efforts to substitute renewable fuels for fossil fuels. Woody biomass (e.g., hog fuel, chips, and densified fuels like briquettes and fuel pellets) is currently a key source of renewable fuels for the production of heat and electricity; research is aggressively pursuing efficient means of converting biomass to liquid transportation fuels such as ethanol for cars or butanol for aviation fuels.

As demand for renewable fuels increases, demand and hence competition for wood fiber will increase as well. Such changes in demand and competition will drive innovation all along the value chain: for example, new systems for harvesting, transporting, and aggregating biomass; development of combined heat and power facilities fueled by biomass in existing forest sector manufacturing facilities; new conversion technologies to produce liquid fuels; and in technologies like torrefaction (treatment of biomass at 220°C–320°C in the absence of oxygen) to produce bio-oil and char for applications like filters and soil amendments. Demand for renewable fuels will drive business systems innovations as well. For example, firms in traditionally unrelated sectors are merging to capitalize on their respective strengths related to growing and processing raw materials and in energy production and distribution. One such example is the joint venture between Weyerhaeuser and Chevron known as Catchlight Energy LLC.

A key shift in the value chain of the forest industry of the future will include development of biorefineries, perhaps by repurposing pulp mills. In this regard, the forest sector's focus for innovation in the future appears to be the same as it has been for decades—"deriving more value from each tree" (PricewaterhouseCoopers 2011). However, in the past, deriving more value from trees focused on increasing yields of traditional products such as lumber, plywood, and pulp. Biorefineries, as discussed in some detail in Chapter 8 of this volume, will focus on "deriving more value from trees" by extracting chemicals from woody biomass that replace petrochemicals. With respect to converting woody biomass to liquid fuels, as is also discussed in Chapter 8, as long as the conversion technologies are unable to compete with costs from cane sugar or starch, the success of biorefineries will depend on increasing product values from the other chemical components of the plants. As such, in addition to liquid biofuels, biorefineries will focus on chemical feedstocks, for example, to make "green" plastics such as for beverage bottles; xylitol, a natural sweetener; phytosterols, used in foods to reduce cholesterol; glucaric acid, a replacement for phosphates in detergents, etc. (see de Jong et al. 2012); and perhaps as a source for cellulose nanocrystals that may be used to reinforce high-performance polymer composites, membranes (e.g., for kidney dialysis), and tissue engineering in biomedical applications, (Simonsen 2012).

16.6.2 Policy as a Driver for Innovation in the Forest Sector

While these changes in demand for fiber are indirectly a result of efforts to address climate change, as discussed in the section on ISs, changes in policy are often the more immediate driver of innovation. Hence, changes in national and regional energy policies will continue to have a significant impact on demand for wood fiber and thereby innovations in the forest sector. However, it is not only new developments in energy policy that will drive future innovations in the sector; changes in building codes will drive innovation as well.

Recent developments in Canada serve as a good example for how changes in building codes impact innovation in the forest sector. In 2009, building codes in the province of British Columbia (BC) were revised to increase the maximum allowable height for wood-framed structures from four to six stories (Anonymous 2009), and the entities driving this change are not stopping at six stories. A 10-story wood-framed building is in the planning stages that would be the tallest such structure in North America (Lee 2012). Further, in 2009, the government of BC passed an act known as *Wood First* that requires projects funded by the province to use wood as the primary construction material (Ho 2009).

These policy changes in BC are also serving to spur demand for a product innovation known as CLT. This product, which is like plywood but made from lumber rather than veneer, was developed in Switzerland in the early 1990s (Crespell and Gagnon 2010). There is a link to sustainability with CLT in that it is said to enable building a home, which can be "…a carbon positive project where more carbon is saved than emitted" (http://www.cstinnovations.ca//benefits.php). And as one additional impact of these policy changes, the first commercial manufacturing facility to produce CLT in North America began operating in BC in 2009 (Lyon 2011). These examples from Canada serve to illustrate the interactions between the societal trends related to sustainability and policy changes, as well as how these forces impact a wide cross section of the economy.

Green building will have a significant impact on innovation in the forest sector—and again, the driver for innovation will begin with changes in policy. Numerous cities, states, public agencies, etc. have developed requirements that new public buildings meet green building standards. In the United States, for example, the US Green Building

Council reported that as of December 2011, "various LEED initiatives including legislation, executive orders, resolutions, ordinances, policies, and incentives are found in 442 localities (384 cities/towns and 58 counties and across 45 states), in 34 state governments (including the Commonwealth of Puerto Rico), in 14 federal agencies or departments, and numerous public school jurisdictions and institutions of higher education across the United States." These policies are resulting in dramatic growth of green building systems, which will continue to drive innovations in materials that require less energy to produce, innovation in adhesives, and finishes related to concerns over indoor air quality, as well as business systems innovations such as in how manufacturers communicate source-of-origin of materials to capitalize on green building systems' credits for "local sourcing" of materials.

Lastly, with respect to the impact of policy on innovation, legislation such as the Lacey Act in the United States and Forest Law Enforcement, Governance, and Trade (FLEGT) in the EU intended to curtail illegal logging will lead to innovations that enable more reliable tracking of the source-of-origin of materials. For example, the Malaysian Forestry Department recently studied the use of RFID tags to track the movement of logs through the supply chain (Friedlos 2009).

To conclude, the forest sector has a long history of innovation focused on efficient utilization of the world's forest resources. Given the continuing increase in global population and rising standards of living in developing nations, the push for sustainability will continue to be the underlying driver of innovation in the forest sector of the future.

References

ABARE. 2010. *Australian Forest and Wood Product Statistics, September and December Quarters 2009*. Canberra, Australian Capital Territory, Australia: Australian Bureau of Agricultural Economics.

Abernathy, W.J. and K. Clark. 1998. *Innovation: Mapping the Winds of Creative Destruction*. Cambridge, MA: Harper Business.

Abernathy, W.J. and J.M. Utterback. 1978. Patterns of innovation in technology. *Technology Review* 80 (7):40–47.

Accoya. 2011. Accoya–modified wood. Accsys Technologies 2011 [cited 30 December 2011]. Available from http://www.accoya.com/

Adair, C. 2004. *Regional Production and Market Outlook Structural Panels and Engineered Wood Products 2004–2009*. Tacoma, WA: APA—The Engineered Wood Association.

Agarwal, S., M.K. Erramilli, and C.S. Dev. 2003. Market orientation and performance in service firms: Role of innovation. *Journal of Services Marketing* 17 (1):68–82.

Anonymous. 2003. *ThermoWood Handbook*. Helsinki, Finland: Finnish ThermoWood Association.

Anonymous. 2004. *Corporate Annual Report*. Helsinki, Finland: Stora Enso.

Anonymous. 2009. Residential mid-rise wood-frame code change. http://www.housing.gov.bc.ca/building/wood_frame/index.htm (Accessed April 18, 2012).

Anonymous. 2012. Protect and defend the US Renewable Fuel Standard, says BP biofuels chief. *Biofuels Digest*, http://www.biofuelsdigest.com/bdigest/2012/04/05/the-game-changer/ (Accessed April 19, 2012).

Attewell, P. 1992. Technology diffusion and organisation learning: The case of business computing. *Organization Science* 3 (1):1–19.

Australia, Montreal Process Implementation Group for. 2008. *Australia's State of the Forests Report 2008*. Canberra, Australian Capital Territory, Australia: Bureau of Rural Sciences.

Balatinecz, J.J. and R.T. Woodhams. 1993. Wood plastic composites: Doing more with less. *Journal of Forestry* 91 (11):22–26.

BERC. (Biomass Energy Resource Center). 2011 [cited December 26 2011]. Available from http://www.biomasscenter.org/

Betz, F. 1998. *Managing Technological Innovation: Competitive Advantage from Change.* ed. D.F. Kocaoglu, Engineering and Technology Management. New York: John Wiley & Sons, Inc.

Brown, S. and K.M. Eisenhardt. 1995. Product development: Past research, present findings and future directions. *The Academy of Management Review* 20 (2):343–363.

Buehlmann, U. and A. Schuler. 2009. The U.S. household furniture industry: Status and opportunities. *Forest Products Journal* 59 (9):20–28.

Bull, L. and I. Ferguson. 2006. Factors influencing the success of wood product innovations in Australia and New Zealand. *Forest Policy and Economics* 8 (7):742–750.

Bumgardner, M.S., R.J. Bush, and C.D. West. 2001. Product development in large furniture companies: A descriptive model with implications for character-marked products. *Wood and Fiber Science* 33 (2):312–313.

Byron, R.N. and J.A. Sayer. 1996. Technological advance and the conservation of resources. *International Journal of Sustainable Development and World Ecology* 3 (3):43–53.

Calantone, R., S.T. Cavusgil, and Y. Zhao. 2002. Learning orientation, firm innovation capability, and firm performance. *Industrial Marketing Management* 31 (6):515–524.

Cao, X. and E. Hansen. 2006. Innovation in China's furniture industry. *Forest Products Journal* 56 (11/12):33–42.

CARB. 2011. *Composite Wood Products ATCM.* California environmental protection agency–air resources board 2011 [cited December 23 2011]. Available from http://arb.ca.gov/toxics/compwood/compwood.htm

Cooper, R. 1990. Stage-gate systems: A new tool for managing new products. *Business Horizons* 33 (3):44–54.

Crespell, P. and S. Gagnon. 2010. *Cross Laminated Timber: A Primer.* Vancouver, British Columbia, Canada: FPInnovations.

Crespell, P. and E. Hansen. 2009. Antecedents to innovativeness in the forest products industry. Review of no. *Journal of Forest Products Business Research* 6 (1):1–18.

Crespell, P., C. Knowles, and E. Hansen. 2006. Innovativeness in the North American softwood sawmilling industry. *Forest Science* 52 (5):568–578.

Damanpour, F. 1991. Organizational innovation: A meta-analysis of effects of determinants and moderators. *Academy of Management Journal* 34 (3):555–590.

Damanpour, F. 1996. Organizational complexity and innovation: Developing and testing multiple contingency models. *Management Science* 42 (5):693–716.

Damanpour, F. and S. Gopalakrishnan. 2001. The dynamics of the adoption of product and process innovation in organisations. *Journal of Management Studies* 38 (1):45–65.

Damanpour, F., K.A. Szabat, and W.M. Evan. 1989. The relationship between types of innovation and organizational performance. *Journal of Management Studies* 26 (6):587–601.

Davis, S.M. 1987. *Future Perfect.* Reading, MA: Addison-Wesley Publishing Company.

de Jong, Ed., A. Higson, P. Walsh, and M. Wellisch. 2012. *Bio-Based Chemicals: Value Added Products from Biorefineries.* Amsterdam, The Netherlands: IEA Bioenergy–Task 42 Biorefinery, pp. 1–34.

Deshpande, R., J.U. Farley, and F.E. Webster, Jr. 1993. Corporate culture, customer orientation, and innovativeness in Japanese firms: A quadrad analysis. *Journal of Marketing* 57 (1):23–27.

Diaz-Balteiro, L., A.C. Herruzo, M. Martinez, and J. González-Pachón. 2006. An analysis of productive efficiency and innovation activity using DEA: An application to Spain's wood-based industry. Review of no. *Forest Policy and Economics* 8 (7):762–773.

Dougherty, D. 1996. Organizing for innovation. In *Handbook of Organization Studies*, eds. S.R. Clegg, C. Hardy, and W.R. Nord. Thousand Oaks, CA: Sage.

Dougherty, D. and C.H. Takacs. 2004. Team play: Heedful interrelating as the boundary for innovation. *Long Range Planning* 37 (6):564–590.

Edquist, C. 1997. *Systems of Innovation: Technologies, Institutions and Organizations.* London, U.K.: Routledge.

Edquist, C. 2001. The system of innovation approach and innovation policy—An account of the state of the art. In *Nelson Winter Conference, DRUID*. Aalborg, Denmark.

Eveland, J.D. 1986. Diffusion, technology transfer and implications: Thinking and talking about change. *Knowledge* 8 (2):303–322.

FAO. 2011. *State of the World's Forests*. Rome, Italy: Food and Agriculture Organisation of the United Nations.

Freeman, C. 1982. *The Economics of Industrial Innovation*. 2nd edn. Cambridge, MA: MIT Press.

Friedlos, D. 2009. Malaysian forestry department studies RFID. http://www.rfidjournal.com/article/view/5350. (Accessed April 18, 2012).

Fruehan, R.J., D.A. Cheu, and D.M. Vislosky. 1999. Steel. In *U.S. Industry in 2000: Studies in Competitive Performance (1999)*, ed. D.C. Mowery. Washington, DC: National Academy Press.

Gebert, D., S. Boerner, and R. Lanwehr. 2003. The risks of autonomy: Empirical evidence for the necessity of a balance management in promoting organizational innovativeness. *Creativity and Innovation Management* 12 (1):41–49.

Han, J.K., N. Kim, and R.K. Srivastava. 1998. Market orientation and organizational performance: Is innovation a missing link? *Journal of Marketing* 62 (4):30–45.

Hansen, E. 2006a. Structural panel industry evolution: Implications for innovation and new product development. *Forest Policy and Economics* 8 (7):774–783.

Hansen, E.N. 2006b. The state of innovation and new product development in the North American lumber and panel industry. *Wood and Fiber Science* 38 (2):325–333.

Hansen, E. 2010. The role of innovation in the forest products industry. *Journal of Forestry* 108 (7):348–353.

Hansen, E. and H. Juslin. 2011. *Strategic Marketing in the Global Forest Industries*. 2nd edn. Corvallis, OR: Self-Published.

Hansen, E., H. Juslin, and C. Knowles. 2007. Innovativeness in the global forest products industry: Exploring new insights. *Canadian Journal of Forest Research* 37 (8):1324–1335.

Hansen, E., C. Knowles, and H. Juslin. 2006. Innovativeness (2 of 3): Hurdles to innovativeness. *Research Briefs* (3), http://www.cof.orst.edu/org/owic/pubs/bizbriefs/hurdles.pdf (Accessed December 29, 2011).

Ho, C. 2009. Act introduced to expand wood use in public buildings. British Columbia Ministry of Forests and Range.

Hovee, E.D. 2005. *Oregon Forest Cluster Analysis*. Portland, OR: Oregon Forest Resources Institute.

Hovgaard, A. and E. Hansen. 2004. Innovativeness in the forest products industry. Review of no. *Forest Products Journal* 54 (1):26–33.

Hull, J.P. 1986. *Science and the Canadian Pulp and Paper Industry, 1903–1933*. Toronto, Ontario, Canada: York University.

Hult, G.T.M., R.F. Hurley, and G.A. Knight. 2004. Innovativeness: Its antecedents and impact on business performance. *Industrial Marketing Management* 33 (5):429–438.

Hurley, R.F. and T.M. Hult. 1998. Innovation, market orientation, and organizational learning: An integration and empirical examination. *Journal of Marketing* 62 (7):42–54.

ITA. International Trade Administration trade database 2009. Available from http://www.tse.export.gov/MapFrameset.aspx?MapPage=NTDMapDisplay.aspx&UniqueURL=teny3f45hzztnb55k44gvobt-2009-11-4-22-0-4 (Accessed January 2, 2012).

Johannessen, J.-A., B. Olsen, and G.T. Lumpkin. 2001. Innovation as newness: What is new, how new and new to whom? *European Journal of Innovation Management* 4 (1):20–31.

Johnson, J.D., M.E. Meyer, J.M. Berkowitz, C.T. Ethington, and V.D. Miller. 1997. Testing two contrasting structural models of innovativeness in a contractual network. *Human Communication Research* 24 (2):320–348.

Kamke, F.A. and H. Sizemore III. 2008. *Viscoelastic Thermal Compression of Wood*. ed. U. P. a. T. Office. Corvallis, OR: Eagle Analytical Company, Inc.

Kaufmann, A. and F. Tödtling. 2002. How effective is innovation support for SMEs? An analysis of the region of Upper Austria. *Technovation* 22 (3):147–159.

Klepper, S. 1996. Entry, exit, growth, and innovation over the product life cycle. *The American Economic Review* 86 (3):562–583.

Knowles, C., E. Hansen, and C. Dibrell. 2008. Measuring firm innovativeness: Development and refinement of the new scale. Review of no. *Journal of Forest Products Business Research* 5 (5):1–25.

Knowles, C., E. Hansen, and S.R. Shook. 2008. Assessing innovativeness in the North American softwood sawmilling industry using three methods. Review of no. *Canadian Journal of Forest Research* 38 (2):363–375.

Kohli, A.K. and B. Jaworski. 1990. Market orientation: The construct, research propositions, and managerial implications. *Journal of Marketing* 54 (2):1–18.

Kubeczko, K. and E. Rametsteiner. 2002. Innovation and entrepreneurship – A new topic for forest related research? Vienna, Austria.

Kubeczko, K., E. Rametsteiner, and G. Weiss. 2006. The role of sectoral and regional innovation systems in supporting innovations in forestry. *Forest Policy and Economics* 8 (7):704–715.

Leavengood, S. 2011. *Identifying Best Quality Management Practices for Achieving Quality and Innovation Performance in the Forest Products Industry*. Portland, OR: Engineering & Technology Management, Portland State University.

Leavengood, S. and T.R. Anderson. 2010. Best practices in quality management for innovation performance. In *53rd International Convention of the Society of Wood Science & Technology*. Geneva, Switzerland.

Lee, J. 2012. North America's tallest wood building to be built in BC: 10-storey building would be tallest such structure on continent, minister says. *The Vancouver Sun*, March 22.

Lihra, T., U. Buehlmann, and R. Beauregard. 2008. Mass customisation of wood furniture as a competitive strategy. *International Journal of Mass Customisation* 2 (3/4):200–215.

Lukas, B.A. and O.C. Ferrell. 2000. The effect of market orientation on product innovation. *Journal of the Academy of Marketing Science* 28 (2):239–247.

Lundvall, B.-Å., B. Johnson, E.S. Andersen, and B. Dalum. 2002. National systems of production, innovation and competence building. *Research Policy* 31 (2):213–231.

Lyon, C. 2011. New Westminster's CST Innovations looks at how wood can replace concrete. *New Westminster News Leader*, April 14, 2011.

Maloney, T.M. 1996. The family of wood composite materials. *Forest Products Journal* 46 (2):19–26.

Maskell, P., T. Pedersen, B. Petersen, and J. Dick-Nielsen. 2006. Learning paths to offshore outsourcing–from cost reduction to knowledge seeking. http://ssrn.com/abstract = 982122 (April 15, 2012).

Myers, S. and D.G. Marquis. 1969. *Successful Industrial Innovation: A Study of Factors Underlying Innovation in Selected Firms*. Washington, DC: National Science Foundation.

Narver, J.C. and S.F. Slater. 1990. The effect of a market orientation on business profitability. *Journal of Marketing* 54 (10):20–35.

Noble, C.H., R.K. Sinha, and A. Kumar. 2002. Market orientation and alternative strategic orientations: A longitudinal assessment of performance implications. *Journal of Marketing* 66 (4):25–39.

Nybakk, E., A. Niskanen, F. Bajric, G. Duduman, D. Feliciano, K. Jablonski, A. Lunnan, L. Sadauskiene, B. Slee, and M. Teder. 2011. Innovation in the wood bio-energy sector in Europe. In *Innovation in Forestry: Territorial and Value Chain Relationships*, eds. G. Weiss, D. Pettenella, P. Ollonqvist, and B. Slee. Oxfordshire, U.K.: CAB International.

Peters, J.S., D.T. Damery, and P. Clouston. 2006. A decade of innovation in particleboard and composite materials: A content analysis of Washington State University's International Particleboard/Composite Materials Symposium Proceedings. Review of no. *Journal of Forest Products Business Research* 3 (1).

PricewaterhouseCoopers. 2011. Growing the future: Exploring new values and directions in the forest, paper & packaging industry. PricewaterhouseCoopers LLP. London, UK.

Quesada, H. and R. Gazo. 2003. Benchmarking study based on critical success factors for household, office and kitchen cabinet wood furniture industries in US. Purdue University. West Lafayette, IN.

Rametsteiner, E. and G.Weiss. 2006. Innovation and innovation policy in forestry: Linking innovation process with systems models. *Forest Policy and Economics* 8 (7):691–703.

Rogers, E.M. 2003. *Diffusion of Innovations*. 5th edn. New York: The Free Press. Original edition, 1962.

Rowell, R.M. 2006. Acetylation. *Forest Products Journal* 56 (9):4–12.

Ruekert, R.W. 1992. Developing a market orientation: An organizational strategy perspective. *International Journal of Research in Marketing* 9 (3):222–245.

Ryan, D. 1993. Engineered lumber: An alternative to old-growth resources. *Journal of Forestry* 91 (11):19–20.

Sayer, J.A., J.K. Vanclay, and R.N. Byron. 1997. *Technologies for Sustainable Forest Management: Challenges for the 21st Century*. Victoria Falls, Zimbabwe: Commonwealth Forestry Congress.

Schaan, S. and F. Anderson. 2002. Innovation in the forest sector. *The Forestry Chronicle* 78 (1):60–63.

Schirmer, J. 2010. Socio-economic characteristics of Victoria's forest industries. Part 1. Report prepared by the Fenner School of Environment and Society for the Victorian Department of Primary Industries. Melbourne, Australia: Department of Primary Industries.

Schuler, A., Craig A., and E. Elias. 2001. Engineered lumber products taking their place in the global market. *Journal of Forestry* 99 (12):28–35.

Schumpeter, J.A. 1939. *Business Cycles: A Historical and Statistical Analysis of the Capitalist Process*. New York: McGraw-Hill Book Company, Inc.

Sherman, L. 2007. Nature's glue: Kaichang Li teaches soybeans to act like shellfish. *Terra: A World of Research and Creativity at Oregon State University* 2 (2):4.

Simonsen, J. 2012. Cellulose nanocrystals research 2012 [cited April 19 2012]. Available from http://www.cof.orst.edu/cof/wse/faculty/simonsen/

Skog, K., P.J. Ince, D.J. Dietzman, and C.D. Ingram. 1995. Wood products technology trends: Changing the face of forestry. *Journal of Forestry* 93 (12):30–33.

Stendahl, M. and A. Roos. 2008. Antecedents and barriers to product innovation—A comparison between innovating and non-innovating strategic business units in the wood industry. *Silva Fennica* 42 (4):659–681.

Stendahl, M., A. Roos, and M. Hugosson. 2007. Product development in the Swedish and Finnish sawmilling industry – A qualitative study of managerial perceptions. *Journal of Forest Products Business Research* 4 (4):1–25.

Stone, I.J., J.G. Benjamin, and J. Leahy. 2011. Applying innovation theory to Maine's logging industry. *Journal of Forestry* 109 (8):462–469.

Trott, P. 2011. *Innovation Management and New Product Development*. 5th edn. New Jersey: Prentice Hall.

Ukrainski, K. and M. Kajanus. 2011. Innovation-related knowledge flows: Comparative analysis of Finnish and Estonian wood sectors. In *Innovation in Forestry: Territorial and Value Chain Relationships*, eds. G. Weiss, D. Pettenella, P. Ollonqvist, and B. Slee. Oxfordshire, U.K.: CAB International.

Utterback, J.M. 1994. *Mastering the Dynamics of Innovation: How Companies Can Seize Opportunities in the Face of Technological Change*. Boston, MA: Harvard Business School Press.

Utterback, J.M. and W.J. Abernathy. 1975. A dynamic model of product and process innovation. *Omega: The International Journal of Management Science* 3 (6):639–656.

Utterback, J.M. and F.F. Suarez. 1993. Innovation, competition and industry structure. *Research Policy* 22 (1):1–21.

Välimäki, H., A. Niskanen, K. Tervonen, and I. Laurila. 2004. Indicators of innovativeness and enterprise competitiveness in the wood products industry in Finland. *Scandinavian Journal of Forest research* 19 (5):90–96.

Van Horne, C., J.-M. Frayret, and D. Poulin. 2006. Creating value with innovation: From centre of expertise to the forest products industry. *Forest Policy and Economics* 8 (7):751–761.

Weiss, G. 2011. Theoretical approaches for the analysis of innovation processes and policies in the forest sector. In *Innovation in Forestry: Territorial and Value Chain Relationships*, eds. G. Weiss, D. Pettenella, P. Ollonqvist, and B. Slee. Oxfordshire, U.K.: CAB International.

Wolfe, R.A. 1994. Organizational innovation: Review, critique and suggested research directions. *Journal of Management Studies* 31 (3):405–431.

Womack, J.P., and D.T. Jones. 1996. *Lean Thinking: Banish Waste and Create Wealth in Your Corporation*. New York: Simon & Schuster.

Womack, J.P., D.T. Jones, and D. Roos. 1990. *The Machine that Changed the World*. New York: Macmillan Publishing Company.

17

Strategic Orientations in the Global Forest Sector

Anne Toppinen, Minli Wan, and Katja Lähtinen

CONTENTS

In this chapter, we explore the strategic orientation* in the global forest sector by first studying strategies for competitive advantage (CA) from three strategic perspectives—proceeding from the market-based view grounded on Porter's generic strategies and five forces of competition to the resource-based view (RBV) and finally to the stakeholder view, which integrates the market-based view, the RBV, and the sociopolitical dimension of a company strategy. We then identify four themes in the history of strategic evolution of the forest industry and examine strategic perspectives regarding both traditional and emerging segments of the forest industry and conclude with a discussion on the possible future paths of strategic orientation in the global forest industry.

17.1 Lenses for Studying Strategy: Starting from Porter and Proceeding via RBV to Stakeholder View

The interest in strategy has been strongly evolving and increasing (see, e.g., Mintzberg 1994) after the pioneering work by Ansoff (1965) on corporate strategic planning. In strategic management research, one of the fundamental missions is to investigate the heterogeneity

---

* Strategic orientation is defined as the strategic directions implemented by a firm to create the proper behaviors for the continuous superior performance of the business (Narver and Slater 1990).

in business performance among firms and CA, which in the theory of strategy is a crucial concept formulated by Porter (1980, 1985). CA is the outcome of a chosen strategy that generates higher increased value for a firm compared to its competitors. Sustainable competitive advantage (SCA) exists if increased value remains and competitors cannot duplicate or imitate the factors that enable the increased value creation (Barney 1991). In recent decades, understanding the sources of SCA has become a major realm in strategic management research (Kim 2008).

The competitiveness of an industry is affected by its structure and the strategic decisions made by managers of individual firms within the industry (Porter 1985). In an individual firm, business success is driven by its organizational structure and the firm-level strategic decisions seeking competitiveness within the business environment (Caves 1980). In studies where both industry- and firm-level effects on firm performance have been examined, both resources and industrial structures have been found to be important for company success (Mauri and Michaels 1998; Spanos and Lioukas 2001; Hawawini et al. 2003).

Strategy can be defined as a set of decisions and actions made by firm managers for using firm-specific resources and capabilities to achieve the highest profits in a certain business environment (Aharoni 1993). The focus of an enterprise strategy should be on defining the means for ensuring competitiveness and positive future development. There exist two major theories of CA that are grounded in economics (Porter 1980, 1985; Conner 1991). The first one is the market-based model, which focuses on achieving CA through Porter's (1980, 1985) three generic strategies—cost leadership, differentiation, and focus—and five forces of competition. This theory of CA is mainly driven by external factors, that is, opportunities, threats, and industry competition (Reed et al. 2000). Conversely, the second model is the resource-based model, which centers on the firm's resources and is driven by the factors that are internal to the organization (Reed et al. 2000). The RBV originates from Penrose's thoughts introduced for the first time in the 1950s (Penrose 1995). In the RBV, firm's internal characteristics based on resources and capabilities employed in business processes (Wernerfelt 1984; Barney 1991; Conner 1991; Penrose 1995; Barney 2001) to implement Porter's generic strategies are considered to be more critical to the determination of strategic action than its external environment (Adewoye et al. 2009). According to the RBV (Barney 1986; Dierickx and Cool 1989; Barney 1991; Conner 1991; Grant 1991, 2005; Penrose 1995), the competitiveness of an individual company in a dynamic business environment in any industry (Brown and Blackmon 2005) is assumed to be based on the availability of heterogeneous firm-specific resources and the capability to coordinate those production factors in a strategically successful way (Helfat and Peteraf 2003).

In general, a firm's resources can be either tangible or intangible.* Internal assessments of the resources and the capabilities controlled by a firm as well as the choice of strategies made within a firm that adjust these firm-specific resources to its business environment may create unique combinations of production factors, which lead to superior business performance (Barney 1986). According to Barney (1991), the most important resources for firms are valuable, rare, imperfectly imitable, and not easily substitutable (VRIN). As such, valuable and rare resources contribute to temporary CA. In order for the resources to provide

---

* Operationalization and measurement of resources and capabilities tend to be the challenging part of the RBV. For example, Lähtinen et al. (2009) introduced a multicriteria decision-making methodology to assess the relative importance of different company-level resources within the RBV framework. They identified the resource pool of the Finnish sawmills to comprise five tangible resources (geographic location, raw material, labor, factory and machinery, finance and strategy) and six intangible resources (management expertise, personnel know-how, collaboration, organization culture, technological know-how, reputation and services).

SCA, they further need to be difficult to imitate and difficult to substitute (Dierickx and Cool 1989; Barney 1991; Amit and Schoemaker 1993; Peteraf 1993; Acedo et al. 2006; Lado et al. 2006; Bonsi et al. 2008). According to the categorization done by Bonsi et al. (2008), intangible resources have the largest contribution to SCA in the forest industry.

Barney's (1991) article in the first *Journal of Management* special issue devoted to resource-based thinking brought about a rapid development of the RBV. After 20 years, the RBV has evolved to be one of the most prominent and powerful theories for understanding organizations. In the 1990s, Conner (1991) forecasted that the RBV was developing into a new theory of the firm with a strong cumulative industrial organization economic heritage comprising also major differences from each of the previous theories. As an outcome of the development during the 1990s and the 2000s, the RBV seems to have reached maturity as a theory (Barney et al. 2011). The development of the RBV has emerged with spin-offs from several theoretical perspectives, that is, the knowledge-based view (Grant 1996), the natural resource-based view (NRBV) (Hart 1995), the dynamic capability (DC) view (Teece et al. 1997), institutional theory (Oliver 1997), and organizational economic perspective (Combs and Ketchen 1999).

The future development of the RBV requires analysis within firm boundaries of the internal processes needed for managing the resources (Kraaijenbrink et al. 2010). A key aspect is the recognition of the criticality of heterogeneous human capital as an underlying mechanism for capabilities (Barney et al. 2011). Foss (2011) addresses microfoundations of the RBV in the context of knowledge-based value creation. Since entrepreneurial behavior is context dependent, in addition to internal resources, further understanding of the boundary conditions of the surrounding environment of the individuals also needs to be taken into account in developing the RBV (Shane and Venkataraman 2000).

In the course of market globalization, the sources of competitiveness have changed from static efficiency and the usage of physical production factors to more dynamic processes requiring continuous learning and innovations (Porter 1994; Teece et al. 1997; Teece 2007). Achieving superior performance requires firms to be flexible and capable in adapting to changing market conditions by unique and rational exploitation of the firm's internal resources (Barney 1991). According to Barney (1997), a firm's sustainable competitiveness is achieved by continuous development of the existing resources and creation of new resources and capabilities in response to dynamic market conditions. Since the traditional RBV is a static view of a dynamic process, the DC perspective that emerged in the 1990s has extended the RBV argument by addressing how the VRIN resources can be created and how the current stock of valuable resources can be refreshed in the changing business environments. Since then, the field of strategic management research has advanced considerably (Ambrosini and Bowman 2009).

The DC perspective emphasizes the firm's ability to integrate, build, and reconfigure internal competences under the rapidly changing and complex external environment (Teece et al. 1997; Verona and Ravasi 2003). DCs include the abilities to detect and assess environmental change, to exploit knowledge, to innovate, to manage across multiple product development schedules, to transcend technology cycles, and to integrate technologies across disparate units. Examples of DCs include creation of new products and alliance formation. The central premise of an offshoot of the DC perspective suggests that firm resources need to fit with the environment and change over time in order to maintain their market relevance and deliver CA. While the RBV tends to focus on the types of resources and the characteristics of these resources that make them strategically important, the DC perspective focuses on how these resources need to change over time to maintain their market relevance (Teece et al. 1997).

To examine the role of the natural environment in a firm's core capability development, Hart (1995) argues that the RBV should be extended to include the opportunities and the constraints the natural environment places on firms and how resources and capabilities rooted in the firm's interaction with its natural environment can lead to CA. The NRBV is an extension of the RBV based on the VRIN presumption simultaneously considering the firm to be interconnected with its natural environment. The NRBV connects environmental challenges and the firm's resources operationalized through three interconnected strategic capabilities—pollution prevention, product stewardship, and sustainable development. The NRBV assumes that these strategic capabilities contribute to CA by lowering production costs or by lowering the use of limited resources (Hart 1995). Hart and Dowell (2011) revisit this earlier approach in light of a number of important developments that have emerged recently in both the RBV and the research on a sustainable enterprise. First, the NRBV of the firm is considered in the light of DCs. Second, the role of the NRBV is examined to understand how firms incorporate environmental sustainability in their quest for CA. Third, the resources and capabilities required to enter and succeed in base-of-the-pyramid (BoP)* (Prahalad and Hammond 2002) markets are discussed (Barney et al. 2011).

Strategic management in different environmental contexts and the way entrepreneurs recognize the business opportunities and access the resources and capabilities needed to exploit those opportunities in the external environment have not so far been analyzed profoundly (Barney et al. 2011). Figure 17.1 describes the strategic management process of a firm from the internal resource and capability perspective, with a simultaneous focus on the external business environment and the natural environment. This figure illustrates that in strategy building, the managers' perceptions of the external business environment largely dictate the tangible and intangible resources that are chosen to be exploited, developed, and protected within a firm (Barney 1986; Dierixck and Cool 1989; Barney 1991; Grant 1991; Fahy 2002).

In the interface between the external business environment and the internal company actions, there are firms' business, knowledge, and financial processes. Business processes are the practical ways in which the use of resources and capabilities of a firm are coordinated *via* activities to achieve the company's strategic goals within the business environment (Zairi 1997; Ray et al. 2004). Business processes include, for instance, the activities that are linked to acquiring raw materials, producing products or services, and delivering products or services to customers (Porter 1985). Knowledge processes of a firm affect how information is acquired, disseminated, interpreted, and used to accomplish organizational goals (Turner and Makhija 2006). Knowledge processes include, for example, the activities that enable embedding knowledge in business processes; representing knowledge in documents, databases, and software; accessing valuable knowledge from outside sources; using accessible knowledge in decision-making; measuring the effects of strategic decision-making; and generating new knowledge (Ruggles 1998). The core task of managers is to design knowledge processes to enhance capability building that supports the ability to understand the internal and external forces affecting company operations (Senge 1990). Finally, the outcomes of business and knowledge processes are materialized in financial processes that provide economic information of a company's activities within its environment (Riahi-Belkaoui 2000). Thus, a company's strategic decisions and performance measurements are closely connected to each other.

---

* Low-income markets with consumer income less than 2.50 USD a day.

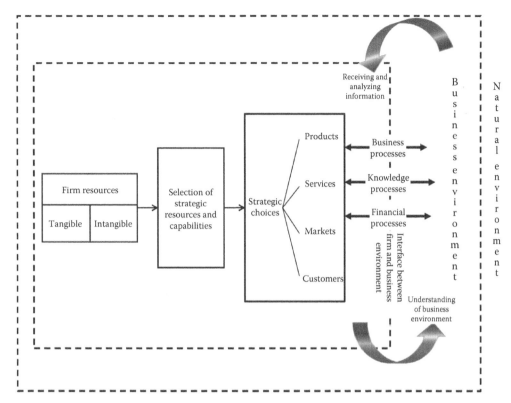

FIGURE 17.1
Interrelationships between the external business environment and the internal strategy building regarding resource usage and capability building. (Modified from Bull, L. and Ferguson, I., *Forest Policy Econ.*, 8(7), 742, 2006; Grant, R.M., *Calif. Manage. Rev.*, 33(3), 114, 1991; Lähtinen, K., *Silva Fennica*, 41(1), 149, 2007.)

In the RBV, there are some aspects that remain underexplored and less developed than others, although it has become one of the dominant perspectives in strategic management with a very large body of work during the previous decades. The early years of the RBV's development were focused on establishing theoretical and empirical relationships between the presence of resources and the development of SCA. More recently, the central issue of where resources come from, that is, a process perspective, has begun to attract attention (Barney et al. 2011). Wernerfelt (2011) approaches the processes of resource acquisition by arguing that the current stock of resources creates asymmetries in competition for new resources. The implication is that firms should expand their resource portfolio by building on their existing resources. Maritan and Peteraf (2011) focus on how the heterogeneous resource positions that lie at the core of the RBV come into existence, building upon two separate mechanisms—resource acquisition in strategic factor markets and internal resource accumulation (Barney et al. 2011).

The RBV of the firm states that firms can earn superior profits and achieve SCA by owning strategic resources without considering the social and ethical dimensions of organizational resources. In response to the need for a more proactive role by state, companies, and communities in a development process aiming at balancing economic growth with environmental sustainability and social cohesion, corporate responsibility issues have attained prominence in the political and business agenda since the early 1990s. Aligning with the RBV of the firm, relations with primary stakeholders* can constitute intangible, socially complex resources that may enhance a firm's ability to outperform its competitors, making stakeholder management a strategic issue for a company (Litz 1996; Hillman and Kleim 2001; Branco and Rodriquez 2006). Thus, an investment in socially responsible activities may have both internal and external benefits by helping a firm develop its resource and capability base. It can be argued that there are numerous benefits to be obtained by respectful and proactive social actions tied to corporate reputation, employee loyalty, and stakeholder commitment. All of these provide each firm with a unique, dynamic, strategic positioning opportunity, and therefore, the role of intangible resources is paramount in formulating and implementing corporate responsibility strategy.

In the market-based view, the focus of strategic actions of the company is on creating superior value for the customer stakeholder group. As a result, the needs of secondary stakeholders† outside economic and market spheres of interest, that is, the broad social and environmental responsibilities required by holistic application of corporate sustainable management, may be considered inadequately (Freeman 1984; Mitchell et al. 2010). The demands for adding social and environmental dimensions to the traditional economic benchmark are reinforced with the increasing societal, market, government, and corporate awareness of the importance of sustainable development and management (Mitchell et al. 2010). Accordingly, there is a need for corporations to broaden their marketing focus to all relevant stakeholders (Ferrell et al. 2010) and create value for a broader range of stakeholders, not only customers.

Compared with the RBV, the stakeholder view of the firm (Freeman 1984) does not separate economics from ethics as most business decisions have an ethical content and an influence on the well-being of firms' stakeholders. The stakeholder view not only integrates both the market-based view and the RBV but also adds the sociopolitical dimension to the examination. Bhattacharyya et al. (2008) suggest that when managers design strategic corporate responsibility initiatives, the firm's interest should be rooted in its potential contribution to the value chain of the firm with regard to the competitive environment and the possibility of creating new business opportunities. Moreover, Galbreath (2009) demonstrates a framework for integrating corporate responsibility with six fundamental and interconnected dimensions of corporate strategy, that is, company mission, identification of strategic issues, markets, customer needs, resources, and CA. McWilliams and Siegel (2011) overview the creation and capture of private and social value by firms that adopt the corporate social responsibility (CSR) strategies, integrating the RBV framework with concepts and tools from economics, notably hedonic pricing, contingent valuation, and the new literature on the economics of industrial organization. Recently, Mitchell et al. (2010) have suggested that adoption of sustainable market orientation (SMO) as a new approach to managing marketing and business strategy offers the potential to produce significant long-term benefits for both primary and secondary corporate stakeholders.

---

* According to Freeman's (1984) seminal proposition, companies have a responsibility to consider all stakeholders that affect or are affected by their actions, and therefore, systematic attention is given to the interests of these stakeholders. Thus, corporate responsibility can add to the bottom line of the firm by improving stakeholder relationships (see also Donaldson and Preston 1995). Primary stakeholders are also called internal stakeholders; they are those without whose continuing participation the company cannot survive, for example, shareholders, employees, customers, suppliers, and government (Clarkson 1995).

† Secondary stakeholders are also called external stakeholders; they are those who may not participate in direct transactions with the company or may be necessary to its survival but otherwise affect or are affected by the company's activities, such as community activists, advocacy groups, religious organizations, and other nongovernmental organizations (Clarkson 1995).

17.2 Historical Evolution of Forest Sector Strategies toward Enhancing Profitability

17.2.1 Introduction

A business strategy covers a number of different and interrelated functions and activities such as manufacturing, operations, research, financing, personnel, and marketing. At the organizational level, marketing is a vital business function necessary in almost any industry and perhaps the most important activity among the knowledge processes of a company with a direct effect on profitability and sales. Especially, with the increasingly fierce competition in the global forest industries, marketing has become an increasingly important factor influencing the competitiveness of companies. Also, as a philosophy, marketing has changed in response to historical developments in a society (Hansen and Juslin 2011), and there is an indication that a properly conceived and implemented marketing strategy can have a significant impact on company performance (Hansen et al. 2006).

SCA, that is, creating sustainable value for customers, is a prerequisite for a company to achieve consistently above-average business performance (Porter 1985; Aaker 1989). Market orientation is rooted in marketing theory as the operationalization of the marketing concept (Liu et al. 2003), and it has been a core foundation of corporate marketing strategy since the mid-1950s (McKitterick 1957; Grönroos 1989; Narver and Slater 1990; Jaworski and Kohli 1993; Hunt and Lambe 2000; Gebhardt et al. 2006). The fundamental work on market orientation started with the publications by Narver and Slater (1990) and Kohli and Jaworski (1990).*

The strategic orientation of the forest industry has evolved in the course of time through four distinct stages: forestry orientation, production orientation, market orientation, and sustainability and increased stakeholder orientation. Market orientation is a business perspective with a focus on customers in the company's operations. Stakeholder orientation is explored as a broader, expanded view of market orientation for marketing strategy (Ferrell et al. 2010). It thus takes into account the interests of all stakeholders for which firms are responsible (Freeman 1984), not only customers.

The forest industry has traditionally relied on a production-oriented management philosophy that emphasizes the cost-efficient production of commodities (Hansen and Juslin 2011). In the 1990s, a combination of pressures, such as increased environmental concerns and competitive threats as a result of globalization, led to a general shift from a production orientation to a more customer-centered market orientation (Beauregard and Bouthillier 1993; Cohen and Kozak 2001; Hansen and Juslin 2011; PWC 2002; Vincent 2002). If a seller wished to continue to create superior value for one's buyer, the seller had to maintain a thorough understanding of the buyer's needs.

In analyzing historical evolution of competitive strategies in the world's leading forest industry companies, Ojala et al. (2007) find that the center of gravity within the forest industry during the 1900s moved from using dominantly upstream strategies, that

---

* As Narver and Slater (1990, 27) state in their seminar paper, "Sellers of commodity products have traditionally created value for buyers by offering lower prices for a given quality, and the retailers and other commercial buyers have shopped actively among the various sellers for the best price-driven value. Today, to some degree, virtually all companies understand that they can create superior value for buyers on a basis other than price. Nevertheless, they differ greatly in their success in implementing non-price-based buyer-value strategies."

is, raw material acquisition and semifinished or standardized products, toward downstream strategies, that is, production of converted or customized products, marketing, and customer orientation. Companies focusing on the production of semifinished or standardized products may be considered to follow upstream-oriented strategies, while companies emphasizing on further processing into higher value-added products may be regarded as implementing more downstream-oriented strategies aiming to have closer contacts and links to customers (Sajasalo 2002). The downstream value is added through product positioning, advertising, marketing channels, and R&D (Chronéer 2003). Overall, Ojala et al. (2007) assess that until the end of the 1990s, only a moderate degree of competition existed in the global forest industry due to constantly growing markets, low degree of internationalization in the leading firms, and the emphasis on business-to-business products with long-term buyer–supplier contracts. In general, especially in the pulp and paper industry, barriers to entry have also been substantial as a result of high investment costs.

In the large-scale wood industry, there is some recent evidence of a dual strategic focus of building capability base to combine leading-edge and innovative solutions with cost-efficient and large-scale production (Korhonen 2006). In addition, an increase in the importance of cooperation networks and strategic alliances in gaining access to strategic resources has been emphasized (Toppinen et al. 2011).

In the following sections, we find it relevant to review the evolution and state of the art of forest sector strategies from four different perspectives: (1) regarding patterns in forestry and production orientation, (2) the process of growing market or customer orientation, (3) from the viewpoint of an increased company internationalization, and (4) increased role of broader stakeholder orientation in corporate-level strategic management.

17.2.2 Strategies Based on Forestry and Production Orientation

Forestry orientation was a dominant strategy in forest industries until the late 1950s. During this period, forestry companies made and sold whatever output they were able to manufacture within the limits of raw material availability. The focus of R&D at this stage was on improving wood procurement from the forests. By the 1960s and 1970s, technological development had enabled the removal of increasing amounts of logs, so the forest industry shifted into a production focus to process readily available logs into lumber, panels, and pulp for paper (Cohen and Kozak 2001). This strong production orientation applied both informational and operational technology to improve productivity and to reduce costs across all aspects of manufacturing. The focus of R&D was on technological advancements to maximize the magnitude of production.

Firms with a production orientation produce what they are most efficient at manufacturing, thereby increasing their production proficiencies but simultaneously narrowing product ranges (Cohen and Kozak 2001). Production orientation assumes that customers are primarily interested in product availability and low prices and that the needs of customers are perceived secondary compared with the need to increase production levels. Managers of production-oriented businesses concentrate on achieving high production efficiency, low costs, and mass distribution (Kotler and Keller 2009). It indicates that the production-oriented companies focus on developing CA based on tangible assets, especially raw materials and physical processes such as production and product distribution (Korhonen and Niemelä 2003). Such an approach is only effective in solutions when customer demand is high, customer needs are simple, and competition is limited (Hansen and Juslin 2011). However, since production orientation focuses on products rather than customers, it may lead to "marketing myopia"* and therefore result in complacency and a loss of knowledge of customers' needs.

In the wood and pulp and paper industry, highly volatile forest product prices have traditionally had the most significant impact on company performance. Economies of scale and scope have been assumed to dictate performance development (Diesen 2007) in addition to the strategic role of investment rivalry in the pulp and paper sector (Christensen and Caves 1997). However, in comparison with the profitability of 30 Scandinavian forest industry companies in 2005, with a few exceptions, niche firms were found to be the most profitable in the industry (Pettersson 2006), and no correlation was found between the pulp and paper industry profitability and firm size during 2003–2005 (Ernst and Young 2007; Uronen 2010).† These findings have thus contradicted the generally maintained assumption of the positive financial outcomes of the economies of scale in the forest industry. Also in retrospect, the synergies of the high level of mergers and acquisitions (M&A) in the 1990s and the early 2000s might have been overestimated, and the strategic fit‡ between the merging companies has been lower and the challenges of successfully integrating distinct company cultures were greater than what was originally anticipated (Siitonen 2003; Ernst and Young 2007). Moreover, the excess supply in the global paper markets causing declining real prices has been a problem in the 2000s.

17.2.3 Strategies Based on Market or Customer Orientation

The marketing concept reflects a philosophy of doing business that is central to firm performance (Lusch and Laczniak 1987; Narver and Slater 1990; Jaworski and Kohli 1996). Instead of mass production, market orientation is characterized as the organization culture that requires customer needs and satisfaction to be placed at the center of business operations (Liu et al. 2003). In addition, the organizational culture that most effectively and efficiently creates necessary behaviors for the creation of superior value for customers and thus continuous superior performance for the business is crucial in market orientation (Narver and Slater 1990; Day 1994). Although the marketing concept and the related construct of market orientation have been fundamental components of business practice for several decades, the forest products industry was late in adopting a marketing philosophy compared with other sectors (Cohen and Kozak 2001).

In terms of Porter's paradigms and the RBV, the strategic choices—if companies have the power of decision in them in the first place—are mostly determined by the external economic environment facing the industry and the strengths and resources of the company. There is an overall trend showing a shift in emphasis from a cost leadership strategy to a differentiation and/or focus strategy (Rich 1986; Bush and Sinclair 1991). Taking the US wood products sector as an example, Rich (1986) utilizes Porter's generic strategies to compare the strategies of the top 50 US forest products companies between the period 1976–1979 and the year 1984. When comparing these two periods, Rich (1986) finds

---

* "Marketing myopia" refers to a short-sighted and inward-looking approach to marketing that focuses on the needs of the company instead of defining the company and its products in terms of the customers' needs and wants (Levitt 1960).
† Contrasting empirical evidence has also been found in Laaksonen-Craig and Toppinen (2008) for the lack of existence of any positive scale economies or firm-specific profits among the largest forest industry companies globally, and regionally most profitable companies have been found to be located in Latin America during that period.
‡ A strategic fit exists if merging companies can fit each other's goals and benefit from each other. The benefits of a good strategic fit include cost reduction, the transfer of knowledge, and skills.

a definite decline in using a cost leadership strategy in the earlier period but an obvious increase in the adoption of a differentiation and/or focus strategies in the later period. He indicates that being a low-cost producer of commodity products worked well during the years of high housing demand and strong overseas markets. But with declining strength in both of these major market areas, a shift to more specialized or differentiated products and markets became necessary. In 1976–1979, production strengths gained importance fueled by a strong housing demand for new home construction, so many companies focused on the cost leadership strategy. With slowing economic growth during the 1980s, the wood products industry was more dependent on the repair and remodeling market and experienced excess production capacity, a sharp drop in raw material prices, and increasing competition from imports resulted in declining prices of finished products. Therefore, many companies had to shift their emphasis from a cost leadership strategy producing low-cost, commodity-type products for ever-expanding markets to a differentiation and/or focus strategy producing more specialized or differentiated products targeting particular market segments. Similar to the findings of Rich (1986), a later study by Bush and Sinclair (1991) also identifies a trend toward the adoption of a differentiation strategy among large US hardwood lumber producers, who attempt to differentiate their products through brand development, proprietary grading, increased customer service, promotional activities, etc.

There are a few results from empirical studies that shed some light on the relationship between strategy and market orientation (Kasper 2005). Narver and Slater (1990) show that the correlation between market orientation and differentiation strategy was much higher than the correlation between market orientation and cost leadership strategy in their empirical study in a major US manufacturing company. This result implies that market orientation is better suited to differentiation strategy than cost leadership strategy as the former closely resembles the marketing concept but the latter is very much related to high volumes and large batches of one type of product (Kasper 2005). It suggests that some companies change their strategies from cost leadership to differentiation during the transition from production orientation to market orientation (Tozanli 1997).

Dramatic changes in resource quality and timber availability have complicated the competitive environment in the forest products industry since the 1990s. The increasingly competitive business environment means that managers need to change their orientation from simply maximizing production efficiencies and minimizing production costs to the market orientation that emphasizes the needs of customers to increase the profitability of the firm. In the wood products industry, driven largely by economic downturns, industrial globalization, changing consumer demographics, increased environmental awareness, and technological advancements (Beauregard and Bouthillier 1993; Cohen and Kozak 2001; PWC 2002; Vincent 2002; Hansen and Juslin 2011), the wood products sector again changed its focus in the latter part of the twentieth century. Since the markets would no longer absorb all that could be produced with a well-entrenched fiber supply and well-established production efficiencies, firms had to start to interact with their customers to remain competitive and prosper. The forest industry found itself exploring and articulating the concept of marketing as a means of expansion, and thus, many companies changed their corporate philosophies by adopting a customer focus (Cohen and Kozak 2001). Moreover, rising prices and price instability have created market opportunities for new engineered wood products and nonwood substitute products. The paradigm shift from volume-based to value-added forest products has occurred through the adoption of marketing-based solutions that produce high-value wood products for customers (Hansen and Juslin 2011). The increasing interest in secondary wood products can be considered a response to this market orientation shift (Kozak and Maness 2001; Vlosky et al. 2003; Haartveit et al. 2004).

During the 1980s, there was a marked increase in the importance of technological strength and decline in the focus on production and raw material availability. Along with the strategy shift occurred an increased emphasis on the marketing-related activities and strengths of marketing, sales force, and distribution. As a result, marketing-related strengths were strongly associated with a differentiation or focus strategy. Presumably the increasing importance of marketing-related activities at the group level represents efforts on the part of top management of multistrategy firms to keep close track of the implementation of each strategy, assuming that different strategies can be identified with different organizational groups within each company. However, as Porter (1980) points out, successful execution of each generic strategy requires different resources, strengths, organizational arrangements, and managerial style. Moreover, the increasing use of strategy combinations, and the evident "stuck-in-the-middle" problems rather than implementation of company-wide single strategies, was also evident (Rich 1986). In addition, Pelham and Wilson (1996) find that the influence of the organizational strategy and structure had less impact on firm performance than did having a market-oriented culture, so they argue that a market-oriented organizational culture can be a very strong resource for developing strategies that lead to increased performance, especially for smaller firms. Similarly, based on data collected from 52 softwood sawmills in the United States, Hansen et al. (2002) find that different size companies pursue different strategies, given the different resources and capabilities of companies depending on their size. Limited resources and better flexibility suggest that small companies are more selective (focused) in their strategies (including product, customer, and market area strategies) than medium or large companies. They also find a clear connection between a specialty product strategy, especially a custom-made product strategy, and CA.

By using one large US-based forest industry company that included 140 different strategic business units (SBUs), Narver and Slater (1990) indicate that market orientation was an important driver for business profitability at that time. Similar to their findings, Hansen et al. (2006) conclude, using a large sample of US forest product companies, that market orientation does have a positive effect on firm performance, except that there was no significant effect on self-reported financial performance regarding customer and product differentiation.

In Finland, Juslin and Tarkkanen (1987) operationalize marketing strategy with regard to products, customers, market areas, and CAs. They conclude that the Finnish forest products industry was at that time moving from commodity products in the direction of special and custom-made products. Several years later, Niemelä (1993) finds, using the same conceptual approach, a clear connection between a commodity product strategy and Porter's cost leadership strategy in analyzing the Finnish and the North American large-scale sawmilling industry.

Strategic resource usage decisions made to enhance competitiveness at sawmills have been addressed within the RBV framework since the 2000s in, for example, Korhonen and Niemelä (2005) and Korhonen (2006). Later, the impacts of managerial decisions on company-level usage of resources and their strategic importance in business performance in the sawmilling sector are found to be significant in the case of Finland (Lähtinen et al. 2009). Recently, Hugosson and McCluskey (2009) conclude similar importance of customer orientation in the case of the Swedish sawmilling industry.

17.2.4 Strategies Based on Expanding Internationalization of the Forest Industry

As a result of production orientation, wood products companies have traditionally produced standardized products, pursued cost leadership strategies, and competed on price.

However, the increased level of internationalization and globalization lately has forced companies to be more market- or customer-oriented and to explore how CAs and increased value creation can take place downstream in the value chain (Alfnes et al. 2006).

The global business environment has changed drastically over the past decades through globalization of markets and in the emergence of new organizational forms such as outsourcing and offshoring activities beyond the home market (Jansson and Sandberg 2008). Since the 1990s, the intensifying international competition and dramatic advances in technology—the main forces of globalization—have changed substantially the nature and operation of the marketplace (Macdonald 1997). The change has not only provided companies with new opportunities but also placed considerable demands on companies to develop new strategies to enhance their competitiveness. From the firm's point of view, key strategic factors to increase competitiveness include diversification of products and markets, expanding firm size and investing in R&D activities, which eventually influence the economic performance of companies. International expansion has been viewed as a strategy pulled by products and production. The companies operating internationally have grown by means of mergers, acquisitions, strategic alliances, and direct investments (Alfnes et al. 2006).

In the case of the forest industry, a desire to ensure high quality and efficient procurement of raw materials, such as round wood or wastepaper, has also motivated companies to spread out globally their production plants. Forest industry companies have become more international since the early 1990s as a result of the rapid increase in foreign direct investments (FDIs), but as a whole the industry is not yet truly globalized (Siitonen 2003). The profitability of globalizing North American companies outperformed that of European companies based on company-level data for the years 1990–1998 (Siitonen 2003). According to the results of the same study, the former obtained better value in stock exchanges than the latter, where investors do not apparently place a premium on companies with a more global size.

In their book on forest industry strategic evolution, Ojala et al. (2007) point out that at the end of the 1990s, the level of internationalization in the pulp and paper industry was among the lowest within the manufacturing industries, with the top five companies possessing less than 15% of world capacity. By comparison, in the same time period, the level of globalization in the automobile industry was very high, with the top five companies occupying about 54% of world capacity (OICA 1999). The growing internationalization of forest industry companies has been reflected clearly in the number of cross-border M&As of both timberland and production capacity since the late 1990s and in the sharp increase in FDIs (Toppinen et al. 2012).

Today, the largest forest companies, for example, have a variable degree of international production as measured by their share of foreign employment; and some of them, such as Swedish SCA or South African Mondi, can be considered as highly internationalized. Regarding the scope of internationalization, some companies, like IP and Stora Enso, have activities in well over 40 countries, but overall, about 80% of the pulp and paper capacity of the largest companies is still located in the home continent (Uronen 2010). Recently, Zhang and Toppinen (2011) investigated the relationship between internationalization and performance in a group of 50 largest pulp and paper companies and found it to be curvilinear, that is, beyond an average of 35% level, increased internationalization, as measured by foreign employment, no longer positively benefits company performance.

According to Laurila and Ropponen (2003), there are additional country-of-origin institutional effects on the patterns and forms of foreign expansion besides the economic effect in the forest industry. This current turbulent competitive situation in the world's forest industry could create new M&A opportunities, for example, with companies looking to expand their limited domestic or international value chains into more global value networks. Ernst and Young (2007) indicate that economic performance is highest in the pulp and paper value chain end (such as in paper converting), and the vital importance of markets and end users and integration of value chains are paramount also for traditional pulp and paper producers.

17.2.5 Strategies Based on Enhanced Stakeholder Orientation

Despite the benefits of being market-oriented that were outlined earlier, some limitations have been found in the context of market orientation (Farrell 2000). From a marketing perspective, a stakeholder orientation requires more of an expansive perspective than is found in current market orientation research (Ferrell et al. 2010). It has been viewed as a broad and long-term philosophy that includes ethics and social responsibility in managerial decisions. In addition, management is expected to use its marketing expertise to improve the welfare of all relevant stakeholders as well as to integrate CSR with marketing strategy and to integrate economic corporate performance with social and environmental principles (Ferrell et al. 2010). By adopting a more in-depth stakeholder orientation, corporate management will move beyond a conventional concentration on microeconomic and functional management prescribed by market orientation and provide more strategic marketing effectiveness through an increased understanding of societal and ecological systems. The ways in which stakeholder orientation affects corporate strategy and the ways in which societal and environmental considerations can be effectively integrated with economic management are crucial to sustain and improve long-term corporate marketing performance (Mitchell et al. 2010).

In the context of the forest industry, organizations also increasingly face a very diverse set of stakeholders, with a broad and often conflicting set of knowledge, demands, and worldviews, some of which are far removed from financial and economic issues (Mikkilä 2006). Internationalization and geographic expansion of operations tend to increase the intensity and diversity of stakeholder pressures in companies' external environments, and the issue of stakeholder orientation is gaining importance based on the ongoing discussion about the social and environmental dimensions of globalization (Zink 2005). There has been a growing concern for a wide array of environmental issues related to forests over the past decades, especially the decrease of global forest area and the climate change and carbon sequestration questions related to forests (Vidal and Kozak 2009). In large forest industry companies, acceptability of operations in the eyes of stakeholders is connected to a certain time and a specific place (Mikkilä 2006; Kourula and Halme 2008). Empirical research on the impact of CSR and sustainability orientation on company performance in the forest industry is an ongoing topic in various universities around the world, and new evidence will be accumulating on the corporate-level strategy–CSR linkage to company performance (e.g., Li et al. 2011).

17.3 Perspectives from the Traditional and Emerging Forest Industries

Recently witnessed progress in polarization of the forest industry between the North and South will likely continue in the future as a result of the foreseen continuation of changes in global economic and demographic conditions. For example, in the pulp and paper industry, the prospects for long-term paper demand in the densely populated emerging markets, especially in China, India, and Southeast Asia, will be much better than that in the mature markets of OECD countries. In the pulp industry, Latin America will be likely to capture the bulk of forest industry FDI. In contrast, consolidation will likely take place in the developed countries, where fiber resources have already been more fully used and consumer markets are more mature. In traditional producing countries, the main strategic challenge is related to industry renewal through new products, services, and business models, as well as to the larger question on the role the forest industry can play in the greening economy. Climate change and rising energy prices have increased environmental and social awareness of forest-based industries, and political programs have been created, especially in the European Union (EU), for higher use of renewable energy in the future (the so-called 20–20–20 target). This might influence the supply of wood and wood-based products as well as forest management and also affect the production and consumption of traditional products.

According to a recent doctoral study (Uronen 2010), most paper and board producers are positioned in the middle of the value chain, where they can only generate about 5% of the total value creation in the paper value chain. Therefore, companies should seriously consider the viability of their current value chain positioning, especially since the prospects for future demand for printing and writing paper are very modest in developed markets. One visible change is that during the last 3–4 years, the pulp and paper industry in both North America and Europe has responded to the declining profits, maturity of the markets, and criticisms in the media by intensifying its R&D efforts. Large-scale technology platforms at the EU level accompanied with national research programs (e.g., Research Program for Finnish Forest Cluster with the aim of generating 50% of value added from new products and services by 2050) have been established. As concluded by Uronen (2010), there are examples from several companies of how R&D activities (related to bioenergy, nanotechnology, etc.) have also been found to receive more internal company funding and top management attention than previously. In the following discussion, we will examine the strategic questions from the viewpoint of emerging bioenergy business at both large- and small-scale levels.

Regarding the forest and energy sectors in the emerging bioenergy production, both the pulp and paper industry and the sawmill industry have shown interest in bioenergy production worldwide in order to create value-added products and benefit from joint products suitable for wood-based energy production. In the pulp and paper industry, the potential for collaboration between energy sectors appears to be particularly strong in the integrated forest biorefinery, which would coproduce pulp and bio-based products (energy and fuels) and would be a promising opportunity for making higher value-added products abreast with an increase in flexibility of production and product assortment, providing new opportunities for companies to gain SCA. Given the synergy benefits associated with this collaboration (or through such collaboration), it seems to outweigh the sole production of transportation fuels; moreover, the potential for increasing value added seems to be remote within the traditional business models of pulp and paper production. On the strategic level, it is considered important for a biorefinery to maximize its value added from the economic resources available to the mill and for novel products to meet international customer needs in different geographic areas.

In a study by Pätäri (2009) on Finland, coproducing traditional pulp and paper products together with bioenergy and biofuels was found to entail adopting more strategies based on economies of scope instead of merely focusing on traditional economies of scale as the only source of CA. In the integrated forest biorefinery concept, potential sources of CA include small-scale bioenergy generation in addition to large-scale bioenergy production. This refers to bioenergy production from versatile fuel sources by smaller firms operating in a distributed manner near the raw material and point of use. In the sawmilling sector, the sawwood-based by-products have been used to produce bioenergy, and partnerships with district heating plants of local communities have been emphasized as a strategic resource to create CA for sawmills' bioenergy business (e.g., Wan et al. 2012). In addition to bioenergy, deeper involvement in the construction sector and service provision for end customers buying buildings constructed in wood in urban areas provides new strategic options for companies (e.g., von Geibler et al. 2010). Especially for companies already possessing capabilities in collaboration with other firms and customer orientation, there are possibilities to be involved in building processes and maintenance of buildings during their whole life cycle.

Also, according to a recent study made in eight dominant forest industry countries in Scandinavia, Latin America, and North America (Hämäläinen et al. 2011), the outlook for technical and raw material choices and the barriers to biorefinery diffusion are very similar, and the volatility and inherent uncertainty of related policies may change factors of competitiveness practically overnight. Despite political uncertainty, many major European companies are openly declaring a strategic shift toward bio-based companies. UPM-Kymmene Biofore Company is a prominent example establishing the first tall-oil-based, second-generation liquid fuel biorefinery in Finland in 2014, and the future will show many new business openings like this if sufficient policy support exists. Another line of strategic renewal concerns integration of pulp, paper, and wood products industry into the value chain of production of bio-based composites and chemicals. This would seem to call for a new wave of both international and cross-sectoral investments, joint ventures and strategic alliances, and other activities that are not yet typical in the context of the global forest industry.

17.4 Future of Strategy in the Forest Industry

As the world becomes more globally interconnected and turbulent, achieving and sustaining CA in today's highly competitive and dynamic market environments is increasingly challenging. Interactivity and complexity are enhanced through a higher level of interconnectedness and interdependencies of market players—also coming outside the forestry domain—within value creation networks (Karpen and Bove 2008). Facing fierce competition in a rapidly changing business environment, managers need to search for new ways to differentiate their companies from those of their competitors.

So, what can we conclude regarding the possible future paths of strategic orientation in the forest industry? Based on the review of literature and our own research and insights, we have identified four themes in the history of strategic evolution of the forest industry (Figure 17.2). Toward the future, we would like to stress the key importance of the following three areas of development as the most relevant avenues for strategic orientation in the forest industry: (1) continuing of the required adjustment strategies coupled with the changing geographic focus in face of changing industry demand and relative competitiveness between different production and consumption regions of the world, (2) addressing the role of industry in the overall goals and agendas toward greening economies and global sustainable development, and (3) foreseeable societal development toward growing

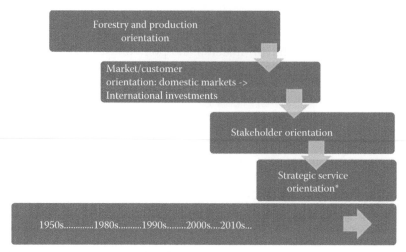

(*Strategic service orientation is based on more theoretical insight than empirical evidence)

FIGURE 17.2
Strategic evolution in the forest industry.

service orientation in business-to-business markets, as a possible fifth paradigm in the historical evolution of forest industry. While the first two areas have been partly covered in this review and are also discussed elsewhere in this edited volume, in this chapter, we focus to a large extent on the role of necessary prerequisites for service orientation and its expected impacts on company performance.

From the view of the first strategic development path, it is worth emphasizing, as has been pointed out by Uronen (2010) in his recent dissertation, that a number of companies in Europe and North America have been profitable even during the financially turbulent years of the early twenty-first century and especially during the recent recession, indicating that either the traditional strategies are still viable under specific industry structural conditions or a successful industry transformation and strategic renewal is indeed possible and has already started to have an effect. Since consumption and production in emerging markets such as China, Brazil, and Russia are growing rapidly, these markets would seem to offer the most important market-driven opportunities and challenges to the traditional Western pulp and paper industry. In local conditions, new businesses such as coproduction of wood-based bioenergy provide new opportunities for strategic transformation both in wood and pulp industries.

From the view of the second strategic path, that is, corporate responsibility perspective, following the conclusions drawn by Etzion (2007), the role of company strategy toward environmental and social issues must be increasingly compatible with a company's general strategy. This seems to be particularly the case in the forest industry with its close natural resource connection and a quest for finding a balance between various local- and global-level stakeholder expectations under a high degree of market and policy uncertainty. Today, there already exist some illuminating examples on how, in the current turbulence in this forest industry operational environment, strategic thinking about corporate responsibility in connection to the quest of sustainable development has become increasingly necessary for developing CA in the forest industry (e.g., Li et al. 2011, and, e.g., the case of Stora Enso Rethink organizational renewal program). All in all, we will find in the future more often in the forest industry that activities under corporate responsibility are no longer considered separately from the firm's strategic management (see also Chapter 10 for further review and discussion on corporate responsibility). In addition, since involvement in CSR is context dependent, taking into account the diverse needs of different stakeholders located in different geographic areas will become more important in the future. This is especially the case if the forest companies are willing to enhance their competitiveness in line with the international political objectives set for enhancing the development of the green economy (e.g., Bär et al. 2011).

Thirdly, with the rise of the service economy, we foresee a continuation of a strategic shift from producing tangible goods to intangible services rendered to customers. Since the uniqueness of service has become a key factor for a firm to differentiate itself from its competitors, many scholars have gradually acknowledged the role of service orientation in helping a firm achieve an SCA (Liang et al. 2010). Within the marketing literature, a rapidly developing and integrated body of thought centered on the service-dominant (S-D) logic (Vargo and Lusch 2004a,b, 2008; Lusch and Vargo 2006; Lusch et al. 2007; Lusch 2011) has significantly contributed to the emerging shift in marketing's prevailing views. In the S-D logic, service is defined as the application of competences for the benefit of customers; customers are operant resources, rather than operand resources,* and they can contribute as value co-creators to the service process (Vargo and Lusch 2004a; Grönroos 2008). Therefore, superior operant resources are regarded as the primary source of CA for a firm (Vargo and Lusch 2004a). In the forest industry context, cocreation of value between producers and customers would inherently require users and consumers to be involved more closely in the product and service development processes. Better fulfillment of customer needs and more efficient provision of product information could facilitate, for example, uptake of greener renewable material-based wood products.

The S-D logic might be the foundation of a paradigm shift in strategic management and marketing. Specifically, the S-D logic proposes a collaborative view of value cocreation from a theoretical perspective (Vargo and Lusch 2004a and b, 2008b; Vargo and Lusch 2008) and potentially sets the stage for full adoption of the marketing concept (Karpen and Bove 2008). In comparison, existing strategic orientations in marketing provide limited guidance for the organizational behaviors that enhance interaction and value cocreation with customers. Therefore, Karpen and Bove (2008) recently propose a concept of strategic service orientation (SSO) consistent with the implications of the S-D logic. Compared with market orientation, the SSO is arguably a stronger source of SCA and firms with an SSO support the social and emotional links between interacting partners (Karpen and Bove 2008). This relational exchange and interaction (Vargo and Lusch 2004a) implies a shift from the traditional notion of firm-controlled customer relationship management toward customer managed relationships (Varey 2002; Karpen and Bove 2008) and therefore increases firms' potential to achieve collaborative benefits and CA (Karpen and Bove 2008).

As a final conclusion, it is thus expected that adoption of the S-D and SSO in the forest industry context could potentially widen the strategic stance of firms and thus allow some proactive companies to find new business opportunities, paving the way toward improved offerings and perhaps an even more responsible way of production, marketing, and consumption. The new business opportunities may also exist at the crossroads of traditional industrial sectors like recently has occurred in bioenergy production that is

---

* Operand resources are usually the tangible and static resources acted upon to produce a product, while operant resources are intangible and dynamic by nature. Employee-based competencies, knowledge, and skills are the examples of operant resources (Constantin and Lusch 1994).

linked to the operations of forest and energy businesses. At the same time, traditional cost minimizing, efficiency seeking way of doing business will not disappear but will certainly remain the "modus operandi" for certain and even a large segment in the industry. But thinking optimistically, adoption of the SSO as a strategic, marketing-led philosophy in the forest-based industries could be an interesting, fresh phenomenon to be seen in our field in the future to aid forest industrial development in the era of sustainable societies and a greening world economy.

References

Aaker, D. A. 1989. Managing assets and skills: The key to sustainable competitive advantage. *California Management Review* 31(2): 91–106.

Acedo, F. J., C. Barroso, and J. L. Galan. 2006. The resource-based theory: Dissemination and main trends. *Strategic Management Journal* 27(6): 621–636.

Adewoye, J. O., J. R. Aworemi, and A. J. Oyedokun. 2009. A study of the effectiveness of IT as a resource based tool deployed by Nigerian banks. *Interdisciplinary Journal of Contemporary Research Business* 1(5): 8–14.

Aharoni, Y. 1993. In search for the unique: Can firm specific advantages be evaluated? *Journal of Management Studies* 30(1): 31–49.

Alfnes, F., F. Asche, T. Bjørndal, S. Baardsen, S. Cantillon, B. Dreyer, R. Døving, A. Følgesvold, S. A. Haugland, and S. E. Haugnes. 2006. *Primary Industries Facing Global Markets: The Supply Chains and Markets for Norwegian Food*, ed. Frank Asche, Oslo, Norway: Copenhagen Business School Press.

Ambrosini, V. and C. Bowman. 2009. What are dynamic capabilities and are they a useful construct in strategic management? *International Journal of Management Reviews* 11(1): 29–49.

Amit, R. and P. J. H. Schoemaker. 1993. Strategic assets and organizational rent. *Strategic Management Journal* 14(1): 33–46.

Ansoff, I. 1965. *Corporate Strategy*. Reading, U.K.: Cox and Wyman Ltd.

Bär, H., K. Jacob, and S. Werland. 2011. Green economy discourses in the Run-Up to Rio 2012. FFU-Report 07-2011, Environmental Policy Research Centre, Freie Universität Berlin, Department of Political and Social Sciences, Berlin, Germany. http://papers.ssrn.com/sol3/papers.cfm?abstract_id=2023052 (accessed July 11, 2012).

Barney, J. B. 1986. Strategic factor markets: Expectations, luck, and business strategy. *Management Science* 32(10): 1231–1241.

Barney, J. B. 1991. Firm resources and sustained competitive advantage. *Journal of Management* 17(1): 99–120.

Barney, J. B. 1997. *Gaining and Sustaining Competitive Advantage*. New York: Addison-Wesley.

Barney, J. B. 2001. Is the resource-based "view" a useful perspective for strategic management research? Yes. *Academy of Management Review* 26(1): 31–46.

Barney, J. B., D. Ketchen, and M. Wright. 2011. The future of resource-based theory: Revitalization or decline? *Journal of Management* 37(5): 1299–1315.

Beauregard, R. and L. Bouthillier. 1993. Crisis in the Quebec forest industry—Problems and possible solutions. *Forestry Chronicle* 69(4): 406–408.

Bhattacharyya, S. S., A. Sahay, A. P. Arora, and A. Chaturvedi. 2008. A toolkit for designing firm level strategic Corporate Social Responsibility (CSR) initiatives. *Social Responsibility Journal* 4(3): 265–282.

Bonsi, R., D. R. Gnyawali, and A. L. Hammett. 2008. Achieving sustained competitive advantage in the forest products firm: The importance of the resource-based view. *Journal of Forest Products Business Research* 5(3): 1–14.

Branco, M. and L. Rodriguez. 2006. Corporate social responsibility and resource-based perspectives. *Journal of Business Ethics* 69(2): 111–132.

Brown, S. and K. Blackmon. 2005. Aligning manufacturing strategy and business-level competitive strategy in new competitive environments: The case for strategic resonance. *Journal of Management Studies* 42(4): 793–815.

Bull, L. and I. Ferguson. 2006. Factors influencing the success of wood product innovations in Australia and New Zealand. *Forest Policy and Economics* 8(7): 742–750.

Bush, R. J. and S. A. Sinclair. 1991. A multivariate model and analysis of competitive strategy in the US hardwood lumber industry. *Forest Science* 37(2): 481–499.

Caves, R. E. 1980. Industrial organization, corporate strategy and structure. *Journal of Economic Literature* 18(1): 64–92.

Christensen, L. R. and R. E. Caves. 1997. Cheap talk and investment rivalry in the pulp and paper industry. *Journal of Industrial Economics* 45(1): 47–73.

Chronéer, D. 2003. Have process industries shifted their centre of gravity during the 90s? *International Journal of Innovation Management* 7(1): 95–129.

Clarkson, M. B. E. 1995. A stakeholder framework for analyzing and evaluating corporate social performance. *Academy of Management Review* 20(1): 92–117.

Cohen, D. H. and R. A. Kozak. 2001. Research and technology: Market-driven innovation in the twenty-first century. *The Forestry Chronicle* 78(1): 108–111.

Combs, J. G. and D. J. Ketchen. 1999. Explaining interfirm cooperation and performance: Toward a reconciliation of predictions from the resource-based view and organizational economics. *Strategic Management Journal* 20(9): 867–888.

Conner, K. R. 1991. A historical comparison of resource-based theory and five schools of thought within industrial organization economics: Do we have a new theory of the firm? *Journal of Management* 17(1): 121–154.

Constantin, J. A. and R. F. Lusch. 1994. *Understanding Resource Management*. Oxford, OH: The Planning Forum.

Day, G. S. 1994. The capabilities of market-driven organizations. *Journal of Marketing* 58(4): 37–52.

Dierickx, I. and K. Cool. 1989. Asset stock accumulation and sustainability of competitive advantage. *Management Science* 34(12): 1504–1511.

Diesen, M. 2007. *Economics of Pulp and Paper Industry*. Helsinki, Finland: Finnish Paper Engineer' Association.

Donaldson, T. and L. E. Preston. 1995. The stakeholder theory of the corporation: Concepts, evidence and implications. *Academy of Management Review* 20(1): 65–91.

Ernst and Young. 2007. At the crossroads—Ernst & Young Global Pulp & Paper Report 2007, Helsinki, Finland.

Etzion, D. 2007. Research on organizations and the natural environment, 1992-present: A review. *Journal of Management* 33(4): 637–664.

Fahy, J. 2002. A resource-based analysis of sustainable competitive advantage in a global environment. *International Business Review* 11(1): 57–78.

Farrell, M. A. 2000. Developing a market-oriented learning organization. *Australian Journal of Management* 25(2): 201–223.

Ferrell, O. C., T. L. Gonzalez-Padron, G. T. M. Hult, and I. Maignan. 2010. From market orientation to stakeholder orientation. *Journal of Public Policy & Marketing* 29(1): 93–96.

Foss, N. 2011. Why micro-foundations for resource-based theory are needed and what they may look like? *Journal of Management* 37(5): 1413–1428.

Freeman, R. E. 1984. *Strategic Management: A Stakeholder Approach*. Boston, MA: Pitman.

Galbreath, J. 2009. Building corporate social responsibility into strategy. *European Business Review* 21(2): 109–127.

Gebhardt, G., G. Carpenter, and J. Sherry. 2006. Creating a market orientation: A longitudinal multi-firm, grounded analysis of cultural transformation. *Journal of Marketing* 70(4): 37–55.

Grant, R. M. 1991. The resource-based theory of competitive advantage: Implications for strategy formulation. *California Management Review* 33(3): 114–135.

Grant, R. M. 1996. Toward a knowledge-based theory of the firm. *Strategic Management Journal* 17(Winter Special Issue): 109–122.

Grant, R. M. 2005. *Contemporary Strategy Analysis* (5th edn.). Malden, MA: Blackwell Publishing.

Grönroos, C. 1989. Defining marketing: A market-oriented approach. *European Journal of Marketing* 23(1): 52–60.

Grönroos, C. 2008. Service-dominant logic revisited: Who create value? And who co-creates? *European Business Review* 20(4): 298–314.

Haartveit, E., R. A. Kozak, and T. C. Maness. 2004. Supply chain management mapping for the forest products industry: Three cases from Western Canada. *Journal of Forest Products Business Research* 1(5): 1–30.

Hämäläinen, S., A. Näyhä, and H. Pesonen. 2011. Forest biorefineries—A business opportunity for the Finnish forest cluster. *Journal of Cleaner Production* 19(6): 1884–1891.

Hansen, E., C. Dibrell, and J. Down. 2006. Market orientation, strategy, and performance in the primary forest industry. *Forest Science* 52(3): 209–220.

Hansen, E. and H. Juslin. 2011. *Strategic Marketing in the Global Forest Industries* (2nd edn.). Eugene, OR: Shelton Turnball.

Hansen, E., J. Seppälä, and H. Juslin. 2002. Marketing strategies of softwood sawmills in western North America. *Forest Products Journal* 52(10): 19–25.

Hart, S. L. 1995. A natural-resource-based view of the firm. *Academy of Management Review* 20(4): 986–1014.

Hart, S. L. and G. Dowell. 2011. A natural-resource-based view of the firm: Fifteen years after. *Journal of Management* 37(5): 1464–1479.

Hawawini, G., V. Subramanian, and P. Verdin, 2003. Is performance driven by industry- or firm-specific factors? A new look at the evidence. *Strategic Management Journal* 24(1): 1–16.

Helfat, C. E. and M. A. Peteraf. 2003. The dynamic resource-based view: Capability lifecycles. *Strategic Management Journal* 24(4): 997–1010.

Hillman, A. J. and G. D. Keim. 2001. Shareholder value, stakeholder management, and social issues: What's the bottom line? *Strategic Management Journal* 22(2): 125–139.

Hunt, S. and C. Lambe. 2000. Marketing's contribution to business strategy: Market orientation, relationship marketing and resource-advantage theory. *International Journal of Management Review* 2(1): 17–43.

Hugosson, M. and D. R. McCluskey. 2009. Marketing competencies of Swedish sawmill firms. *Journal of Forest Products Business Research* 6: 31.

Jansson, H. and S. Sandberg. 2008. Internationalization of small and medium sized enterprises in the Baltic Sea Region. *Journal of International Management* 14(1): 65–77.

Jaworski, B. J. and A. K. Kohli. 1993. Market orientation: Antecedents and consequences. *Journal of Marketing* 57(3): 53–70.

Jaworski, B. J. and A. K. Kohli. 1996. Market orientation: Review, refinement, roadmap. *Journal of Market Focused Management* 1(2): 119–35.

Juslin, H. and T. Tarkkanen. 1987. Marketing strategies of the Finnish forest industries. *Communicationes Instituti Forestalis Fenniae* 143: 61.

Karpen, I. O. and L. L. Bove. 2008. Linking S-D logic and marketing practice: Toward a strategic service orientation. *Otago Forum* 2 (2008)—Academic Papers. http://www.business.otago.ac.nz/marketing/events/OtagoForum/Final%20forum%20papers/Otago%20Forum%20Paper%2013_Karpen.pdf (accessed March 12, 2012).

Kasper, H. 2005. The culture of market oriented organizations. http://arno.unimaas.nl/show.cgi?fid = 3912 (accessed March 9, 2012).

Kim, Y. 2008. Corporate responses to climate change: The resource-based view. Oikos Ph.D. summer academy 2008. Entrepreneurial strategies for sustainability. Oikos Foundation for Economy and Ecology, St. Gallen, Switzerland. http://www.oikos-international.org/fileadmin/oikos-international/international/oikos_PhD_summer_academy/Papers_2008/Kim_Paper.pdf (accessed April 30, 2012).

Kohli, A. K. and B. Jaworski. 1990. Market orientation: The construct, research propositions, and managerial implications. *Journal of Marketing* 54(2): 1–18.

Korhonen, S. 2006. A capability based view of organizational renewal: Combining opportunity and advantage seeking growth in large, established European and North American wood industry companies. Ph.D. dissertation, University of Helsinki, Helsinki, Finland.

Korhonen, S. and J. S. Niemelä. 2003. Strategy analysis of the leading European and North American wood-industry companies in 1998–2001. Series B: 30. Research Report. University of Helsinki, Institute for Rural Research and Training, Seinäjoki.

Korhonen, S. and J. S. Niemelä. 2005. A conceptual analysis of capabilities: Identifying and classifying sources of competitive advantage in the wood industry. *The Finnish Journal of Business Economics* 54(1): 11–47.

Kotler, P. and K. L. Keller. 2009. *Marketing Management.* London, U.K.: Pearson Prentice Hall.

Kourula, A. and M. Halme. 2008. Types of corporate responsibility and engagement with nongovernmental organizations: An exploration of business and societal outcomes. *Corporate Governance: The International Journal of Business in Society* 8(4): 557–570.

Kozak, R. A. and T. C. Maness. 2001. Quality assurance for value-added wood producers in British Columbia. *Forest Products Journal* 51(6): 47–55.

Kraaijenbrink, J., J. C. Spender, and A. J. Groen. 2010. The resource-based view: A review and assessment of its critiques. *Journal of Management* 36(1): 349–372.

Laaksonen-Craig, S. and A. Toppinen. 2008. Profit persistence in globalizing forest industry companies. *International Forestry Review* 10(4): 608–618.

Lado, A. A., N. G. Boyd, P. Wright, and M. Kroll. 2006. Paradox and theorizing within the resource-based view. *Academy of Management Review* 31(1): 115–131.

Lähtinen, K. 2007. Linking resource-based view with business economics of woodworking industry: Earlier findings and future insights. *Silva Fennica* 41(1): 149–165.

Lähtinen, K., A. Toppinen, P. Leskinen, and A. Haara. 2009. Resource usage decisions and business success: A case study of Finnish large-and medium-sized sawmills. *Journal of Forest Products Business Research* 6(3): 1–18.

Laurila, J. and M. Ropponen. 2003. Institutional conditioning of foreign expansion: Some evidence from Finnish-based paper industry firms, 1994–2000. *Journal of Management Studies* 40(3): 725–751.

Levitt, T. C. 1960. Marketing myopia. *Harvard Business Review* 38(4): 45–56.

Li, N., A. Toppinen, A. Tuppura, K. Puumalainen, and M. Hujala. 2011. Determinants of sustainability disclosure in the global forest industry. *Electronic Journal of Business Ethics and Organization Studies* 16(1): 33–40.

Liang, R., H. Tseng, and Y. Lee. 2010. Impact of service orientation on frontline employee service performance and consumer response. *International Journal of Marketing Studies* 2(2): 67–74.

Litz, R. A. 1996. A resource-based view of the socially responsible firm: Stakeholder interdependence, ethical awareness, and issue responsiveness as strategic assets. *Journal of Business Ethics* 15(12): 1355–1363.

Liu, S. S., X. Luo, and Y. Shi. 2003. Market oriented organizations in an emerging economy: A study of missing links. *Journal of Business Research* 56(6): 481–491.

Lusch, R. F. 2011. Reframing supply chain management: A service-dominant logic perspective. *Journal of Supply Chain Management* 47(1): 14–18.

Lusch, R. F. and G. R. Laczniak. 1987. The evolving marketing concept, competitive intensity and organizational performance. *Journal of the Academy of Marketing Sciences* 15(3): 1–11.

Lusch, R. F. and S. L. Vargo. 2006. Service dominant logic: Reactions, reflections, and refinements. *Marketing Theory* 6(3): 281–288.

Lusch, R. F., S. L. Vargo, and M. O'Brien. 2007. Competing through service: Insights from service-dominant logic. *Journal of Retailing* 83(1): 5–18.

Macdonald, D. 1997. *Industrial Relations and Globalization: Challenges for Employers and Their Organizations.* Turin, Italy: International Labour Organisation ACT/EMP Publications.

Maritan, C. and M. A. Peteraf. 2011. Building a bridge between resource acquisition and resource accumulation. *Journal of Management* 37: 1374–1389.

Mauri, A. J. and M. P. Michaels. 1998. Firm and industry effects within strategic management: An empirical examination. *Strategic Management Journal* 19(3): 211–219.

McKitterick, J. 1957. What is the marketing management concept? In *The Frontiers of Marketing Thought and Science*, ed. F. M. Bass, pp. 71–81. Chicago, IL: American Marketing Association.

McWilliams, A. and D. Siegel. 2011. Creating and capturing value: Strategic corporate social responsibility, resource based theory, and sustainable competitive advantage. *Journal of Management* 37(5): 1480–1495.

Mikkilä, M. 2006. The many faces of responsibility: Acceptability of the global pulp and paper industry in various societies. Ph.D. dissertation, University of Joensuu, Joensuu, Finland.

Mintzberg, H. 1994. *The Rise and Fall of Strategic Planning*. New York: FT Prentice Hall. 458 pp.

Mitchell, R. W., B. Woolliscroft, and J. Higham. 2010. Sustainable market orientation: A new approach to managing marketing strategy. *Journal of Macromarketing* 30(2): 160–170.

Narver, J. and S. Slater. 1990. The effect of market orientation on a business profitability. *Journal of Marketing* 54(4): 20–35.

Niemelä. J. 1993. Marketing-oriented strategy concept and its empirical testing with large sawmills. Ph.D. dissertation, University of Helsinki, Helsinki, Finland.

OICA. 1999. International organization of motor vehicle manufacturers production statistics. World motor vehicle production by manufacturer in 1999. http://oica.net/wp-content/uploads/2007/06/cl99cons2.pdf (accessed March 12, 2012).

Ojala, J., J. Lamberg, A. Ahola, and A. Melander. 2007. The ephemera of success: Strategy, structure and performance in the forestry industries. In *The Evolution of Competitive Strategies in Global Forest Industries: Comparative Perspectives*, eds., J. Lamberg, J. Näsi, J. Ojala, and P. Sajasalo. World forests: Dordrecht, the Netherlands: Springer.

Oliver, C. 1997. Sustainable competitive advantage: Combining institutional and resource-based views. *Strategic Management Journal* 18(9): 697–713.

Pätäri, S. 2009. On value creation at an industrial intersection—Bioenergy in the forest and energy sectors. Ph.D. dissertation, Lappeenranta University of Technology, Lappeenranta, Finland.

Pelham, A. M. and D. T. Wilson. 1996. A longitudinal study of the impact of market structure, firm structure, strategy and market orientation culture on dimensions of small-firm performance. *Journal of the Academy of Marketing Science* 24(1): 27–43.

Penrose, E. T. 1995. *The Theory of the Growth of the Firm* (3rd edn.). Oxford, NY: Oxford University Press (reprint of 1959).

Peteraf, M. A. 1993. The cornerstone of competitive advantage: A resource-based view. *Strategic Management Journal* 14(3): 179–191.

Pettersson, M. 2006. Mest lönsamt vara liten och nischad. *Nordisk Papperstidning* 5: 8–9 (in Swedish).

Porter, M. E. 1980. *Competitive Strategy: Techniques for Analyzing Industries and Competitors*. New York: Free Press.

Porter, M. E. 1985. *Competitive Advantage: Creating and Sustaining Superior Performance*. New York: Free Press.

Porter, M. E. 1994. The role of location in competition. *Journal of the Economics of Business* 1(1): 35–39.

Prahalad, C. K. and A. Hammond. 2002. Serving the world's poor, profitably. *Harvard Business Review* 80(9): 48–57.

PWC. 2002. *Global Forest and Paper Industry—Survey of 2001 Results*. Vancouver, British Columbia, Canada: PricewaterhouseCoopers.

Ray, G., J. B. Barney, and W. Muhanna. 2004. Capabilities, business processes and competitive advantage: Choosing the dependent variable in empirical tests of resource-based view. *Strategic Management Journal* 25(1): 23–37.

Reed, R., D. J. Lemak, and N. P. Mero. 2000. Total quality management and sustainable competitive advantage. *Journal of Quality Management* 5(1): 5–26.

Riahi-Belkaoui, A. 2000. *Accounting Theory* (4th edn.). London, U.K.: Business Press, Thomson Learning.

Rich, S. U. 1986. Recent shifts in competitive strategies in the U.S. forest products industry and the increased importance of key marketing functions. *Forest Products Journal* 36(7/8): 33–44.

Ruggles, R. 1998. The state of the notion: Knowledge management in practice. *California Management Review* 40(3): 80–89.

Sajasalo, P. 2002. Change in the extent and form of internationalization-the Finnish forest industry from the mid-1980s to 2000. *Journal of International Business Research* 1(1): 109–134.

Senge, P. M. 1990. *The Fifth Discipline: The Art and Practice of the Learning Organization.* London, U.K.: Random House.

Shane, S. and S. Venkataraman. 2000. The promise of entrepreneurship as a field of research. *Academy of Management Review* 25(1): 217–226.

Siitonen, S. 2003. *Impact of Globalisation and Regionalisation Strategies on the Performance of the World's Pulp and Paper Companies.* Helsinki School of Economics, Helsinki, Finland A-225. 260 pp. + appendices.

Spanos, Y. E. and S. Lioukas. 2001. An examination into the causal logic of rent generation: Contrasting Porter's competitive strategy framework and the resource-based perspective. *Strategic Management Journal* 22(10): 907–934.

Teece, D. J. 2007. Explicating dynamic capabilities: The nature and microfoundations of (sustainable) enterprise performance. *Strategic Management Journal* 28(13): 1319–1350.

Teece, D. J., G. Pisano, and A. Shuen. 1997. Dynamic capabilities and strategic management. *Strategic Management Journal* 18(7): 509–533.

Toppinen, A., K. Lähtinen, L. Leskinen, and N. Österman. 2011. Business networks as a source for competitiveness of the medium-sized Finnish sawmills. *Silva Fennica* 45(4): 743–759.

Toppinen, A., N. Li., A. Tuppura, and Y. Xiong. 2012. Corporate responsibility and strategic group in the forest-based industry: Exploratory analysis based on the global reporting initiative (GRI) framework. *Corporate Social Responsibility and Environmental Management* 19(4): 191–205.

Tozanli, S. 1997. European dairy oligopoly and the performances of large processing firms. In *A Case Study of Structural Change: The EU Dairy Industry*, eds., Tozanli, S., and Gilpin J. Reading, MA: University of Reading.

Turner, K. L. and M. V. Makhija. 2006. The role of organizational controls in managing knowledge. *Academy of Management Review* 31(1): 197–217.

Uronen, T. 2010. On the transformation processes of the global pulp and paper industry and their implications for corporate strategies—A European perspective. Ph.D. dissertation, Helsinki University of Technology, Helsinki, Finland.

Varey, R. J. 2002. *Relationship Marketing: Dialogue and Networks in the E-Commerce Era.* Chichester, U.K.: John Wiley & Sons.

Vargo, S. L. and R. F. Lusch. 2004a. Evolving to a new dominant logic for marketing. *Journal of Marketing* 68(1): 1–17.

Vargo, S. L. and R. F. Lusch. 2004b. The four service marketing myths: Remnants of a goods-based manufacturing model. *Journal of Service Research* 6(4): 324–335.

Vargo, S. L. and R. F. Lusch. 2008. Service-dominant logic: Continuing the evolution. *Journal of the Academic Marketing Science* 36(1): 1–10.

Verona, G. and D. Ravasi. 2003. Unbundling dynamic capabilities: An exploratory study of continuous product innovation. *Industrial and Corporate Change* 12(3): 577–606.

Vidal, N. G. and R. A. Kozak. 2009. From forest certification to corporate responsibility: Adapting to changing global competitive factors. *Proceedings of the XIII World Forestry Congress 2009*: *Forests in Development—A Vital Balance*, Buenos Aires, Argentina, August 2009.

Vincent, M. 2002. The Canadian forest sector and international competition (in French: Le Secteur Forestier Canadien Et La Compétition Internationale). Personal Communication. Chief Economist, Forest Industry Council of Quebec. Quebec City, Canada.

Vlosky, R. P., R. Gazo, and D. Cassens. 2003. Certification involvement by selected United States value-added solid wood products sectors. *Wood and Fiber Science* 35(4): 560–569.

von Geibler, J., K. Kristof, and K. Bienge. 2010. Sustainability assessment of entire forest value chains: Integrating stakeholder perspectives and indicators in decision support tools. *Ecological Modelling* 221(18): 2206–2214.

Wan, M., K. Lähtinen, A. Toppinen, and M. Toivio. 2012. Opportunities and challenges in emerging bioenergy business: Case of Finnish sawmilling industry. *Journal of Forest Engineering* 23(2): 13 pp.

Wernerfelt, B. 1984. A resource-based view of the firm. *Strategic Management Journal* 5(2): 171–180.

Wernerfelt, B. 2011. The use of resources in resource acquisition. *Journal of Management* 37: 1369–1373.

Zairi, M. 1997. Business process management: A boundaryless approach to modern competitiveness. *Business Process Management* 3(1): 1355–2503.

Zhang, Y. and A. Toppinen. 2011. Internationalization and financial performance in the global forest industry. *International Forestry Review* 13(1): 96–105.

Zink, K. J. 2005. Stakeholder orientation and corporate social responsibility as a precondition for sustainability. *Total Quality Management* 16(8–9): 1041–1052.

Section V

Bringing It All Together

18

What Now, Mr. Jones? Some Thoughts about Today's Forest Sector and Tomorrow's Great Leap Forward

Robert A. Kozak

CONTENTS

18.1 Something Is Happening Here

> Because something is happening here
> But you don't know what it is
> Do you, Mr. Jones?
>
> —Bob Dylan, "Ballad of a Thin Man" (*Highway 61 Revisited*), 1965

If today is like any other day on Planet Earth, in the time that it takes to read this chapter, we will have permanently lost dozens of acres of forest lands, not to mention one or two species, all to feed our voracious collective appetites. Moreover, our population will have grown by thousands, many of whom will go to sleep tonight without food or clean water.

This is not meant to sound alarmist. Rather, it is a clarion call to the forest sector: a call to action. The time to act is now, not when it is too late. And who better to do so than the forest sector?

Forests, as we know, cover a huge portion of this planet. Forest ecosystems are important repositories for biodiversity and genetic resources and they provide important regulatory functions with respect to climate, water, and erosion, to name a few examples. Yet, the future of forests, and the well-being of millions of people who depend on them, hangs precariously in the balance.

As the economic engine responsible for providing bountiful goods and services to society as a whole, including food, shelter, medicine, and energy, the forest sector is ideally positioned, obliged even, to take a leadership role on these sorts of complex and multifaceted issues. Why? Because, in many ways, the forest sector is light-years ahead of the rest of the world in thinking through these sorts of questions. Who else deals with such complex spatial and temporal scales? What other sectors try to balance social, environmental, and economic needs day in and day out? And who else must deal with uncertainty, trade-offs, and varying futures in the way that the forest sector must? As the noted wildlife biologist, Fred Bunnell, once bemused, "Forestry is not rocket science. It's much more complicated."

Arguably, there is no force—be it climate change, pests, disease, fire, poverty, and so on—that has had as big an impact on the current and future states of our forests as business has. In other words, if the forest sector—and, specifically, the people behind this massive economic engine—is a major part of the problem, then it must, therefore, be a key part of the solution. This is no simple task. The world is uncertain, markets are continually evolving, global competition is becoming increasingly fierce, and the rules of engagement for business interests are fluid and, oftentimes, perplexing. At the same time, it behooves the forest sector to be responsible stewards of one of Planet Earth's most delicate and vital natural resources—forests. This requires a sector with a good deal of imagination and the ability to put forward bold and innovative programs "that capture the complexity of forest ecosystems and the needs of communities that depend on them" (Kozak 2005, 62). Is the forest sector there yet? No. Could it be? Definitely.

This chapter explores and discusses a number of the pressing issues that the forest sector must contend with in order to move forward in a meaningful, sensible, and vital manner. While the intent of this chapter is not to summarize the book *per se*, it does incorporate and build upon a number of the recurring themes found within its pages as a means of illustrating the issues that the forest sector is facing today and the challenges that it must address in the future. These revolve mainly around the notion that our main thrust as global citizens ought to be to maintain the ecological integrity of Planet Earth and ensure the well-being of future generations. The forest sector can do its part by striving toward an ethic rooted in conservation and better adopting the principles of what is increasingly becoming understood as a "conservation-based economy."

18.2 Back to Basics: What Is a Conservation-Based Economy, Anyways?

> I have read many definitions of what is a conservationist, and written not a few myself, but I suspect that the best one is written not with a pen, but with an axe. It is a matter of what a man thinks about while chopping, or while deciding what to chop. A conservationist is one who is humbly aware that with each stroke he is writing his signature on the face of his land.
>
> **—Aldo Leopold, November entry,** *A Sand County Almanac,* **1949**

If we are talking about moving toward a conservation-based economy, perhaps the best place to start is with a definition, or at least a common understanding, of the term "conservation." This is not an uncomplicated task, muddied by the fact that there exist countless definitions of this somewhat nebulous term. This complexity is educed in the earlier excerpt from Aldo Leopold's *A Sand County Almanac,* a seminal masterpiece in formulating our modern-day thinking around conservation and the conservation-based economy. But more important, perhaps, is the observation that the "best" definition of conservation is written "with an axe" (Leopold 1949, 73).

In order to better situate this paradox, it is perhaps worth noting that the quotation comes from the November entry wherein Leopold contemplates the autumnal harvest, how it is that he consumes and, ultimately, what it means to be a conservationist. In the end, he concludes that conservation is not so much about preserving our resources but more a consideration of how we consume. It is about understanding and appreciating and respecting how we use the natural resources that are available to us. In many ways, business (and by proxy, society) has lost sight of this.

A conservation-based economy, then, says that businesses must consider consumption that is responsible, not maximized, and that a truly balanced and accountable industry must be consistent with this conservation ethic. To say that we cannot do so is defeatist, an admission that, collectively, we do not have the smarts to figure out this conundrum, this "wicked problem" (Rittel and Webber 1973). But the reality is that business is inextricably linked to extracting natural resources. Concurrently, there is a pressing need to maintain those resources for future generations. Reconciling this paradox is becoming an increasingly tall order—not to mention an increasingly urgent one—as the population of Planet Earth grows well beyond its carrying capacity.

18.3 The Big Picture: Global Population and Consumption

Earth [pritvi] provides enough to satisfy every man's need but not for every man's greed.

—Mahatma Gandhi, 1947

In describing the "big picture," perhaps the best place to start is with global population growth, something that has been quantified and discussed numerous times and will only be touched upon here. For millennia, the human population of Planet Earth grew at a steady rate, reaching about 500 million at the time of the Renaissance. During the age of Enlightenment in the 1800s, a milestone of one billion humans was reached (Bongaarts and Bulatao 2000). Since then, population has grown to over seven billion, mostly concentrated in developing regions of the world (Bloom 2011).

This growth has aptly been described as explosive, meaning that we appear to be tracking toward infinity (Nielsen 2006; Harmon 2011). This is worrisome because Planet Earth simply cannot support an infinite number of people—"a breakdown in one form or another will have to occur" (Nielsen 2006, 13). Scientists reckon that the carrying capacity of our planet is somewhere between 8 and 12 billion people (Cohen 1995). The trouble is that most estimates of human population over the next century—depending on what assumptions are made about birth rates, life expectancies, the state of political and social unrest, and so on—far exceed these levels (Bongaarts and Bulatao 2000; Nielsen 2006).

This is a cause for major concern. What it essentially means is that each subsequent generation will have to deal with harder and harder choices between fewer and fewer options and that the time to react is now, not when it's too late. But it gets even more complicated when one considers our collective and voracious patterns of consumption, the proverbial elephant in the room.

Not only is population growing at an alarming rate but so too is the rate at which we consume resources that are becoming more and more scarce (notably, assumed rates of consumption are one of the key determinants of the planet's carrying capacity [Nielsen 2006]). And forest products are by no means immune to this trend. One need only look at the consumption of wood products in the North American housing sector to realize that this is true. While housing starts in the region are very much a reflection of the economy, the one trend that is undeniable is that housing sizes are most definitively on the rise, increasing by about one-third since 1980 (United States Census Bureau 2013). This reflection of our consumptive nature should raise a number of questions, if not eyebrows. Do we really need to live in massive houses? Why are homes no longer designed to last hundreds of years? Should we not be promoting recycling, reuse, and design for disassembly in housing instead opting to landfill construction waste after only a few decades of use? And so on. These are hard-hitting questions—challenging the *status quo* may have potentially serious economic consequences—but it is imperative that we ask them right now (Kozak 2005).

The unfortunate reality, though, is that not that many people are posing these sorts of questions, and we, as a society, seem to be overwhelmingly complacent, lacking the resolve to tackle these very real issues and make things better for future generations who will have to live in an increasingly crowded world. Surely, in this 24-7 digital age of unyielding data and information, consumers are aware of the problems facing Planet Earth. But why has nothing changed? Why do we continue to consume more and more? This point cannot be overstated. Somewhere along the way, we seem to have supplanted our sense of global citizenship to make way for an ethic of greed. Consumption has become our new *raison d'être*, complete with its own mottos: "whoever dies with the most toys wins" and "keeping up with the Joneses." The trouble is that, unlike global citizens, consumers have no obligations or responsibilities to anything or anyone beyond feeding their own ravenous appetites.

18.4 Globalization: The Double-Edged Sword

> These are the days of miracle and wonder
> And don't cry baby don't cry
> Don't cry, don't cry
>
> **—Paul Simon, "The Boy in the Bubble"** (*Graceland*), **1986**

It is little wonder that globalization is so frequently discussed and debated. Since the fall of the Berlin wall, the "triumph" of capitalism as the only game in town, and the bonanza of comparatively cheap oil that has been available to fuel this transnational experiment, globalization has been an unrelenting, omnipresent force. It is everywhere, in everything we see, feel, hear, taste, and touch. It is the familiar double-edged sword—an opportunity for some through opened up markets and/or the ability to manufacture goods more cheaply halfway across the world, a threat to others in the form of increased competition from regions with lower costs of production (Kozak 2005).

Examples abound in the forest products industry—many documented throughout this book—of how globalization has manifested itself, altered the forestry landscape, and impacted the sector: the emergence of the BRIC countries (Brazil, Russia, India, China) as major players in the production of forest products; the rise of China as the "factory" of the world; the subsequent decline of secondary wood products manufacturing like furniture in the United States and elsewhere; the shift in production of lower value commodity wood products (logs, dimension lumber, and pulp and paper) to the Southern hemisphere and its so-called wall of wood; the evolution of global supply chains and the concomitant ability to outsource production in lower cost regions; the increased access to new, lucrative markets for all manner of wood products further and further afield; the dynamic and evolving global trade flows of forest products as a result of all of the above; and so on.

For better or for worse, globalization is here to stay, at least for the foreseeable future. But what is it? Again, countless definitions of globalization exist, but, fundamentally, it means that this pair of trousers is made in Calcutta and that shirt is made in Jakarta and the outfit costs $20 less at the big box mega-outlet than at the "mom and pop" store that stood there before it.

Where forestry is concerned, this has huge implications. First off, the forest sector—value-added and commodity producers alike—now faces stiff competition from new international entrants, the likes of which it has never seen. Second, success in the new global economy and participation in international markets are predicated on forest-based businesses adopting a renewed focus, new sets of business and cultural skills, and heightened levels of strategic thinking. Third, the corporatization and homogenization of the retail sector has had tremendous impacts on countless forest-dependent communities throughout the world, which, for obvious reasons, tend to be located in rural regions. It is the "mom and pop" stores that sponsored the neighborhood baseball teams and sat on our local boards of commerce, and their demise—surgically, with the double-edged sword—drastically alters the social fabric, dynamics, and vitality of small, rural communities everywhere. Last, but certainly not least, we need to pose some very serious questions, especially in light of the fact that many forested regions of the world tend to go hand in hand with extreme poverty. Are we willing to accept the fact that the main reason globalization works as a construct is that multinational corporations are able to exploit cheap labor in parts of the world with less costly material inputs and lower environmental standards? And, moreover, what is the right thing to do given that these developing economies want to—in fact, have a right to—participate in the global economy?

18.5 Sustainability: The Thin Veneer of Truth

> They took all the trees
> Put 'em in a tree museum
> And they charged the people
> A dollar and a half just to see 'em
>
> —**Joni Mitchell, "Big Yellow Taxi" (*Ladies of the Canyon*), 1970**

Of course, one cannot have a meaningful dialogue on issues like population growth, consumption, and globalization, and ultimately achieve a conservation-based economy without beginning with, or at least stumbling upon, the notion of sustainability. Sustainability is a ubiquitous term. In fact, it has been trotted out in many chapters of this book—on activism, global forest policies, green marketing, strategy, innovation, and corporate social responsibility (CSR), to name a few examples. And while it is not the intent of this present chapter to define this term, it is worth recounting Seager's (2008, 447) observation that "Sustainability might best be defined as an ethical concept that things should be better in the future than they are at present. Like other ethical concepts such as fairness or justice, sustainability is best interpreted conceptually rather than technically." Basically, sustainability is a panacea, a means of redressing the "intergenerational tyranny" that is now taking place, wherein our natural environment is in precipitous decline and we are leaving our children a legacy of depletion and harm.

For millennia, we have been perturbing ecosystems and drawing down from finite reserves of natural capital that blanket Planet Earth (Kozak 2005). This should come as no surprise—humans need to consume nature's bounty, be it for food, shelter, medicine, clothing, or energy—in order to survive as a species. Where sustainability is concerned, the important historical moment comes during the Neolithic age, when humans first made the transformation from nomadic hunter/gatherers to settled agriculturalists. At the risk of oversimplification, this spawned a series of technological advances over a period of 10,000 years or so, which allowed society to produce goods at faster and faster rates and higher and higher volumes. This so-called expansionist paradigm posits that human beings—the cerebral creatures that we are—should be able to "master" nature and that technology is the tool that liberates us. And given that population growth is "explosive," the ability to manipulate nature for our benefit could not have come at better time (Nielsen 2006).

But the expansionist paradigm is not without its problems. First, are our technical innovations always in the best interest of society and the environment? For instance, in order to meet our growing demand for commodity products like dimension lumber and pulp, we have become increasingly reliant upon intensively managed plantation forestry, in some cases, seeded with genetically modified trees (Food and Agricultural Organization of the United Nations 2010) and subjected to fertilizers, pesticides, herbicides, and massive irrigation systems (Richardson 1993; Ecobichon 2001; Coyle and Coleman 2005). Are we going to accept these anthropogenic abstractions of forests as viable substitutes for natural forests, not to mention the countless species that reside within them (Kozak 2005)? This stamp of human progress has, and will continue to have, huge implications for the ecological integrity of the entire planet. Second, the fact that technology has allowed us to produce from nature at ever increasing rates ultimately encourages us to accelerate our consumptive patterns (Kozak 2005). And consuming we are. The estimate of our global ecological footprint for 2012 is 1.5, which is to say that we would need 1.5 Earth-like planets to meet our demands for natural resources if everyone continues to enjoy their current standards of living. Not surprisingly, these estimates vary by country and are strongly correlated with wealth and consumption. For example, ecological footprint estimates are typically between 4 and 7 in North America and Europe (with the United States at 7.2), 2.1 in China, and less than 1, on average, in Africa, but growing over time (WWF 2012; Global Footprint Network 2013).

These are menacing data points, but the important take-home message is that society is simply not living within its means. Not only is sustainability a ubiquitous term, it is a highly misappropriated term. We can talk all we want about the *sustainable* forest management of individual landscapes, certified lumber and toilet paper that come from *sustainably* harvested stands, *sustainable* production of secondary wood products, *sustainable* supply chains, and so on—and there is a thin veneer of truth there—but let's be mindful of the fact that all this chatter may represent nothing more than a "house of cards" (Kozak 2005). This is not to discourage such efforts but rather to point out that these are merely *less unsustainable* strategies, driven largely by the profit motive and doing little or nothing in the way of maintaining the ecological integrity of Planet Earth and ensuring the well-being of future generations (Ehrenfeld and Hoffman 2013). There is no middle ground here. Sustainability is an absolute. It must be pervasive. We cannot have pockets of sustainability. Either we are sustainable or we are not.

Sometimes, in fact, these incremental steps toward sustainability can be counterproductive. It is worth offering up a few, arguably naïve, but telling illustrations. Take, for instance, Leadership in Energy and Environmental Design (LEED) certification of "green" buildings. The aim of more sustainable designs for buildings is a noble and serviceable goal, but the points-based approach that comprises the system, in some ways, confers upon designers the freedom to be unsustainable. In theory, a building's furnishings could be made of an exotic species from Southeast Asia and come from a region where the (carbon sequestering, biodiverse) forests have been mowed down and the communities decimated. But this atrocity can be easily negated with the installation of the right kind of heating, ventilation, and air conditioning (HVAC) system. Is this sustainable? Or take the example of a large trucking fleet that decides to "green" its supply chain by using biodiesel. Does this expunge the problem of child labor being used in the production of goods that sit on the truck bed? Similarly, when a large multinational forestry conglomerate operating in Western Africa fulfills its CSR mandate by building a school for local children, what good does this do if there are no teachers?

Third-party certification of sustainably managed forests provides a more nuanced, but equally compelling, example. Why does most of the thinking (and research) in the area of forest certification revolve around possible price premiums that consumers would pay for certified wood? In theory, certification provides a means of capturing the "externalities" of production, the environmental and social costs that would otherwise be incurred by society from harvesting forests unsustainably. But why should these costs be passed on to the consumer? Is this not an incentive for forest products companies to behave badly and eke out more market share by selling cheaper products? Does it not make more sense to charge an environmental levy—in effect, a price premium—on wood that is not certified (Kozak 2005)? Last but certainly not least, the promise of biofuels from renewable resources like trees is an interesting, potentially game-changing notion that has gained much traction in the past decade. But if creating a conservation-based economy is fundamentally about respecting our natural resources, exactly how respectful is it to deconstruct a tree down to its rudimentary chemistry to produce a low-value commodity good that is used to fuel (literally and figuratively) our consumptive patterns? And, moreover, how would this impact society's competing demands for land use on what is already an overcrowded planet?

The fact that sustainability has now become so commonplace in our everyday lexicon may actually do more harm than good. We run the risk of being lulled into a state of complacency and acceptance: "if we say it enough times, it must be true." This is dangerous. The appropriate question to be asking ourselves at this moment in time—and business needs to be a big part of the solution—is how do you build sustainability into a model that is fundamentally unsustainable?

18.6 New Products...

> You're going to reap just what you sow.
>
> **—Lou Reed, "Perfect Day" (*Transformer*), 1972**

In the midst of all of the challenges and uncertainty related to the sustainability of Planet Earth, it is difficult to quarrel with the fact that these are certainly interesting times for the forest sector. Against this backdrop, it is worth reinforcing the fact that forests are a renewable resource and, therefore, have the potential to contribute to the goals of sustainability and the conservation-based economy in a meaningful and important way. The list of products derived from forests is immense (see, e.g., Chamberlain et al. 1998; Conners 2002), and today's forest sector is characterized by the introduction of a wide range of new and exciting products (see, e.g., Forest Products Association of Canada 2010). Consequently, there are a virtually limitless number of paths that the forest sector of the future can pursue.

For lack of a better taxonomy, this vast basket of products includes commodity products and their differentiated cousins (e.g., dimension lumber, pulp and paper products, plywood, oriented strand board), secondary or value-added wood products (e.g., furniture and cabinetry, prefabricated homes, specialty papers), innovative products (e.g., rayon, nanocrystalline cellulose, nutraceuticals), subsistence products (e.g., firewood), and nontimber forest products (e.g., fruits and nuts, medicinal plants, florals, rubber). One last product class is not a product at all but rather refers to the ecosystem services and benefits that forests provide—clean water, biodiversity habitats, recreation opportunities, and carbon sequestration to mitigate the forces of climate change. As these provisioning services become increasingly valued (and, consequently, monetized in the marketplace), they present a new opportunity for the forest sector. It is not unforeseeable, for example, to imagine some natural forests around the world being less a part of an industrial wood products sector but more as providers of ecosystem services. The challenge for the forest industry lies in generating profits from "products" that are not as tangible, and markets that are ethereal, or at least much more difficult to grasp.

Many of these products—especially commodity goods like dimension lumber and pulp and paper—could best be characterized as mature. Consequently, manufacturers of these products are finding themselves operating within a highly competitive industrial complex, with new entrants from around the world, notably the BRIC countries, aggressively eking out market share. Not surprisingly, as prices for commodity products spiral downward, regions with clear comparative cost advantages are seemingly getting the upper hand (Martin and Porter 2000) over more traditional producers who now find themselves a bit "out of step" (Kozak 2005). It is becoming more and more difficult, for example, for a Nordic producer operating in a natural forest landscape to compete with a Southern hemisphere producer operating on an intensively grown forest plantation. Some businesses that insist on maintaining a low-value, high-volume commodity focus are finding themselves in a proverbial "race to the bottom," heedlessly pursuing markets that are fleeting and increasingly unprofitable (Kozak 2005).

Perhaps the most exciting recent trend is the emergence of a multitude of new and novel forest products. Notionally, many of these products are not, in fact, new but throwbacks to a time when humans used forests for countless purposes in imaginative and innovative ways. Wood from trees has always been used for the production of energy, for example, whether it be by burning logs to create heat or by deconstructing wood fiber to its barest essentials to create biofuels (Kozak 2005).

Many of these new products seem to reconcile well with the need to create a conservation-based economy. Not coincidentally, this is occurring as the disciplines of "green" or "environmental" marketing and branding gain acceptance as important and strategic business functions (Charter et al. 2002), and market segments of "green" consumers—ranging from "fence-sitters" to the "hardcore"—continue to grow (GfK Roper Consulting 2007). This trend has manifested itself in the forest sector in many ways. Wood products are now commonly marketed for their environmental attributes, such as the fact that they come from a natural and renewable resource (e.g., wood flooring) or for their carbon sequestering potential (e.g., cross-laminated timber). Certified wood products (e.g., dimension lumber, paper products) have now become the industry standard and the *de facto* rule of engagement for companies wishing to access international markets and large retail outlets. In the meantime, markets for certified wood products continue to grow as consumers acquire knowledge on sustainability issues and begin to understand the significant role that third-party certification can play. Environmental product declarations (EPDs) for communicating the environmental impacts of systems based on life cycle assessment data (e.g., large-scale wood buildings) continue to make inroads into the marketplace, especially in the domain of green or sustainable buildings, which continue to offer a plethora of opportunities to wood products manufacturers interested in marketing to environmentally friendly building specifiers (e.g., architects and structural engineers) and progressive homeowners. Finally, the forest sector has become increasingly reliant upon and/or experimental with wood furnish from waste streams—forest residues, sawmill byproducts, recycled materials—to produce a variety of innovative products (e.g., engineered wood panels, bioethanol, wood pellets).

This latest trend has evolved into what is commonly becoming known as the "biorefinery" concept as a means of confronting the challenges of plummeting global pulp and paper markets and addressing excess capacity in pulp mills. The idea that pulp mills can be reimagined to produce thermal bioenergy products, liquid biofuels, biochemicals, bioplastics, nanocrystalline cellulose, pharmaceuticals, nutraceuticals, textiles, and other lignocellulosic biomaterials is appealing, to be sure (see, e.g., Forest Products Association of Canada 2010), but it warrants a cautionary note. Many governments are upholding these new and novel products as yet another panacea to cure the woes of faltering commodity wood sectors. This may well be the path forward, but much of the technology has yet to be scaled up, and some of it is not yet economically viable. It is easy to get caught up in these sorts of "sexy" innovations, but the reality in the forest sector is that, sometimes, all it takes to be successful is to make and sell something as exhilaratingly simple as a table and chair.

It is also worth noting that, as innovative as they are, these new wood products and services will not sell themselves. As with the more traditional forest sector, the tenets of conventional marketing wisdom still apply. New product development programs must be approached with thoughtfulness, prudence, and due diligence. In this age of intense competition and an increased focus on environmental and social issues, marketing—including messaging and branding—is fundamental to the success of a business, and the importance of tools like the "total product concept" and the "product life cycle" cannot be overstated (Hansen and Juslin 2011). They must be incorporated into the strategic thinking of an organization at every level to understand how environmental attributes can best be used as a means of competitive differentiation, how diversification and continual renewal can help to moderate the inevitable forces of market decline, and, increasingly, (this is the tricky one) how their products can contribute positively to the social well-being and ecological integrity of Planet Earth.

18.7 ...Same Old Business Models

And if you tolerate this
Then your children will be next.

—Manic Street Preachers, "If You Tolerate This Your Children Will Be Next"
(*This Is My Truth Tell Me Yours*), 1998

Despite a reinvigorated sense of purpose for the forest sector as a result of the potential to explore and market a range of new products and services (witness the recent espousal of a new rhetorical term, "bioeconomy," to refer to the broad range of forest-based activities), the dominant business paradigm still revolves around exploitation, consumption, and profit maximization. Without a quantum shift in the way business operates, we will never truly achieve a conservation-based economy. This is true for any industry, but the forest sector, which has, arguably, made modest gains toward the goal of sustainability, has so much at stake.

The forest sector, like any other industrial segment, is rooted in a system of neoclassical economics, which is predicated upon an assumption of unlimited resources (Daly and Farley 2010). It breeds and favors a Fordist approach, wherein large volumes of low-value goods are mass produced. In this light, it is little wonder that the forest sector landscape—policy, dialogue, and even research—is eclipsed by large, multinational corporations producing commodity products. The important point, though, is that this model—and its spurious assumptions—simply cannot work in a conservation-based economy.

This hegemony of big business in the forest sector is illogical, unfounded, and misleading. In reality, small- and medium-sized enterprises (SMEs)—employing less than 250 individuals but more typically less than 10—are key contributors to the global forest enterprise, both in developed and developing economies (Kozak 2007). It has been estimated that SMEs in the forest sector contribute approximately $US 130 billion of gross value added globally (Macqueen 2004; Mayers 2006) and provide employment for more than 20 million people (Macqueen et al. 2006; Mayers 2006). Importantly, these figures do not take into account employment in the informal sector, which, while difficult to quantify, has been estimated at upward of 140 million individuals, mostly working at the microenterprise level in developing regions (Mayers 2006). In the final analysis, SMEs employ more than half of the forest sector workers in most countries, with proportions exceeding 80% in many developing economies (Mayers 2006).

Clearly, SMEs are an important piece of the forest sector mosaic, and their interests must be acknowledged, be it by research programs, in policy reform, or as part of the general discourse. Notably, the smaller scale of SMEs translates into nimbleness, oftentimes leading to highly innovative business solutions in the forms of new product development, progressive sustainability programs, rapid market responses, high-tech gadgetry, and inventive supply chain strategies, to name a few. The recent evolution of the secondary wood products sector provides a telling illustration, specifically its adoption of "mass customization," "localization," and "service orientation" approaches to meet the exacting demands of an increasingly fickle marketplace.

Importantly, SMEs are also being upheld as a viable poverty alleviation strategy in developing regions of the world that are highly forest dependent (Kozak 2007). This becomes all the more pressing as the world witnesses significant reforms in forest tenure, with rights of access to forest resources slowly devolving from large, multinational concessionaires into the hands of communities and indigenous peoples (Sunderlin et al. 2008). The advantages of this type of pro-poor approach are many, but most revolve around generating wealth and employment that stays within impoverished communities and creating meaningful business interests that are locally vested and, therefore, more likely to operate in an ecologically and socially relevant manner (Kozak 2007). Challenges are equally numerous, as interventions seek to create enabling business environments and promote favorable forest policies that allow SMEs to flourish. Many of the solutions being put forward include capacity building programs, legalizing and empowering the informal sector, capitalizing on growing domestic markets (which flies in the face of globalization), and the exploration and promotion of alternative business models, like community forest enterprises, business associations, and strategic alliances with other supply chain actors and stakeholders (Kozak 2009).

Change in the forest sector is a tall order, especially for the "big business" commodity segment. This is an entrenched, old-world industry, anxiously clinging to its resolve and identity in an increasingly competitive and globalized world. The forest sector is oftentimes characterized as a low-tech "sunset" industry. This goes a long way in explaining the precipitous declines in forestry enrolments at universities and colleges around the world or the rebranding of forestry schools as "environmental science" programs (or some variation thereof). This sort of attitude is not entirely baseless. The traditional forest sector is a notorious laggard, happily focusing on a low-cost production orientation, and late to the table in implementing a modern-day customer orientation (Hansen and Juslin 2011). What marketing programs exist are generally not very novel or groundbreaking, and there is what can only be qualified as a "hipness" gap—something of a communication breakdown—between industry leaders and some of their more media-savvy and culturally attuned audiences (architects, structural engineers, builders, interior designers, industrial designers, homeowners, publishers, and so on).

Even the recent adoption of CSR (and its antecedent, forest certification), while a step in the right direction, is not without its problems. To an outsider, a forest company's CSR practices—philanthropic activities in local communities, "greening" supply chains, reducing manufacturing waste, improved stakeholder engagement, partnering with environmental nongovernmental organizations (ENGOs), and so on—may appear noble and selfless. But the reality is that the drivers that lead businesses to go beyond their legal and economic responsibilities into the realm of ethical and discretionary responsibilities (Carroll 1979) have as much, or more, to do with bettering their economic positions: increased sales and market shares, cost savings, increased investor appeal, brand management, enhanced corporate image, and risk management, to name a few (Lee and Kotler 2011). In other words, CSR practices are merely market responses. In today's world, they are simply not justified unless they also make good, prudent business sense. Is this necessarily a bad thing? Certainly not. But it does suggest that companies, in general, will not engage in socially or environmentally responsible practices—the foundation of a conservation-based economy—without the carrot of a "bottom-line" benefit. The Nobel Prize winning economist, Milton Friedman, goes one step further and argues that corporations are legally obliged to put the financial interests of their owners ahead of everything else and that it is, in fact, illegal to engage in activities that would do otherwise; "The social responsibility of business is to increase its profits" (Friedman 1970, 1).

18.8 The Great Leap Forward

> Here comes the future and you can't run from it
> If you've got a blacklist, I want to be on it
> (Waiting for the great leap forwards)

> **—Billy Bragg, "Waiting for the Great Leap Forwards" (*Worker's Playtime*), 1988**

It is unclear whether humankind will ever be able to nurture a truly sustainable Planet Earth or whether the forest sector will adopt and adhere to the principles of a conservation-based economy. But if we do not, what will the future hold for us? Natural resources are finite, and we are taking from nature faster than we are replacing it. We do not need high-level mathematics to understand that this simply cannot go on forever. Somehow, we need to reconcile this spatial and temporal tension of sustainability. Perhaps the expansionist paradigm will be our "white knight" and we (rather, our children) will muddle our way through the problem with a series of "techno-fixes" (Kozak 2005). But is this a chance that we, as a society, are willing to take? And if we head down the wrong path, then what?

Again, this chapter is meant to be a call to action. Solutions necessarily require all of us, collectively, to get involved, communicate ideas, and debate plausible solutions. Again, there is much to learn from the forest sector, which has had to contend with these sorts of issues for a very long time. One need only look at the forest sector's progress in things like forest certification, sustainable forest management, CSR, green marketing, pollution abatement, energy efficiency, novel bio-based products, value-added products, strategic alliances with environmental groups, and green building to realize that this is true. These actions are driven largely by the profit motive, which understandably will never go away. But does it need to be so up front? In ancient Greece, and up until Medieval times, the social contract was very much about the "commons," and the role of business was to ensure the well-being of communities, not the creation of wealth. This is not to suggest that we step back in time but rather that we ponder the potential role of business in this evolving and increasingly complex world. Is it possible for businesses to be less profit-driven, but to do good for good's sake, and by doing so, earn a profit?

We find ourselves in the midst of a massive challenge. We are at a point in time where we need to take some chances, and perhaps a leap of faith, by questioning our fundamental economic assumptions and systems, not to mention our policy regimes: the proverbial "paradigm shift" that we all hear about. This is nothing new. In 1942, Schumpeter's seminal text, "Capitalism, Socialism and Democracy," put forward the notion of "creative destructionism," wherein economic systems are continually being reinvented as a result of the eradication of prior economic structures (Schumpeter 1942). Where to begin is the question. Fortunately, many authors have attempted, through their writings, to enter the fray and catalyze this lofty and dizzying debate (see, e.g., Hawken et al. 1999; Brown 2001).

Paul Hawken's "The Ecology of Commerce" is particularly notable, especially within the context of a natural resource sector like forestry. In it, he discusses three potential scenarios unfolding for Planet Earth: (1) we continue to draw down our natural capital and hope for the best; (2) we draw down our natural capital, but at a reduced rate, by using resources more sensibly and efficiently; and (3) we take no more than we need, we try not to harm the environment, we waste nothing, and we make amends whenever and wherever we can (Hawken 1993). This last scenario, which he dubs the "restorative economy," represents a paradigm shift in so much as restoring degraded ecosystems to their fullest biological potential—in effect, exceeding sustainability—becomes an inherent part of the business process (Hawken 1993). This may seem Utopian (Kozak 2005), but the remarkable realization is that the forest sector is tantalizingly close to achieving this ideal. Will the forest sector get over this last hurdle on its own and truly create a conservation-based economy? If not, what will it take? Policy reform? Incentives? The promise of market success?

In all likelihood, quantum change in the forest sector will, in fact, be driven by people—company owners and employees, ENGO representatives, government bureaucrats, policy makers, union members, community residents, supply chain actors, customers, and any other interested stakeholder or citizen. Noam Chomsky, the great American linguist and philosopher, talks about the "paradox of the slave owner." There can be no question that the slave owner was part of a grotesque, hateful institution. Yet, a slave owner may also be a decent person: nurturing, kind, principled, responsible, and so on. The point is that, in answering "big picture" questions on the path toward sustainability, or on how to achieve a conservation-based economy, it is vital and necessary to separate the institution from the individual. The institution of business will not change unless its hand is forced. It is up to individuals, compelled by their moral prerogative, to drive change by placing social and environmental issues at the forefront of business discourse. Such individuals—smart, strong-willed, audacious, forward-thinking individuals—exist in today's forest sector. It is up to them to address tomorrow's challenges for the sake of Planet Earth.

Acknowledgments

I would like to thank Rajat Panwar, Denise Allen, and my comrades in the FACT Lab for their encouragement, ideas, help, and inspiration. Special thanks also to the many friends who make an appearance, now and then, on dog-eared pages and scratched vinyl.

> ...the problem is not to fight Wall Street. The problem is why does the system need Wall Street to function?
>
> **—Slavoj Žižek, 2011**

References

Bloom, D. 2011. 7 Billion and counting. *Science* 333(6042):562–569.

Bongaarts, J. and R. Bulatao (Eds.). 2000. *Beyond Six Million: Forecasting the World's Population. Panel on Population Projects, Committee on Population, Commission on Behavioral and Social Sciences and Education, and National Research Council.* National Academy Press, Washington, DC.

Brown, L. 2001. *Eco-Economy: Building an Economy for the Earth.* W.W. Norton and Company, New York.

Carroll, A. 1979. A three-dimensional conceptual model of corporate performance. *Academy of Management Review* 4(4):497–505.

Chamberlain, J., R. Bush, and A. Hammet. 1998. Non-timber forest products: The other forest products. *Forest Products Journal* 48(10):10–19.

Charter, M., K. Peattie, J. Ottman, and M. Polonsky. 2002. *Marketing and Sustainability.* Centre for Business Relationships, Accountability, Sustainability and Society (BRASS), in association with The Centre for Sustainable Design, Swindon, U.K.

Cohen, J. 1995. Population growth and earth's human carrying capacity. *Science* 269(1995):341–346.

Conners, T. 2002. Products made from wood. University of Kentucky Cooperative Extension Service. Department of Forestry Fact Sheet FORFS 02–02.

Coyle, D. and M. Coleman. 2005. Forest production responses to irrigation and fertilization are not explained by shifts in Allocation. *Forest Ecology and Management* 208(1–3):137–152.

Daly, H. and J. Farley. 2010. *Ecological Economics: Principles and Applications* (2nd edn.). Island Press, Washington, DC.

Ecobichon, D. 2001. Pesticide use in developing countries. *Toxicology* 160(1–3):27–33.

Ehrenfeld, J. and A. Hoffman. 2013. *Flourishing: A Frank Conversation about Sustainability*. Stanford University Press, Stanford, CA.

Food and Agricultural Organization of the United Nations. 2010. *Forests and Genetically Modified Trees*. International Union of Forest Research Organizations and Agricultural Organization of the United Nations, Rome, Italy.

Forest Products Association of Canada. 2010. *Transforming Canada's Forest Products Industry: Summary of the Findings from the Future Bio-Pathways Project*. Forest Products Association of Canada, Ottawa, Ontario, Canada.

Friedman, M. 1970. The social responsibility of business is to increase its profits. *New York Times Magazine*, September 13, 1970.

GfK Roper Consulting. 2007. *GfK Roper Green Gauge (R) Study*. GfK Roper Consulting, New York.

Global Footprint Network. 2013. Footprint basics. http://www.footprintnetwork.org/en/index. php/GFN/page/basics_introduction/. Accessed April, 2013.

Hansen, E. and H. Juslin. 2011. *Strategic Marketing in the Global Forest Industries* (2nd edn.). Self-published, Corvallis, OR.

Harmon, K. 2011. 7 Billion and still growing: Explosive population growth might have helped us displace neandertals. *Scientific American*, July 28, 2011.

Hawken, P. 1993. *The Ecology of Commerce: A Declaration of Sustainability*. Harper Business, New York.

Hawken, P., A. Lovins and L. Hunter Lovins. 1999. *Natural Capitalism: Creating the Next Industrial Revolution*. Little, Brown and Company, New York.

Kozak, R. 2005. Research and resource dependent communities: A world of possibilities. *BC Journal of Ecosystems and Management* 6(2):55–62.

Kozak, R. 2007. *Small and Medium Forest Enterprises: Instruments of Change in the Developing World*. Rights and Resources Initiative, Washington, DC.

Kozak, R. 2009. Alternative business models for forest-dependent communities in Africa: A pragmatic consideration of small-scale enterprises and a path forward. *Madagascar Conservation and Development* 4(2):76–81.

Lee, N. and P. Kotler. 2011. *Social Marketing: Influencing Behaviors for Good* (4th edn.). Sage Publications, Thousand Oaks, CA.

Leopold, A. 1949. *A Sand County Almanac (And Sketches Here and There)*. Oxford University Press, Oxford, U.K.

Macqueen, D. 2004. Associations of small and medium forest enterprise: An initial review of issues for local livelihoods and sustainability. International Institute for Environment and Development (IIED), Briefing Paper. London, U.K.

Macqueen, D., S. Bose, S. Bukula, C. Kazoora, S. Ousman, N. Porro, and H. Weyerhaeuser. 2006. Working together: Forest-linked small and medium enterprise associations and collective action. International Institute for Environment and Development (IIED), Gatekeeper Series 125. London, U.K.

Martin, R. and M. Porter. 2000. Canadian competitiveness: Nine years after the crossroads. *Proceedings of the Centre for the Study of Living Standards Conference on the Canada-US Manufacturing Productivity Gap*. Ottawa, Ontario, Canada.

Mayers, J. 2006. Small- and medium-sized forestry enterprises. International Tropical Timber Organization (ITTO) Tropical Forest Update 16/2(2006):10–11.

Nielsen, R. 2006. *The Little Green Handbook*. Picador, New York.

Richardson, B. 1993. Vegetation management practices in plantation forests of Australia and New Zealand. *Canadian Journal of Forest Research* 23(10):1989–2005.

Rittel, H. and M. Webber. 1973. Dilemmas in a general theory of planning. *Policy Sciences* 4(1973):155–169.

Schumpeter, J. 1942. *Capitalism, Socialism and Democracy*. Harper, New York.

Seager, T. 2008. The sustainability spectrum and the sciences of sustainability. *Business Strategy and the Environment* 17(2008):444–453.

Sunderlin, W., J. Hatcher, and M. Liddle. 2008. *From Exclusion to Ownership? Challenges and Opportunities in Advancing Forest Tenure Reform.* Rights and Resources Initiative, Washington, DC.

United States Census Bureau. 2013. Median and average square feet of floor area in new single-family houses completed by location. http://www.census.gov/const/C25Ann/sftotalmedavgsqft. pdf. Accessed April, 2013.

WWF. 2012. *Africa Ecological Footprint Report: Green Infrastructure for Africa's Ecological Security.* WWF-World Wildlife Fund, Gland, Switzerland.

Index